Linear Algebra
and
Differential Equations

Gary L. Peterson

James S. Sochacki
James Madison University

Boston • San Francisco • New York • London • Toronto • Sydney • Tokyo
Singapore • Madrid • Mexico City • Munich • Paris • Cape Town • Hong Kong • Montreal

Senior Acquisitions Editor: Laurie Rosatone
Marketing Manager: Michael Boezi
Senior Marketing Coordinator: Sara Anderson
Project Manager: Jennifer Albanese
Senior Production Supervisor: Peggy McMahon
Editorial Production Services: Barbara Pendergast
Manufacturing Buyer: Evelyn Beaton
Senior Prepress Supervisor: Caroline Fell
Technical Art Supervisor: Joe Vetere
Composition and Technical Art Rendering: Techsetters, Inc.
Cover Design: Leslie Haimes
Cover Image: © Digital Vision

Library of Congress Cataloging-in-Publication Data

Peterson, Gary L.
    Linear algebra and differential equations / Gary L. Peterson, James S. Sochacki.
      p.   cm.
    Includes indexes.
    ISBN 0-201-66212-4
    1. Algebras, Linear.   2. Differential equations.   I. Sochacki, James S.   II. Title.
QA184.2.P48 2002
512′5–dc21

2001024930

ISBN 0-201-66212-4

1 2 3 4 5 6 7 8 9 10 — CRW — 04030201

Preface

The idea of an integrated linear algebra and differential equations course is an attractive one to many mathematics faculty. One significant reason for this is that, while linear algebra has widespread application throughout mathematics, the traditional one-term linear algebra course leaves little time for these applications—especially those dealing with vector spaces and eigenvalues and eigenvectors. Meanwhile, linear algebra concepts such as bases of vector spaces and eigenvalues and eigenvectors arise in the traditional introductory ordinary differential equations course in the study of linear differential equations and systems of linear differential equations. We began offering such an integrated course at James Madison University several years ago. Our experiences in teaching this course along with our unhappiness with existing texts led us to write this book.

As we tried various existing texts in our course, we realized that a better organization was in order. The most typical approach has been to start with the beginning material of an introductory ordinary differential equations text, then cover the standard material of a linear algebra course, and finally return to the study of differential equations. Because of the disjointed nature and inefficiency of such an arrangement, we started to experiment with different arrangements as we taught out of these texts that evolved to the one reflected by our table of contents. Preliminary drafts of our book with this arrangement have been successfully class tested at James Madison University for several years.

TEXT ORGANIZATION

To obtain a better integrated treatment of linear algebra and differential equations, our arrangement begins with two chapters on linear algebra, Chapter 1 on matrices and determinants and Chapter 2 on vector spaces. We then turn our attention to differential equations in the next two chapters. Chapter 3 is our introductory chapter on differential equations with a primary focus on first order equations. Students have encountered a good bit of this material already in their calculus courses. Chapter 4 then links together the subject areas of linear algebra and differential equations with the study of linear differential equations. In Chapter 5 we return to linear algebra with the study of linear transformations and eigenvalues and eigenvectors. Chapter 5 then naturally dovetails with Chapter 6 on systems of differential equations. We conclude with Chapter 7 on Laplace transforms, which includes systems of differential equations, Chapter 8 on power series solutions to differential equations, and Chapter 9 on inner product spaces. The following diagram illustrates the primary dependencies of the chapters of this book.

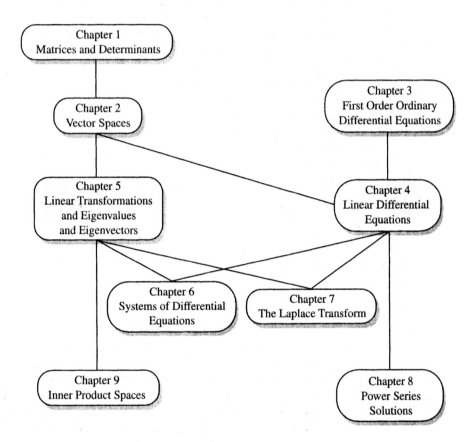

PRESENTATION

In writing this text, we have striven to write a book that gives the student a solid foundation in both linear algebra and ordinary differential equations. We have kept the writing style to the point and student friendly while adhering to conventional mathematical style. Theoretic developments are as complete as possible. Almost all proofs of theorems have been included with a few exceptions, such as the proofs of the existence and uniqueness of solutions to higher order and systems of linear differential equation initial value problems in Chapters 4 and 6 and the existence of Jordan canonical form in Chapter 5. In a number of instances we have strategically placed those proofs that, in our experience, many instructors are forced to omit so that they can be skipped with minimum disruption to the flow of the presentation. We have taken care to include an ample number of examples, exercises, and applications in this text. For motivational purposes, we begin each chapter with an introduction relating its contents to either knowledge already possessed by the student or an application.

COURSE ARRANGEMENTS

While our motivation has been to write a text that would be suitable for a one-term course, this book certainly contains more than enough material for a two-term sequence of courses on linear algebra and differential equations consisting of either two integrated courses or a linear algebra course followed by a differential equations course. For those teaching a one-term course, we describe our integrated linear algebra and differential equations course at James Madison University. This course is a four-credit one-semester course with approximately 56 class meetings and is centered around Chapters 1–6. Most students taking this course either have completed the calculus sequence or are concurrently in the final semester of calculus. We suggest instructors of our course follow the following approximate 48-day schedule:

Chapter 1 (10 days): 2 days each on Sections 1.1–1.3, 1 day on Section 1.4, 3 days on Sections 1.5 and 1.6

Chapter 2 (9 days): 2 days each on Sections 2.1–2.4 and 1 day on Section 2.5

Chapter 3 (7 days): 4 days on Sections 3.1–3.4, 2 days on Section 3.6, and 1 day on Section 3.7

Chapter 4 (9 days): 2 days on each of Sections 4.1–4.3, 1 day on Section 4.4, and 2 days on Section 4.5

Chapter 5 (7 days): 2 days on Sections 5.1 and 5.2, 2 days each on Sections 5.3 and 5.4, and 1 day on Section 5.5

Chapter 6 (6 days): 1 day on Section 6.1, 2 days on Section 6.2, 1 day on Sections 6.3 and 6.4, and 2 days on Sections 6.5 and 6.6.

After adding time for testing, this schedule leaves an ample number of days to cover selected material from Chapters 7–9 or to devote more time to material in Chapters 1–6. Those interested in a two-term sequence of integrated linear algebra and differential equations courses should have little difficulty constructing a syllabus for such a sequence. A sequence of courses consisting of a linear algebra course followed by a differential equations course can be built around Chapters 1, 2, 5, and 9 for the linear algebra course and Chapters 3, 4, 6, 7, and 8 for the differential equations course.

MATHEMATICAL SOFTWARE

This text is written so that it can be used both by those who wish to make use of mathematical software packages for linear algebra and differential equations (such as Maple or Mathematica) and by those who do not wish to do so. For those who wish to do so, exercises involving the use of an appropriate software package are included throughout the text. To facilitate the use of these packages, we have concluded that it is best to include discussions with a particular package rather than trying to give vague generic descriptions. We have chosen Maple as our primary accompanying software package for this textbook. For those who wish to use Mathematica or MATLAB instead of Maple, a supplementary *Technology Resource Manual* on the use of these packages for this book is available from Addison-Wesley (ISBN: 0-201-75815-6). Those wishing to use other software packages should have little difficulty incorporating the use of these other packages. In a few instances, instructions on how to use Maple for Maple users appear within the text. But most of the time we simply indicate the appropriate Maple commands in the exercises with the expectation that either the student using Maple will consult Maple's help menu to see how to use the commands or the instructor will add supplements on how to apply Maple's commands.

ACKNOWLEDGMENTS

It is our pleasure to thank the many people who have helped shape this book. We are grateful to the students and faculty at James Madison University who put up with the preliminary drafts of this text and offered numerous improvements. We have received many valuable suggestions from our reviewers, including:

> Paul Eenigenburg, Western Michigan University
>
> Gary Howell, Florida Institute of Technology
>
> David Johnson, Lehigh University
>
> Douglas Meade, University of South Carolina
>
> Ronald Miech, University of California, Los Angeles
>
> Michael Nasab, Long Beach City College
>
> Clifford Queen, Lehigh University
>
> Murray Schechter, Lehigh University
>
> Barbara Shipman, University of Texas at Arlington
>
> William Stout, Salve Regina University
>
> Russell Thompson, Utah State University
>
> James Turner, Calvin College (formerly at Purdue University)

Finally, we wish to thank the staff of and individuals associated with Addison-Wesley including sponsoring editor Laurie Rosatone, project manager Jennifer Albanese, assistant editors Ellen Keohane and Susan Laferriere, production supervisor Peggy McMahon, copyediting and production coordinator Barbara Pendergast, technical art supervisor Joe Vetere, and mathematics accuracy checker Paul Lorczak whose efforts kept this project on track and brought it to fruition. To all of you, we owe a debt of gratitude.

G. L. P.
J. S. S.
Harrisonburg, VA

Contents

First Order Ordinary Differential Equations   **111**

3.1   Introduction to Differential Equations   112
3.2   Separable Differential Equations   120
3.3   Exact Differential Equations   124
3.4   Linear Differential Equations   130
3.5   More Techniques for Solving First Order Differential Equations   136
3.6   Modeling with Differential Equations   144
3.7   Reduction of Order   153
3.8   The Theory of First Order Differential Equations   157
3.9   Numerical Solutions of Ordinary Differential Equations   168

4

Linear Differential Equations   **179**

4.1   The Theory of Higher Order Linear Differential Equations   179
4.2   Homogeneous Constant Coefficient Linear Differential Equations   189
4.3   The Method of Undetermined Coefficients   203
4.4   The Method of Variation of Parameters   211
4.5   Some Applications of Higher Order Differential Equations   217

5

Linear Transformations and Eigenvalues and Eigenvectors   **231**

5.1   Linear Transformations   231
5.2   The Algebra of Linear Transformations; Differential Operators and Differential Equations   245
5.3   Matrices for Linear Transformations   253
5.4   Eigenvalues and Eigenvectors of Matrices   269
5.5   Similar Matrices, Diagonalization, and Jordan Canonical Form   278
5.6   Eigenvectors and Eigenvalues of Linear Transformations   287

6

Systems of Differential Equations   293

7

The Laplace Transform   345

8

Power Series Solutions to Linear Differential Equations   375

9

Matrices and Determinants

Beginning with your first algebra course you have encountered problems such as the following:

> *A boat traveling at a constant speed in a river with a constant current speed can travel 48 miles downstream in 4 hours. The same trip upstream takes 6 hours. What is the speed of the boat in still water and what is the speed of the current?*

Here is one way we can solve this problem: Let x be the speed of the boat in still water and y be the speed of the current. Since the speed of the boat going downstream is $x + y$, we have

$$4(x + y) = 48 \quad \text{or} \quad x + y = 12.$$

Since the speed of the boat going upstream is $x - y$,

$$6(x - y) = 48 \quad \text{or} \quad x - y = 8.$$

Thus we can determine the speed of the boat in still water and the speed of the current by solving the system of equations:

$$x + y = 12$$
$$x - y = 8.$$

Doing so (try it), we find the speed of the boat in still water is $x = 10$ miles per hour and the speed of the current is $y = 2$ miles per hour.

The system of equations we have just considered is an example of a system of linear equations, and you have encountered many such linear systems over the years. Of course, you probably have come to realize that the larger the system—that is, the more variables

1

and/or equations in the system—the more difficult it often is to solve the system. For instance, suppose we needed to find the partial fraction decomposition of

$$\frac{1}{(x^2+1)(x^2+4)},$$

which, as you saw in calculus, is used to integrate this expression. (In our study of the Laplace transform in Chapter 7, we will see another place where finding partial fraction decompositions of expressions such as this arises.) This partial fraction decomposition has the form

$$\frac{1}{(x^2+1)(x^2+4)} = \frac{Ax+B}{x^2+1} + \frac{Cx+D}{x^2+4},$$

and finding it involves solving a system of four linear equations in the four unknowns A, B, C, and D, which takes more time and effort to solve than the problem with the boat. There is no limit to the size of linear systems that arise in practice. It is not unheard of to encounter systems of linear equations with tens, hundreds, or even thousands of unknowns and equations.

The larger the linear system, the easier it is to get lost in your work if you are not careful. Because of this, we are going to begin this chapter by showing you a systematic way of solving linear systems of equations so that, if you follow this approach, you will always be led to the correct solutions of a given linear system. Our approach will involve representing linear systems of equations by a type of expression called a matrix. After you have seen this particular use of matrices (it will be just one of many more to come) in Section 1.1, we will go on to study matrices in their own right in the rest of this chapter. We begin with a discussion of some of the basics.

1.1   SYSTEMS OF LINEAR EQUATIONS

A **linear equation** in the variables or unknowns $x_1, x_2, \ldots, x_n$ is an equation that can be written in the form

$$\boxed{a_1x_1 + a_2x_2 + \cdots + a_nx_n = b}$$

where $a_1, a_2, \ldots, a_n, b$ are constants. For instance,

$$2x - 3y = 1$$

is a linear equation in the variables x and y,

$$3x - y + 2z = 8$$

is a linear equation in the variables x, y, and z, and

$$-x_1 + 5x_2 - \pi x_3 + \sqrt{2}x_4 - 9x_5 = e^2$$

is a linear equation in the variables x_1, x_2, x_3, x_4, and x_5. The graph of a linear equation in two variables such as $2x - 3y = 1$ is a line in the xy-plane, and the graph of a linear equation in three variables such as $3x - y + 2z = 8$ is a plane in 3-space.

When considered together, a collection of linear equations

$$
\begin{aligned}
a_{11}x_1 + a_{12}x_2 + \cdots + a_{1n}x_n &= b_1 \\
a_{21}x_1 + a_{22}x_2 + \cdots + a_{2n}x_n &= b_2 \\
&\vdots \\
a_{m1}x_1 + a_{m2}x_2 + \cdots + a_{mn}x_n &= b_m
\end{aligned}
$$

is called a **system of linear equations.** For instance,

$$
\begin{aligned}
x - y + z &= 0 \\
2x - 3y + 4z &= -2 \\
-2x - y + z &= 7
\end{aligned}
$$

is a system of three linear equations in three variables.

A **solution** to a system of equations with variables $x_1, x_2, \ldots, x_n$ consists of values of $x_1, x_2, \ldots, x_n$ that satisfy each equation in the system. From your first algebra course you should recall that the solutions to a system of two linear equations in x and y,

$$
\begin{aligned}
a_{11}x + a_{12}y &= b_1 \\
a_{21}x + a_{22}y &= b_2,
\end{aligned}
$$

are the points at which the graphs of the lines given by these two equations intersect. Consequently, such a system will have exactly one solution if the graphs intersect in a single point, will have infinitely many solutions if the graphs are the same line, and will have no solution if the graphs are parallel. As we shall see, this in fact holds for all systems of linear equations; that is, a linear system either has exactly one solution, infinitely many solutions, or no solutions.

The main purpose of this section is to present the **Gauss-Jordan elimination method**,[1] a systematic way for solving systems of linear equations that will always lead us to solutions of the system. The Gauss-Jordan method involves the repeated use of three basic transformations on a system. We shall call the following transformations **elementary operations.**

1.   Interchange two equations in the system.

2.   Multiply an equation by a nonzero number.

3.   Replace an equation by itself plus a multiple of another equation.

Two systems of equations are said to be **equivalent** if they have the same solutions. It is not difficult to see that applying an elementary operation to a system produces an equivalent system.

---

[1] Named in honor of Karl Friedrich Gauss (1777–1855), who is one of the greatest mathematicians of all time and is often referred to as the "prince of mathematics," and Wilhelm Jordan (1842–1899), a German engineer.

To illustrate the Gauss-Jordan elimination method, consider the system:

$$x - y + z = 0$$
$$2x - 3y + 4z = -2$$
$$-2x - y + z = 7.$$

We are going to use elementary operations to transform this system to one of the form

$$x \qquad\qquad = *$$
$$y \qquad\quad = *$$
$$z \quad = *$$

where each $*$ is a constant from which we have the solution. To this end, let us first replace the second equation by itself plus -2 times the first equation (or subtracting 2 times the first equation from the second) and replace the third equation by itself plus 2 times the first equation to eliminate x from the second and third equations. (We are doing two elementary operations simultaneously here.) This gives us the system

$$x - y + z = 0$$
$$-y + 2z = -2$$
$$-3y + 3z = 7.$$

Next, let us use the second equation to eliminate y in the first and third equations by replacing the first equation by itself minus the second equation and replacing the third equation by itself plus -3 times the second equation, obtaining

$$x \qquad -z = 2$$
$$-y + 2z = -2$$
$$-3z = 13.$$

Now we are going to use the third equation to eliminate z from the first two equations by multiplying the first equation by 3 and then subtracting the third equation from it (we actually are doing two elementary operations here) and multiplying the second equation by 3 and then adding 2 times the third equation to it (here too we are doing two elementary operations). This gives us the system:

$$3x \qquad\qquad = -7$$
$$-3y \quad = 20$$
$$3z = -13.$$

Finally, dividing the first equation by 3 (or multiplying it by 1/3), dividing the second equation by -3, and dividing the third equation by 3, we have our system in the promised form as

$$x \qquad\qquad = -\frac{7}{3}$$
$$y \qquad = -\frac{20}{3}$$
$$z \quad = -\frac{13}{3},$$

which tells us the solution.

You might notice that we only really need to keep track of the coefficients as we transform our system. To keep track of them, we will indicate a system such as

$$x - y + z = 0$$
$$2x - 3y + 4z = -2$$
$$-2x - y + z = 7$$

by the following array of numbers:

$$\begin{bmatrix} 1 & -1 & 1 & \vdots & 0 \\ 2 & -3 & 4 & \vdots & -2 \\ -2 & -1 & 1 & \vdots & 7 \end{bmatrix}.$$

This array is called the **augmented matrix** for the system. The entries appearing to the left of the dashed vertical line are the coefficients of the variables as they appear in the system. This part of the augmented matrix is called the **coefficient matrix** of the system. The numbers to the right of the dashed vertical line are the constants on the right-hand side of the system as they appear in the system. In general, the augmented matrix for the system

$$a_{11}x_1 + a_{12}x_2 + \cdots + a_{1n}x_n = b_1$$
$$a_{21}x_1 + a_{22}x_2 + \cdots + a_{2n}x_n = b_2$$
$$\vdots$$
$$a_{m1}x_1 + a_{m2}x_2 + \cdots + a_{mn}x_n = b_m$$

is

$$\begin{bmatrix} a_{11} & a_{12} & \cdots & a_{1n} & \vdots & b_1 \\ a_{21} & a_{22} & \cdots & a_{2n} & \vdots & b_2 \\ \vdots & \vdots & & \vdots & \vdots & \vdots \\ a_{m1} & a_{m2} & \cdots & a_{mn} & \vdots & b_m \end{bmatrix}.$$

The portion of the augmented matrix to the left of the dashed line with entries a_{ij} is the coefficient matrix of the system.

Corresponding to the elementary operations for systems of equations are elementary row operations that we perform on the augmented matrix for a linear system. These are as follows.

1. Interchange two rows.[2]

2. Multiply a row by a nonzero number.

3. Replace a row by itself plus a multiple of another row.

---

[2] A line of numbers going across the matrix from left to right is called a **row**; a line of numbers going down the matrix is called a **column.**

As our first formal example of this section, we are going to redo the work we did in solving the system

$$x - y + z = 0$$
$$2x - 3y + 4z = -2$$
$$-2x - y + z = 7$$

with augmented matrices.

EXAMPLE 1   Solve the system:

$$x - y + z = 0$$
$$2x - 3y + 4z = -2$$
$$-2x - y + z = 7.$$

Solution   Our work will consist of four steps. In the first step, we shall use the first row and row operations to make all other entries in the first column zero. In the second step, we shall use the second row to make all other entries in the second column zero. In the third step, we shall use the third row to make all other entries in the third column zero. In the fourth step, we shall make the nonzero entries in the coefficient matrix 1 at which point we will be able to read off our solution. To aid you in following the steps, an expression such as $R_2 - 2R_1$ next to the second row indicates that we are replacing the second row by itself plus -2 times the first row; an expression such as $R_1/3$ next to the first row indicates we are dividing this row by 3. Arrows are used to indicate the progression of our steps.

$$\begin{bmatrix} 1 & -1 & 1 & \vdots & 0 \\ 2 & -3 & 4 & \vdots & -2 \\ -2 & -1 & 1 & \vdots & 7 \end{bmatrix} \begin{array}{l} \\ R_2 - 2R_1 \\ R_3 + 2R_1 \end{array}$$

$$\rightarrow \begin{bmatrix} 1 & -1 & 1 & \vdots & 0 \\ 0 & -1 & 2 & \vdots & -2 \\ 0 & -3 & 3 & \vdots & 7 \end{bmatrix} \begin{array}{l} R_1 - R_2 \\ \\ R_3 - 3R_2 \end{array}$$

$$\rightarrow \begin{bmatrix} 1 & 0 & -1 & \vdots & 2 \\ 0 & -1 & 2 & \vdots & -2 \\ 0 & 0 & -3 & \vdots & 13 \end{bmatrix} \begin{array}{l} 3R_1 - R_3 \\ 3R_2 + 2R_3 \end{array} \rightarrow \begin{bmatrix} 3 & 0 & 0 & \vdots & -7 \\ 0 & -3 & 0 & \vdots & 20 \\ 0 & 0 & -3 & \vdots & 13 \end{bmatrix} \begin{array}{l} R_1/3 \\ -R_2/3 \\ -R_3/3 \end{array}$$

$$\rightarrow \begin{bmatrix} 1 & 0 & 0 & \vdots & -7/3 \\ 0 & 1 & 0 & \vdots & -20/3 \\ 0 & 0 & 1 & \vdots & -13/3 \end{bmatrix}$$

The solution is then $x = -7/3$, $y = -20/3$, $z = -13/3$.   ●

In Gauss-Jordan elimination, we use elementary row operations on the augmented matrix of the system to transform it so that the final coefficient matrix has a form called **reduced row-echelon form** with the following properties.

1. Any rows of zeros (called **zero rows**) appear at the bottom.
2. The first nonzero entry of a nonzero row is 1 (called a **leading 1**).
3. The leading 1 of a nonzero row appears to the right of the leading 1 of any preceding row.
4. All the other entries of a column containing a leading 1 are zero.

Looking back at Example 1, you will see that the coefficient matrix in our final augmented matrix is in reduced row-echelon form. Once we have the coefficient matrix in reduced row-echelon form, the solutions to the system are easily determined.

Let us do some more examples.

EXAMPLE 2 Solve the system:

$$x_1 + x_2 - x_3 + 2x_4 = 1$$
$$x_1 + x_2 + x_4 = 2$$
$$x_1 + 2x_2 - 4x_3 = 1$$
$$2x_1 + x_2 + 2x_3 + 5x_4 = 1.$$

Solution We try to proceed as we did in Example 1. Notice, however, that we will have to modify our approach here. The symbol $R_2 \leftrightarrow R_3$ after the first step is used to indicate that we are interchanging the second and third rows.

$$\begin{bmatrix} 1 & 1 & -1 & 2 & \vdots & 1 \\ 1 & 1 & 0 & 1 & \vdots & 2 \\ 1 & 2 & -4 & 0 & \vdots & 1 \\ 2 & 1 & 2 & 5 & \vdots & 1 \end{bmatrix} \begin{matrix} \\ R_2 - R_1 \\ R_3 - R_1 \\ R_4 - 2R_1 \end{matrix}$$

$$\rightarrow \begin{bmatrix} 1 & 1 & -1 & 2 & \vdots & 1 \\ 0 & 0 & 1 & -1 & \vdots & 1 \\ 0 & 1 & -3 & -2 & \vdots & 0 \\ 0 & -1 & 4 & 1 & \vdots & -1 \end{bmatrix} \begin{matrix} \\ R_2 \leftrightarrow R_3 \\ \\ \end{matrix}$$

$$\rightarrow \begin{bmatrix} 1 & 1 & -1 & 2 & \vdots & 1 \\ 0 & 1 & -3 & -2 & \vdots & 0 \\ 0 & 0 & 1 & -1 & \vdots & 1 \\ 0 & -1 & 4 & 1 & \vdots & -1 \end{bmatrix} \begin{matrix} R_1 - R_2 \\ \\ \\ R_4 + R_2 \end{matrix}$$

$$
\rightarrow \begin{bmatrix} 1 & 0 & 2 & 4 & \vdots & 1 \\ 0 & 1 & -3 & -2 & \vdots & 0 \\ 0 & 0 & 1 & -1 & \vdots & 1 \\ 0 & 0 & 1 & -1 & \vdots & -1 \end{bmatrix} \begin{matrix} R_1 - 2R_3 \\ R_2 + 3R_3 \\ \\ R_4 - R_3 \end{matrix} \rightarrow \begin{bmatrix} 1 & 0 & 0 & 6 & \vdots & -1 \\ 0 & 1 & 0 & -5 & \vdots & 3 \\ 0 & 0 & 1 & -1 & \vdots & 1 \\ 0 & 0 & 0 & 0 & \vdots & -2 \end{bmatrix}
$$

We now have the coefficient matrix in reduced row-echelon form. Our final augmented matrix represents the system

$$
\begin{aligned}
x_1 + 6x_4 &= -1 \\
x_2 - 5x_4 &= 3 \\
x_3 - x_4 &= 1 \\
0 &= -2,
\end{aligned}
$$

which is equivalent to our original system. Since this last system contains the false equation $0 = -2$, it has no solutions. Hence our original system has no solutions. ●

EXAMPLE 3   Solve the system:

$$
\begin{aligned}
2x + 3y - z &= 3 \\
-x - y + 3z &= 0 \\
x + 2y + 2z &= 3 \\
y + 5z &= 3.
\end{aligned}
$$

Solution   We first reduce the augmented matrix for this system so that its coefficient matrix is in reduced row-echelon form.

$$
\begin{bmatrix} 2 & 3 & -1 & \vdots & 3 \\ -1 & -1 & 3 & \vdots & 0 \\ 1 & 2 & 2 & \vdots & 3 \\ 0 & 1 & 5 & \vdots & 3 \end{bmatrix} \begin{matrix} \\ 2R_2 + R_1 \\ 2R_3 - R_1 \\ \\ \end{matrix}
$$

$$
\rightarrow \begin{bmatrix} 2 & 3 & -1 & \vdots & 3 \\ 0 & 1 & 5 & \vdots & 3 \\ 0 & 1 & 5 & \vdots & 3 \\ 0 & 1 & 5 & \vdots & 3 \end{bmatrix} \begin{matrix} R_1 - 3R_2 \\ \\ R_3 - R_2 \\ R_4 - R_2 \end{matrix}
$$

$$
\rightarrow \begin{bmatrix} 2 & 0 & -16 & \vdots & -6 \\ 0 & 1 & 5 & \vdots & 3 \\ 0 & 0 & 0 & \vdots & 0 \\ 0 & 0 & 0 & \vdots & 0 \end{bmatrix} \begin{matrix} R_1/2 \\ \\ \\ \end{matrix} \rightarrow \begin{bmatrix} 1 & 0 & -8 & \vdots & -3 \\ 0 & 1 & 5 & \vdots & 3 \\ 0 & 0 & 0 & \vdots & 0 \\ 0 & 0 & 0 & \vdots & 0 \end{bmatrix}.
$$

This final augmented matrix represents the equivalent system:

$$x - 8z = -3$$
$$y + 5z = 3$$
$$0 = 0$$
$$0 = 0.$$

Solving the first two equations for x and y in terms of z, we can say that our solutions have the form

$$x = -3 + 8z, \qquad y = 3 - 5z$$

where z is any real number. In particular, we have infinitely many solutions in this example. (Any choice of z gives us a solution. If $z = 0$, we have $x = -3$, $y = 3$, $z = 0$ as a solution; if $z = 1$, we have $x = 5$, $y = -2$, $z = 1$ as a solution; if $z = \sqrt{17}$, we have $x = -3 + 8\sqrt{17}$, $y = 3 - 5\sqrt{17}$, $z = \sqrt{17}$ as a solution; and so on.) In a case such as this, we refer to z as the **free variable** and x and y as the **dependent** variables in our solutions. When specifying our solutions to systems like this, we will follow the convention of using variables that correspond to leading ones as dependent variables and those that do not as free variables. It is not necessary to specify our solutions this way, however. For instance, in this example we could solve for z in terms of x, obtaining

$$z = \frac{x}{8} + \frac{3}{8}$$

and

$$y = 3 - 5z = 3 - 5\left(\frac{x}{8} + \frac{3}{8}\right) = -\frac{5x}{8} + \frac{9}{8},$$

giving us the solutions with x as the free variable and y and z as the dependent variables. ●

EXAMPLE 4 Solve the system:

$$4x_1 - 8x_2 - x_3 + x_4 + 3x_5 = 0$$
$$5x_1 - 10x_2 - x_3 + 2x_4 + 3x_5 = 0$$
$$3x_1 - 6x_2 - x_3 + x_4 + 2x_5 = 0.$$

Solution We again begin by reducing the augmented matrix to the point where its coefficient matrix is in reduced row-echelon form:

$$\left[\begin{array}{ccccc|c} 4 & -8 & -1 & 1 & 3 & 0 \\ 5 & -10 & -1 & 2 & 3 & 0 \\ 3 & -6 & -1 & 1 & 2 & 0 \end{array}\right] \begin{array}{l} \\ 4R_2 - 5R_1 \\ 4R_3 - 3R_1 \end{array}$$

$$\rightarrow \begin{bmatrix} 4 & -8 & -1 & 1 & 3 & \vdots & 0 \\ 0 & 0 & 1 & 3 & -3 & \vdots & 0 \\ 0 & 0 & -1 & 1 & -1 & \vdots & 0 \end{bmatrix} \begin{matrix} R_1 + R_2 \\ \\ R_3 + R_2 \end{matrix}$$

$$\rightarrow \begin{bmatrix} 4 & -8 & 0 & 4 & 0 & \vdots & 0 \\ 0 & 0 & 1 & 3 & -3 & \vdots & 0 \\ 0 & 0 & 0 & 4 & -4 & \vdots & 0 \end{bmatrix} \begin{matrix} R_1/4 \\ \\ R_3/4 \end{matrix}$$

$$\rightarrow \begin{bmatrix} 1 & -2 & 0 & 1 & 0 & \vdots & 0 \\ 0 & 0 & 1 & 3 & -3 & \vdots & 0 \\ 0 & 0 & 0 & 1 & -1 & \vdots & 0 \end{bmatrix} \begin{matrix} R_1 - R_3 \\ R_2 - 3R_3 \end{matrix}$$

$$\rightarrow \begin{bmatrix} 1 & -2 & 0 & 0 & 1 & \vdots & 0 \\ 0 & 0 & 1 & 0 & 0 & \vdots & 0 \\ 0 & 0 & 0 & 1 & -1 & \vdots & 0 \end{bmatrix}.$$

We now have arrived at the equivalent system

$$x_1 - 2x_2 + x_5 = 0$$
$$x_3 = 0$$
$$x_4 - x_5 = 0,$$

which has solutions

$$x_1 = 2x_2 - x_5, \qquad x_3 = 0, \qquad x_4 = x_5$$

with x_2 and x_5 as the free variables and x_1 and x_4 as the dependent variables.   ●

Systems of equations that have solutions such as those in Examples 1, 3, and 4 are called **consistent systems**; those that do not have solutions as occurred in Example 2 are called **inconsistent systems.** Notice that an inconsistent system is easily recognized once the coefficient matrix of its augmented matrix is put in reduced row-echelon form: There will be a row with zeros in the coefficient matrix with nonzero entry in the right-hand entry of this row. If we do not have this, the system is consistent. Consistent systems break down into two types. Once the coefficient matrix of the augmented matrix is put in reduced row-echelon form, the number of nonzero rows in the coefficient matrix is always less than or equal to the number of columns of the coefficient matrix. (That is, there will never be more nonzero rows than columns when the coefficient matrix is in reduced row-echelon form. Why is this the case?) If there are fewer nonzero rows than columns, as we had in Examples 3 and 4, the system will have infinitely many solutions. If we have as many nonzero rows as columns, as occurred in Example 1, we have exactly one solution. Recall that it was mentioned at the beginning of this section that every system of linear equations either has exactly one solution, infinitely many solutions, or no solutions. Now we can see why this is true.

A system of linear equations that can be written in the form

$$
\begin{aligned}
a_{11}x_1 + a_{12}x_2 + \cdots + a_{1n}x_n &= 0 \\
a_{21}x_1 + a_{22}x_2 + \cdots + a_{2n}x_n &= 0 \\
&\vdots \\
a_{m1}x_1 + a_{m2}x_2 + \cdots + a_{mn}x_n &= 0
\end{aligned}
\tag{1}
$$

is called a **homogeneous system.** The system of equations in Example 4 is homogeneous. Notice that

$$ x_1 = 0, \qquad x_2 = 0, \qquad \ldots, \qquad x_n = 0 $$

is a solution to the homogeneous system in Equations (1). This is called the **trivial solution** of the homogeneous system. Because homogeneous systems always have a trivial solution, they are never inconsistent systems. Homogeneous systems will occur frequently in our future work and we will often be interested in whether such a system has solutions other than the trivial one, which we naturally call **nontrivial solutions.** The system in Example 4 has nontrivial solutions. For instance, we would obtain one (among the infinitely many such nontrivial solutions) by letting $x_2 = 1$ and $x_5 = 2$, in which case we have the nontrivial solution $x_1 = 0$, $x_2 = 1$, $x_3 = 0$, $x_4 = 2$, $x_5 = 2$. Actually, we can tell ahead of time that the system in Example 4 has nontrivial solutions. Because this system has fewer equations than variables, the reduced row-echelon form of the coefficient matrix will have fewer nonzero rows than columns and hence must have infinitely many solutions (only one of which is the trivial solution) and consequently must have infinitely many nontrivial solutions. This reasoning applies to any homogeneous system with fewer equations than variables, and hence we have the following theorem.

THEOREM 1.1   A homogeneous system of m linear equations in n variables with $m < n$ has infinitely many nontrivial solutions.

Of course, if a homogeneous system has at least as many equations as variables such as the systems

$$
\begin{aligned}
x + y + z &= 0 \\
x - y - z &= 0 \\
2x + y + z &= 0
\end{aligned}
\qquad \text{and} \qquad
\begin{aligned}
2x + y + z &= 0 \\
x - 2y - z &= 0 \\
3x - y &= 0 \\
4x - 3y - z &= 0
\end{aligned}
$$

we would have to do some work toward solving these systems before we would be able to see whether they have nontrivial solutions. We shall do this for the second system a bit later.

Gaussian elimination, which is another systematic approach for solving linear systems, is similar to the approach we have been using but does not require that all the other entries of the column containing a leading 1 be zero. That is, it uses row operations to transform the augmented matrix so that the coefficient matrix has the following form:

1. Any zero rows appear at the bottom.

2. The first nonzero entry of a nonzero row is 1.

3. The leading 1 of a nonzero row appears to the right of the leading 1 of any preceding row.

Such a form is called a **row-echelon form** for the coefficient matrix. In essence, we do not eliminate (make zero) entries above the leading 1s in Gaussian elimination. Here is how this approach can be applied to the system in Example 1.

$$\left[\begin{array}{ccc:c} 1 & -1 & 1 & 0 \\ 2 & -3 & 4 & -2 \\ -2 & -1 & 1 & 7 \end{array}\right] \begin{array}{l} \\ R_2 - 2R_1 \\ R_3 + 2R_1 \end{array}$$

$$\rightarrow \left[\begin{array}{ccc:c} 1 & -1 & 1 & 0 \\ 0 & -1 & 2 & -2 \\ 0 & -3 & 3 & 7 \end{array}\right] \begin{array}{l} \\ \\ R_3 - 3R_2 \end{array}$$

$$\rightarrow \left[\begin{array}{ccc:c} 1 & -1 & 1 & 0 \\ 0 & -1 & 2 & -2 \\ 0 & 0 & -3 & 13 \end{array}\right] \begin{array}{l} \\ -R_2 \\ -R_3/3 \end{array} \rightarrow \left[\begin{array}{ccc:c} 1 & -1 & 1 & 0 \\ 0 & 1 & -2 & 2 \\ 0 & 0 & 1 & -13/3 \end{array}\right]$$

We now have the coefficient matrix in a row-echelon form and use this result to find the solutions. The third row tells us

$$z = -\frac{13}{3}.$$

The values of the remaining variables are found by a process called **back substitution.** From the second row, we have the equation

$$y - 2z = 2$$

from which we can find y:

$$y + \frac{26}{3} = 2$$

$$y = -\frac{20}{3}.$$

Finally, the first row represents the equation

$$x - y + z = 0$$

from which we can find x:

$$x + \frac{20}{3} - \frac{13}{3} = 0$$

$$x = -\frac{7}{3}.$$

On the plus side, Gaussian elimination requires fewer row operations. But on the minus side, the work is sometimes messy when doing the back substitutions. Often, we find ourselves having to deal with fractions even if our original system involves only integers. The back substitutions are also cumbersome to do when dealing with systems that have infinitely many solutions. Try the Gaussian elimination procedure in Example 3 or 4 if you would like to see how it goes.

As a rule we will tend to use Gauss-Jordan elimination when we have to find the solutions to a linear system in this text. Sometimes, however, we will not have to completely solve a system and will use Gaussian elimination since it will involve less work. The next example illustrates an instance of this. In fact, in this example we will not even have to bother completing Gaussian elimination by making the leading entries one.

EXAMPLE 5   Determine the values of a, b, and c so that the system

$$x - y + 2z = a$$
$$2x + y - z = b$$
$$x + 2y - 3z = c$$

has solutions.

Solution   We begin doing row operations as follows.

$$\begin{bmatrix} 1 & -1 & 2 & \vdots & a \\ 2 & 1 & -1 & \vdots & b \\ 1 & 2 & -3 & \vdots & c \end{bmatrix} \begin{matrix} \\ R_2 - 2R_1 \\ R_3 - R_1 \end{matrix} \rightarrow \begin{bmatrix} 1 & -1 & 2 & \vdots & a \\ 0 & 3 & -5 & \vdots & b - 2a \\ 0 & 3 & -5 & \vdots & c - a \end{bmatrix} \begin{matrix} \\ \\ R_3 - R_2 \end{matrix}$$

$$\rightarrow \begin{bmatrix} 1 & -1 & 2 & \vdots & a \\ 0 & 3 & -5 & \vdots & b - 2a \\ 0 & 0 & 0 & \vdots & a - b + c \end{bmatrix}$$

Now we can see that this system has solutions if and only if a, b, and c satisfy the equation

$$a - b + c = 0.$$ ●

Another place where we will sometimes use an abbreviated version of Gaussian elimination is when we are trying to see if a homogeneous system has nontrivial solutions.

EXAMPLE 6   Determine if the system

$$2x + y + z = 0$$
$$x - 2y - z = 0$$
$$3x - y = 0$$
$$4x - 3y - z = 0$$

has nontrivial solutions.

Solution   Perform row operations:

$$
\begin{bmatrix}
2 & 1 & 1 & \vdots & 0 \\
1 & -2 & -1 & \vdots & 0 \\
3 & -1 & 0 & \vdots & 0 \\
4 & -3 & -1 & \vdots & 0
\end{bmatrix}
\begin{matrix} \\ 2R_2 - R_1 \\ 2R_3 - 3R_1 \\ R_4 - 2R_1 \end{matrix}
\rightarrow
\begin{bmatrix}
2 & 1 & 1 & \vdots & 0 \\
0 & -5 & -3 & \vdots & 0 \\
0 & -5 & -3 & \vdots & 0 \\
0 & -5 & -3 & \vdots & 0
\end{bmatrix}
\begin{matrix} \\ \\ R_3 - R_2 \\ R_4 - R_2 \end{matrix}
$$

$$
\rightarrow
\begin{bmatrix}
2 & 1 & 1 & \vdots & 0 \\
0 & -5 & -3 & \vdots & 0 \\
0 & 0 & 0 & \vdots & 0 \\
0 & 0 & 0 & \vdots & 0
\end{bmatrix}.
$$

It is now apparent that this system has nontrivial solutions. In fact, you should be able to see this after the first set of row operations. ●

It is not difficult to write computer programs for solving systems of linear equations using the Gauss-Jordan or Gaussian elimination methods. Thus it is not surprising that there are computer software packages for solving systems of linear systems.[3] Maple is one among several available mathematical software packages that can be used to find the solutions of linear systems of equations.

In the preface we mentioned that we will use Maple as our accompanying software package within this text. The use of Maple is at the discretion of your instructor. Some may use it, others may prefer to use a different software package, and yet others may choose to not use any such package (and give an excellent and complete course). For those intructors who wish to use Maple—or for students who are independently interested in gaining some knowledge of its capabilities—we will include occasional remarks about how to use it when we deem it appropriate. On many other occasions we will not include any remarks and will simply provide some exercises asking you to use indicated Maple commands. In these cases, you are expected to look up the command in the Help menu under Topic Search to see how to use it. This is one place where we will include a few remarks to get you started. For those who wish to use the software packages Mathematica or MATLAB, the accompanying *Technology Resource Manual* contains corresponding commands for these software packages.

Here we explain how to use Maple to find solutions to linear systems. One way to do this is to use the *linsolve* command. To use this command in a Maple worksheet, you will first have to load Maple's linear algebra package by typing

```
with(linalg);
```

at the command prompt > and then hitting the enter key. After doing this, you will get a list of Maple's linear algebra commands. To solve the system in Example 1, first enter the coefficient matrix of the system by typing

```
A:= matrix([[1,-1,1],[2,-3,4],[-2,-1,1]]);
```

---

[3] Often these packages employ methods that are more efficient than Gauss-Jordan or Gaussian elimination, but we will not concern ourselves with these issues in this text.

at the command prompt and then hitting the enter key. (The symbol := is used in Maple for indicating that we are defining A to be the coefficient matrix we type on the right.) The constants on the right-hand side of the system are typed and entered as

$$b:=\text{vector}([0,-2,7]);$$

at the command prompt. Finally, type and enter

$$\text{linsolve}(A,b);$$

at the commmand prompt and Maple will give us the solution as

$$\left[\frac{-7}{3},\frac{-20}{3},\frac{-13}{3}\right]^4.$$

Doing the same set of steps for the system in Example 2 results in no output, indicating there is no solution. Doing them in Example 3 yields the output

$$[-3+8\_t_1, 3-5\_t_1, \_t_1],$$

which is Maple's way of indicating our solutions in Example 3 with t_1 in place of z. In Example 4, these steps yield

$$[2\_t_1-\_t_2, \_t_1, 0, \_t_2, \_t_2].$$

EXERCISES 1.1

Solve the systems of equations in Exercises 1–16.

1.
$$x+y-z=0$$
$$2x+3y-2z=6$$
$$x+2y+2z=10$$

2.
$$2x+y-2z=0$$
$$2x-y-2z=0$$
$$x+2y-4z=0$$

3.
$$2x+3y-4z=3$$
$$2x+3y-2z=3$$
$$4x+6y-2z=7$$

4.
$$3x+y-2z=3$$
$$x-8y-14z=-14$$
$$x+2y+z=2$$

5.
$$x+3z=0$$
$$2x+y-z=0$$
$$4x+y+5z=0$$

6.
$$2x+3y+z=4$$
$$x+9y-4z=2$$
$$x-y+2z=3$$

7.
$$3x_1+x_2-3x_3-x_4=6$$
$$x_1+x_2-2x_3+x_4=0$$
$$3x_1+2x_2-4x_3+x_4=5$$
$$x_1+2x_2-3x_3+3x_4=4$$

8.
$$x_1+x_2-x_3+2x_4=1$$
$$x_1+x_2-x_3-x_4=-1$$
$$x_1+2x_2+x_3+2x_4=-1$$
$$2x_1+2x_2+x_3+x_4=2$$

9.
$$x_1+2x_2-3x_3+4x_4=2$$
$$2x_1-4x_2+6x_3-5x_4=10$$
$$x_1-6x_2+9x_3-9x_4=8$$
$$3x_1-2x_2+4x_3-x_4=12$$

---

[4] Software packages such as Maple often will have several ways of doing things. This is the case for solving systems of linear equations. One variant is to enter b as a matrix with one column by typing and entering

$$b:=\text{matrix}([[0],[-2],[7]]);$$

When we then type and enter

$$\text{linsolve}(A,b);$$

our solution is given in column form. Another way is to use Maple's *solve* command for solving equations and systems of equations. (With this approach it is not necessary to load Maple's linear algebra package.) To do it this way for the system in Example 1, we type and enter

$$\text{solve}(\{x-y+z=0, 2*x-3*y+4*z=-2, -2*x-y+z=7\}, \{x,y,z\});$$

10.
$$x_1 - x_2 + x_3 + x_4 - x_5 = 0$$
$$2x_1 - x_2 + 2x_3 - x_4 + 3x_5 = 0$$
$$2x_1 - x_2 - 2x_4 + x_5 = 0$$
$$x_1 + x_2 - x_3 - x_4 + 2x_5 = 0$$
$$2x_1 + 4x_3 + x_4 + 3x_5 = 0$$

11.
$$x + 2y + z = -2$$
$$2x + 2y - 2z = 3$$

12.
$$2x - 4y + 6z = 2$$
$$-3x + 6y - 9z = 3$$

13.
$$x - 2y = 2$$
$$x + 8y = -4$$
$$2x + y = 1$$

14.
$$2x + 3y = 5$$
$$2x + y = 2$$
$$x - 2y = 1$$

15.
$$2x_1 - x_2 - x_3 + x_4 + x_5 = 0$$
$$x_1 - x_2 + x_3 + 2x_4 - 3x_5 = 0$$
$$3x_1 - 2x_2 - x_3 - x_4 + 2x_5 = 0$$

16.
$$x_1 - 3x_2 + x_3 - x_4 - x_5 = 1$$
$$2x_1 + x_2 - x_3 + 2x_4 + x_5 = 2$$
$$-x_1 + 3x_2 - x_3 - 2x_4 - x_5 = 3$$
$$2x_1 + x_2 - x_3 - x_4 - x_5 = 6$$

Determine conditions on a, b, and c so that the systems of equations in Exercises 17 and 18 have solutions.

17.
$$2x - y + 3z = a$$
$$x - 3y + 2z = b$$
$$x + 2y + z = c$$

18.
$$x + 2y - z = a$$
$$x + y - 2z = b$$
$$2x + y - 3z = c$$

Determine conditions on a, b, c, and d so that the systems of equations in Exercises 19 and 20 have solutions.

19.
$$x_1 + x_2 + x_3 - x_4 = a$$
$$x_1 - x_2 - x_3 + x_4 = b$$
$$x_1 + x_2 + x_3 + x_4 = c$$
$$x_1 - x_2 + x_3 + x_4 = d$$

20.
$$x_1 - x_2 + x_3 + x_4 = a$$
$$x_1 + x_2 - 2x_3 + 3x_4 = b$$
$$3x_1 - 2x_2 + 3x_3 - 2x_4 = c$$
$$2x_2 - 3x_3 + 2x_4 = d$$

Determine if the homogeneous systems of linear equations in Exercises 21–24 have nontrivial solutions. You do not have to solve the systems.

21.
$$9x - 2y + 17z = 0$$
$$13x + 81y - 27z = 0$$

22.
$$99x_1 + \pi x_2 - \sqrt{5}x_3 = 0$$
$$2x_1 + (\sin 1)x_2 + 2x_4 = 0$$
$$3.38x_1 - ex_3 + (\ln 2)x_4 = 0$$

23.
$$x - y + z = 0$$
$$2x + y + 2z = 0$$
$$3x - 5y + 3z = 0$$

24.
$$x + y + 2z = 0$$
$$3x - y - 2z = 0$$
$$2x - 2y - 4z = 0$$
$$x + 3y + 6z = 0$$

25. We have seen that homogeneous linear systems with fewer equations than variables always have infinitely many solutions. What possibilities can arise for non-homogeneous linear systems with fewer equations than variables? Explain your answer.

26. Give an example of a system of linear equations with more equations than variables that illustrates each of the following possibilities: Has exactly one solution, has infinitely many solutions, and has no solution.

27. Describe graphically the possible solutions to a system of two linear equations in x, y, and z.

28. Describe graphically the possible solutions to a system of three linear equations in x, y, and z.

Use Maple or another appropriate software package to solve the systems of equations in Exercises 29–32. If you are using Mathematica or MATLAB, see the *Technology Resource Manual* for appropriate commands. (To become more comfortable with the software package you are using, you may wish to practice using it to solve some of the smaller systems in Exercises 1–16 before doing these.)

29.
$$7x_1 - 3x_2 + 5x_3 - 8x_4 + 2x_5 = 13$$
$$12x_1 + 4x_2 - 16x_3 - 9x_4 + 7x_5 = 21$$
$$-22x_1 - 8x_2 + 25x_3 - 16x_4 - 8x_5 = 47$$
$$-52x_1 - 40x_2 + 118x_3 - 37x_4 - 29x_5 = 62$$

30.
$$46x_1 + 82x_2 - 26x_3 + 44x_4 = 122$$
$$69x_1 + 101x_2 + 43x_3 + 30x_4 = 261$$
$$-437x_1 - 735x_2 + 335x_3 + 437x_4 = -406$$
$$299x_1 + 379x_2 - 631x_3 - 2501x_4 = -4146$$
$$1863x_1 + 2804x_2 + 62x_3 - 1983x_4 = 4857$$
$$1748x_1 + 2291x_2 - 461x_3 - 9863x_4 = 4166$$

31. $62x_1 + 82x_2 + 26x_3 - 4x_4$
$$+ 32x_5 + 34x_6 - 2x_7 - 4x_8 = 0$$
$93x_1 + 123x_2 + 67x_3 - 36x_4$
$$+ 106x_5 + 51x_6 + 31x_7 - 188x_8 = 0$$
$-589x_1 - 779x_2 - 303x_3 + 647x_4$
$$- 330x_5 - 323x_6 - 256x_7 - 246x_8 = 0$$
$403x_1 + 533x_2 + 365x_3 - 2493x_4$
$$+ 263x_5 + 50x_6 + 981x_7 + 1345x_8 = 0$$
$2511x_1 + 3321x_2 + 1711x_3 - 2636x_4$
$$+ 2358x_5 + 1357x_6 + 1457x_7 - 2323x_8 = 0$$
$2356x_1 + 3116x_2 + 2038x_3 - 6828x_4$
$$+ 2418x_5 + 1936x_6 + 3596x_7 - 357x_8 = 0$$
32. $3.3x_1 + 3.3x_2 + 12.1x_3 + 2.2x_4$
$$+ 45.1x_5 + 7.7x_6 + 12.1x_7$$
$$+ 35.2x_8 + 1.1x_9 = -3.3$$
$3x_1 + 3x_2 + 15.8x_3 - 4x_4$
$$+ 61.4x_5 + 82x_6 + 5x_7$$
$$+ 21.2x_8 + 5.8x_9 = -0.6$$
$$(continued)$$

$-3.3x_1 - 3.3x_2 - 16.1x_3 + 1.8x_4$
$$- 61.1x_5 - 9.7x_6 - 10.1x_7$$
$$- 28.2x_8 - 4.2x_9 = 7.3$$
$3x_1 + 3x_2 + 15x_3$
$$+ 56.3x_5 + 8.4x_6 + 13.7x_7$$
$$+ 30.3x_8 + 9.8x_9 = -9.9$$
$3x_1 + 3x_2 + 11x_3 + 3x_4$
$$+ 37x_5 + 19.5x_6 + 14x_7$$
$$+ 30.5x_8 - 7.5x_9 = -17$$
$-3x_1 - 3x_2 - 11x_3 - 3x_4$
$$- 41.1x_5 - 3.8x_6 - 5.9x_7$$
$$- 34.1x_8 + 16.4x_9 = 38.3$$
$-2.2x_4 + 5.2x_5 - 4.2x_6$
$$- 11.6x_7 - 1.4x_8 + 31.2x_9 = 48.2$$
$4.2x_1 + 4.2x_2 + 19.4x_3 - 3.2x_4$
$$+ 76.4x_5 - 0.2x_6 + 3.4x_7$$
$$+ 35.8x_8 - 9.6x_9 = -23.2$$

1.2 MATRICES AND MATRIX OPERATIONS

In the previous section we introduced augmented matrices for systems of linear equations as a convenient way of representing these systems. This is one of many uses of matrices. In this section we will look at matrices from a general point of view.

We should be explicit about exactly what a matrix is, so let us begin with a definition. A **matrix** is a rectangular array of objects called the **entries** of the matrix. (For us, the objects will be numbers, but they do not have to be. For example, we could have matrices whose entries are automobiles or members of a marching band.) We write matrices down by enclosing their entries within brackets (some use parentheses instead) and, if we wish to give a matrix a name, we will do so by using capital letters such as A, B, or C. Here are some examples of matrices:

$$A = \begin{bmatrix} 1 & 2 & 3 \\ 4 & 5 & 6 \end{bmatrix}, \quad B = \begin{bmatrix} -7 & 4 & 4 & 0 & 3 \end{bmatrix},$$

$$C = \begin{bmatrix} 0 \\ -1 \\ 1/2 \\ 4 \end{bmatrix}, \quad D = \begin{bmatrix} 0 & -2 & \pi & 8 \\ -1 & 12 & 3/8 & \ln 2 \\ 1 & 1 & 1 & -1 \\ \sqrt{2} & -7 & 0.9 & -391/629 \end{bmatrix}$$

Augmented matrices of systems of linear equations have these forms if we delete the dashed line. In fact, the dashed line is included merely as a convenience to help distinguish the left- and right-hand sides of the equations. If a matrix has m rows (which go across) and n columns (which go up and down), we say the **size** (or **dimensions**) of the matrix is (or are) $m \times n$ (read "m by n"). Thus, for the matrices just given, A is a 2×3 matrix, B is a 1×5 matrix, C is a 4×1 matrix, and D is a 4×4 matrix. A matrix such as B that has one row is called a **row matrix** or **row vector**; a matrix such as C that has one column is called a **column matrix** or **column vector.** Matrices that have the same number of rows as columns (that is, $n \times n$ matrices) are called **square matrices.** The matrix D is an example of a square matrix.

As you would expect, we consider two matrices A and B to be equal, written $A = B$, if they have the same size and entries. For example,

$$\begin{bmatrix} -1 & 2 \\ 1 & 12 \end{bmatrix} = \begin{bmatrix} -1 & 8/4 \\ 2 - 1 & 3 \cdot 4 \end{bmatrix}$$

while

$$\begin{bmatrix} -1 & 2 \\ 1 & 12 \end{bmatrix} \neq \begin{bmatrix} 5 & 2 \\ 1 & 12 \end{bmatrix} \quad \text{and} \quad \begin{bmatrix} 1 \\ 2 \end{bmatrix} \neq \begin{bmatrix} 1 & 2 \end{bmatrix}.$$

The general form of an $m \times n$ matrix A is

$$A = \begin{bmatrix} a_{11} & a_{12} & a_{13} & \cdots & a_{1n} \\ a_{21} & a_{22} & a_{23} & \cdots & a_{2n} \\ a_{31} & a_{32} & a_{33} & \cdots & a_{3n} \\ \vdots & \vdots & \vdots & & \vdots \\ a_{m1} & a_{m2} & a_{m3} & \cdots & a_{mn} \end{bmatrix}. \tag{1}$$

Notice that in this notation the first subscript i of an entry a_{ij} is the row in which the entry appears and the second subscript j is the column in which it appears. To save writing, we shall often indicate a matrix such as this by simply writing

$$A = [a_{ij}].$$

If we wish to single out the ij-entry of a matrix A, we will write

$$\text{ent}_{ij}(A).$$

For instance, if B is the matrix

$$B = \begin{bmatrix} -1 & 2 & 1 \\ 5 & 4 & -9 \\ 3 & -4 & 7 \end{bmatrix},$$

then

$$\text{ent}_{23}(B) = -9.$$

If $A = [a_{ij}]$ is an $n \times n$ matrix, the entries $a_{11}, a_{22}, \ldots, a_{nn}$ are called the **diagonal entries** of A. The matrix B has diagonal entries $-1, 4, 7$.

We will use the symbol $\mathbb{R}$ to denote the set of real numbers. The set of $m \times n$ matrices with entries from $\mathbb{R}$ will be denoted

$$M_{m \times n}(\mathbb{R}).$$

Thus, for example, in set notation

$$M_{2 \times 2}(\mathbb{R}) = \left\{ \begin{bmatrix} a_{11} & a_{12} \\ a_{21} & a_{22} \end{bmatrix} \mid a_{11}, a_{12}, a_{21}, a_{22} \in \mathbb{R} \right\}.^5$$

You have encountered two-dimensional vectors in two-dimensional space (which we will denote by $\mathbb{R}^2$) and three-dimensional vectors in three-dimensional space (which we will denote by $\mathbb{R}^3$) in previous courses. One standard notation for indicating such vectors is to use ordered pairs (a, b) for two-dimensional vectors and ordered triples (a, b, c) for three-dimensional vectors. Notice that these ordered pairs and triples are in fact row matrices or row vectors. However, we will be notationally better off if we use column matrices for two- and three-dimensional vectors. We also will identify the set of two-dimensional vectors with $\mathbb{R}^2$ and the set of three-dimensional vectors with $\mathbb{R}^3$; in other words,

$$\mathbb{R}^2 = M_{2 \times 1}(\mathbb{R}) = \left\{ \begin{bmatrix} a \\ b \end{bmatrix} \mid a, b \in \mathbb{R} \right\}$$

and

$$\mathbb{R}^3 = M_{3 \times 1}(\mathbb{R}) = \left\{ \begin{bmatrix} a \\ b \\ c \end{bmatrix} \mid a, b, c \in \mathbb{R} \right\}$$

in this book. More generally, the set of $n \times 1$ column matrices $M_{n \times 1}(\mathbb{R})$ will be denoted $\mathbb{R}^n$ and we will refer to the elements of $\mathbb{R}^n$ as **vectors in $\mathbb{R}^n$** or **n-dimensional vectors**.

We next turn our attention to the "arithmetic" of matrices beginning with the operations of **addition** and a multiplication by numbers called **scalar multiplication**.[6] If A and B are matrices of the same size, we add A and B by adding their corresponding

---

[5] In set notation, the vertical bar, |, denotes "such that" (some use a colon, :, instead of a vertical bar) and the symbol $\in$ denotes "element of" (or "member of"). One way of reading

$$\left\{ \begin{bmatrix} a_{11} & a_{12} \\ a_{21} & a_{22} \end{bmatrix} \mid a_{11}, a_{12}, a_{21}, a_{22} \in \mathbb{R} \right\}$$

is as "the set of matrices

$$\begin{bmatrix} a_{11} & a_{12} \\ a_{21} & a_{22} \end{bmatrix}$$

such that $a_{11}, a_{12}, a_{21}, a_{22}$ are elements of the set of real numbers."

[6] These two operations are extensions of the ones you already know for vectors in $\mathbb{R}^2$ or $\mathbb{R}^3$ to matrices in general.

entries; that is, if

$$A = [a_{ij}] \quad \text{and} \quad B = [b_{ij}]$$

are matrices in $M_{m \times n}(\mathbb{R})$, the sum of A and B is the $m \times n$ matrix

$$A + B = [a_{ij} + b_{ij}].$$

For instance, if

$$A = \begin{bmatrix} 1 & 2 \\ 3 & 4 \\ 5 & 6 \end{bmatrix} \quad \text{and} \quad B = \begin{bmatrix} 8 & 9 \\ 10 & 11 \\ 12 & 13 \end{bmatrix},$$

then

$$A + B = \begin{bmatrix} 1+8 & 2+9 \\ 3+10 & 4+11 \\ 5+12 & 6+13 \end{bmatrix} = \begin{bmatrix} 9 & 11 \\ 13 & 15 \\ 17 & 19 \end{bmatrix}.$$

Note that we have only defined sums of matrices of the same size. The sum of matrices of different sizes is undefined. For example, the sum

$$\begin{bmatrix} -2 & 5 \\ 3 & 1 \end{bmatrix} + \begin{bmatrix} 3 & 0 & -2 \end{bmatrix}$$

is undefined. If c is a real number (which we call a **scalar** in this setting) and $A = [a_{ij}]$ is an $m \times n$ matrix, the **scalar product** cA is the $m \times n$ matrix obtained by multiplying c times each entry of A:

$$cA = c[a_{ij}] = [ca_{ij}].$$

For example,

$$5 \begin{bmatrix} 1 & 2 \\ 3 & 4 \end{bmatrix} = \begin{bmatrix} 5 \cdot 1 & 5 \cdot 2 \\ 5 \cdot 3 & 5 \cdot 4 \end{bmatrix} = \begin{bmatrix} 5 & 10 \\ 15 & 20 \end{bmatrix}.$$

The following theorem lists some elementary properties involving addition and scalar multiplication of matrices.

THEOREM 1.2   If A, B, and C are matrices of the same size and if c and d are scalars, then:

1. $A + B = B + A$ (commutative law of addition).
2. $A + (B + C) = (A + B) + C$ (associative law of addition).
3. $c(dA) = (cd)A$.
4. $c(A + B) = cA + cB$.
5. $(c + d)A = cA + dA$.

Proof   We prove these equalities by showing that the matrices on each side have the same entries. Let us prove parts (1) and (4) here. The proofs of the remaining parts will be left as exercises (Exercise 24). For notational purposes, we set

$$A = [a_{ij}] \quad \text{and} \quad B = [b_{ij}].$$

Part (1) follows since

$$\text{ent}_{ij}(A + B) = a_{ij} + b_{ij} = b_{ij} + a_{ij} = \text{ent}_{ij}(B + A).$$

To obtain part (4),

$$\text{ent}_{ij}(c(A + B)) = c(a_{ij} + b_{ij}) = ca_{ij} + cb_{ij} = \text{ent}_{ij}(cA + cB). \qquad \bullet$$

One special type of matrix is the set of **zero matrices**. The $m \times n$ zero matrix, denoted $O_{m \times n}$, is the $m \times n$ matrix that has all of its entries zero. For example,

$$O_{2 \times 2} = \begin{bmatrix} 0 & 0 \\ 0 & 0 \end{bmatrix} \quad \text{and} \quad O_{4 \times 3} = \begin{bmatrix} 0 & 0 & 0 \\ 0 & 0 & 0 \\ 0 & 0 & 0 \\ 0 & 0 & 0 \end{bmatrix}.$$

Notice that if A is an $m \times n$ matrix, then:

1. $A + O_{m \times n} = A$.
2. $0 \cdot A = O_{m \times n}$.

We often will indicate a zero matrix by simply writing O. (To avoid confusion with the number zero, we put this in boldface print in this book.) For instance, we might write the first property as $A + O = A$. The second property could be written as $0 \cdot A = O$.

The **negative** of a matrix $A = [a_{ij}]$, denoted $-A$, is the matrix whose entries are the negatives of those of A:

$$\boxed{-A = [-a_{ij}].}$$

Notice that

$$-A = (-1)A \quad \text{and} \quad A + (-A) = O.$$

Subtraction of matrices A and B of the same size can be defined in terms of adding the negative of B:

$$\boxed{A - B = A + (-B).}$$

Of course, notice that $A - B$ could also be found by subtracting the entries of B from the corresponding entries of A.

Up to this point, all of the operations we have introduced on matrices should seem relatively natural. Our final operation will be matrix multiplication, which upon first glance may not seem to be the natural way to multiply matrices. However, the manner of multiplying matrices you are about to see is the one that we will need as we use matrix multiplication in our future work.

Here is how we do matrix multiplication: Suppose that $A = [a_{ij}]$ is an $m \times n$ matrix and $B = [b_{ij}]$ is an $n \times l$ matrix. The product of A and B is defined to be the $m \times l$ matrix

$$\boxed{AB = [p_{ij}]}$$

where

$$\boxed{p_{ij} = a_{i1}b_{1j} + a_{i2}b_{2j} + a_{i3}b_{3j} + \cdots + a_{in}b_{nj} = \sum_{k=1}^{n} a_{ik}b_{kj}.}$$

In other words, for each $1 \leq i \leq m$ and $1 \leq j \leq l$ the ij-entry of AB is found by multiplying each entry of row i of A times its corresponding entry of column j of B and then summing these products.

Here is an example illustrating our matrix multiplication procedure.

EXAMPLE 1   Find the product AB for

$$A = \begin{bmatrix} 1 & 2 \\ 3 & 4 \end{bmatrix} \quad \text{and} \quad B = \begin{bmatrix} 5 & 6 \\ 7 & 8 \end{bmatrix}.$$

Solution   The product AB is

$$AB = \begin{bmatrix} 1 & 2 \\ 3 & 4 \end{bmatrix} \begin{bmatrix} 5 & 6 \\ 7 & 8 \end{bmatrix}$$

$$= \begin{bmatrix} 1 \cdot 5 + 2 \cdot 7 & 1 \cdot 6 + 2 \cdot 8 \\ 3 \cdot 5 + 4 \cdot 7 & 3 \cdot 6 + 4 \cdot 8 \end{bmatrix} = \begin{bmatrix} 19 & 22 \\ 43 & 50 \end{bmatrix}. \qquad \bullet$$

Once you practice this sum of row entries times column entries a few times, you should find yourself getting the hang of it.[7] Let us do another example of matrix multiplication.

EXAMPLE 2   Find the product CD for

$$C = \begin{bmatrix} -1 & 2 & -3 \\ 0 & -1 & 1 \\ 4 & 2 & -1 \end{bmatrix} \quad \text{and} \quad D = \begin{bmatrix} 1 & -2 \\ -3 & 4 \\ 1 & 1 \end{bmatrix}.$$

---

[7] You might find it convenient to note that the ij-entry of AB is much like the dot product of the vector formed by row i of A with the vector formed by column j of B. We will discuss dot products more fully in Chapter 9.

Solution

$$CD = \begin{bmatrix} -1 & 2 & -3 \\ 0 & -1 & 1 \\ 4 & 2 & -1 \end{bmatrix} \begin{bmatrix} 1 & -2 \\ -3 & 4 \\ 1 & 1 \end{bmatrix}$$

$$= \begin{bmatrix} (-1)\cdot1+2(-3)-3\cdot1 & (-1)(-2)+2\cdot4-3\cdot1 \\ 0\cdot1-1(-3)+1\cdot1 & 0(-2)-1\cdot4+1\cdot1 \\ 4\cdot1+2(-3)-1\cdot1 & 4(-2)+2\cdot4-1\cdot1 \end{bmatrix} = \begin{bmatrix} -10 & 7 \\ 4 & -3 \\ -3 & -1 \end{bmatrix}$$

Notice that for the product AB of two matrices A and B to be defined, it is necessary that the number of columns of A be the same as the number of rows of B. If this is not the case, the product is not defined. For instance, the product DC for the matrices in Example 2 is not defined. In particular, CD is not the same as DC. This is an illustration of the fact that *matrix multiplication is not commutative;* that is, AB is not in general the same as BA for matrices A and B. Sometimes these products are not the same because one is defined while the other is not, as the matrices C and D illustrate. But even if both products are defined, it is often the case that they are not the same. If you compute the product BA for the matrices in Example 1, you will find (try it)

$$BA = \begin{bmatrix} 23 & 34 \\ 31 & 46 \end{bmatrix},$$

which is not the same as AB.[8] In the case when $AB = BA$ for two matrices A and B, we say A and B **commute.**

While matrix multiplication is not commutative, some properties that you are used to having for multiplication of numbers do carry over to matrices when the products are defined.

THEOREM 1.3   Provided that the indicated sums and products are defined, the following properties hold where A, B, and C are matrices and d is a scalar.

1. $A(BC) = (AB)C$ (associative law of multiplication)
2. $A(B + C) = AB + AC$ (left-hand distributive law)
3. $(A + B)C = AC + BC$ (right-hand distributive law)
4. $d(AB) = (dA)B = A(dB)$

Proof   We will prove the first two parts here and leave proofs of the remaining parts as exercises (Exercise 25). For notational purposes, suppose

$$A = [a_{ij}], \quad B = [b_{ij}], \quad \text{and} \quad C = [c_{ij}].$$

---

[8] This is not the first time you have encountered an example of a noncommutative operation. Composition of functions is noncommutative. The cross product of two three-dimensional vectors is another example of a noncommutative operation.

To prove part (1), we also have to introduce some notation for the sizes of A, B, and C. Suppose A is an $m \times n$ matrix, B is an $n \times l$ matrix, and C is an $l \times h$ matrix. Both $A(BC)$ and $(AB)C$ are $m \times h$ matrices. (Why?) To see that these products are the same, we work out the ij-entry of each. For $A(BC)$, this is

$$\text{ent}_{ij}(A(BC)) = \sum_{k=1}^{n} a_{ik}\text{ent}_{kj}(BC) = \sum_{k=1}^{n} a_{ik}\left(\sum_{q=1}^{l} b_{kq}c_{qj}\right) = \sum_{k=1}^{n}\left(\sum_{q=1}^{l} a_{ik}b_{kq}c_{qj}\right).$$

Carrying out the same steps for $(AB)C$,

$$\text{ent}_{ij}((AB)C) = \sum_{q=1}^{l} \text{ent}_{iq}(AB)c_{qj} = \sum_{q=1}^{l}\left(\sum_{k=1}^{n} a_{ik}b_{kq}\right)c_{qj} = \sum_{q=1}^{l}\left(\sum_{k=1}^{n} a_{ik}b_{kq}c_{qj}\right).$$

Since the summations over k and q are interchangeable, we see that the ij-entries of $A(BC)$ and $(AB)C$ are the same and hence $A(BC) = (AB)C$.

To prove part (2), we again introduce notation for the sizes of our matrices. Suppose A is an $m \times n$ matrix and B and C are $n \times l$ matrices. Both $A(B + C)$ and $AB + AC$ are $m \times l$ matrices. We have

$$\text{ent}_{ij}(A(B + C)) = \sum_{k=1}^{n} a_{ik}(\text{ent}_{kj}(B + C)) = \sum_{k=1}^{n} a_{ik}(b_{kj} + c_{kj})$$

$$= \sum_{k=1}^{n}(a_{ik}b_{kj} + a_{ik}c_{kj})$$

and

$$\text{ent}_{ij}(AB + AC) = \text{ent}_{ij}(AB) + \text{ent}_{ij}(AC) = \sum_{k=1}^{n} a_{ik}b_{kj} + \sum_{k=1}^{n} a_{ik}c_{kj}$$

$$= \sum_{k=1}^{n}(a_{ik}b_{kj} + a_{ik}c_{kj}).$$

Thus $A(B + C) = AB + AC$ since they have the same entries. ●

If A is a square matrix, we can define positive integer powers of A in the same manner as we do for real numbers; that is,

$$A^1 = A, \quad A^2 = AA, \quad \text{and} \quad A^3 = A^2A = AAA, \ldots.$$

Such powers are not defined, however, if A is not a square matrix. (Why is this the case?) If A is an $m \times n$ matrix, it is easy to see that

$$O_{l \times m}A = O_{l \times n} \quad \text{and} \quad AO_{n \times l} = O_{m \times l}.$$

Besides zero matrices, another special type of matrices is the set of **identity matrices**. The $n \times n$ identity matrix, denoted I_n, has diagonal entries 1 and all other entries 0. For example,

$$I_2 = \begin{bmatrix} 1 & 0 \\ 0 & 1 \end{bmatrix} \quad \text{and} \quad I_3 = \begin{bmatrix} 1 & 0 & 0 \\ 0 & 1 & 0 \\ 0 & 0 & 1 \end{bmatrix}.$$

Identity matrices play the role the number 1 plays for the real numbers with respect to multiplication in the sense that

$$I_m A = A \quad \text{and} \quad A I_n = A$$

for any $m \times n$ matrix A. (Convince yourself of these statements.)

One use (among many more to come) of matrix multiplication arises in connection with systems of linear equations. Given a system of linear equations

$$a_{11}x_1 + a_{12}x_2 + \cdots + a_{1n}x_n = b_1$$
$$a_{21}x_1 + a_{22}x_2 + \cdots + a_{2n}x_n = b_2$$
$$\vdots$$
$$a_{m1}x_1 + a_{m2}x_2 + \cdots + a_{mn}x_n = b_m,$$

we will let A denote the coefficient matrix of this system,

$$A = \begin{bmatrix} a_{11} & a_{12} & \cdots & a_{1n} \\ a_{21} & a_{22} & \cdots & a_{2n} \\ \vdots & \vdots & & \vdots \\ a_{m1} & a_{m2} & \cdots & a_{mn} \end{bmatrix},$$

X denote the column of variables,

$$X = \begin{bmatrix} x_1 \\ x_2 \\ \vdots \\ x_n \end{bmatrix},$$

and B denote the column,

$$B = \begin{bmatrix} b_1 \\ b_2 \\ \vdots \\ b_m \end{bmatrix}.$$

Observe that our system can then be conveniently written as the **matrix equation**

$$\boxed{AX = B.}$$

For instance, the system

$$2x - y + 4z = 1$$
$$x - 7y + z = 3$$
$$-x + 2y + z = 2$$

would be written

$$\begin{bmatrix} 2 & -1 & 4 \\ 1 & -7 & 1 \\ -1 & 2 & 1 \end{bmatrix} \begin{bmatrix} x \\ y \\ z \end{bmatrix} = \begin{bmatrix} 1 \\ 3 \\ 2 \end{bmatrix}$$

as a matrix equation. Notice that a homogeneous linear system takes on the form $AX = O$ when written as a matrix equation.

EXERCISES 1.2

In Exercises 1–18, either perform the indicated operations or state that the expression is undefined where A, B, C, D, E, and F are the matrices:

$$A = \begin{bmatrix} 1 & 2 \\ 3 & -1 \\ 2 & -1 \end{bmatrix}, \quad B = \begin{bmatrix} 2 & -1 \\ -3 & -2 \\ 0 & 4 \end{bmatrix},$$

$$C = \begin{bmatrix} 2 & -1 \\ 1 & 5 \end{bmatrix}, \quad D = \begin{bmatrix} 0 & 1 \\ 3 & -1 \end{bmatrix},$$

$$E = \begin{bmatrix} 1 & -3 & 5 \\ 2 & 1 & -1 \\ 1 & 1 & 0 \end{bmatrix}, \quad F = \begin{bmatrix} 1 & -1 & 4 \\ 2 & -3 & 6 \\ 1 & 0 & 1 \end{bmatrix}.$$

1. $A + B$ **2.** $D - C$ **3.** $2B$

4. $-\frac{3}{4}F$ **5.** $A - 4B$ **6.** $3D + 2C$

7. CD **8.** DC **9.** EF

10. FE **11.** AE **12.** EA

13. $(E + F)A$ **14.** $B(C + D)$ **15.** $3AC$

16. $F(-2B)$ **17.** C^2 **18.** A^3

Write the systems of equations in Exercises 19 and 20 in the matrix form $AX = B$.

19. $\quad 2x - y + 4z = 1$
$\qquad x + y - z = 4$
$\qquad\quad y + 3z = 5$
$\qquad x + y = 2$

20. $\quad x_1 - 3x_2 + x_3 - 5x_4 = 2$
$\qquad x_1 + x_2 - x_3 + x_4 = 1$
$\qquad x_1 - x_2 - x_3 + 6x_4 = 6$

Write the matrix equations as systems of equations in Exercises 21 and 22.

21. $\begin{bmatrix} 2 & -2 & 5 & 7 \\ 4 & 5 & -11 & 3 \end{bmatrix} \begin{bmatrix} x_1 \\ x_2 \\ x_3 \\ x_4 \end{bmatrix} = \begin{bmatrix} 12 \\ -3 \end{bmatrix}$

22. $\begin{bmatrix} 2 & 2 & -11 \\ 0 & -1 & -5 \\ 2 & -3 & 0 \end{bmatrix} \begin{bmatrix} x \\ y \\ z \end{bmatrix} = \begin{bmatrix} 51 \\ -33 \\ 1/2 \end{bmatrix}$

23. Suppose that A and B are $n \times n$ matrices.
     a) Show that $(A + B)^2 = A^2 + AB + BA + B^2$.
     b) Explain why $(A + B)^2$ is not equal to $A^2 + 2AB + B^2$ in general.

24. Prove the following parts of Theorem 1.2.
     a) Part (2)
     b) Part (3)
     c) Part (5)

25. Prove the following parts of Theorem 1.3.
     a) Part (3)
     b) Part (4)

26. Suppose A is an $m \times n$ matrix and B is an $n \times l$ matrix. Further, suppose that A has a row of zeros. Does AB have a row of zeros? Why or why not? Does this also hold if B has a row of zeros? Why or why not?

27. Suppose A is an $m \times n$ matrix and B is an $n \times l$ matrix. Further, suppose that B has a column of zeros. Does AB have a column of zeros? Why or why not? Does this also hold if A has a column of zeros? Why or why not?

28. Give an example of two matrices A and B for which $AB = O$ with $A \neq O$ and $B \neq O$.

29. a) Suppose that A is the row vector

$$A = \begin{bmatrix} a_1 & a_2 & \cdots & a_n \end{bmatrix}$$

and B is an $n \times l$ matrix. View B as the column of row vectors

$$B = \begin{bmatrix} B_1 \\ B_2 \\ \vdots \\ B_n \end{bmatrix}$$

where $B_1, B_2, \ldots, B_n$ are the rows of B. Show that

$$AB = a_1 B_1 + a_2 B_2 + \cdots + a_n B_n.$$

b) Use the result of part (a) to find AB for

$$A = \begin{bmatrix} -2 & 1 & 6 \end{bmatrix} \quad \text{and} \quad B = \begin{bmatrix} -1 & 1 & 0 \\ 2 & 1 & 1 \\ 4 & -1 & 2 \end{bmatrix}.$$

30. a) Suppose that B is the column vector

$$B = \begin{bmatrix} b_1 \\ b_2 \\ \vdots \\ b_n \end{bmatrix}$$

and A is an $m \times n$ matrix. View A as the row of column vectors

$$A = \begin{bmatrix} A_1 & A_2 & \cdots & A_n \end{bmatrix}$$

where $A_1, A_2, \ldots, A_n$ are the columns of A. Show that

$$AB = b_1 A_1 + b_2 A_2 + \cdots + b_n A_n.$$

b) Use the result of part (a) to find AB for

$$A = \begin{bmatrix} 3 & 2 & -1 \\ 0 & 3 & 5 \\ 1 & 1 & 2 \end{bmatrix} \quad \text{and} \quad B = \begin{bmatrix} 2 \\ -1 \\ 3 \end{bmatrix}.$$

31. The **trace** of a square matrix A, denoted $\text{tr}(A)$, is the sum of the diagonal entries of A. Find $\text{tr}(A)$ for

$$A = \begin{bmatrix} 5 & 0 & -4 \\ 2 & -11 & 6 \\ 2 & 10 & 3 \end{bmatrix}.$$

32. Prove the following where A and B are square matrices of the same size and c is a scalar.

a) $\text{tr}(A + B) = \text{tr}(A) + \text{tr}(B)$

b) $\text{tr}(cA) = c\,\text{tr}(A)$

c) $\text{tr}(AB) = \text{tr}(BA)$

The *matrix* command introduced in the previous section is one way of entering matrices on a Maple worksheet. Maple uses the *evalm* command along with +, -, *, &*, and ∧ to find sums, differences, scalar products, matrix products, and matrix powers, respectively. For instance, to find $A - B + 4C + AB - C^3$ where A, B, and C are matrices already entered on a Maple worksheet, we would type and enter

```
evalm(A-B+4*C+A&*B-C∧3);
```

at the command prompt. A scalar product cA also may be found with the *scalarmul* command by typing

```
scalarmul(A,c);
```

at the command prompt. Products of two or more matrices can be found by using the *multiply* command. For instance, typing and entering

```
multiply(B,A,C);
```

will give us the product BAC. Use these Maple commands or appropriate commands in another suitable software package (keep in mind that corresponding Mathematica and MATLAB commands can be found in the *Technology Resource Manual*) to find the indicated expression (if possible) where

$$A = \begin{bmatrix} 4 & -2 & 16 & 27 & -11 \\ 9 & 43 & 9 & -8 & -1 \\ 34 & 20 & -3 & 0 & 21 \\ -5 & 4 & 4 & 7 & 41 \\ 0 & 12 & -2 & -2 & 3 \end{bmatrix},$$

$$B = \begin{bmatrix} 1 & 1 & 0 & 0 & 0 \\ 2 & 2 & 2 & 0 & 0 \\ 0 & 3 & 3 & 3 & 0 \\ 0 & 0 & 4 & 4 & 4 \\ 0 & 0 & 0 & 5 & 5 \end{bmatrix}, \quad \text{and}$$

$$C = \begin{bmatrix} -2 & 1 & 0 & 3 & 0 \\ 0 & 1 & 2 & 1 & -1 \\ 3 & -1 & -1 & 1 & 1 \\ -1 & 0 & 2 & 2 & -3 \end{bmatrix}$$

in Exercises 33–40.

33. $A - 2B$

34. $5A + 6C$

35. ABC

36. $CB + C$

37. $(A + B)^2$

38. $4A + CB$

39. $4CA - 5CB - 2C$

40. $B^2 - 4AB + 2A^2$

1.3   INVERSES OF MATRICES

If A is an $n \times n$ matrix, we say that an $n \times n$ matrix B is an **inverse** of A if

$$\boxed{AB = BA = I}$$

where I is the $n \times n$ identity matrix. To illustrate, the matrix

$$A = \begin{bmatrix} 1 & 2 \\ 3 & 5 \end{bmatrix}$$

has the matrix

$$B = \begin{bmatrix} -5 & 2 \\ 3 & -1 \end{bmatrix}$$

as an inverse since

$$AB = \begin{bmatrix} 1 & 2 \\ 3 & 5 \end{bmatrix} \begin{bmatrix} -5 & 2 \\ 3 & -1 \end{bmatrix} = \begin{bmatrix} 1 & 0 \\ 0 & 1 \end{bmatrix}$$

and

$$BA = \begin{bmatrix} -5 & 2 \\ 3 & -1 \end{bmatrix} \begin{bmatrix} 1 & 2 \\ 3 & 5 \end{bmatrix} = \begin{bmatrix} 1 & 0 \\ 0 & 1 \end{bmatrix}.$$

(How we obtain B will be seen later.)

Not all square matrices have inverses. Certainly, square zero matrices do not have inverses. (The products OB and BO are O, not I.) But even a nonzero square matrix may fail to have an inverse. As a simple example, the matrix

$$A = \begin{bmatrix} 1 & 0 \\ 0 & 0 \end{bmatrix}$$

cannot have an inverse since for any 2×2 matrix

$$B = \begin{bmatrix} a & b \\ c & d \end{bmatrix},$$

we have

$$AB = \begin{bmatrix} 1 & 0 \\ 0 & 0 \end{bmatrix} \begin{bmatrix} a & b \\ c & d \end{bmatrix} = \begin{bmatrix} a & b \\ 0 & 0 \end{bmatrix} \neq \begin{bmatrix} 1 & 0 \\ 0 & 1 \end{bmatrix}.$$

Square matrices that have inverses are called **invertible** or **nonsingular** matrices; those that do not have inverses are called **noninvertible** or **singular** matrices.

When a matrix has an inverse, it has only one inverse.

THEOREM 1.4   If A is an invertible matrix, then the inverse of A is unique.

Proof   Suppose that A did have two inverses B and C. Consider the product BAC. If we group B and A together,

$$BAC = (BA)C = IC = C$$

since $BA = I$. If we group A and C together,

$$BAC = B(AC) = BI = B$$

since $AC = I$. Thus,

$$C = B \qquad \qquad \bullet$$

The uniqueness of the inverse of an invertible matrix A allows us to speak of *the* inverse of A rather than *an* inverse of A. It also allows us to introduce a symbol for the inverse of A. Henceforth we shall denote the inverse of A by

$$\boxed{A^{-1}}$$

in much the same manner as we use the exponent -1 for denoting inverses of functions.[9]

Let us now turn our attention to a method for finding inverses of square matrices. Consider again the matrix

$$A = \begin{bmatrix} 1 & 2 \\ 3 & 5 \end{bmatrix}.$$

Let us think of the inverse of A as an unknown matrix

$$A^{-1} = \begin{bmatrix} x_{11} & x_{12} \\ x_{21} & x_{22} \end{bmatrix}.$$

We want to find the entries so that

$$AA^{-1} = \begin{bmatrix} 1 & 2 \\ 3 & 5 \end{bmatrix} \begin{bmatrix} x_{11} & x_{12} \\ x_{21} & x_{22} \end{bmatrix} = \begin{bmatrix} x_{11} + 2x_{21} & x_{12} + 2x_{22} \\ 3x_{11} + 5x_{21} & 3x_{12} + 5x_{22} \end{bmatrix} = \begin{bmatrix} 1 & 0 \\ 0 & 1 \end{bmatrix}.$$

This gives us a system of equations in x_{11} and x_{21},

$$x_{11} + 2x_{21} = 1$$
$$3x_{11} + 5x_{21} = 0,$$

and a system in x_{12} and x_{22},

$$x_{12} + 2x_{22} = 0$$
$$3x_{12} + 5x_{22} = 1.$$

---

[9] Do note that A^{-1} does not stand for $1/A$ any more than $\sin^{-1} x$ stands for $1/\sin x$; indeed writing $1/A$ for a matrix A amounts to writing nonsense.

We then will have a unique matrix A^{-1} so that $AA^{-1} = I$ if and only if each of these systems of equations has a unique solution (which occurs if and only if the reduced row-echelon form of A is I). Let us solve these systems to see if this is the case. To save writing, notice that since both of these systems have the same coefficient matrices, any set of row operations that leads to the solution of one system leads to the solution of the other system too. Thus we can simultaneously solve these two systems by forming the augmented matrix

$$[A|I] = \begin{bmatrix} 1 & 2 & \vdots & 1 & 0 \\ 3 & 5 & \vdots & 0 & 1 \end{bmatrix}$$

and then using row operations to reduce its left-hand portion A to reduced row-echelon form (which will be I):

$$\rightarrow \begin{bmatrix} 1 & 2 & \vdots & 1 & 0 \\ 0 & -1 & \vdots & -3 & 1 \end{bmatrix} \rightarrow \begin{bmatrix} 1 & 0 & \vdots & -5 & 2 \\ 0 & 1 & \vdots & 3 & -1 \end{bmatrix}.$$

(We have not indicated the row operations here. Can you determine the ones we used?) The right-hand portion of our final augmented matrix tells us $x_{11} = -5$, $x_{21} = 3$, $x_{12} = 2$, $x_{22} = -1$, and hence

$$A^{-1} = \begin{bmatrix} -5 & 2 \\ 3 & -1 \end{bmatrix}.$$

We must be honest, however. There is a gap in our development here. The procedure we have just illustrated produces a matrix B so that $AB = I$. (We describe this by saying B is a right-hand inverse.) But the inverse of A must also have the property that $BA = I$. (When $BA = I$, we say B is a left-hand inverse.) You can check that the right-hand inverse we have just found for the given 2×2 matrix A is also a left-hand inverse and hence is A^{-1}. Shortly we will fill in this left-hand inverse gap. Once we do so, we then will have that the inverse of a square matrix A (if any) can be found by the following procedure:

1. Form the augmented matrix $[A|I]$ where I is the identity matrix with the same size as A.

2. Use row operations to reduce the left-hand portion into reduced row-echelon form.

3. If the augmented matrix after step 2 has the form $[I|B]$, then $B = A^{-1}$; if it does not have this form (or equivalently, if the reduced row-echelon form of A contains a zero row), A does not have an inverse.

Examples 1 and 2 illustrate our procedure for finding inverses.

EXAMPLE 1   If possible, find the inverse of the following matrix.

$$\begin{bmatrix} 2 & 1 & 3 \\ 2 & 1 & 1 \\ 4 & 5 & 1 \end{bmatrix}$$

Solution   We first form the augmented matrix and then apply row operations:

$$\left[\begin{array}{ccc|ccc} 2 & 1 & 3 & 1 & 0 & 0 \\ 2 & 1 & 1 & 0 & 1 & 0 \\ 4 & 5 & 1 & 0 & 0 & 1 \end{array}\right] \rightarrow \left[\begin{array}{ccc|ccc} 2 & 1 & 3 & 1 & 0 & 0 \\ 0 & 0 & -2 & -1 & 1 & 0 \\ 0 & 3 & -5 & -2 & 0 & 1 \end{array}\right]$$

$$\rightarrow \left[\begin{array}{ccc|ccc} 2 & 1 & 3 & 1 & 0 & 0 \\ 0 & 3 & -5 & -2 & 0 & 1 \\ 0 & 0 & -2 & -1 & 1 & 0 \end{array}\right] \rightarrow \left[\begin{array}{ccc|ccc} 6 & 0 & 14 & 5 & 0 & -1 \\ 0 & 3 & -5 & -2 & 0 & 1 \\ 0 & 0 & -2 & -1 & 1 & 0 \end{array}\right]$$

$$\rightarrow \left[\begin{array}{ccc|ccc} 6 & 0 & 0 & -2 & 7 & -1 \\ 0 & 6 & 0 & 1 & -5 & 2 \\ 0 & 0 & -2 & -1 & 1 & 0 \end{array}\right] \rightarrow \left[\begin{array}{ccc|ccc} 1 & 0 & 0 & -1/3 & 7/6 & -1/6 \\ 0 & 1 & 0 & 1/6 & -5/6 & 1/3 \\ 0 & 0 & 1 & 1/2 & -1/2 & 0 \end{array}\right].$$

Thus the inverse of this matrix is

$$\left[\begin{array}{ccc} -1/3 & 7/6 & -1/6 \\ 1/6 & -5/6 & 1/3 \\ 1/2 & -1/2 & 0 \end{array}\right].$$

●

EXAMPLE 2   If possible, find the inverse of the following matrix.

$$\left[\begin{array}{ccc} 1 & -2 & 2 \\ 2 & -3 & 1 \\ 1 & -1 & -1 \end{array}\right]$$

Solution   We again form the augmented matrix and apply row operations:

$$\left[\begin{array}{ccc|ccc} 1 & -2 & 2 & 1 & 0 & 0 \\ 2 & -3 & 1 & 0 & 1 & 0 \\ 1 & -1 & -1 & 0 & 0 & 1 \end{array}\right] \rightarrow \left[\begin{array}{ccc|ccc} 1 & -2 & 2 & 1 & 0 & 0 \\ 0 & 1 & -3 & -2 & 1 & 0 \\ 0 & 1 & -3 & -1 & 0 & 1 \end{array}\right].$$

At this point it is apparent that the left-hand portion cannot be reduced to I and hence the matrix in this example does not have an inverse.   ●

When the inverse of a square matrix A is known, we can easily find the solutions to a system of linear equations

$$AX = B.$$

If we multiply this matrix equation by A^{-1} on the left, we have

$$A^{-1}AX = A^{-1}B$$

and hence the solution is given by

$$X = A^{-1}B.$$

We use this approach to solve the system in the next example.

EXAMPLE 3   Solve the system

$$2x + y + 3z = 6$$
$$2x + y + z = -12$$
$$4x + 5y + z = 3.$$

Solution   From Example 1, we have that the inverse of the coefficient matrix

$$A = \begin{bmatrix} 2 & 1 & 3 \\ 2 & 1 & 1 \\ 4 & 5 & 1 \end{bmatrix}$$

of this system is

$$\begin{bmatrix} -1/3 & 7/6 & -1/6 \\ 1/6 & -5/6 & 1/3 \\ 1/2 & -1/2 & 0 \end{bmatrix}.$$

The solution is then given by

$$X = \begin{bmatrix} x \\ y \\ z \end{bmatrix} = A^{-1}B = \begin{bmatrix} -1/3 & 7/6 & -1/6 \\ 1/6 & -5/6 & 1/3 \\ 1/2 & -1/2 & 0 \end{bmatrix} \begin{bmatrix} 6 \\ -12 \\ 3 \end{bmatrix} = \begin{bmatrix} -33/2 \\ 12 \\ 9 \end{bmatrix};$$

that is, $x = -33/2$, $y = 12$, $z = 9$.   ●

The following theorem gives us a characterization of when a system of n linear equations in n unknowns has a unique solution.

THEOREM 1.5   A system $AX = B$ of n linear equations in n unknowns has a unique solution if and only if A is invertible.

Proof   If A is invertible, the solutions to the system are given by $X = A^{-1}B$ and hence are unique. Conversely, suppose $AX = B$ has a unique solution. Considering the result of Gauss-Jordan elimination on the system, it follows that the reduced row-echelon form of A is I. Hence A is invertible.   ●

We now develop some mathematics that will justify why $B = A^{-1}$ when we are able to reduce $[A|I]$ to $[I|B]$.

Matrices obtained from an identity matrix I by applying an elementary row operation to I are called **elementary matrices.** We classify elementary matrices into the following three types.

Type 1: An elementary matrix obtained by interchanging two rows of I

Type 2: An elementary matrix obtained by multiplying a row of I by a nonzero number

Type 3: An elementary matrix obtained by replacing a row of I by itself plus a multiple of another row of I

Some examples of elementary matrices of each of these respective types obtained from the 2×2 identity matrix are

$$E_1 = \begin{bmatrix} 0 & 1 \\ 1 & 0 \end{bmatrix} \quad \text{(interchange rows 1 and 2)},$$

$$E_2 = \begin{bmatrix} 2 & 0 \\ 0 & 1 \end{bmatrix} \quad \text{(multiply row 1 by 2)},$$

$$E_3 = \begin{bmatrix} 1 & 0 \\ 3 & 1 \end{bmatrix} \quad \text{(add 3 times row 1 to row 2)}.$$

An interesting fact is that multiplication by elementary matrices on the left of another matrix performs the corresponding row operation on the other matrix. Notice how this works when we multiply E_1, E_2, and E_3 times a 2×2 matrix:

$$\begin{bmatrix} 0 & 1 \\ 1 & 0 \end{bmatrix}\begin{bmatrix} a & b \\ c & d \end{bmatrix} = \begin{bmatrix} c & d \\ a & b \end{bmatrix}$$

$$\begin{bmatrix} 2 & 0 \\ 0 & 1 \end{bmatrix}\begin{bmatrix} a & b \\ c & d \end{bmatrix} = \begin{bmatrix} 2a & 2b \\ c & d \end{bmatrix}$$

$$\begin{bmatrix} 1 & 0 \\ 3 & 1 \end{bmatrix}\begin{bmatrix} a & b \\ c & d \end{bmatrix} = \begin{bmatrix} a & b \\ 3a+c & 3b+d \end{bmatrix}.$$

These illustrate the following theorem whose proof we leave as an exercise (Exercise 13).

THEOREM 1.6 Suppose that A is an $m \times n$ matrix and E is an $m \times m$ elementary matrix.

1. If E is obtained by interchanging rows i and j of I, then EA is the matrix obtained from A by interchanging rows i and j of A.

2. If E is obtained by multiplying row i of I by c, then EA is the matrix obtained from A by multiplying row i of A by c.

3. If E is obtained by replacing row i of I by itself plus c times row j of I, then EA is the matrix obtained from A by replacing row i of A by itself plus c times row j of A.

Of course, we would not use multiplication by elementary matrices to perform row operations—certainly we would just do the row operations! Nevertheless, they do serve as a useful theoretical tool from time to time. Our first instance of this involves seeing why our procedure for finding inverses does in fact produce the inverse. Look at our procedure in the following way. We begin with the augmented matrix $[A|I]$ and use elementary row operations to reduce it to $[I|B]$. Suppose this takes k elementary row operations and $E_1, E_2, \ldots, E_k$ are the elementary matrices that perform the successive row operations. Since performing these elementary operations on $[A|I]$ is the same as performing them on A and I individually, it follows that

$$E_k \cdots E_2 E_1 [A|I] = [E_k \cdots E_2 E_1 A | E_k \cdots E_2 E_1 I] = [I|B].$$

From the right-hand portion of this augmented matrix, we see

$$B = E_k \cdots E_2 E_1.$$

From the left-hand portion, we see

$$E_k \cdots E_2 E_1 A = BA = I.$$

Thus B is not only the right-hand inverse of A as we saw from conception of our method for finding inverses, but is the necessary left-hand inverse too.

Let us proceed to further develop the theory of invertible matrices. We begin with the following theorem.

THEOREM 1.7   If A and B are invertible matrices of the same size, then AB is invertible and

$$(AB)^{-1} = B^{-1} A^{-1}.$$

Proof   It suffices to show that $B^{-1} A^{-1}$ is the inverse of AB. This we do by showing the necessary products are I:

$$ABB^{-1} A^{-1} = AIA^{-1} = AA^{-1} = I$$

and

$$B^{-1} A^{-1} AB = B^{-1} IB = B^{-1} B = I. \qquad \bullet$$

Notice that $(AB)^{-1}$ is not $A^{-1} B^{-1}$. The result of Theorem 1.7 generalizes to products of invertible matrices with more factors as follows: If $A_1, A_2, \ldots, A_n$ are invertible matrices of the same size, then

$$(A_1 A_2 \cdots A_n)^{-1} = A_n^{-1} \cdots A_2^{-1} A_1^{-1}$$

since

$$(A_1 A_2 \cdots A_n)^{-1} = (A_2 A_3 \cdots A_n)^{-1} A_1^{-1}$$
$$= (A_3 \cdots A_n)^{-1} A_2^{-1} A_1^{-1} = \ldots = A_n^{-1} \cdots A_2^{-1} A_1^{-1}.$$

Next, we consider the invertibility of elementary matrices.

THEOREM 1.8   If E is an elementary matrix, then E is invertible and:

  1. If E is obtained by interchanging two rows of I, then $E^{-1} = E$;

2. If E is obtained by multiplying row i of I by a nonzero scalar c, then E^{-1} is the matrix obtained by multiplying row i of I by $1/c$;

3. If E is obtained by replacing row i of I by itself plus c times row j of I, then E^{-1} is the matrix obtained by replacing row i of I by itself plus $-c$ times row j of I.

Proof In each part, let B denote the described matrix. We can then prove each part by showing that $EB = I$ and $BE = I$. This can be done by either directly calculating these products or by using Theorem 1.6. We leave these details as exercises (Exercise 14). ●

Up to this point we have been careful to show that B is both a right-hand inverse (that is, $AB = I$) and a left-hand inverse (that is, $BA = I$) when verifying that a square matrix B is the inverse of a square matrix A. There are places where a right-hand inverse need not be a left-hand inverse or vice versa. The next theorem tells us that this is not the case for square matrices, however.

THEOREM 1.9 Suppose that A and B are $n \times n$ matrices such that either $AB = I$ or $BA = I$. Then A is an invertible matrix and $A^{-1} = B$.

Proof Let us prove this in the case when $AB = I$; the case $BA = I$ will be left as an exercise (Exercise 17). Suppose A is not invertible. Since the reduced row-echelon form of A is not I, there are then elementary matrices $E_1, E_2, \ldots, E_m$ so that $E_1 E_2 \cdots E_m A$ contains a zero row. Consequently

$$E_1 E_2 \cdots E_m AB = E_1 E_2 \cdots E_m \tag{1}$$

contains a zero row. But a matrix with a zero row is not invertible (Exercise 16) and hence $E_1 E_2 \cdots E_m$ is not invertible. This gives us a contradiction since each E_i is invertible (Theorem 1.8) and products of invertible matrices are invertible (Theorem 1.7). Now that we know A is invertible, we can choose $E_1, E_2, \ldots, E_m$ so that

$$E_1 E_2 \cdots E_m A = I.$$

This along with Equation (1) gives us that $B = E_1 E_2 \cdots E_m = A^{-1}$. ●

Because of Theorem 1.9, from now on we will only have to verify one of $AB = I$ or $BA = I$ to see if a square matrix B is the inverse of a square matrix A.

Our final result gives a characterization of invertible matrices in terms of elementary matrices.

THEOREM 1.10 A square matrix A is invertible if and only if A is a product of elementary matrices.

Proof If A is a product of elementary matrices, then A is invertible by Theorems 1.7 and 1.8. Conversely, if A is invertible there are elementary matrices $E_1, E_2, \ldots E_m$ so that

$$E_1 E_2 \cdots E_m A = I.$$

Thus,

$$A = E_m^{-1} \cdots E_2^{-1} E_1^{-1} E_1 E_2 \cdots E_m A = E_m^{-1} \cdots E_2^{-1} E_1^{-1}.$$

Since each E_i^{-1} is an elementary matrix by Theorem 1.8, A is a product of elementary matrices.   ●

EXERCISES 1.3

For each of the following matrices, either find the inverse of the matrix or determine that the matrix is not invertible.

1. $\begin{bmatrix} 1 & 2 \\ -3 & 1 \end{bmatrix}$

2. $\begin{bmatrix} 2 & -6 \\ -3 & 9 \end{bmatrix}$

3. $\begin{bmatrix} 1 & -2 & 3 \\ 2 & -1 & 4 \\ 1 & 1 & 1 \end{bmatrix}$

4. $\begin{bmatrix} 2 & -1 & 3 \\ 1 & 1 & -2 \\ 1 & 1 & 5 \end{bmatrix}$

5. $\begin{bmatrix} 0 & -2 & 1 \\ 2 & 4 & -1 \\ 2 & 1 & 2 \end{bmatrix}$

6. $\begin{bmatrix} 0 & -1 & 3 \\ 0 & -4 & 1 \\ 2 & -1 & 3 \end{bmatrix}$

7. $\begin{bmatrix} 1 & -1 & 1 & 2 \\ 1 & 2 & -1 & -1 \\ 1 & -4 & 1 & 5 \\ 3 & 1 & 1 & 6 \end{bmatrix}$

8. $\begin{bmatrix} 1 & -2 & -1 & 1 & 0 \\ -1 & 1 & 0 & 1 & 1 \\ 1 & -1 & -1 & 2 & 1 \\ 1 & -1 & 1 & 4 & 2 \\ 2 & 0 & -1 & 8 & 4 \end{bmatrix}$

9. Use the inverse of the matrix in Exercise 5 to solve the system.

$$-2y + z = 2$$
$$2x + 4y - z = -1$$
$$2x + y + 2z = 5$$

10. Use the inverse of the matrix in Exercise 7 to solve the system.

$$x_1 - x_2 + x_3 + 2x_4 = 3$$
$$x_1 + 2x_2 - x_3 - x_4 = 5$$
$$x_1 - 4x_2 + x_3 + 5x_4 = 1$$
$$3x_1 + x_2 + x_3 + 6x_4 = 2$$

11. For $A = \begin{bmatrix} 1 & 2 \\ 3 & 4 \end{bmatrix}$, find an elementary matrix E so that:

a) $EA = \begin{bmatrix} 1 & 2 \\ 6 & 8 \end{bmatrix}$.

b) $EA = \begin{bmatrix} 7 & 10 \\ 3 & 4 \end{bmatrix}$.

c) $EA = \begin{bmatrix} 3 & 4 \\ 1 & 2 \end{bmatrix}$.

12. Express the matrix in Exercise 5 as a product of elementary matrices.

13. Prove the following parts of Theorem 1.6.
   a) Part (1)
   b) Part (2)
   c) Part (3)

14. Prove the following parts of Theorem 1.8.
   a) Part (1)
   b) Part (2)
   c) Part (3)

15. Show that if A is an invertible matrix, then so is A^{-1} and $(A^{-1})^{-1} = A$.

16. Show that a square matrix containing a zero row or a zero column is not invertible.

17. Complete the proof of Theorem 1.9 by showing if $BA = I$, then A is invertible and $B = A^{-1}$.

18. Suppose that A is a noninvertible square matrix. Show that the homogeneous system $AX = 0$ has nontrivial solutions.

19. Suppose that A is an invertible matrix and m is a positive integer. Show that $(A^m)^{-1} = (A^{-1})^m$.

20. a) Suppose that A is an invertible $n \times n$ matrix and B and C are $n \times l$ matrices such that $AB = AC$. Show that $B = C$.

   b) Give an example to show that the result of part (a) need not hold if A is not invertible.

21. Suppose that A and B are $n \times n$ matrices such that AB is invertible. Show that A and B are invertible.

Use the *inverse* command in Maple or the appropriate command in another suitable software package to find the inverses of the matrices in Exercises 22–25 (if possible).

22.
$$\begin{bmatrix} 14 & -25 & 39 & 29 & 6 \\ 58 & -41 & 88 & 24 & 18 \\ 15 & -6 & 31 & -23 & 12 \\ -3 & -22 & -25 & 73 & -24 \\ 3 & 6 & 12 & -24 & 9 \end{bmatrix}$$

23.
$$\begin{bmatrix} 8 & -21 & 14 & 26 & -3 \\ 7 & -28 & -6 & 66 & -18 \\ 23 & -12 & 45 & -13 & 15 \\ -9 & 14 & -10 & 26 & 3 \\ 2 & 4 & 8 & -16 & 6 \end{bmatrix}$$

24.
$$\begin{bmatrix} 13.2 & -11 & 13 & 27 \\ -4.4 & -10.2 & 5 & -9 \\ 15 & -17.6 & 21.2 & 32.4 \\ -10 & 12 & -14 & -18 \end{bmatrix}$$

25.[10]
$$\begin{bmatrix} 18 & -24 & 25 & 27 \\ -12 - 2\pi & 14 - \pi & -17 & -27 \\ 15 + 3\sqrt{3} & -18 - 4\sqrt{3} & 21 + 4\sqrt{3} & 27 \\ -10 & 12 & -14 & -18 \end{bmatrix}$$

26. Use Maple or another appropriate software package to find $P^{-1}AP$ where

$$A = \begin{bmatrix} -46 & 192 & 36 & -23 & -84 \\ -122 & 437 & 73 & -45 & -194 \\ 45 & -191 & -37 & 22 & 84 \\ -120 & 438 & 74 & -48 & -193 \\ -200 & 686 & 110 & -67 & -306 \end{bmatrix},$$

$$P = \begin{bmatrix} 1 & -1 & 2 & 0 & -1 \\ 1 & 0 & 5 & -1 & 3 \\ -1 & 1 & -3 & 2 & 4 \\ 1 & 0 & 4 & 2 & 4 \\ 1 & 1 & 8 & -2 & 8 \end{bmatrix}.$$

1.4   SPECIAL MATRICES AND ADDITIONAL PROPERTIES OF MATRICES

You already have seen some special types of matrices: zero matrices, identity matrices, and elementary matrices. There are some other special forms of matrices that will come up in our future work. One such type are **diagonal matrices,** which are square matrices whose off diagonal entries are zero. The matrix

$$A = \begin{bmatrix} 2 & 0 & 0 & 0 \\ 0 & -4 & 0 & 0 \\ 0 & 0 & 7 & 0 \\ 0 & 0 & 0 & 3 \end{bmatrix}$$

is an example of a diagonal matrix. We will write

$$\boxed{\text{diag}(d_1, d_2, \ldots, d_n)}$$

[10]In Maple a square root such as $\sqrt{3}$ is indicated by typing `sqrt(3)`; π is indicated by typing `pi`. Products require a multiplication star so that $-12 - 2\pi$ is typed as `-12-2*pi`; likewise, $15 + 3\sqrt{3}$ is typed as `15+3*sqrt(3)`.

to indicate an $n \times n$ diagonal matrix

$$\begin{bmatrix} d_1 & 0 & 0 & \cdots & 0 \\ 0 & d_2 & 0 & \cdots & 0 \\ \vdots & \vdots & \vdots & & \vdots \\ 0 & 0 & 0 & \cdots & d_n \end{bmatrix}.$$

For instance, the diagonal matrix A would be written as

$$A = \mathrm{diag}(2, -4, 7, 3)$$

in this notation. Some easily verified properties of diagonal matrices whose proofs we leave as exercises (Exercise 21) are listed in Theorem 1.11.

THEOREM 1.11   Suppose that A and B are $n \times n$ diagonal matrices

$$A = \mathrm{diag}(a_1, a_2, \ldots, a_n) \quad \text{and} \quad B = \mathrm{diag}(b_1, b_2, \ldots, b_n).$$

1. $A + B = \mathrm{diag}(a_1 + b_1, a_2 + b_2, \ldots, a_n + b_n)$.
2. $AB = \mathrm{diag}(a_1 b_1, a_2 b_2, \ldots, a_n b_n)$.
3. A is invertible if and only if each $a_i \neq 0$. Further, if each $a_i \neq 0$,
    $A^{-1} = \mathrm{diag}(1/a_1, 1/a_2, \ldots, 1/a_n)$.

Two other special types of square matrices are **triangular matrices,** which come in two forms: **upper triangular matrices** in which all entries below the diagonal are zero and **lower triangular matrices** in which all entries above the diagonal are zero. The matrix

$$\begin{bmatrix} -2 & 3 & 1 \\ 0 & 4 & -2 \\ 0 & 0 & 5 \end{bmatrix}$$

is an example of an upper triangular matrix, and the matrix

$$\begin{bmatrix} 1 & 0 & 0 \\ 2 & 3 & 0 \\ 6 & 0 & 5 \end{bmatrix}$$

is an example of a lower triangular matrix. Some easily seen properties of triangular matrices whose proofs are left as exercises (Exercise 22) are listed in Theorem 1.12.

THEOREM 1.12   Suppose that A and B are $n \times n$ triangular matrices.

1. If A and B are both upper triangular, then so is $A + B$; if A and B are both lower triangular, then so is $A + B$.

2. If A and B are both upper triangular, then so is AB; if A and B are both lower triangular, then so is AB.

3. A is invertible if and only if each of the diagonal entries of A is nonzero.

The **transpose** of a matrix A, denoted A^T, is the matrix obtained by interchanging the rows and columns of A; to put it another way, if $A = [a_{ij}]$ is an $m \times n$ matrix, then A^T is the $n \times m$ matrix with entries

$$\boxed{\text{ent}_{ij}(A^T) = a_{ji}.}$$

For instance, the transpose of the matrix

$$A = \begin{bmatrix} 1 & 2 & 3 \\ 4 & 5 & 6 \end{bmatrix}$$

is

$$A^T = \begin{bmatrix} 1 & 4 \\ 2 & 5 \\ 3 & 6 \end{bmatrix}.$$

In the next theorem we list some basic properties of transposes of matrices.

THEOREM 1.13 If A and B are matrices so that the indicated sum or product is defined and c is a scalar, then:

1. $(A^T)^T = A$.
2. $(A + B)^T = A^T + B^T$.
3. $(cA)^T = cA^T$.
4. $(AB)^T = B^T A^T$.
5. $(A^T)^{-1} = (A^{-1})^T$.

Proof Part (4) is the most difficult to prove. We will prove it here and leave proofs of the remaining parts as exercises (Exercise 23). Suppose A is an $m \times n$ matrix and B is an $n \times l$ matrix. The matrix $(AB)^T$ is an $l \times m$ matrix whose ij-entry is

$$\text{ent}_{ij}((AB)^T) = \text{ent}_{ji}(AB) = \sum_{k=1}^{n} a_{jk} b_{ki}. \tag{1}$$

$B^T A^T$ is also an $l \times m$ matrix whose ij-entry is

$$\text{ent}_{ij}(B^T A^T) = \sum_{k=1}^{n} \text{ent}_{ik}(B^T)\text{ent}_{kj}(A^T) = \sum_{k=1}^{n} b_{ki} a_{jk}. \tag{2}$$

As the results in Equations (1) and (2) are the same, we have $(AB)^T = B^T A^T$. ●

A matrix A is called a **symmetric matrix** if

$$A^T = A.$$

The matrix

$$\begin{bmatrix} 1 & 2 & -3 \\ 2 & -1 & 4 \\ -3 & 4 & 0 \end{bmatrix}$$

is a symmetric matrix; the matrix

$$\begin{bmatrix} 0 & 5 \\ 1 & -1 \end{bmatrix}$$

is not symmetric. Notice that a symmetric matrix must necessarily be a square matrix. We leave the proofs of the properties of symmetric matrices listed in the following theorem as exercises (Exercise 24).

THEOREM 1.14   Suppose A and B are matrices of the same size.

1. If A and B are symmetric matrices, then so is $A + B$.
2. If A is a symmetric matrix, then so is cA for any scalar c.
3. $A^T A$ and $A A^T$ are symmetric matrices.
4. If A is an invertible symmetric matrix, then A^{-1} is a symmetric matrix.

We often have applied a finite number of elementary row operations to a matrix A obtaining a matrix B. In this setting, we say that the matrix A is **row equivalent** to the matrix B. We frequently will use this terminology making statements such as "a square matrix A is invertible if and only if A is row equivalent to I" or "a system of linear equations $AX = B$ has no solution if and only if the augmented matrix $[A|B]$ is row equivalent to an augmented matrix containing a row that consists of zero entries in the left-hand portion and a nonzero entry in the right-hand portion." To give a couple more illustrations, notice that the terminology of reduced row-echelon form that we introduced for coefficient matrices of linear systems can be applied to any matrix; that is, a matrix is in reduced row-echelon form if:

1. Any zero rows appear at the bottom.
2. The first nonzero entry of a nonzero row is 1.
3. The leading 1 of a nonzero row appears to the right of the leading 1 of any preceding row.
4. All other entries of a column containing a leading 1 are zero.

Using ideas we shall develop in the next chapter, it can be shown that the reduced row-echelon form of a matrix is unique, so we may speak of the reduced row-echelon form of a matrix and say "a matrix is row equivalent to its reduced row-echelon form." When we do not require property 4, the matrix is in row-echelon form. Row-echelon form is not unique, so we would have to say "a matrix is row equivalent to any of its row-echelon forms."

The notion of row equivalence gives us a relationship between matrices of the same size that possesses the properties listed in Theorem 1.15.

THEOREM 1.15

1. Every matrix A is row equivalent to itself.

2. If a matrix A is row equivalent to a matrix B, then B is row equivalent to A.

3. If a matrix A is row equivalent to a matrix B and B is row equivalent to a matrix C, then A is row equivalent to C.

We will leave the proof of Theorem 1.15 as another exercise (Exercise 25). In Theorem 1.15, the first property is called the **reflexive property,** the second is called the **symmetric property,** and the third is called the **transitive property** of row equivalence. A relation that has all three of these properties is called an **equivalence relation.** Equivalence relations are important types of relations occurring frequently throughout mathematics. A couple of other important equivalence relations you have encountered before are congruence and similarity of triangles. Not all relations are equivalence relations. The inequality $<$ on the set of real numbers $\mathbb{R}$ is not an equivalence relation since it is neither reflexive nor symmetric (although it is transitive).

We conclude this section by pointing out that just as we perform elementary row operations, it is also possible to perform **elementary column operations** on a matrix. As you might expect, these are the following:

1. Interchange two columns.

2. Multiply a column by a nonzero number.

3. Replace a column by itself plus a multiple of another column.

When we apply a finite number of elementary column operations to a matrix A obtaining a matrix B, we say A is **column equivalent** to B.

Many (we authors included) are so used to performing row operations that they find it awkward to perform column operations. For the most part, we will avoid using column operations in this book. But we will see one place in the next chapter where column operations arise. If you too feel uncomfortable doing them, notice that column operations may be performed on a matrix A by first performing the corresponding row operations on A^T and then transposing again.

EXERCISES 1.4

Let A be the matrix

$$A = \begin{bmatrix} -1 & 0 & 0 \\ 0 & -2 & 0 \\ 0 & 0 & 1 \end{bmatrix}.$$

Find:

1. A^2

2. A^{-1}

3. A^5

4. $(A^{-1})^4$

Let A and B be the matrices

$$A = \begin{bmatrix} 1 & 2 & 0 \\ 0 & -1 & 3 \\ 0 & 0 & 1 \end{bmatrix} \quad \text{and} \quad B = \begin{bmatrix} 6 & -1 & 3 \\ 0 & 4 & -1 \\ 0 & 0 & 2 \end{bmatrix}.$$

Find:

5. AB **6.** A^{-1}.

Let A and B be the matrices

$$A = \begin{bmatrix} 1 & 2 & -3 \\ 1 & -2 & 1 \end{bmatrix} \quad \text{and} \quad B = \begin{bmatrix} -2 & 1 \\ 3 & 5 \\ -4 & 1 \end{bmatrix}.$$

If possible, find:

7. A^T.   **8.** B^T.

9. $A^T + 4B$.   **10.** $2A - 5B^T$.

11. $(AB)^T$.   **12.** $B^T A^T$.

13. $A^T B^T$.   **14.** $A^T A$.

Let A and B be the matrices

$$A = \begin{bmatrix} 1 & -2 & 3 \\ -2 & 0 & 4 \\ 3 & 4 & 5 \end{bmatrix} \quad \text{and} \quad B = \begin{bmatrix} 3 & 0 & -1 \\ 1 & 4 & 2 \\ -1 & 2 & 1 \end{bmatrix}.$$

Determine which of the following are symmetric matrices in Exercises 15–20.

15. A **16.** B

17. $A + B$ **18.** A^{-1}

19. BB^T **20.** $B^T B$

21. Prove the following parts of Theorem 1.11.

     a) Part (1)

     b) Part (2)

     c) Part (3)

22. Prove the following parts of Theorem 1.12.

     a) Part (1)

     b) Part (2)

     c) Part (3)

23. Prove the following parts of Theorem 1.13.

     a) Part (1)   **b)** Part (2)

     c) Part (3)   **d)** Part (5)

24. Prove the following parts of Theorem 1.14.

     a) Part (1)   **b)** Part (2)

     c) Part (3)   **d)** Part (4)

25. Prove the following parts of Theorem 1.15.

     a) Part (1)

     b) Part (2)

     c) Part (3)

26. Show that two matrices A and B are row equivalent if and only if there is an invertible matrix C so that $CA = B$.

27. Show that two matrices A and B are row equivalent if and only if they have the same reduced row-echelon form.

28. Use the result of Exercise 27 to show that the matrices

$$\begin{bmatrix} 1 & 2 & -1 \\ 3 & -1 & 2 \\ 1 & -5 & 4 \end{bmatrix} \quad \text{and}$$

$$\begin{bmatrix} 2 & -3 & 3 \\ 1 & -5 & 4 \\ -2 & -11 & 7 \end{bmatrix}$$

are row equivalent.

29. Show that any two invertible matrices of the same size are row equivalent.

30. Just as we speak of the reduced row-echelon form of a matrix, we may speak of the reduced column-echelon form of a matrix. Write a statement that describes the form of a reduced column-echelon matrix.

31. Find the reduced column-echelon form of the matrix

$$\begin{bmatrix} -3 & -1 & 4 \\ 2 & 3 & -1 \\ 1 & -2 & -3 \end{bmatrix}.$$

32. Find A^3 for

$$A = \begin{bmatrix} 0 & 1 & 2 \\ 0 & 0 & 3 \\ 0 & 0 & 0 \end{bmatrix}.$$

33. A square matrix A is called a **nilpotent** matrix if there is a positive integer m so that $A^m = 0$. Prove that if A is a triangular $n \times n$ matrix whose diagonal entries are all zero, then A is a nilpotent matrix by showing $A^n = 0$.

34. The Maple command for finding the transpose of a matrix is *transpose*. Use Maple or another appropriate software package to find $AA^T - 3A^T$ for

$$A = \begin{bmatrix} 2 & 1 & -3 & 4 & 1 \\ -2 & 1 & 1 & 0 & 7 \\ 6 & 2 & -8 & 4 & 9 \\ -2 & 0 & 2 & -6 & -2 \\ 10 & 3 & -13 & 14 & 12 \end{bmatrix}.$$

35. a) Either of the Maple commands *rref* or *gaussjord* can be used to find the reduced row-echelon form of a matrix. Use them or corresponding commands in another appropriate software package to find the reduced row-echelon form of matrix A in Exercise 34.

b) Apply the *gausselim* command of Maple to matrix A. Describe the form of the answer Maple is giving us.

c) Is A an invertible matrix?

36. How could Maple commands or commands of another appropriate software package be used to find the reduced column-echelon form of a matrix? Use them to find the reduced column-echelon form of matrix A in Exercise 34.

1.5   DETERMINANTS

You already may have had some exposure to determinants. For instance, you might have encountered them for finding cross products of three-dimensional vectors. Or perhaps you have learned a method for finding solutions to some systems of linear equations involving determinants called Cramer's rule. (We shall discuss this rule in the next section.) The Jacobian of a transformation is yet another example of a determinant you might have encountered. Even if you have had some prior experience with determinants, however, it is likely that your knowledge of them is not thorough. The purpose of this and the remaining sections of this chapter is to a give a thorough treatment of determinants.

There are a number of equivalent ways of defining determinants. We are going to begin with a process called a cofactor expansion approach. Suppose that A is a square matrix:

$$A = \begin{bmatrix} a_{11} & a_{12} & \cdots & a_{1n} \\ a_{21} & a_{22} & \cdots & a_{2n} \\ \vdots & \vdots & & \vdots \\ a_{n1} & a_{n2} & \cdots & a_{nn} \end{bmatrix}.$$

The **minor** of the entry a_{ij} of A, denoted M_{ij}, is the matrix obtained from A by deleting row i and column j from A.[11] (Of course, this only makes sense if $n \geq 2$.) For instance, if A is a 3×3 matrix,

$$A = \begin{bmatrix} a_{11} & a_{12} & a_{13} \\ a_{21} & a_{22} & a_{23} \\ a_{31} & a_{32} & a_{33} \end{bmatrix},$$

[11]There are two different ways in which the concept of a minor is commonly defined. Quite a few books as well as Maple take our approach. Many other books, however, prefer to define the minor of a_{ij} as the determinant of the matrix obtained from A by deleting row i and column j. In other words, minors in these other books are the determinants of our minors.

some minors of A are:

$$M_{11} = \begin{bmatrix} a_{22} & a_{23} \\ a_{32} & a_{33} \end{bmatrix}, \qquad M_{12} = \begin{bmatrix} a_{21} & a_{23} \\ a_{31} & a_{33} \end{bmatrix}, \qquad M_{32} = \begin{bmatrix} a_{11} & a_{13} \\ a_{21} & a_{23} \end{bmatrix}.$$

We are going to define the determinant of an $n \times n$ matrix A (also called an $n \times n$ determinant), denoted $\det(A)$, with what is called an inductive definition as follows: If A is a 1×1 matrix,

$$A = [a_{11}],$$

we define the determinant of A to be its entry,

$$\det(A) = a_{11}.$$

If $n \geq 2$, the determinant of $A = [a_{ij}]$ is defined to be

$$\det(A) = a_{11} \det(M_{11}) - a_{12} \det(M_{12}) + a_{13} \det(M_{13}) - \cdots + (-1)^{1+n} a_{1n} \det(M_{1n})$$

$$= \sum_{j=1}^{n} (-1)^{1+j} a_{1j} \det(M_{1j}).$$

In effect, we have reduced $\det(A)$ to the determinants of the smaller matrices M_{1j}, which (by repeating this reduction procedure if necessary) we already know how to find.

To illustrate, if A is a 2×2 matrix,

$$A = \begin{bmatrix} a_{11} & a_{12} \\ a_{21} & a_{22} \end{bmatrix},$$

our definition tells us that

$$\det(A) = a_{11} \det([a_{22}]) - a_{12} \det([a_{21}]) = a_{11}a_{22} - a_{12}a_{21}.$$

The products $a_{11}a_{22}$ and $a_{12}a_{21}$ are often referred to as cross products; the determinant of a 2×2 matrix can then be easily remembered as being the cross product of the diagonal entries minus the cross product of the off diagonal entries. Determinants of $n \times n$ matrices when $n \geq 2$ are also indicated by putting vertical bars around the entries of the matrix.[12] If we do this for a 2×2 matrix, our determinant formula becomes

$$\begin{vmatrix} a_{11} & a_{12} \\ a_{21} & a_{22} \end{vmatrix} = a_{11}a_{22} - a_{12}a_{21}. \tag{1}$$

---

[12] We do not do this for a 1×1 matrix $A = [a_{11}]$ to avoid confusion with the absolute value of a_{11}.

For instance,

$$\begin{vmatrix} 2 & -3 \\ 5 & 6 \end{vmatrix} = 2 \cdot 6 - (-3) \cdot 5 = 27.$$

For a 3×3 matrix, our definition tells us

$$\begin{vmatrix} a_{11} & a_{12} & a_{13} \\ a_{21} & a_{22} & a_{23} \\ a_{31} & a_{32} & a_{33} \end{vmatrix} = a_{11} \begin{vmatrix} a_{22} & a_{23} \\ a_{32} & a_{33} \end{vmatrix} - a_{12} \begin{vmatrix} a_{21} & a_{23} \\ a_{31} & a_{33} \end{vmatrix} + a_{13} \begin{vmatrix} a_{21} & a_{22} \\ a_{31} & a_{32} \end{vmatrix}$$

where the remaining 2×2 determinants can be found by the formula in Equation (1). For instance,

$$\begin{vmatrix} 2 & 3 & -2 \\ -1 & 6 & 3 \\ 4 & -2 & 1 \end{vmatrix} = 2 \begin{vmatrix} 6 & 3 \\ -2 & 1 \end{vmatrix} - 3 \begin{vmatrix} -1 & 3 \\ 4 & 1 \end{vmatrix} + (-2) \begin{vmatrix} -1 & 6 \\ 4 & -2 \end{vmatrix}$$
$$= 2(6 + 6) - 3(-1 - 12) - 2(2 - 24) = 107.$$

Continuing, a 4×4 determinant can be found by forming the alternating sum and difference of the entries of the first row of the matrix times the determinants of their respective minors, which are 3×3 determinants and so on.

The **cofactor** of an entry a_{ij} of an $n \times n$ matrix A where $n \geq 2$, denoted C_{ij}, is

$$\boxed{C_{ij} = (-1)^{i+j} \det(M_{ij})}$$

where M_{ij} is the minor of a_{ij}. Some cofactors of a 3×3 matrix,

$$A = \begin{bmatrix} a_{11} & a_{12} & a_{13} \\ a_{21} & a_{22} & a_{23} \\ a_{31} & a_{32} & a_{33} \end{bmatrix},$$

are:

$$C_{11} = (-1)^{1+1} \begin{vmatrix} a_{22} & a_{23} \\ a_{32} & a_{33} \end{vmatrix} = \begin{vmatrix} a_{22} & a_{23} \\ a_{32} & a_{33} \end{vmatrix}$$

$$C_{12} = (-1)^{1+2} \begin{vmatrix} a_{21} & a_{23} \\ a_{31} & a_{33} \end{vmatrix} = - \begin{vmatrix} a_{21} & a_{23} \\ a_{31} & a_{33} \end{vmatrix}.$$

$$C_{32} = (-1)^{3+2} \begin{vmatrix} a_{11} & a_{13} \\ a_{21} & a_{23} \end{vmatrix} = - \begin{vmatrix} a_{11} & a_{13} \\ a_{21} & a_{23} \end{vmatrix}$$

The signs of the cofactors can be easily remembered by noting that they form the checkerboard pattern:

$$
\begin{array}{ccccc}
+ & - & + & - & \cdots \\
- & + & - & + & \cdots \\
+ & - & + & - & \cdots \\
\vdots & \vdots & \vdots & \vdots &
\end{array}
$$

Our formula for finding the determinant of an $n \times n$ matrix $A = [a_{ij}]$ where $n \geq 2$ can be written in terms of cofactors as

$$
\det(A) = \sum_{j=1}^{n} a_{1j} C_{1j},
$$

which we will refer to as the *cofactor expansion about the first row* or simply the *expansion about the first row*. What is remarkable is that the same procedure may be followed for any row or column.

THEOREM 1.16   If $A = [a_{ij}]$ is an $n \times n$ matrix with $n \geq 2$, then for any $1 \leq i \leq n$

$$
\boxed{\det(A) = \sum_{j=1}^{n} a_{ij} C_{ij} \quad \text{(cofactor expansion about the ith row)}}
$$

or any $1 \leq j \leq n$,

$$
\boxed{\det(A) = \sum_{i=1}^{n} a_{ij} C_{ij} \quad \text{(cofactor expansion about the jth column).}}
$$

To illustrate this theorem, earlier we had found

$$
\begin{vmatrix}
2 & 3 & -2 \\
-1 & 6 & 3 \\
4 & -2 & 1
\end{vmatrix} = 107
$$

by using the cofactor expansion about the first row. Notice that we get the same result if, for instance, we expand about the third row,

$$
\begin{vmatrix}
2 & 3 & -2 \\
-1 & 6 & 3 \\
4 & -2 & 1
\end{vmatrix} = 4 \begin{vmatrix} 3 & -2 \\ 6 & 3 \end{vmatrix} - (-2) \begin{vmatrix} 2 & -2 \\ -1 & 3 \end{vmatrix} + \begin{vmatrix} 2 & 3 \\ -1 & 6 \end{vmatrix}
$$

$$
= 4(21) + 2(4) + 15 = 107,
$$

or the second column,

$$
\begin{vmatrix}
2 & 3 & -2 \\
-1 & 6 & 3 \\
4 & -2 & 1
\end{vmatrix} = -3 \begin{vmatrix} -1 & 3 \\ 4 & 1 \end{vmatrix} + 6 \begin{vmatrix} 2 & -2 \\ 4 & 1 \end{vmatrix} - (-2) \begin{vmatrix} 2 & -2 \\ -1 & 3 \end{vmatrix}
$$

$$
= -3(-13) + 6(10) + 2(4) = 107.
$$

So that you first gain an overview of the theory of determinants, we are going to postpone many of the proofs of our results about determinants to the end of this chapter in Section 1.7. The proof of Theorem 1.16 is the first of these we postpone.

You may raise the question: Why is it important to be able to expand a determinant about any row or column? One reason is that sometimes we can make the work easier by choosing a particular row or column. Consider the following example.

EXAMPLE 1   Evaluate the determinant

$$\begin{vmatrix} 7 & -3 & 0 & 4 \\ 0 & 1 & 0 & 3 \\ 2 & 1 & -2 & -5 \\ 0 & 4 & 0 & 6 \end{vmatrix}.$$

Solution   Since the third column contains three zeros, let us begin by expanding about it obtaining

$$-2 \begin{vmatrix} 7 & -3 & 4 \\ 0 & 1 & 3 \\ 0 & 4 & 6 \end{vmatrix}.$$

The remaining 3×3 determinant is now quickly found by expanding about its first column. Doing so, we get our answer:

$$(-2) \cdot 7 \begin{vmatrix} 1 & 3 \\ 4 & 6 \end{vmatrix} = 84. \qquad \bullet$$

The fact that Theorem 1.16 allows us to expand about any row or column gives us some cases in which determinants can be quickly found. One of these is described in the following corollary.

COROLLARY 1.17   If an $n \times n$ matrix A has a zero row or zero column, then $\det(A) = 0$.

Proof   This result is immediate if A is a 1×1 matrix, so assume $n \geq 2$. Expanding the determinant about the row or column whose entries are all zero, the result follows. $\quad \bullet$

Corollary 1.18 describes another case in which we can quickly see the value of a determinant.

COROLLARY 1.18   The determinant of a triangular matrix is the product of its diagonal entries.

Proof   We will do the upper triangular case here. The lower triangular case will be left as an exercise (Exercise 15). Suppose A is an upper triangular matrix:

$$A = \begin{bmatrix} a_{11} & a_{12} & a_{13} & \cdots & a_{1n} \\ 0 & a_{22} & a_{23} & \cdots & a_{2n} \\ 0 & 0 & a_{33} & \cdots & a_{3n} \\ \vdots & \vdots & \vdots & & \vdots \\ 0 & 0 & 0 & \cdots & a_{nn} \end{bmatrix}.$$

Again the result is immediate if $n = 1$, so assume $n \geq 2$. Expanding $\det(A)$ about the first column, we have

$$\det(A) = a_{11} \begin{vmatrix} a_{22} & a_{23} & \cdots & a_{2n} \\ 0 & a_{33} & \cdots & a_{3n} \\ \vdots & \vdots & & \vdots \\ 0 & 0 & \cdots & a_{nn} \end{vmatrix}.$$

If we continue to expand each remaining determinant about the first column, we obtain

$$\det(A) = a_{11}a_{22} \cdots a_{nn}$$

as desired.[13]   ●

In Section 1.7 we shall use the fact that the determinant of a square matrix can be found by performing a cofactor expansion about either its first row or first column to obtain the result stated in Theorem 1.19.

THEOREM 1.19   If A is an $n \times n$ matrix, then

$$\det(A^T) = \det(A).$$

As the square matrices grow in size, the calculations of determinants using cofactor expansions become lengthy. We next develop a more efficient method for calculating determinants of large square matrices involving row operations. To use this approach, we will have to know the effects of elementary row operations on a determinant. These effects are listed in the following theorem.

THEOREM 1.20   Suppose that $A = [a_{ij}]$ is an $n \times n$ matrix with $n \geq 2$.

1. If B is a matrix obtained from A by interchanging two rows of A, then $\det(B) = -\det(A)$.

2. If B is a matrix obtained from A by multiplying a row of A by a scalar c, then $\det(B) = c \det(A)$.

3. If B is a matrix obtained from A by replacing a row of A by itself plus a multiple of another row of A, then $\det(B) = \det(A)$.

Because $\det(A) = \det(A^T)$, we can replace the elementary row operation by the corresponding elementary column operation in each part of Theorem 1.20 and obtain

---

[13] If you are familiar with mathematical induction, you will notice that this proof could be more effectively written by using induction on n.

the same result. The proof of Theorem 1.20 is another one that we postpone until Section 1.7.

We use row operations to calculate the determinant of a square matrix A in a manner similar to the way we use them to solve a system of linear equations with Gaussian elimination: Use row operations to reduce to row-echelon form with the exception of making the leading entries one. This reduced matrix is an upper triangular matrix whose determinant is easily found by Corollary 1.18.

Of course, when we apply the row operations, we must be careful to compensate for their effects. The following example illustrates how we may do this.

EXAMPLE 2   Find the determinant of the matrix

$$A = \begin{bmatrix} 1 & -1 & 2 & 3 \\ 2 & 1 & 2 & 1 \\ 1 & 1 & -1 & -2 \\ 1 & -1 & 1 & 4 \end{bmatrix}.$$

Solution   We begin with a first set of row operations toward the goal of getting A into an upper triangular form. To help you follow our work, we have indicated the row operations that will be performed.

$$\det(A) = \begin{vmatrix} 1 & -1 & 2 & 3 \\ 2 & 1 & 2 & 1 \\ 1 & 1 & -1 & -2 \\ 1 & -1 & 1 & 4 \end{vmatrix} \begin{matrix} \\ R_2 - 2R_1 \\ R_3 - R_1 \\ R_4 - R_1 \end{matrix}$$

By part (3) of Theorem 1.20, performing these elementary row operations does not affect the determinant and hence

$$\det(A) = \begin{vmatrix} 1 & -1 & 2 & 3 \\ 0 & 3 & -2 & -5 \\ 0 & 2 & -3 & -5 \\ 0 & 0 & -1 & 1 \end{vmatrix} \begin{matrix} \\ \\ 3R_3 - 2R_2 \\ \\ \end{matrix}$$

where again we have indicated the row operation we will perform in the next step. This row operation is a combination of two elementary row operations: (1) multiplying the third row by 3 and (2) adding -2 times the second row. The second of these has no effect, but the first changes the determinant by a factor of 3 by part (2) of Theorem 1.20. Notice how we multiply by 1/3 to compensate, obtaining

$$\det(A) = \frac{1}{3} \begin{vmatrix} 1 & -1 & 2 & 3 \\ 0 & 3 & -2 & -5 \\ 0 & 0 & -5 & -5 \\ 0 & 0 & -1 & 1 \end{vmatrix} \begin{matrix} \\ \\ \\ R_3 \leftrightarrow R_4 \end{matrix}.$$

The next indicated row operation we will perform changes the sign by part (1) of Theorem 1.20. Observe how we compensate:

$$
\det(A) = -\frac{1}{3}
\begin{vmatrix}
1 & -1 & 2 & 3 \\
0 & 3 & -2 & -5 \\
0 & 0 & -1 & 1 \\
0 & 0 & -5 & -5
\end{vmatrix}
\begin{matrix}
\\ \\ \\ R_4 - 5R_3
\end{matrix}
.
$$

Performing our last indicated elementary row operation, which has no effect on the determinant, we have a desired upper triangular form from which we easily compute the determinant:

$$
\det(A) = -\frac{1}{3}
\begin{vmatrix}
1 & -1 & 2 & 3 \\
0 & 3 & -2 & -5 \\
0 & 0 & -1 & 1 \\
0 & 0 & 0 & -10
\end{vmatrix}
= -\frac{1}{3} \cdot 1 \cdot 3(-1)(-10) = -10.
$$

●

EXERCISES 1.5

Find det(A) for

$$
A = \begin{bmatrix}
2 & -1 & 3 \\
4 & 1 & -2 \\
-3 & 2 & 1
\end{bmatrix}
$$

by expanding about the indicated row or column in Exercises 1–6.

1. Row 1   **2.** Row 2   **3.** Row 3

4. Column 1   **5.** Column 2   **6.** Column 3

Find the determinants in Exercises 7–10 by expanding about appropriate rows or columns.

7.
$$
\begin{vmatrix}
-3 & 0 & 4 \\
2 & -1 & 3 \\
4 & 0 & 5
\end{vmatrix}
$$

8.
$$
\begin{vmatrix}
2 & -1 & 5 & 6 \\
0 & 3 & 4 & 0 \\
0 & 1 & 5 & 2 \\
0 & 1 & -3 & 0
\end{vmatrix}
$$

9.
$$
\begin{vmatrix}
4 & 3 & 2 & 1 \\
-2 & 5 & -1 & -2 \\
0 & 1 & 0 & 0 \\
0 & 2 & 0 & -2
\end{vmatrix}
$$

10.
$$
\begin{vmatrix}
0 & 3 & 2 & 5 & 0 \\
0 & -2 & 0 & 0 & 0 \\
2 & 4 & 0 & 1 & 1 \\
3 & 7 & 3 & -2 & 0 \\
4 & 5 & 0 & 1 & 1
\end{vmatrix}
$$

Use row operations to find the determinants in Exercises 11–14.

11.
$$
\begin{vmatrix}
1 & -2 & 1 \\
2 & 1 & 3 \\
-1 & 4 & 5
\end{vmatrix}
$$

12.
$$
\begin{vmatrix}
2 & -1 & 3 & 1 \\
-1 & 2 & -1 & 4 \\
1 & -1 & 3 & 1 \\
3 & 2 & -1 & 5
\end{vmatrix}
$$

13.
$$
\begin{vmatrix}
1 & -2 & 1 & -1 \\
2 & -4 & 3 & 2 \\
5 & -11 & 2 & -6 \\
1 & -1 & 1 & 3
\end{vmatrix}
$$

14.
$$\begin{vmatrix} 1 & -1 & 2 & 1 & 3 \\ 1 & 2 & -1 & -1 & 1 \\ 2 & -1 & 3 & 1 & 4 \\ -1 & 1 & 4 & -1 & 2 \\ 3 & -4 & 5 & 2 & 8 \end{vmatrix}$$

15. Complete the proof of Corollary 1.18 by showing that the determinant of a lower triangular matrix is the product of its diagonal entries.

16. Suppose that A is a square matrix in which one of the rows is a scalar multiple of another. Show that $\det(A) = 0$. Does the same result hold if A has one of its columns being a scalar multiple of another? Why or why not?

The Maple command for finding the determinant of a square matrix is *det*. Use this Maple command or the corresponding command in another appropriate software package to find the determinants in Exercises 17 and 18.

17.
$$\begin{vmatrix} 7 & -11 & 4 & 6 & 3 \\ -5 & 5 & 4 & -2 & 3 \\ 6 & 12 & -14 & 3 & 5 \\ 13 & -1 & 3 & 2 & -4 \\ 3 & -2 & 8 & -7 & 6 \end{vmatrix}$$

18.
$$\begin{vmatrix} \pi - 3 & 7 & -4 & 6 \\ 12 & \pi - 4 & 2 & -5 \\ 3 & -8 & \pi - 8 & 7 \\ 6 & 4 & -5 & \pi + 2 \end{vmatrix}$$

1.6   FURTHER PROPERTIES OF DETERMINANTS

In this section we shall see some additional properties of determinants. Our first such property gives us a relationship between the value of the determinant of a square matrix and its invertibility.

THEOREM 1.21   A square matrix A is invertible if and only if $\det(A) \neq 0$.

Proof   Suppose that A is invertible. Then A is row equivalent to I. From Theorem 1.20, we can see that the determinants of two row equivalent matrices are nonzero multiples of one another. Thus $\det(A)$ is a nonzero multiple of $\det(I) = 1$ and hence $\det(A) \neq 0$. Conversely, suppose $\det(A) \neq 0$. Were A not invertible, the reduced row-echelon form of A, let us call this matrix B, would have a zero row. But then $\det(B) = 0$ by Corollary 1.17. Since $\det(A)$ is a multiple of $\det(B)$, we obtain $\det(A) = 0$, which is a contradiction.   ●

Theorem 1.21 is useful for determining if a square matrix A is invertible when $\det(A)$ can be quickly calculated. For instance, since

$$\begin{vmatrix} -2 & 3 & 0 \\ 4 & 10 & 2 \\ -5 & 7 & 0 \end{vmatrix} = -2 \begin{vmatrix} -2 & 3 \\ -5 & 7 \end{vmatrix} = -2 \neq 0,$$

the matrix

$$\begin{bmatrix} -2 & 3 & 0 \\ 4 & 10 & 2 \\ -5 & 7 & 0 \end{bmatrix}$$

is invertible.

Our next major objective will be to obtain a result about the determinant of a product of matrices. To reach this objective, we first prove a couple of lemmas about elementary matrices.

LEMMA 1.22   Suppose that E is an elementary matrix.

1. If E is obtained from I by interchanging two rows of I, then $\det(E) = -1$.
2. If E is obtained from I by multiplying a row of I by a nonzero scalar c, then $\det(E) = c$.
3. If E is obtained from I by replacing a row of I by itself plus a multiple of another row of I, then $\det(E) = 1$.

Proof   These are all consequences of Theorem 1.20. For example, part (1) follows because $\det(E) = -\det(I) = -1$ by part (1) of Theorem 1.20. We leave the proofs of the remaining two parts as exercises (Exercise 13).   ●

LEMMA 1.23   If A is an $n \times n$ matrix and E is an $n \times n$ elementary matrix, then

$$\det(EA) = \det(E)\det(A).$$

More generally, if $E_1, E_2, \ldots, E_m$ are $n \times n$ elementary matrices, then

$$\det(E_1 E_2 \cdots E_m A) = \det(E_1 E_2 \cdots E_m)\det(A).$$

Proof   The first part is an immediate consequence of Theorem 1.6, which tells us that left multiplication by an elementary matrix performs an elementary row operation, Theorem 1.20, which tells us the effect of an elementary row operation on a determinant, and Lemma 1.22. The second part follows by repeated use of the first part:

$$\det(E_1 E_2 \cdots E_m A) = \det(E_1)\det(E_2 \cdots E_m A)$$

$$\vdots$$

$$= \det(E_1)\det(E_2)\cdots\det(E_m)\det(A)$$
$$= \det(E_1 E_2)\det(E_3)\cdots\det(E_m)\det(A)$$

$$\vdots$$

$$= \det(E_1 E_2 \cdots E_m)\det(A).$$ ●

Now we are ready to prove Theorem 1.24.

THEOREM 1.24   If A and B are $n \times n$ matrices,

$$\det(AB) = \det(A)\det(B).$$

Proof   We will break our proof into two cases: one in which A is invertible and the other in which it is not. If A is invertible, then A is a product of elementary matrices by Theorem 1.10 and the result now follows from Lemma 1.23.

Suppose A is not invertible. Then $\det(A) = 0$ by Theorem 1.21 and consequently

$$\det(A)\det(B) = 0.$$

Since A is not invertible, A is row equivalent to a matrix with a zero row. Thus there are elementary matrices $E_1, E_2, \ldots, E_m$ so that $E_1 E_2 \cdots E_m A$ has a zero row. Then $E_1 E_2 \cdots E_m AB$ has a zero row and consequently

$$\det(E_1 E_2 \cdots E_m AB) = 0.$$

Hence

$$\det(E_1 E_2 \cdots E_m)\det(AB) = 0,$$

which implies

$$\det(AB) = 0$$

since $\det(E_1 E_2 \cdots E_m) \neq 0$, and we again have the desired result that $\det(AB) = \det(A)\det(B)$. ●

As a consequence of Theorem 1.24, we have the following corollary.

COROLLARY 1.25   If A is an invertible matrix, $\det(A^{-1}) = 1/\det(A)$.

Proof   Since

$$\det(A^{-1})\det(A) = \det(A^{-1}A) = \det(I) = 1,$$

it follows that $\det(A^{-1}) = 1/\det(A)$. ●

If $A = [a_{ij}]$ is an $n \times n$ matrix, the $n \times n$ matrix with entries the cofactors of A,

$$\begin{bmatrix} C_{11} & C_{12} & \cdots & C_{1n} \\ C_{21} & C_{22} & \cdots & C_{2n} \\ \vdots & \vdots & & \vdots \\ C_{n1} & C_{n2} & \cdots & C_{nn} \end{bmatrix},$$

is called the **cofactor matrix** of A. The transpose of this cofactor matrix is called the **adjoint** of A and is denoted $\text{adj}(A)$; that is,

$$\text{adj}(A) = \begin{bmatrix} C_{11} & C_{21} & \cdots & C_{n1} \\ C_{12} & C_{22} & \cdots & C_{n2} \\ \vdots & \vdots & & \vdots \\ C_{1n} & C_{2n} & \cdots & C_{nn} \end{bmatrix}.$$

For example, the cofactor matrix of

$$A = \begin{bmatrix} 1 & 2 \\ 3 & 4 \end{bmatrix}$$

is

$$\begin{bmatrix} 4 & -3 \\ -2 & 1 \end{bmatrix}$$

and

$$\mathrm{adj}(A) = \begin{bmatrix} 4 & -2 \\ -3 & 1 \end{bmatrix}.$$

A curious feature of the adjoint of a square matrix is Theorem 1.26.

THEOREM 1.26   If A is a square matrix,

$$A\,\mathrm{adj}(A) = \mathrm{adj}(A)A = \det(A)I.$$

The proof of Theorem 1.26 is another one we postpone until the next section. Notice this theorem is telling us that $\mathrm{adj}(A)$ is almost the inverse of A. Indeed, we have Corollary 1.27.

COROLLARY 1.27   If A is an invertible matrix, then

$$A^{-1} = \frac{1}{\det(A)}\mathrm{adj}(A).$$

For instance, for the matrix

$$A = \begin{bmatrix} 1 & 2 \\ 3 & 4 \end{bmatrix}$$

just prior to Theorem 1.26, Corollary 1.27 tells us that

$$A^{-1} = \frac{1}{\det(A)}\mathrm{adj}(A) = \frac{1}{-2}\begin{bmatrix} 4 & -2 \\ -3 & 1 \end{bmatrix} = \begin{bmatrix} -2 & 1 \\ 3/2 & -1/2 \end{bmatrix}.$$

As a rule, however, Corollary 1.27 is not an efficient way of finding inverses of matrices because of all the determinants that must be calculated. The approach used in Section 1.3 is usually the better way to go. This adjoint method for finding inverses is used primarily as a theoretical tool.

To develop our final property of this section, consider a linear system with two equations and two unknowns:

$$a_{11}x + a_{12}y = b_1$$
$$a_{21}x + a_{22}y = b_2.$$

Let us start to solve this system. We could use our matrix method, but we will not bother for such a small system. Instead, let us eliminate y by multiplying the first equation by a_{22} and subtracting a_{12} times the second equation. This gives us

$$(a_{11}a_{22} - a_{21}a_{12})x = a_{22}b_1 - a_{12}b_2.$$

If $a_{11}a_{22} - a_{12}a_{21} \neq 0$, we then have

$$x = \frac{a_{22}b_1 - a_{12}b_2}{a_{11}a_{22} - a_{12}a_{21}}.$$

Carrying out similar steps to find y (try it), we find

$$y = \frac{a_{11}b_2 - a_{21}b_1}{a_{11}a_{22} - a_{12}a_{21}}.$$

Our formulas for x and y can be conveniently expressed in terms of determinants. If we let A denote the coefficient matrix of the system,

$$A = \begin{bmatrix} a_{11} & a_{12} \\ a_{21} & a_{22} \end{bmatrix},$$

and A_1 and A_2 be the matrices obtained from A by replacing the first and second columns, respectively, of A by the column

$$B = \begin{bmatrix} b_1 \\ b_2 \end{bmatrix}$$

so that

$$A_1 = \begin{bmatrix} b_1 & a_{12} \\ b_2 & a_{22} \end{bmatrix} \quad \text{and} \quad A_2 = \begin{bmatrix} a_{11} & b_1 \\ a_{21} & b_2 \end{bmatrix},$$

then

$$x = \frac{\det(A_1)}{\det(A)}, \qquad y = \frac{\det(A_2)}{\det(A)}.$$

We have just discovered what is known as **Cramer's rule,** named after Gabriel Cramer (1704–1752). Variations of this rule were apparently known prior to Cramer, but his name became attached to it when it appeared in his 1750 work *Introduction à l'analyse des lignes courbes algébriques.* The rule extends to any system of n linear equations in n unknowns provided the determinant of the coefficient matrix is nonzero (or equivalently, provided the coefficient matrix is invertible).

THEOREM 1.28   **(Cramer's Rule)** Suppose that $AX = B$ is a system of n linear equations in n unknowns such that $\det(A) \neq 0$. Let A_1 be the matrix obtained from A by replacing the first column of A by B, A_2 be the matrix obtained from A by replacing the second column of A by $B, \ldots, A_n$ be the matrix obtained from A by replacing the nth column of A by B. Then

$$\boxed{x_1 = \frac{\det(A_1)}{\det(A)}, \qquad x_2 = \frac{\det(A_2)}{\det(A)}, \qquad \cdots \qquad x_n = \frac{\det(A_n)}{\det(A)}.}$$

The proof of Cramer's rule for a general positive integer n will be given in the next section. Let us look at an example using it.

EXAMPLE 1   Use Cramer's rule to solve the system

$$x + y - z = 2$$
$$2x - y + z = 3$$
$$x - 2y + z = 1.$$

Solution   We first calculate the necessary determinants (the details of which we leave out).

$$\det(A) = \begin{vmatrix} 1 & 1 & -1 \\ 2 & -1 & 1 \\ 1 & -2 & 1 \end{vmatrix} = 3$$

$$\det(A_1) = \begin{vmatrix} 2 & 1 & -1 \\ 3 & -1 & 1 \\ 1 & -2 & 1 \end{vmatrix} = 5$$

$$\det(A_2) = \begin{vmatrix} 1 & 2 & -1 \\ 2 & 3 & 1 \\ 1 & 1 & 1 \end{vmatrix} = 1$$

$$\det(A_3) = \begin{vmatrix} 1 & 1 & 2 \\ 2 & -1 & 3 \\ 1 & -2 & 1 \end{vmatrix} = 0$$

Our solution is then:

$$x = \frac{5}{3}, \qquad y = \frac{1}{3}, \qquad z = \frac{0}{3} = 0.$$ ●

As is the case with the adjoint method for finding inverses, Cramer's rule is usually not a very efficient way to solve a linear system. The Gauss-Jordan or Gaussian elimination methods are normally much better. Note too that Cramer's rule may not be used should the system not have the same number of equations as unknowns or, if it does have as many equations as unknowns, should the coefficient matrix not be invertible. One place where it is often convenient to use Cramer's rule, however, is if the coefficients involve functions such as in the following example.

EXAMPLE 2   Solve the following system for x and y:

$$xe^{2t} \sin t - ye^{2t} \cos t = 1$$
$$2xe^{2t} \cos t + 2ye^{2t} \sin t = t.$$

Solution   In this example, we have:

$$\det(A) = \begin{vmatrix} e^{2t} \sin t & -e^{2t} \cos t \\ 2e^{2t} \cos t & 2e^{2t} \sin t \end{vmatrix} = 2e^{4t} \sin^2 t + 2e^{4t} \cos^2 t = 2e^{4t}$$

$$\det(A_1) = \begin{vmatrix} 1 & -e^{2t}\cos t \\ t & 2e^{2t}\sin t \end{vmatrix} = 2e^{2t}\sin t + te^{2t}\cos t$$

$$\det(A_2) = \begin{vmatrix} e^{2t}\sin t & 1 \\ 2e^{2t}\cos t & t \end{vmatrix} = te^{2t}\sin t - 2e^{2t}\cos t$$

$$x = e^{-2t}\sin t + \frac{1}{2}te^{-2t}\cos t, \qquad y = \frac{1}{2}te^{-2t}\sin t - e^{-2t}\cos t \qquad \bullet$$

EXERCISES 1.6

Use Theorem 1.21 to determine whether the following matrices are invertible.

1. $\begin{bmatrix} 6 & -3 \\ -4 & 2 \end{bmatrix}$ **2.** $\begin{bmatrix} 5 & -1 \\ 3 & 4 \end{bmatrix}$

3. $\begin{bmatrix} -1 & 0 & 2 \\ 1 & 1 & -1 \\ 3 & 0 & 1 \end{bmatrix}$ **4.** $\begin{bmatrix} 2 & -1 & -3 \\ 1 & 1 & 3 \\ 6 & 0 & 0 \end{bmatrix}$

Use the adjoint method to find the inverse of the following matrices.

5. $\begin{bmatrix} 1 & 3 \\ -2 & 1 \end{bmatrix}$ **6.** $\begin{bmatrix} -2 & 3 \\ 1 & -2 \end{bmatrix}$

Use Cramer's rule to solve the following systems.

7. $3x - 4y = 1$ **8.** $7x + y = 4$
  $2x + 3y = 2$ $2x - 5y = 8$

9. $3x - y + z = 1$ **10.** $5x - 4y + z = 2$
  $2x + y - 3z = 3$ $2x - 3y - 2z = 4$
  $x - 2y + z = 7$ $3x + y + 3z = 2$

11. Use Cramer's rule to solve the following system for x and y.

$$xe^t \sin 2t + ye^t \cos 2t = t$$
$$2xe^t \cos 2t - 2ye^t \sin 2t = t^2$$

12. Use Cramer's rule to solve the following system for x, y, and z.

$$e^t x + e^{2t} y + e^{-t} z = 1$$
$$e^t x + 2e^{2t} y - e^{-t} z = t$$
$$e^t x + 4e^{2t} y + e^{-t} z = t^2$$

13. Prove the following parts of Lemma 1.22.

  a) Part (2)

  b) Part (3)

14. a) An invertible matrix A with integer entries is said to be **unimodular** if A^{-1} also has integer entries. Show that if A is a square matrix with integer entries such that $\det(A) = \pm 1$, then A is a unimodular matrix.

  b) Prove the converse of the result in part (a); i.e., prove that if A is a unimodular matrix, then $\det(A) = \pm 1$.

15. a) Find the determinants of the following matrices.

$$A = \begin{bmatrix} 3 & -2 \\ 1 & 4 \end{bmatrix} \quad \text{and} \quad B = \begin{bmatrix} 1 & 2 \\ -2 & 3 \end{bmatrix}$$

  b) Find $\det(AB)$, $\det(A^{-1})$, and $\det(B^T A^{-1})$ without finding AB, A^{-1}, or $B^T A^{-1}$.

  c) Show that $\det(A + B)$ is not the same as $\det(A) + \det(B)$.

16. Show that if A and B are square matrices of the same size, then $\det(AB) = \det(BA)$.

17. a) Either of the commands *adj* or *adjoint* can be used in Maple to find the adjoint of a square matrix. Use either one of these Maple commands or corresponding commands in another appropriate software package to find the adjoint of the matrix

$$A = \begin{bmatrix} 1.2 & 2 & -3.1 & -2 \\ 1 & 1.2 & -2 & 2.6 \\ -2.1 & 3.7 & 1 & -4 \\ 2.3 & 4 & -3 & 6.5 \end{bmatrix}.$$

b) Use your software package and the result of part (a) to find $adj(A)A$.

c) By part (b), what is the value of $\det(A)$?

1.7   PROOFS OF THEOREMS ON DETERMINANTS

In this section we will prove those results about determinants whose proofs were omitted in the previous two sections. Many of these proofs will use the technique of mathematical induction, a technique of proof with which we will assume you are familiar.

Recall that we defined the determinant of an $n \times n$ matrix $A = [a_{ij}]$ as the cofactor expansion about the first row:

$$\det(A) = \sum_{j=1}^{n} a_{1j}C_{1j} = \sum_{j=1}^{n}(-1)^{1+j}a_{1j}\det(M_{1j}).$$

As we prove some of our results, we will sometimes have minors of minors; that is, we will have matrices obtained by deleting two rows and two columns from A. For notational purposes, let us use

$$M(ij, kl)$$

to denote the matrix obtained from A by deleting rows i and k ($i \neq k$) and columns j and l ($j \neq l$).

Our first theorem about determinants (Theorem 1.16) was that we could expand about any row or column. As a first step toward obtaining this result, we show that we can expand about any row.

LEMMA 1.29   If $A = [a_{ij}]$ is an $n \times n$ matrix with $n \geq 2$, then for any i, $1 \leq i \leq n$,

$$\det(A) = \sum_{j=1}^{n} a_{ij}C_{ij}.$$

Proof   The verification is easy for $n = 2$ and is left as an exercise (Exercise 1). Assume the result is valid for all $k \times k$ matrices and suppose that $A = [a_{ij}]$ is a $(k+1) \times (k+1)$ matrix. There is nothing to show if $i = 1$, so assume $i > 1$. By definition,

$$\det(A) = \sum_{j=1}^{k+1}(-1)^{1+j}a_{1j}\det(M_{1j}).$$

Using the induction hypothesis, we may expand each $\det(M_{1j})$ about its row and obtain

$$
\det(A) = \sum_{j=1}^{k+1}(-1)^{1+j}a_{1j}\left\{\sum_{l=1}^{j-1}(-1)^{i-1+l}a_{il}\det(M(1j,il))\right.
$$

$$
\left. + \sum_{l=j+1}^{k+1}(-1)^{i-1+l-1}a_{il}\det(M(1j,il))\right\}
$$

$$
= \sum_{j=1}^{k+1}\sum_{l=1}^{j-1}(-1)^{i+j+l}a_{1j}a_{il}\det(M(1j,il))
$$

$$
+ \sum_{j=1}^{k+1}\sum_{l=j+1}^{k+1}(-1)^{i+j+l-1}a_{1j}a_{il}\det(M(1j,il))
$$

(On the second summation, $l-1$ occurs in the exponent of -1 instead of l since the lth column of A becomes the $(l-1)$st column of M_{1j} when $l > j$.) Now consider the cofactor expansion about the ith row, which we write as

$$
\sum_{l=1}^{k+1}(-1)^{i+l}a_{il}\det(M_{il}).
$$

Let us expand each $\det(M_{il})$ about the first row:

$$
\sum_{l=1}^{k+1}(-1)^{i+l}a_{il}\left\{\sum_{j=1}^{l-1}(-1)^{1+j}a_{1j}\det(M(il,1j))\right.
$$

$$
\left. + \sum_{j=l+1}^{k+1}(-1)^{1+j-1}a_{1j}\det(M(il,1j))\right\}
$$

$$
= \sum_{l=1}^{k+1}\sum_{j=1}^{l-1}(-1)^{i+j+l+1}a_{il}a_{1j}\det(M(il,1j))
$$

$$
+ \sum_{l=1}^{k+1}\sum_{j=l+1}^{k+1}(-1)^{i+j+l}a_{il}a_{1j}\det(M(il,1j)).
$$

While they may look different, the results in Equations (1) and (2) are the same. To see why, consider a term with a j and an l. If $j > l$, this term in Equation (1) is

$$
(-1)^{i+j+l}a_{1j}a_{il}\det(M(1j,il)),
$$

which is exactly the same term as we have in Equation (2) for $j > l$. We leave it as an exercise (Exercise 2) to show that the terms with $j < l$ in Equations (1) and (2) are the same. ●

We next show that we can expand about the first column.

LEMMA 1.30 If $A = [a_{ij}]$ is an $n \times n$ matrix with $n \geq 2$, then

$$\det(A) = \sum_{i=1}^{n} a_{i1} C_{i1}.$$

Proof We again use induction on n. We will let you verify this for $n = 2$ (Exercise 3). Assume this result holds for all $k \times k$ matrices and let $A = [a_{ij}]$ be a $(k+1) \times (k+1)$ matrix. By definition

$$\det(A) = \sum_{j=1}^{k+1} (-1)^{1+j} a_{1j} \det(M_{1j})$$

$$= a_{11} \det(M_{11}) + \sum_{j=2}^{k+1} (-1)^{1+j} a_{1j} \det(M_{1j}).$$

Using the induction hypothesis, we expand each $\det(M_{1j})$ about its first column for $j \geq 2$ and obtain

$$\det(A) = a_{11} \det(M_{11}) + \sum_{j=2}^{k+1} (-1)^{1+j} a_{1j} \left\{ \sum_{i=2}^{k+1} (-1)^{1+i-1} a_{i1} \det(M(1j, i1)) \right\} \quad (3)$$

$$= a_{11} \det(M_{11}) + \sum_{j=2}^{k+1} \sum_{i=2}^{k+1} (-1)^{1+j+i} a_{1j} a_{i1} \det(M(1j, i1)).$$

Writing the cofactor expansion about the first column as

$$\sum_{i=1}^{k+1} (-1)^{i+1} a_{i1} \det(M_{i1}) = a_{11} \det(M_{11}) + \sum_{i=2}^{k+1} (-1)^{i+1} a_{i1} \det(M_{i1})$$

and then expanding each $\det(M_{i1})$ for $i \geq 2$ about its first row, we obtain

$$a_{11} \det(M_{11}) + \sum_{i=2}^{k+1} (-1)^{i+1} a_{i1} \left\{ \sum_{j=2}^{k+1} (-1)^{1+j-1} a_{1j} \det(M(i1, 1j)) \right\} \quad (4)$$

$$= a_{11} \det(M_{11}) + \sum_{i=2}^{k+1} \sum_{j=2}^{k+1} (-1)^{i+1+j} a_{i1} a_{1j} \det(M(i1, 1j))$$

Since the results of Equations (3) and (4) are the same, our proof is complete. ●

Before completing the proof of Theorem 1.16, we use Lemma 1.30 to prove Theorem 1.19. For convenience, let us restate Theorem 1.19 as Theorem 1.31.

THEOREM 1.31 If A is an $n \times n$ matrix,

$$\det(A^T) = \det(A).$$

Proof Here we use induction too, only we may start with $n = 1$ where the result is trivial for a 1×1 matrix $A = [a_{11}]$. Assume the result holds for any $k \times k$ matrix and let $A = [a_{ij}]$

be a $(k+1) \times (k+1)$ matrix. Note that the $j1$-entry of A^T is a_{1j} and its minor is M_{1j}^T where M_{1j} is the minor of the entry a_{1j} of A. Thus if we expand $\det(A^T)$ about the first column,

$$\det(A^T) = \sum_{j=1}^{k+1}(-1)^{j+1}a_{1j}\det(M_{1j}^T).$$

By the induction hypothesis, $\det(M_{1j}^T) = \det(M_{1j})$ and hence

$$\det(A^T) = \sum_{j=1}^{k+1}(-1)^{1+j}a_{1j}\det(M_{1j}) = \det(A). \qquad \bullet$$

To complete the proof of Theorem 1.16, we must show that for any $1 \le j \le n$,

$$\det(A) = \sum_{i=1}^{n} a_{ij}C_{ij} = \sum_{i=1}^{n}(-1)^{i+j}a_{ij}\det(M_{ij})$$

where $A = [a_{ij}]$ is an $n \times n$ matrix with $n \ge 2$. To obtain this, we first expand $\det(A^T)$ about row j (which we may do by Lemma 1.29). This gives us

$$\det(A^T) = \sum_{i=1}^{n}(-1)^{j+i}a_{ij}\det(M_{ij}^T).$$

Now applying the result of Theorem 1.31 to $\det(A^T)$ and each $\det(M_{ij}^T)$, we obtain the desired result.

Another result we have not proved that we now prove is Theorem 1.20, which we restate as Theorem 1.32.

THEOREM 1.32 Suppose that $A = [a_{ij}]$ is an $n \times n$ matrix with $n \ge 2$.

1. If B is a matrix obtained from A by interchanging two rows of A, then $\det(B) = -\det(A)$.
2. If B is a matrix obtained from A by multiplying a row of A by a scalar c, then $\det(B) = c\det(A)$.
3. If B is a matrix obtained from A by replacing a row of A by itself plus a multiple of another row of A, then $\det(B) = \det(A)$.

Proof

1. We proceed by induction on n leaving the first case with $n = 2$ as an exercise (Exercise 4.) Assume part (1) holds for all $k \times k$ matrices and let $A = [a_{ij}]$ be a $(k+1) \times (k+1)$ matrix. Suppose that B is obtained from A by interchanging rows i and l. We are going to expand $\det(B)$ about a row other than row i or row l. Pick such a row. Let us call this the mth row. We have $\text{ent}_{mj}(B) = a_{mj}$. If we interchange the rows of the minor M_{mj} of the entry a_{mj} of A that come from rows i and l of A, we obtain the minor of $\text{ent}_{mj}(B)$ in B. By the

induction hypothesis, the determinant of this minor of B is $-\det(M_{mj})$. Thus,

$$\det(B) = \sum_{j=1}^{k+1}(-1)^{m+j}a_{mj}(-\det(M_{mj})) = -\det(A).$$

2. Suppose that B is obtained from A by multiplying row i of A by c. The minor of $\text{ent}_{ij}(B)$ of B is the same as the minor M_{ij} of the entry a_{ij} of A. Hence if we expand about the ith row,

$$\det(B) = \sum_{j=1}^{n}(-1)^{i+j}ca_{ij}\det(M_{ij}) = c\det(A).$$

Before proving part (3), we prove the following lemma.

LEMMA 1.33   If A is an $n \times n$ matrix where $n \geq 2$ with two rows that have the same entries, then $\det(A) = 0$.

Proof   Suppose that row i and row j of A have the same entries. Let B be the matrix obtained from A by interchanging rows i and j. On the one hand, by part (1) of Theorem 1.32, we have

$$\det(B) = -\det(A).$$

On the other hand, $B = A$ and hence

$$\det(B) = \det(A).$$

Thus $\det(A) = -\det(A)$, which implies $\det(A) = 0$.   ●

Proof of   Suppose that B is obtained from A by replacing row i of A by itself plus c times row
Theorem 1.32, l of A. Then $\text{ent}_{ij}(B) = a_{ij} + ca_{lj}$ and the minor of $\text{ent}_{ij}(B)$ of B is the same as the
Part (3)   minor M_{ij} of the entry a_{ij} of A. If we expand $\det(B)$ about row i,

$$\det(B) = \sum_{j=1}^{n}(-1)^{i+j}(a_{ij} + ca_{mj})\det(M_{ij})$$

$$= \sum_{j=1}^{n}(-1)^{i+j}a_{ij}\det(M_{ij}) + c\sum_{j=1}^{n}(-1)^{i+j}a_{mj}\det(M_{ij}).$$

The second sum is the same as the determinant of the matrix obtained from A by replacing the ith row of A by row m and hence is zero by Lemma 1.33. Thus

$$\det(B) = \sum_{j=1}^{n}(-1)^{i+j}a_{ij}\det(M_{ij}) = \det(A).$$

The proof of our result about the product of a square matrix and its adjoint (Theorem 1.26) is another place where Lemma 1.33 is used. We restate this result as Theorem 1.34.

THEOREM 1.34 If A is a square matrix, then

$$A \operatorname{adj}(A) = \operatorname{adj}(A)A = \det(A)I.$$

Proof We will prove that $A \operatorname{adj}(A) = \det(A)I$ and leave the proof for $\operatorname{adj}(A)A$ as an exercise (Exercise 5). Suppose A is an $n \times n$ matrix. Notice that

$$\operatorname{ent}_{ij}(A \operatorname{adj}(A)) = \sum_{k=1}^{n} a_{ik}C_{jk}.$$

If $i = j$,

$$\operatorname{ent}_{ii}(A \operatorname{adj}(A)) = \sum_{k=1}^{n} a_{ik}C_{ik} = \det(A).$$

If $i \neq j$,

$$\operatorname{ent}_{ij}(A \operatorname{adj}(A)) = \sum_{k=1}^{n} a_{ik}C_{jk}$$

is the determinant of the matrix obtained from A by replacing the jth row of A by the ith row of A. Since this determinant contains two rows with the same entries, we have

$$\operatorname{ent}_{ij}(A \operatorname{adj}(A)) = \sum_{k=1}^{n} a_{ik}C_{jk} = 0$$

when $i \neq j$. This gives us $A \operatorname{adj}(A) = \det(A)I$. ●

The final result about determinants we have yet to prove is Cramer's rule, restated as Theorem 1.35.

THEOREM 1.35 Suppose that $AX = B$ is a system of n linear equations in n unknowns such that $\det(A) \neq 0$. Let A_1 be the matrix obtained from A by replacing the first column of A by B, A_2 be the matrix obtained from A by replacing the second column of A by $B, \ldots, A_n$ be the matrix obtained from A by replacing the nth column of A by B. Then

$$x_1 = \frac{\det(A_1)}{\det(A)}, \qquad x_2 = \frac{\det(A_2)}{\det(A)}, \qquad \ldots, \qquad \frac{\det(A_n)}{\det(A)}.$$

Proof Since

$$A^{-1} = \frac{1}{\det(A)}\operatorname{adj}(A),$$

we have

$$X = A^{-1}B = \frac{1}{\det(A)}\operatorname{adj}(A)B.$$

Thus for each $1 \leq i \leq n$,

$$x_i = \frac{1}{\det(A)} \sum_{k=1}^{n} C_{ki} b_k.$$

The summation is exactly the determinant of A_i expanded about the ith column and hence

$$x_i = \frac{\det(A_i)}{\det(A)}. \qquad \bullet$$

EXERCISES 1.7

1. Prove Lemma 1.29 for $n = 2$.

2. Show that the terms with $j < l$ in Equations (1) and (2) are the same.

3. Prove Lemma 1.30 for $n = 2$.

4. Prove part (1) of Theorem 1.32 for $n = 2$.

5. Prove that $\text{adj}(A)A = \det(A)I$.

6. If A is an $n \times n$ matrix and c is a scalar, show that $\det(cA) = c^n \det(A)$.

7. A matrix of the form

$$V = \begin{bmatrix} 1 & 1 & 1 & \cdots & 1 \\ x_1 & x_2 & x_3 & \cdots & x_n \\ x_1^2 & x_2^2 & x_3^2 & \cdots & x_n^2 \\ \vdots & \vdots & \vdots & & \vdots \\ x_1^{n-1} & x_2^{n-1} & x_3^{n-1} & \cdots & x_n^{n-1} \end{bmatrix}$$

is called a **Vandermonde** matrix.[14]

a) Show that if $n = 2$,

$$\det(V) = x_2 - x_1.$$

b) Use row operations to show that if $n = 3$,

$$\det(V) = (x_2 - x_1)(x_3 - x_1)(x_3 - x_2).$$

c) Use row operations to show that if $n = 4$,

$$\det(V) = (x_2 - x_1)(x_3 - x_1)(x_4 - x_1)(x_3 - x_2)$$
$$(x_4 - x_2)(x_4 - x_3).$$

d) In general,

$$\det(V) = \prod_{j=2}^{n} \left\{ \prod_{i=1}^{j-1} (x_j - x_i) \right\}.$$

Prove this result.

---

[14] Named for Alexandre Théophile Vandermonde (1735–1796) who studied the theory of equations and determinants.

Vector Spaces

Your first encounter with vectors in two or three dimensions likely was for modeling physical situations. For example, winds blowing with speeds of 5 and 10 miles per hour 45° east of north may be illustrated by the velocity vectors **u** and **v** in Figure 2.1 drawn as directed line segments of lengths 5 and 10 units pointing 45° east of north. A force pushing a block up an inclined plane might be illustrated by drawing a force vector **F** as in Figure 2.2. In your calculus courses you should have encountered many uses of vectors in two and three dimensions in the study of equations of lines and planes, tangent and normal vectors to curves, and gradients, just to name a few.

Figure 2.1

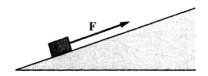

Figure 2.2

Were someone to ask you to briefly tell them about vectors you might well respond by saying simply that a vector $\mathbf{v}$ is a directed line segment in two or three dimensions. If we place a vector so that its initial point is at the origin (choice of the initial point does not matter since we are only trying to indicate magnitude and direction with a vector) and its terminal point is (a, b) in two dimensions or (a, b, c) in three dimensions, we can denote the vector by its terminal point as

$$\mathbf{v} = (a, b) \quad \text{or} \quad \mathbf{v} = (a, b, c)$$

in two or three dimensions, respectively. (Other standard notations you might have used instead are $\mathbf{v} = \langle a, b \rangle$ or $\mathbf{v} = a\mathbf{i} + b\mathbf{j}$ in two dimensions and $\mathbf{v} = \langle a, b, c \rangle$ or $\mathbf{v} = a\mathbf{i} + b\mathbf{j} + c\mathbf{k}$ in three dimensions. In Chapter 1 we mentioned that in this text we will write our vectors in two, three, and even n dimensions as column matrices or column vectors.) Vectors are added by the rules

$$(a_1, b_1) + (a_2, b_2) = (a_1 + a_2, b_1 + b_2)$$

or

$$(a_1, b_1, c_1) + (a_2, b_2, c_2) = (a_1 + a_2, b_1 + b_2, c_1 + c_2)$$

and we have a scalar multiplication defined as

$$k(a, b) = (ka, kb) \quad \text{or} \quad k(a, b, c) = (ka, kb, kc)$$

(which, of course, are special cases of matrix addition and scalar multiplication).

We could continue discussing things such as the geometric impact of vector addition (the parallelogram rule), the geometric impact of scalar multiplication (stretching, shrinking, and reflecting), dot products, cross products, and so on, but that is not our purpose here. Our purpose is to study sets of vectors forming a type of structure called a vector space from an algebraic point of view rather than a geometric one. To us a vector space will be a set on which we have defined an addition and a scalar multiplication satisfying certain properties. Two old friends, vectors in two dimensions and vectors in three dimensions, are two examples of vector spaces, but they are not the only ones as you are about to see.

2.1   VECTOR SPACES

As just mentioned, a vector space will be a set of objects on which we have an addition and a scalar multiplication satisfying properties. The formal definition of a vector space is as follows.

DEFINITION   A nonempty set V is called a **vector space** if there are operations of addition and scalar multiplication on V such that the following eight properties are satisfied:

1. $u + v = v + u$ for all u and v in V.

2. $u + (v + w) = (u + v) + w$ for all u, v, and w in V.

3. There is an element 0 in V so that $v + 0 = v$ for all v in V.
4. For each v in V there is an element $-v$ in V so that $v + (-v) = 0$.
5. $c(u + v) = cu + cv$ for all real numbers c and for all u and v in V.
6. $(c + d)v = cv + dv$ for all real numbers c and d and for all v in V.
7. $c(dv) = (cd)v$ for all real numbers c and d and for all v in V.
8. $1 \cdot v = v$ for all v in V.

The eight properties of a vector space are also called the *laws*, *axioms*, or *postulates* of a vector space. The elements of the set V when V is a vector space are called the **vectors** of V and, as we have done already with matrices, real numbers are called **scalars** in connection with the scalar multiplication on V.[1] Actually, not all vector spaces are formed using real numbers for the scalars. Later we shall work with some vector spaces where the complex numbers are used as scalars. But for now, all scalars will be real numbers.

Some terminology is associated with the vector space properties. Property 1 is called the **commutative law of addition,** and property 2 is called the **associative law of addition.** The element 0 of V in property 3 is called an **additive identity** or a **zero vector,**[2] and the element $-v$ of V in property 4 is called an **additive inverse** or a **negative** of the vector v. Because of commutativity of addition, we could have equally well put our zero and negative vectors on the left in the equations in properties 3 and 4, writing them as

$$0 + v = v \quad \text{and} \quad -v + v = 0.$$

Properties 5 and 6 are distributive properties: Property 5 is a left-hand distributive property saying that scalar multiplication distributes over vector addition, and property 6 is a right-hand distributive property saying that scalar multiplication distributes over scalar addition. Property 7 is an associative property for scalar multiplication.

Let us now look at some examples of vector spaces.

EXAMPLE 1   From our matrix addition and scalar multiplication properties in Chapter 1, we immediately see the set of $n \times 1$ column vectors or n-dimensional vectors $\mathbb{R}^n$ satisfies the eight properties of a vector space under our addition and scalar multiplication of column vectors,

$$\begin{bmatrix} x_1 \\ x_2 \\ \vdots \\ x_n \end{bmatrix} + \begin{bmatrix} y_1 \\ y_2 \\ \vdots \\ y_n \end{bmatrix} = \begin{bmatrix} x_1 + y_1 \\ x_2 + y_2 \\ \vdots \\ x_n + y_n \end{bmatrix} \quad \text{and} \quad c \begin{bmatrix} x_1 \\ x_2 \\ \vdots \\ x_n \end{bmatrix} = \begin{bmatrix} cx_1 \\ cx_2 \\ \vdots \\ cx_n \end{bmatrix}.$$

---

[1] In print, vectors are often set in boldface type and, in handwritten work, marked with an arrow over the top to distinguish them from scalars. Such confusion will not arise in this text, however, since we will reserve the lowercase letters u, v, and w for vectors. Scalars usually will be denoted by letters such as a, b, c, and d.

[2] In cases where zero vectors could be confused with the scalar zero, we will put zero vectors in boldface print as **0** as we did for zero matrices in Chapter 1.

Hence $\mathbb{R}^n$ (of which vectors in two and three dimensions are special cases when we write these vectors in column form) is a vector space for each positive integer n. ●

Row vectors of a fixed length n would also form a vector space under our addition and scalar multiplication of row vectors. More generally, so would matrices of a fixed size under matrix addition and scalar multiplication. Let us make this our second example.

EXAMPLE 2 The set of $m \times n$ matrices $M_{m \times n}(\mathbb{R})$ satisfies the eight properties of a vector space under matrix addition and scalar multiplication and hence is a vector space. ●

Making our definition of a vector space as general as we have done will prove valuable to us in the future. For instance, various sets of real-valued functions[3] form vector spaces as is illustrated in the next example. Because of this, sets of real-valued functions will have many properties similar to those enjoyed by our matrix vector spaces, which we shall exploit later in our study of differential equations.

EXAMPLE 3 Let $F(a, b)$ denote the set of all real-valued functions defined on the open interval (a, b).[4] We can define an addition on $F(a, b)$ as follows: If f and g are two functions in $F(a, b)$, we let $f + g$ be the function defined on (a, b) by

$$(f + g)(x) = f(x) + g(x).$$

We can also define a scalar multiplication on $F(a, b)$: If c is a real number and f is a function in $F(a, b)$, we let cf be the function defined on (a, b) by

$$(cf)(x) = cf(x).$$

Show that $F(a, b)$ is a vector space under this addition and this scalar multiplication.

Solution We verify that the eight properties of a vector space are satisfied.

1. Do we have equality of the function $f + g$ and $g + f$ for all f and g in $F(a, b)$? To see, we have to check if $(f + g)(x)$ is the same as $(g + f)(x)$ for any x in (a, b). If x is in (a, b),

    $$(f + g)(x) = f(x) + g(x) = g(x) + f(x) = (g + f)(x)$$

    (the second equality holds since addition of the real numbers $f(x)$ and $g(x)$ is commutative) and we do have $f + g = g + f$.

2. Do we have equality of the functions $f + (g + h)$ and $(f + g) + h$ for any f, g, and h in $F(a, b)$? We proceed in much the same manner as in the previous

---

[3] A real-valued function is a function whose range is contained in the set of real numbers. For instance, the function f from $\mathbb{R}$ to $\mathbb{R}$ defined by $f(x) = x^2$ is real-valued; so is the function f from $\mathbb{R}^2$ to $\mathbb{R}$ defined by $f(x, y) = x^2 + y^2$. For a fixed positive integer n the determinant is a real-valued function from $M_{n \times n}(\mathbb{R})$ to $\mathbb{R}$.

[4] For instance, the functions given by $f(x) = x^2$, $f(x) = |x|$, $f(x) = \sin x$, $f(x) = e^x$, and the greatest integer function would be elements of $F(-\infty, \infty)$; these five functions along with the functions defined by $f(x) = 1/x$, $f(x) = \ln x$, and $f(x) = \cot x$ would be elements of $F(0, \pi)$.

part. For any x in (a, b),

$$(f + (g + h))(x) = f(x) + (g + h)(x) = f(x) + (g(x) + h(x))$$
$$= (f(x) + g(x)) + h(x) = (f + g)(x) + h(x)$$
$$= ((f + g) + h)(x)$$

(the third equality holds since addition of the real numbers $f(x)$, $g(x)$, and $h(x)$ is associative) and we have $f + (g + h) = (f + g) + h$.

3. Do we have a zero vector? How about if we use the constant function that is 0 for each x, which we shall also denote by 0 and call the *zero function*? That is, the zero function is

$$0(x) = 0.$$

The zero function is then an element of $F(a, b)$, and for any f in $F(a, b)$ and x in (a, b),

$$(f + 0)(x) = f(x) + 0(x) = f(x) + 0 = f(x).$$

Hence $f + 0 = f$ and the zero function serves as a zero vector for $F(a, b)$.

4. What could we use for the negative of a function f in $F(a, b)$? How about the function $-f$ defined by

$$(-f)(x) = -f(x)?$$

The function $-f$ is in $F(a, b)$, and for any x in (a, b) we have

$$(f + (-f))(x) = f(x) + (-f)(x) = f(x) + (-f(x)) = 0 = 0(x).$$

Hence $-f$ serves as a negative of f.

5. This and the remaining properties are verified in much the same manner as the first two, so we will go a little faster now. For any real number c and any functions f and g in $F(a, b)$,

$$(c(f + g))(x) = c(f + g)(x) = c(f(x) + g(x)) = cf(x) + cg(x)$$
$$= (cf)(x) + (cg)(x)$$

for any x in (a, b), and hence $c(f + g) = cf + cg$.

6. For any real numbers c and d and any function f in $F(a, b)$,

$$((c + d)f)(x) = (c + d)f(x) = cf(x) + df(x) = (cf)(x) + (df)(x)$$

for any x in (a, b), and hence $(c + d)f = cf + df$.

Lest we be accused of doing everything for you, we will let you verify the last two properties as an exercise (Exercise 1). ●

There is nothing special about using an open interval in Example 3. The set of real-valued functions defined on a closed interval $[a, b]$, which we will denote by $F[a, b]$, also forms a vector space under the addition and scalar multiplication of functions we used in Example 3. More generally, the set of real-valued functions defined on any set

S, which we will denote by $F(S)$, is a vector space under the types of addition and scalar multiplication of functions we used in Example 3.

The next example illustrates that not every set on which is defined an addition and a scalar multiplication is a vector space.

EXAMPLE 4   On the set of pairs of real numbers (x, y), define an addition by

$$(x_1, y_1) + (x_2, y_2) = (x_1 + x_2 + 1, y_1 + y_2)$$

and a scalar multiplication by

$$c(x, y) = (cx, cy).$$

Determine if this is a vector space.

Solution   Let us start checking the eight properties.

1. We have

$$(x_1, y_1) + (x_2, y_2) = (x_1 + x_2 + 1, y_1 + y_2)$$

and

$$(x_2, y_2) + (x_1, y_1) = (x_2 + x_1 + 1, y_2 + y_1).$$

Since these ordered pairs are the same, addition is commutative.

2. For three pairs (x_1, y_1), (x_2, y_2), and (x_3, y_3) of real numbers, we have

$$(x_1, y_1) + ((x_2, y_2) + (x_3, y_3)) = (x_1, y_1) + (x_2 + x_3 + 1, y_2 + y_3)$$
$$= (x_1 + x_2 + x_3 + 2, y_1 + y_2 + y_3)$$

while

$$((x_1, y_1) + (x_2, y_2)) + (x_3, y_3) = (x_1 + x_2 + 1, y_1 + y_2) + (x_3, y_3)$$
$$= (x_1 + x_2 + x_3 + 2, y_1 + y_2 + y_3).$$

Since these ordered pairs are again the same, addition is associative.

3. The pair $(-1, 0)$ serves as an additive identity here since

$$(x, y) + (-1, 0) = (x + (-1) + 1, y + 0) = (x, y)$$

so we have property 3. (Notice that an additive identity does not have to be $(0, 0)$!)

4. An additive inverse of (x, y) is $(-x - 2, -y)$ (and is not $(-x, -y)$!) since

$$(x, y) + (-x - 2, -y) = (x + (-x - 2) + 1, y + (-y)) = (-1, 0)$$

and hence we have property 4.

5. Since

$$c((x_1, y_1) + (x_2, y_2)) = c(x_1 + x_2 + 1, y_1 + y_2) = (cx_1 + cx_2 + c, cy_1 + cy_2)$$

while

$$c(x_1, y_1) + c(x_2, y_2) = (cx_1, cy_1) + (cx_2, cy_2) = (cx_1 + cx_2 + 1, cy_1 + cy_2),$$

we see that property 5 does not hold for every real number c. (In fact, it holds only when $c = 1$.) Thus, this is not a vector space. Once we see one property that does not hold we are done with determining whether we have a vector space in this example. Were we to continue going through the properties, we would also find that property 6 does not hold. (Try it.) Properties 7 and 8 will hold. (Verify this.) ●

Do not let Example 4 mislead you into thinking that unusual operations for addition or scalar multiplication will not produce a vector space. Consider the next example.

EXAMPLE 5 Let $\mathbb{R}^+$ denote the set of positive real numbers. Define addition on $\mathbb{R}^+$ by

$$x \oplus y = xy$$

and scalar multiplication by

$$c \odot x = x^c$$

where x and y are in $\mathbb{R}^+$ and c is a real number. (We use the symbols $\oplus$ and $\odot$ to avoid confusion with usual addition and multiplication.) Determine if $\mathbb{R}^+$ is a vector space under this addition and scalar multiplication.

Solution

1. Since

$$x \oplus y = xy = yx = y \oplus x$$

   this addition is commutative.
2. Since

$$x \oplus (y \oplus z) = x(yz) = (xy)z = (x \oplus y) \oplus z$$

   this addition is associative.
3. The positive real number 1 is an additive identity since

$$x \oplus 1 = x \cdot 1 = x.$$

4. An additive inverse of x is $1/x$. (You verify this one.)
5. Since

$$c \odot (x \oplus y) = c \odot (xy) = (xy)^c = x^c y^c = x^c \oplus y^c = c \odot x \oplus c \odot y$$

   we have property 5.

We will let you verify that

6. $(c + d) \odot x = c \odot x \oplus d \odot x$,

7. $c \odot (d \odot x) = (cd) \odot x$, and

8. $1 \odot x = x$

as an exercise (Exercise 2) and hence we have a vector space.   ●

We conclude this section by discussing some properties of vector spaces. The first theorem deals with the uniqueness of zero and negative vectors.

THEOREM 2.1   Suppose that V is a vector space.

1. A zero vector of V is unique.

2. A negative of a vector v in V is unique.

Proof

1. Suppose that 0 and $0'$ are zero vectors of V. On the one hand, since 0 is a zero vector, we have that

$$0' + 0 = 0'.$$

On the other hand,

$$0' + 0 = 0$$

since $0'$ is a zero vector. Thus we see $0' = 0$.

2. Suppose that $-v$ and $-v'$ are negatives of v. Notice that

$$-v' + (v + (-v)) = -v' + 0 = -v' = (-v' + v) + (-v) = 0 + (-v) = -v,$$

and hence we see $-v = -v'$.   ●

Because of Theorem 2.1, we may now say *the* zero vector instead of *a* zero vector and *the* negative of a vector instead of *a* negative of a vector.

Theorem 2.2 contains some more properties that we shall use often.

THEOREM 2.2   Let V be a vector space.

1. For any vector v in V, $0 \cdot v = \mathbf{0}$.[5]

2. For any real number c, $c\mathbf{0} = \mathbf{0}$.

3. For any vector v in V, $(-1)v = -v$.

Proof   We will prove parts (1) and (3) and leave the proof of part (2) as an exercise (Exercise 10).

1. One way to prove this is to first notice that since

$$0 \cdot v = (0 + 0)v = 0 \cdot v + 0 \cdot v$$

---

[5] Here is a place where we have put the zero vector in boldface print to distinguish it from the scalar zero.

we have

$$0 \cdot v = 0 \cdot v + 0 \cdot v.$$

Adding $-(0 \cdot v)$ to each side of the preceding equation, we obtain

$$0 \cdot v + (-(0 \cdot v)) = 0 \cdot v + 0 \cdot v + (-(0 \cdot v))$$

from which the desired equation

$$\mathbf{0} = 0 \cdot v$$

now follows.

2. Noting on the one hand that

$$(1 + (-1))v = 1 \cdot v + (-1)v = v + (-1)v$$

and on the other hand that

$$(1 + (-1))v = 0 \cdot v = \mathbf{0}$$

by part (1), we have

$$v + (-1)v = \mathbf{0}.$$

Adding $-v$ to each side of the preceding equation, we obtain

$$-v + v + (-1)v = -v + \mathbf{0}$$

and hence

$$(-1)v = -v. \qquad \bullet$$

Finally, we point out that we can define **subtraction** on a vector space V by setting the difference of two vectors u and v in V to be

$$u - v = u + (-v).$$

You might notice that we could have equally well subtracted the vector whenever we added its negative in the proofs of parts (1) and (3) in Theorem 2.2.

EXERCISES 2.1

1. Complete Example 3 by showing that properties 7 and 8 of a vector space hold.

2. Complete Example 5 by showing that properties 6, 7, and 8 of a vector space hold.

3. In each of the following, determine whether the indicated addition and scalar multiplication on ordered pairs of real numbers yields a vector space. For those that are not vector spaces, determine which properties of a vector space fail to hold.

a) $(x_1, y_1) + (x_2, y_2) = (x_1, y_2),$
    $c(x, y) = (cx, cy)$

b) $(x_1, y_1) + (x_2, y_2) = (x_1 + x_2, y_1 + y_2),$
    $c(x, y) = (c + x, c + y)$

c) $(x_1, y_1) + (x_2, y_2) = (x_1 + y_2, x_2 + y_1),$
    $c(x, y) = (cx, cy)$

4. In each of the following, determine whether the indicated addition and scalar multiplication of ordered triples of real numbers yields a vector space. For those that are not vector spaces, determine which properties of a vector space fail to hold.

   a) $(x_1, y_1, z_1) + (x_2, y_2, z_2) =$
   $(x_1 + x_2, y_1 + y_2, z_1 + z_2),$
   $c(x, y, z) = (cx, y, cz)$

   b) $(x_1, y_1, z_1) + (x_2, y_2, z_2) =$
   $(z_1 + z_2, y_1 + y_2, x_1 + x_2),$
   $c(x, y, z) = (cx, cy, cz)$

   c) $(x_1, y_1, z_1) + (x_2, y_2, z_2) =$
   $(x_1 + x_2, y_1 + y_2 - 2, z_1 + z_2),$
   $c(x, y, z) = (cx, y, z)$

5. Show that the set of ordered pairs of positive real numbers is a vector space under the addition and scalar multiplication

$$(x_1, y_1) + (x_2, y_2) = (x_1 x_2, y_1 y_2), \quad c(x, y) = (x^c, y^c).$$

6. Does the set of complex numbers under the addition and scalar multiplication

$$(a + bi) + (c + di) = (a + c) + (b + d)i,$$
$$c(a + bi) = ca + cbi$$

   where $a, b, c,$ and d are real numbers form a vector space? If not, why not?

7. Let C denote the set of all convergent sequences of real numbers $\{a_n\}$. Is C a vector space under the

   addition and scalar multiplication

$$\{a_n\} + \{b_n\} = \{a_n + b_n\}, c\{a_n\} = \{ca_n\}?$$

   If not, why not?

8. Let S denote the set of all convergent series of real numbers $\sum_{n=1}^{\infty} a_n$. Is S a vector space under the addition and scalar multiplication

$$\sum_{n=1}^{\infty} a_n + \sum_{n=1}^{\infty} b_n = \sum_{n=1}^{\infty} (a_n + b_n),$$

$$c \sum_{n=1}^{\infty} a_n = \sum_{n=1}^{\infty} ca_n?$$

   If not, why not?

9. Let V be a set consisting of a single element z. Define addition and scalar multiplication on V by

$$z + z = z, \quad cz = z.$$

   Show that V is a vector space. Such a vector space is called a **zero vector space.**

10. Prove part (2) of Theorem 2.2.

11. Prove that if c is a real number and v is a vector in a vector space V such that $cv = \mathbf{0}$, then either $c = 0$ or $v = \mathbf{0}$.

12. Show that subtraction is not an associative operation on a vector space.

2.2 SUBSPACES AND SPANNING SETS

We begin this section with subspaces. Roughly speaking, by a subspace we mean a vector space sitting within a larger vector space. The following definition states this precisely.

> **DEFINITION** A subset W of a vector space V is called a **subspace** of V if W is itself a vector space under the addition and scalar multiplication of V restricted to W.

EXAMPLE 1 Let W be the set of all column vectors of the form

$$\begin{bmatrix} x \\ y \\ 0 \end{bmatrix}.$$

The set W is a subset of $\mathbb{R}^3$. In fact, W is a subspace of $\mathbb{R}^3$. To see this, first notice that the addition of $\mathbb{R}^3$ on elements of W gives us an addition on W: For two elements

$$\begin{bmatrix} x_1 \\ y_1 \\ 0 \end{bmatrix} \quad \text{and} \quad \begin{bmatrix} x_2 \\ y_2 \\ 0 \end{bmatrix}$$

of W, we have the sum

$$\begin{bmatrix} x_1 \\ y_1 \\ 0 \end{bmatrix} + \begin{bmatrix} x_2 \\ y_2 \\ 0 \end{bmatrix} = \begin{bmatrix} x_1 + x_2 \\ y_1 + y_2 \\ 0 \end{bmatrix},$$

which is also an element of W. The fact that the sum of two elements of W is again an element of W is usually described by saying that W is **closed under addition.** Next notice that the scalar multiplication of $\mathbb{R}^3$ gives us a multiplication on W: If c is a scalar and

$$\begin{bmatrix} x \\ y \\ 0 \end{bmatrix}$$

is an element of W, then

$$c \begin{bmatrix} x \\ y \\ 0 \end{bmatrix} = \begin{bmatrix} cx \\ cy \\ 0 \end{bmatrix}$$

is an element of W. Here we say W is **closed under scalar multiplication.** So we have two of the ingredients we need (an addition and a scalar multiplication) for W to be a vector space.

Let us move on to the eight properties. Since addition on $\mathbb{R}^3$ is commutative and associative, it certainly is on W too since the elements of W are elements of $\mathbb{R}^3$. Hence properties 1 and 2 of a vector space hold for W. Property 3 holds since

$$\begin{bmatrix} 0 \\ 0 \\ 0 \end{bmatrix}$$

is an element of W. Because columns of the form

$$\begin{bmatrix} -x \\ -y \\ 0 \end{bmatrix}$$

are in W, property 4 holds. As was the case with commutativity and associativity of addition, the scalar multiplication properties 5–8 will carry over from $\mathbb{R}^3$ to the subset W. Hence W is a vector space. ●

Looking back at Example 1, notice that properties 1, 2, and 5–7 are immediately inherited by any subset of a vector space, so we really do not need to check for them. In fact, the next theorem tells us that the closure properties are really the crucial ones in determining whether a nonempty subset of a vector space is a subspace.

THEOREM 2.3   Let W be a nonempty subset of a vector space V. Then W is a subspace of V if and only if for all u and w in W and for all scalars c, $u + w$ is in W and cu is in W.

Proof   If W is a subspace, W is a vector space and hence we immediately have W is closed under addition and scalar multiplication. (Otherwise, W would not have an addition or scalar multiplication.) The main part of this proof is then to show the converse: If W is closed under addition and scalar multiplication, then W is a subspace. As already noted, properties 1, 2, and 5–8 carry over to W from V, so we only have to do some work to get properties 3 and 4. To get 3, pick an element v in W. (We can do this because W is nonempty.) Since W is closed under scalar multiplication, $(-1)v = -v$ is in W. Now since W is closed under addition,

$$v + (-v) = 0$$

lies in W and we have property 3. The multiplying by -1 trick also gives us negatives: For any u in W, $(-1)u = -u$ is in W by the closure under scalar multiplication and hence property 4 holds.   ●

Let us do some more examples determining whether subsets of vector spaces are subspaces, but now applying Theorem 2.3 by only checking the closure properties.

EXAMPLE 2   Do the vectors of the form

$$\begin{bmatrix} x \\ 1 \end{bmatrix}$$

form a subspace of $\mathbb{R}^2$?

Solution   Since the sum of two such vectors,

$$\begin{bmatrix} x_1 \\ 1 \end{bmatrix} + \begin{bmatrix} x_2 \\ 1 \end{bmatrix} = \begin{bmatrix} x_1 + x_2 \\ 2 \end{bmatrix},$$

has 2 not 1 for its second entry, the set of such vectors is not closed under addition and hence is not a subspace. It is also easily seen that this set of vectors is not closed under scalar multiplication.   ●

EXAMPLE 3   Do the vectors of the form

$$\begin{bmatrix} x \\ y \\ x - 2y \end{bmatrix}$$

form a subspace of $\mathbb{R}^3$?

Solution   Adding two such vectors, we obtain a vector of the same form:

$$\begin{bmatrix} x_1 \\ y_1 \\ x_1 - 2y_1 \end{bmatrix} + \begin{bmatrix} x_2 \\ y_2 \\ x_2 - 2y_2 \end{bmatrix} = \begin{bmatrix} x_1 + x_2 \\ y_1 + y_2 \\ x_1 - 2y_1 + x_2 - 2y_2 \end{bmatrix}$$

$$= \begin{bmatrix} x_1 + x_2 \\ y_1 + y_2 \\ x_1 + x_2 - 2(y_1 + y_2) \end{bmatrix}.$$

Hence we have closure under addition. We also have closure under scalar multiplication since

$$c \begin{bmatrix} x \\ y \\ x - 2y \end{bmatrix} = \begin{bmatrix} cx \\ cy \\ cx - 2cy \end{bmatrix}.$$

Thus these vectors do form a subspace.   ●

Solutions to systems of homogeneous linear equations form subspaces. Indeed this will be such an important fact for us that we record it as a theorem.

THEOREM 2.4   If A is an $m \times n$ matrix, then the solutions to the system of homogeneous linear equations $AX = 0$ is a subspace of $\mathbb{R}^n$.

Proof   First notice that the set of solutions contains the trivial solution $X = 0$ and hence is a nonempty subset of $\mathbb{R}^n$. If X_1 and X_2 are two solutions of $AX = 0$, then $AX_1 = 0$ and $AX_2 = 0$ so that

$$A(X_1 + X_2) = AX_1 + AX_2 = 0 + 0 = 0$$

and hence the set of solutions is closed under addition. If X is a solution and c is a scalar, then

$$A(cX) = cAX = c0 = 0$$

and hence the set of solutions is closed under scalar multiplication. Thus the set of solutions to $AX = 0$ is a subspace of $\mathbb{R}^n$.   ●

In Section 2.1, we noted that sets of real-valued functions on intervals form vector spaces. There are numerous examples of subspaces of such function spaces that will come up in our future work listed in Examples 4–10.

EXAMPLE 4   Let $C(a, b)$ denote the set of continuous real-valued functions on the open interval (a, b), which is a nonempty subset of $F(a, b)$. From calculus, we know that sums of continuous functions and constant multiples of continuous functions are continuous. Hence $C(a, b)$ is closed under addition and scalar multiplication of functions and is a subspace of $F(a, b)$.   ●

EXAMPLE 5   Let $D(a, b)$ denote the set of differentiable functions on (a, b). From calculus we know that $D(a, b)$ is a nonempty subset of $C(a, b)$ that is closed under addition and scalar multiplication of functions. Hence $D(a, b)$ is a subspace of $C(a, b)$.   ●

EXAMPLE 6   Generalizing Example 5, for each positive integer n, let $D^n(a, b)$ denote the set of functions that have an nth derivative on (a, b). We have that $D^1(a, b) = D(a, b)$ and each $D^{n+1}(a, b)$ is a subspace of $D^n(a, b)$.   ●

EXAMPLE 7   For each nonnegative integer n, we will use $C^n(a, b)$ to denote the set of all functions that have a continuous nth derivative on (a, b). Notice that $C^0(a, b) = C(a, b)$, each $C^{n+1}(a, b)$ is a subspace of $C^n(a, b)$, and $C^n(a, b)$ is a subspace of $D^n(a, b)$ for each $n \geq 1$.   ●

EXAMPLE 8   We will let $C^\infty(a, b)$ denote the set of functions that have a continuous nth derivative for every nonnegative integer n. The set $C^\infty(a, b)$ is a subspace of $C^n(a, b)$ for every nonnegative integer n.   ●

EXAMPLE 9   We will let P denote the set of all polynomials; that is, P consists of all expressions $p(x)$ of the form

$$p(x) = a_n x^n + a_{n-1} x^{n-1} + \cdots + a_1 x + a_0$$

where n is a nonnegative integer and each a_i is a real number. Each such polynomial $p(x)$ gives us a function p that is an element of $C^\infty(-\infty, \infty)$. Identifying the polynomial $p(x)$ with the function p, we may view P as being a subset of $C^\infty(-\infty, \infty)$. Since polynomials are closed under addition and scalar multiplication, P is a subspace of $C^\infty(-\infty, \infty)$.   ●

EXAMPLE 10   For each nonnegative integer k, we will let P_k denote the set of all polynomials of degree less than or equal to k along with the polynomial 0. In particular, P_0 is the set of all constant functions $p(x) = a$, P_1 is the set of all linear functions $p(x) = mx + b$, and P_2 is the set of all functions of the form $p(x) = ax^2 + bx + c$. Each P_k is a subspace of P. Also, P_0 is a subspace of P_1, P_1 is a subspace of P_2, and so on.   ●

We could equally well use other types of intervals in Examples 4–8. When doing so, we will adjust the notation accordingly. For example, $C[a, b]$ will denote the set of continuous functions on the closed interval $[a, b]$, which is a subspace of $F[a, b]$.

We next turn our attention to spanning sets. If V is a vector space and $v_1, v_2, \ldots, v_n$ are vectors in V, an expression of the form

$$c_1 v_1 + c_2 v_2 + \cdots + c_n v_n$$

where $c_1, c_2, \ldots, c_n$ are scalars is called a **linear combination** of $v_1, v_2, \ldots, v_n$. Given a fixed collection of vectors, Theorem 2.5 tells us the set of their linear combinations forms a subspace.

THEOREM 2.5   If V is a vector space and $v_1, v_2, \ldots, v_n$ are vectors in V, then the set of all linear combinations of $v_1, v_2, \ldots, v_n$ is a subspace of V.

Proof Consider two linear combinations

$$c_1 v_1 + c_2 v_2 + \cdots + c_n v_n \quad \text{and} \quad d_1 v_1 + d_2 v_2 + \cdots + d_n v_n$$

of $v_1, v_2, \ldots, v_n$. As

$$c_1 v_1 + c_2 v_2 + \cdots + c_n v_n + d_1 v_1 + d_2 v_2 + \cdots + d_n v_n$$
$$= (c_1 + d_1)v_1 + (c_2 + d_2)v_2 + \cdots + (c_n + d_n)v_n$$

is a linear combination of $v_1, v_2, \ldots, v_n$, we have closure under addition. Also, for any scalar c, the fact that

$$c(c_1 v_1 + c_2 v_2 + \cdots + c_n v_n) = cc_1 v_1 + cc_2 v_2 + \cdots + cc_n v_n$$

shows we have closure under scalar multiplication and completes our proof. ●

The subspace of a vector space V consisting of all linear combinations of vectors $v_1, v_2, \ldots, v_n$ of V will henceforth be called the **subspace of V spanned by** $v_1, v_2, \ldots, v_n$ and will be denoted

$$\boxed{\text{Span}\{v_1, v_2, \ldots, v_n\}.}$$

In Examples 11 and 12 we determine if a vector lies in the subspace spanned by some vectors.

EXAMPLE 11 Is the vector

$$\begin{bmatrix} 2 \\ -5 \\ 1 \\ 10 \end{bmatrix} \quad \text{in Span} \left\{ \begin{bmatrix} 1 \\ -1 \\ 2 \\ 3 \end{bmatrix}, \begin{bmatrix} 1 \\ -2 \\ -1 \\ 2 \end{bmatrix}, \begin{bmatrix} -1 \\ 0 \\ 1 \\ 3 \end{bmatrix} \right\} ?$$

Solution We need to see if we can find scalars c_1, c_2, c_3 so that

$$c_1 \begin{bmatrix} 1 \\ -1 \\ 2 \\ 3 \end{bmatrix} + c_2 \begin{bmatrix} 1 \\ -2 \\ -1 \\ 2 \end{bmatrix} + c_3 \begin{bmatrix} -1 \\ 0 \\ 1 \\ 3 \end{bmatrix} = \begin{bmatrix} 2 \\ -5 \\ 1 \\ 10 \end{bmatrix}.$$

Comparing entries in these columns, we arrive at the system

$$c_1 + c_2 - c_3 = 2$$
$$-c_1 - 2c_2 = -5$$
$$2c_1 - c_2 + c_3 = 1$$
$$3c_1 + 2c_2 + 3c_3 = 10$$

and our answer will be yes or no depending on whether or not this system has solutions. Reducing the augmented matrix for this system until it becomes clear whether or not we have a solution,

$$\begin{bmatrix} 1 & 1 & -1 & \vdots & 2 \\ -1 & -2 & 0 & \vdots & -5 \\ 2 & -1 & 1 & \vdots & 1 \\ 3 & 2 & 3 & \vdots & 10 \end{bmatrix} \rightarrow \begin{bmatrix} 1 & 1 & -1 & \vdots & 2 \\ 0 & -1 & -1 & \vdots & -3 \\ 0 & -3 & 3 & \vdots & -3 \\ 0 & -1 & 6 & \vdots & 4 \end{bmatrix}$$

$$\rightarrow \begin{bmatrix} 1 & 1 & -1 & \vdots & 2 \\ 0 & -1 & -1 & \vdots & -3 \\ 0 & 0 & 6 & \vdots & 6 \\ 0 & 0 & 7 & \vdots & 7 \end{bmatrix},$$

we see the system does have a solution and hence our answer to this problem is yes. ●

EXAMPLE 12   Is $2x^2 + x + 1$ in Span$\{x^2 + x, x^2 - 1, x + 1\}$?

Solution   By comparing coefficients of x^2, x, and the constant terms, we see that there are scalars c_1, c_2, c_3 so that

$$c_1(x^2 + x) + c_2(x^2 - 1) + c_3(x + 1) = 2x^2 + x + 1$$

if and only if the system of equations

$$c_1 + c_2 = 2$$
$$c_1 + c_3 = 1$$
$$-c_2 + c_3 = 1$$

has solutions. Starting to reduce the augmented matrix for this system,

$$\begin{bmatrix} 1 & 1 & 0 & \vdots & 2 \\ 1 & 0 & 1 & \vdots & 1 \\ 0 & -1 & 1 & \vdots & 1 \end{bmatrix} \rightarrow \begin{bmatrix} 1 & 1 & 0 & \vdots & 2 \\ 0 & -1 & 1 & \vdots & -1 \\ 0 & -1 & 1 & \vdots & 1 \end{bmatrix},$$

we can see that we have arrived at a system with no solution and hence our answer to this problem is no. ●

We say that the vectors $v_1, v_2, \ldots, v_n$ of a vector space V **span** V if Span$\{v_1, v_2, \ldots, v_n\} = V$. To put it another way, $v_1, v_2, \ldots, v_n$ span V if every vector in V is a linear combination of $v_1, v_2, \ldots, v_n$. In our final two examples of this section we determine if some given vectors span the given vector space.

EXAMPLE 13   Do

$$\begin{bmatrix} 1 \\ -2 \end{bmatrix}, \begin{bmatrix} 2 \\ -4 \end{bmatrix}$$

span $\mathbb{R}^2$?

Solution For an arbitrary vector

$$\begin{bmatrix} a \\ b \end{bmatrix}$$

of $\mathbb{R}^2$, we must determine whether there are scalars c_1, c_2 so that

$$c_1 \begin{bmatrix} 1 \\ -2 \end{bmatrix} + c_2 \begin{bmatrix} 2 \\ -4 \end{bmatrix} = \begin{bmatrix} a \\ b \end{bmatrix}.$$

Reducing the augmented matrix for the resulting system of equations,

$$\begin{bmatrix} 1 & 2 & \vdots & a \\ -2 & -4 & \vdots & b \end{bmatrix} \rightarrow \begin{bmatrix} 1 & 2 & \vdots & a \\ 0 & 0 & \vdots & b+2a \end{bmatrix},$$

we see that the system does not have a solution for all a and b and hence the answer to this problem is no. ●

EXAMPLE 14 Do $x^2 + x - 3$, $x - 5$, 3 span P_2?

Solution Here we must determine whether for an arbitrary element $ax^2 + bx + c$ there are scalars c_1, c_2, c_3 so that

$$c_1(x^2 + x - 3) + c_2(x - 5) + c_3 \cdot 3 = ax^2 + bx + c.$$

Comparing coefficients, we are led to the system

$$c_1 = a$$
$$c_1 + c_2 = b$$
$$-3c_1 - 5c_2 + 3c_3 = c,$$

which obviously has a solution. Thus the answer to this problem is yes. ●

EXERCISES 2.2

1. Determine which of the following sets of vectors are subspaces of $\mathbb{R}^2$.

a) All vectors of the form $\begin{bmatrix} 0 \\ y \end{bmatrix}$

b) All vectors of the form $\begin{bmatrix} x \\ 3x \end{bmatrix}$

c) All vectors of the form $\begin{bmatrix} x \\ 2 - 5x \end{bmatrix}$

d) All vectors $\begin{bmatrix} x \\ y \end{bmatrix}$ where $x + y = 0$

2. Determine which of the following sets of vectors are subspaces of $\mathbb{R}^3$.

a) All vectors of the form $\begin{bmatrix} x \\ y \\ y - 4x \end{bmatrix}$

b) All vectors of the form $\begin{bmatrix} y + z + 1 \\ y \\ z \end{bmatrix}$

c) All vectors $\begin{bmatrix} x \\ y \\ z \end{bmatrix}$ where $z = x + y$

d) All vectors $\begin{bmatrix} x \\ y \\ z \end{bmatrix}$ where $z = x^2 + y^2$

3. Determine which of the following sets of functions are subspaces of $F[a, b]$.

   a) All functions f in $F[a, b]$ for which $f(a) = 0$
   b) All functions f in $F[a, b]$ for which $f(a) = 1$
   c) All functions f in $C[a, b]$ for which
      $\int_a^b f(x)\, dx = 0$
   d) All functions f in $D[a, b]$ for which
      $f'(x) = f(x)$
   e) All functions f in $D[a, b]$ for which $f'(x) = e^x$

4. Determine which of the following sets of $n \times n$ matrices are subspaces of $M_{n \times n}(\mathbb{R})$.

   a) The $n \times n$ diagonal matrices
   b) The $n \times n$ upper triangular matrices
   c) The $n \times n$ symmetric matrices
   d) The $n \times n$ matrices of determinant zero
   e) The $n \times n$ invertible matrices

5. If A is an $m \times n$ matrix and B is a nonzero element of $\mathbb{R}^m$, do the solutions to the system $AX = B$ form a subspace of $\mathbb{R}^n$? Why or why not?

6. Complex numbers $a + bi$ where a and b are integers are called *Gaussian integers*. Do the Gaussian integers form a subspace of the vector space of complex numbers? Why or why not?

7. Do the sequences that converge to zero form a subspace of the vector space of convergent sequences? How about the sequences that converge to a rational number?

8. Do the series that converge to a positive number form a subspace of the vector space of convergent series? How about the series that converge absolutely?

9. Is $\begin{bmatrix} 2 \\ 1 \end{bmatrix}$ in $\text{Span}\left\{ \begin{bmatrix} -1 \\ 1 \end{bmatrix}, \begin{bmatrix} 2 \\ 3 \end{bmatrix} \right\}$?

10. Is $\begin{bmatrix} -4 \\ 3 \end{bmatrix}$ in $\text{Span}\left\{ \begin{bmatrix} 3 \\ -3 \end{bmatrix}, \begin{bmatrix} -4 \\ 4 \end{bmatrix}, \begin{bmatrix} 2 \\ 2 \end{bmatrix} \right\}$?

11. Is $\begin{bmatrix} 1 \\ -5 \\ -3 \end{bmatrix}$ in $\text{Span}\left\{ \begin{bmatrix} 1 \\ -1 \\ 0 \end{bmatrix}, \begin{bmatrix} 1 \\ 1 \\ 1 \end{bmatrix}, \begin{bmatrix} 2 \\ 0 \\ 1 \end{bmatrix} \right\}$?

12. Is $\begin{bmatrix} 3 & 5 \\ 3 & 4 \end{bmatrix}$ in

$\text{Span}\left\{ \begin{bmatrix} 1 & 1 \\ 0 & 1 \end{bmatrix}, \begin{bmatrix} 1 & 1 \\ 1 & 0 \end{bmatrix}, \begin{bmatrix} 1 & 0 \\ 0 & -1 \end{bmatrix} \right\}$?

13. Is $3x^2$ in $\text{Span}\{x^2 - x, x^2 + x + 1, x^2 - 1\}$?

14. Is $\sin(x + \pi/4)$ in $\text{Span}\{\sin x, \cos x\}$?

15. Determine if $\begin{bmatrix} 1 \\ 1 \end{bmatrix}, \begin{bmatrix} 2 \\ 1 \end{bmatrix}, \begin{bmatrix} -1 \\ 1 \end{bmatrix}$ span $\mathbb{R}^2$.

16. Determine if $\begin{bmatrix} 1 \\ 1 \\ 0 \end{bmatrix}, \begin{bmatrix} 1 \\ -1 \\ -1 \end{bmatrix}, \begin{bmatrix} 1 \\ 3 \\ 1 \end{bmatrix}$ span $\mathbb{R}^3$.

17. Determine if $\begin{bmatrix} 1 \\ 0 \\ 1 \\ 1 \end{bmatrix}, \begin{bmatrix} -1 \\ -1 \\ 1 \\ 1 \end{bmatrix}, \begin{bmatrix} 0 \\ 1 \\ 1 \\ 0 \end{bmatrix}$ span $\mathbb{R}^4$.

18. Determine if $\begin{bmatrix} 1 & 0 \\ 0 & 1 \end{bmatrix}, \begin{bmatrix} 0 & 1 \\ -1 & 0 \end{bmatrix}$,

$\begin{bmatrix} 1 & 0 \\ 0 & -1 \end{bmatrix}, \begin{bmatrix} 1 & 1 \\ -1 & 0 \end{bmatrix}$ span $M_{2 \times 2}(\mathbb{R})$.

19. Determine if $x^2 - 1, x^2 + 1, x^2 + x$ span P_2.

20. Determine if $x^3 + x^2, x^2 + x, x + 1$ span P_3.

Use the system of linear equation solving capabilities of Maple or another appropriate software package in Exercises 21–24.

21. Determine if $\begin{bmatrix} 1 \\ 2 \\ -2 \\ 0 \\ -1 \\ 4 \end{bmatrix}$ is in

$\text{Span}\left\{ \begin{bmatrix} 4 \\ -2 \\ 1 \\ 2 \\ 3 \\ 1 \end{bmatrix}, \begin{bmatrix} 2 \\ -2 \\ 1 \\ -5 \\ 7 \\ 1 \end{bmatrix}, \begin{bmatrix} -17 \\ 22 \\ -1 \\ 0 \\ 2 \\ 1 \end{bmatrix}, \begin{bmatrix} -31 \\ -3 \\ -8 \\ 1 \\ 1 \\ 0 \end{bmatrix} \right\}$

$$\left\{\begin{bmatrix} 17 \\ 44 \\ 1 \\ -22 \\ 11 \\ 1.9 \end{bmatrix}, \begin{bmatrix} 1 \\ 0 \\ 1 \\ 15 \\ -3 \\ 3 \end{bmatrix}\right\}.$$

22. Determine if $x^4 + x^2 + 1$ is in

Span$\{x^4 - x^3 + 3x - 4, x^4 - x^3 - x^2 + x - 4,$
$x^3 + x^2 - 3x + 3,$
$x^4 + x^3 - 2x^2 + 4x - 8,$
$5x^4 - 7x^3 - 2x^2 - x + 9, 2x^4 - 7x^3 + 1\}.$

23. Determine if

$$x^5 - x^4 + x^3 + x,$$
$$x^4 - x^3 + 2x - 4,$$
$$x^5 - 5x^3 + 6x^2 - 8x + 2,$$
$$x^5 + \tfrac{2}{3}x^4 - x^3 + 2x^2 + 3x - 1,$$
$$-2x^3 - 4x^2 + 3x - 9,$$
$$x^4 - 3x^3 + \pi x^2 - 2x + 1$$

span P_5.

24. Determine if

$$\begin{bmatrix} 2 & -1 & 4 \\ -1 & 3 & 3 \end{bmatrix}, \begin{bmatrix} 0 & -1 & -1 \\ 2 & 2 & 2 \end{bmatrix},$$

$$\begin{bmatrix} 5 & -2 & -1 \\ 2 & 6 & -3 \end{bmatrix}, \begin{bmatrix} \sqrt{3} & 7 & -1 \\ 2 & 1 & \pi \end{bmatrix},$$

$$\begin{bmatrix} -7 & 9 & 21 \\ 14 & -22 & 8 \end{bmatrix}, \begin{bmatrix} 0 & 3 & -1 \\ 1 & 1 & 2 \end{bmatrix}$$

span $M_{2\times3}(\mathbb{R})$.

25. Suppose that $v_1, v_2, \ldots, v_k$ are vectors in $\mathbb{R}^n$. How can we tell from a row-echelon form of the matrix

$$A = [v_1 \quad v_2 \quad \ldots \quad v_k]$$

if $v_1, v_2, \ldots, v_k$ span $\mathbb{R}^n$?

26. Use the answer to Exercise 25 and one of the *gausselim, gaussjord,* or *rref* commands of Maple or corresponding commands in another appropriate software package to determine if the vectors

$$\begin{bmatrix} 1 \\ -1 \\ 2 \\ 4 \\ 5 \end{bmatrix}, \begin{bmatrix} 0 \\ 1 \\ 1 \\ -3 \\ \sqrt{2} \end{bmatrix}, \begin{bmatrix} -2 \\ 2 \\ 5 \\ -3 \\ 1 \end{bmatrix},$$

$$\begin{bmatrix} -3 \\ 8 \\ 7 \\ -9 \\ 11 \end{bmatrix}, \begin{bmatrix} 1 \\ 1 \\ 1 \\ -4 \\ -4 \end{bmatrix}$$

span $\mathbb{R}^5$.

2.3 LINEAR INDEPENDENCE AND BASES

The concept of a spanning set that we encountered in the previous section is a very fundamental one in theory of vector spaces involving linear combinations. Another is the concept of linear independence and its opposite, linear dependence. These concepts are defined as follows.

> **DEFINITION** Suppose that $v_1, v_2, \ldots, v_n$ are vectors in a vector space V. We say that $v_1, v_2, \ldots, v_n$ are **linearly dependent** if there are scalars $c_1, c_2, \ldots, c_n$ *not all zero* so that
>
> $$c_1 v_1 + c_2 v_2 + \cdots + c_n v_n = 0.$$
>
> If $v_1, v_2, \ldots, v_n$ are not linearly dependent, we say $v_1, v_2, \ldots, v_n$ are **linearly independent**.

We can always get a linear combination of vectors $v_1, v_2, \ldots, v_n$ equal to the zero vector by using zero for each scalar:

$$0 \cdot v_1 + 0 \cdot v_2 + \cdots + 0 \cdot v_n = 0.$$

Carrying over the terminology we used for solutions of homogenous systems of linear equations, let us call this the **trivial linear combination** of $v_1, v_2, \ldots, v_n$. We then could equally well say $v_1, v_2, \ldots, v_n$ are linearly dependent if there is a nontrivial linear combination of them equal to the zero vector; saying $v_1, v_2, \ldots, v_n$ are linearly independent would mean that the trivial linear combination is the only linear combination of $v_1, v_2, \ldots, v_n$ equal to the zero vector.

EXAMPLE 1   Are the vectors

$$\begin{bmatrix} 1 \\ 2 \\ 3 \end{bmatrix}, \begin{bmatrix} 3 \\ 2 \\ 1 \end{bmatrix}, \begin{bmatrix} -1 \\ 2 \\ 5 \end{bmatrix}$$

linearly dependent or linearly independent?

Solution   Consider a linear combination of these vectors equal to the zero vector of $\mathbb{R}^3$:

$$c_1 \begin{bmatrix} 1 \\ 2 \\ 3 \end{bmatrix} + c_2 \begin{bmatrix} 3 \\ 2 \\ 1 \end{bmatrix} + c_3 \begin{bmatrix} -1 \\ 2 \\ 5 \end{bmatrix} = \begin{bmatrix} 0 \\ 0 \\ 0 \end{bmatrix}.$$

This leads us to the system of equations:

$$c_1 + 3c_2 - c_3 = 0$$
$$2c_1 + 2c_2 + 2c_3 = 0$$
$$3c_1 + c_2 + 5c_3 = 0.$$

Beginning to apply row operations to the augmented matrix for this system,

$$\begin{bmatrix} 1 & 3 & -1 & \vdots & 0 \\ 2 & 2 & 2 & \vdots & 0 \\ 3 & 1 & 5 & \vdots & 0 \end{bmatrix} \rightarrow \begin{bmatrix} 1 & 3 & -1 & \vdots & 0 \\ 0 & -4 & 4 & \vdots & 0 \\ 0 & -8 & 8 & \vdots & 0 \end{bmatrix},$$

we can see that our system has nontrivial solutions. Thus there are nontrivial linear combinations of our three vectors equal to the zero vector and hence they are linearly dependent.   ●

EXAMPLE 2   Are $x^2 + 1$, $x^2 - x + 1$, $x + 2$ linearly dependent or linearly independent?

Solution   Suppose

$$c_1(x^2 + 1) + c_2(x^2 - x + 1) + c_3(x + 2) = 0.$$

Comparing coefficients of x^2, x and the constant terms on each side of the preceding equation, we obtain the system:

$$c_1 + c_2 = 0$$
$$-c_2 + c_3 = 0$$
$$c_1 + c_2 + 2c_3 = 0.$$

We do not need to set up an augmented matrix here. Notice that subtracting the first equation from the third will give us $c_3 = 0$. The second equation then tells us $c_2 = 0$ from which we can now see $c_1 = 0$ from the first equation. Since our system has only the trivial solution, it follows that the three given polynomials are linearly independent. ●

The next theorem gives us another characterization of linear dependence.

THEOREM 2.6   Suppose $v_1, v_2, \ldots, v_n$ are vectors in a vector space V. Then $v_1, v_2, \ldots, v_n$ are linearly dependent if and only if one of $v_1, v_2, \ldots, v_n$ is a linear combination of the others.

Proof   Suppose $v_1, v_2, \ldots, v_n$ are linearly dependent. Then there are scalars $c_1, c_2, \ldots, c_n$ not all zero so that

$$c_1 v_1 + c_2 v_2 + \cdots + c_n v_n = 0. \tag{1}$$

Suppose that $c_i \neq 0$. Then we may solve Equation (1) for v_i as follows

$$c_i v_i = -c_1 v_1 - \cdots - c_{i-1} v_{i-1} - c_{i+1} v_{i+1} - \cdots - c_n v_n$$
$$v_i = -\frac{c_1}{c_i} v_1 - \cdots - \frac{c_{i-1}}{c_i} v_{i-1} - \frac{c_{i+1}}{c_i} v_{i+1} - \cdots - \frac{c_n}{c_i} v_n$$

and hence obtain that v_i is a linear combination of $v_1, \ldots, v_{i-1}, v_{i+1}, \ldots, v_n$.

To prove the converse, suppose one of $v_1, v_2, \ldots, v_n$, let us call it v_i, is a linear combination of the other $v_1, v_2, \ldots, v_n$:

$$v_i = c_1 v_1 + \cdots + c_{i-1} v_{i-1} + c_{i+1} v_{i+1} + \cdots + c_n v_n.$$

Then

$$-c_1 v_1 - \cdots - c_{i-1} v_{i-1} + v_i - c_{i+1} v_{i+1} - \cdots - c_n v_n = 0.$$

This gives us a nontrivial linear combination of $v_1, v_2, \ldots, v_n$ equal to the zero vector since the scalar with v_i is 1 and hence $v_1, v_2, \ldots, v_n$ are linearly dependent. ●

If we view having one vector as a linear combination of others as a dependence of the vector on the others, Theorem 2.6 in some sense gives us a more natural way to think of linear dependence. Unfortunately, it is not as practical in general. Were we to use it to check if a set of vectors $v_1, v_2, \ldots, v_n$ are linearly dependent, we could start by checking if v_1 is a linear combination of $v_2, \ldots, v_n$. If so, we would have linear dependence. But if not, we would then have to look at v_2 and see if v_2 is a linear combination of $v_1, v_3, \ldots, v_n$. If so, we again would have linear dependence. If not, we would move on to v_3 and so on. Notice that checking to see if there is a nontrivial linear combination of $v_1, v_2, \ldots, v_n$ equal to the zero vector is much more efficient.

One exception is in the case of two vectors v_1 and v_2, for having one vector a linear combination of the other is the same as saying one vector is a scalar multiple of the other, which is often easily seen by inspection. For example,

$$\begin{bmatrix} 1 \\ -2 \end{bmatrix} \text{ and } \begin{bmatrix} 3 \\ -6 \end{bmatrix}$$

are linearly dependent since the second vector is 3 times the first (or the first is 1/3 times the second). The polynomials $x^2 + x$ and $x^2 - 1$ are linearly independent since neither is a scalar multiple of the other.

We next introduce the concept of a basis.

DEFINITION   We say that the vectors $v_1, v_2, \ldots, v_n$ of a vector space V are a **basis** for V if both of the following two conditions are satisfied:

1. $v_1, v_2, \ldots, v_n$ are linearly independent.
2. $v_1, v_2, \ldots, v_n$ span V.

EXAMPLE 3   The vectors

$$e_1 = \begin{bmatrix} 1 \\ 0 \end{bmatrix}, \qquad e_2 = \begin{bmatrix} 0 \\ 1 \end{bmatrix}$$

are easily seen to be linearly independent since if

$$c_1 e_1 + c_2 e_2 = \begin{bmatrix} c_1 \\ c_2 \end{bmatrix} = \begin{bmatrix} 0 \\ 0 \end{bmatrix},$$

then $c_1 = 0$ and $c_2 = 0$. They also span $\mathbb{R}^2$ since

$$a e_1 + b e_2 = \begin{bmatrix} a \\ b \end{bmatrix}.$$

Thus e_1, e_2 form a basis for $\mathbb{R}^2$.   ●

EXAMPLE 4   As in Example 3, it is easily seen that the three vectors

$$e_1 = \begin{bmatrix} 1 \\ 0 \\ 0 \end{bmatrix}, \qquad e_2 = \begin{bmatrix} 0 \\ 1 \\ 0 \end{bmatrix}, \qquad e_3 = \begin{bmatrix} 0 \\ 0 \\ 1 \end{bmatrix}$$

both are linearly independent and span $\mathbb{R}^3$. Hence e_1, e_2, e_3 form a basis for $\mathbb{R}^3$.   ●

EXAMPLE 5   Generalizing Examples 3 and 4 to $\mathbb{R}^n$, let us use e_i to denote the vector in $\mathbb{R}^n$ that has 1 in the ith position and 0s elsewhere:

$$e_1 = \begin{bmatrix} 1 \\ 0 \\ 0 \\ \vdots \\ 0 \\ 0 \end{bmatrix}, \qquad e_2 = \begin{bmatrix} 0 \\ 1 \\ 0 \\ \vdots \\ 0 \\ 0 \end{bmatrix}, \qquad \dots, \qquad e_n = \begin{bmatrix} 0 \\ 0 \\ 0 \\ \vdots \\ 0 \\ 1 \end{bmatrix}.$$

Here again the vectors $e_1, e_2, \dots, e_n$ are easily seen to both be linearly independent and span $\mathbb{R}^n$ and consequently form a basis for $\mathbb{R}^n$.[6]   ●

EXAMPLE 6   Generalizing even further, the $m \times n$ matrices E_{ij} that have 1 in the ij-position and 0s elsewhere for $i = 1, 2, \dots, m$ and $j = 1, 2, \dots, n$ are linearly independent and span $M_{m \times n}(\mathbb{R})$. Hence they are a basis for $M_{m \times n}(\mathbb{R})$. For instance,

$$E_{11} = \begin{bmatrix} 1 & 0 \\ 0 & 0 \end{bmatrix}, \qquad E_{12} = \begin{bmatrix} 0 & 1 \\ 0 & 0 \end{bmatrix}, \qquad E_{21} = \begin{bmatrix} 0 & 0 \\ 1 & 0 \end{bmatrix}, \qquad E_{22} = \begin{bmatrix} 0 & 0 \\ 0 & 1 \end{bmatrix}$$

form a basis for $M_{2 \times 2}(\mathbb{R})$.   ●

EXAMPLE 7   For a nonnegative integer n, the $n + 1$ polynomials

$$x^n, x^{n-1}, \dots, x, 1$$

are linearly independent since if

$$c_1 x^n + c_2 x^{n-1} + \dots + c_n x + c_{n+1} \cdot 1 = 0,$$

then $c_1 = 0, c_2 = 0, \dots, c_n = 0, c_{n+1} = 0$. Further, they span P_n; indeed we typically write the polynomials in P_n as linear combinations of $x^n, x^{n-1}, \dots, x, 1$ when we write them as $a_n x^n + a_{n-1} x^{n-1} + \dots + a_1 x + a_0$. Hence $x^n, x^{n-1}, \dots, x, 1$ form a basis for P_n.   ●

The bases given in each of Examples 3–7 are natural bases to use and are called the **standard bases** for each of these respective vector spaces. Standard bases are not the only bases, however, for these vector spaces. Consider Examples 8 and 9.

EXAMPLE 8   Show that

$$\begin{bmatrix} 1 \\ 0 \\ 1 \end{bmatrix}, \begin{bmatrix} 1 \\ 1 \\ 1 \end{bmatrix}, \begin{bmatrix} -1 \\ 1 \\ 1 \end{bmatrix}$$

form a basis for $\mathbb{R}^3$.

---

[6] The vectors e_1 and e_2 of $\mathbb{R}^2$ in Example 3 are often denoted by **i** and **j**, respectively; the vectors e_1, e_2, and e_3 of $\mathbb{R}^3$ in Example 4 are often denoted by **i, j,** and **k**, respectively.

Solution   Let us first show that these three vectors are linearly independent. Suppose

$$
c_1 \begin{bmatrix} 1 \\ 0 \\ 1 \end{bmatrix} + c_2 \begin{bmatrix} 1 \\ 1 \\ 1 \end{bmatrix} + c_3 \begin{bmatrix} -1 \\ 1 \\ 1 \end{bmatrix} = \begin{bmatrix} 0 \\ 0 \\ 0 \end{bmatrix}.
$$

Setting up and reducing the augmented matrix for the resulting homogeneous system,

$$
\begin{bmatrix} 1 & 1 & -1 & \vdots & 0 \\ 0 & 1 & 1 & \vdots & 0 \\ 1 & 1 & 1 & \vdots & 0 \end{bmatrix} \rightarrow \begin{bmatrix} 1 & 1 & -1 & \vdots & 0 \\ 0 & 1 & 1 & \vdots & 0 \\ 0 & 0 & 2 & \vdots & 0 \end{bmatrix},
$$

we see that we have the trivial solution $c_1 = 0$, $c_2 = 0$, $c_3 = 0$. Hence these vectors are linearly independent. Similar work applied to the system of equations resulting from the vector equation

$$
c_1 \begin{bmatrix} 1 \\ 0 \\ 1 \end{bmatrix} + c_2 \begin{bmatrix} 1 \\ 1 \\ 1 \end{bmatrix} + c_3 \begin{bmatrix} -1 \\ 1 \\ 1 \end{bmatrix} = \begin{bmatrix} a \\ b \\ c \end{bmatrix}
$$

shows us that our three vectors span $\mathbb{R}^3$. Thus these vectors do form a basis for $\mathbb{R}^3$.   ●

EXAMPLE 9   Show that $x^2 + x - 3, x - 5, 3$ form a basis for P_2.

Solution   Let us first check to see if these three polynomials are linearly independent. If

$$
c_1(x^2 + x - 3) + c_2(x - 5) + c_3 \cdot 3 = 0,
$$

we have the homogeneous system:

$$
c_1 = 0
$$
$$
c_1 + c_2 = 0
$$
$$
-3c_1 - 5c_2 + 3c_3 = 0.
$$

Since this system has only the trivial solution, the three polynomials are linearly independent. Likewise, the system of equations resulting from

$$
c_1(x^2 + x - 3) + c_2(x - 5) + 3c_3 = ax^2 + bx + c
$$

has a solution and hence our three polynomials span P_2. (Indeed, if you have a good memory, you will note that we already did the spanning part in the last example of the previous section.) Thus we have shown $x^2 + x - 3, x - 5, 3$ form a basis for $\mathbb{R}^3$.   ●

Keep in mind that we must have both linear independence and spanning to have a basis. If either one (or both) fails to hold, we do not have a basis.

EXAMPLE 10   Do $x^2 + x - 1, x^2 - x + 1$ form a basis for P_2?

Solution   Since neither polynomial is a scalar multiple of the other, these two polynomials are linearly independent. However, they do not span P_2. To see this, observe that the system of equations resulting from

$$c_1(x^2 + x - 1) + c_2(x^2 - x + 1) = ax^2 + bx + c$$

is

$$c_1 + c_2 = a$$
$$c_1 - c_2 = b$$
$$-c_1 + c_2 = c,$$

which does not have a solution for all $a, b,$ and c since adding the second and third equations gives us

$$0 = b + c.$$

Since $x^2 + x - 1, x^2 - x + 1$ do not span P_2, they do not form a basis for P_2. ●

EXAMPLE 11   Do

$$\begin{bmatrix} 1 \\ 1 \end{bmatrix}, \begin{bmatrix} -1 \\ 1 \end{bmatrix}, \begin{bmatrix} 2 \\ 3 \end{bmatrix}$$

form a basis for $\mathbb{R}^2$?

Solution   Let us first see if these vectors are linearly independent. If

$$c_1 \begin{bmatrix} 1 \\ 1 \end{bmatrix} + c_2 \begin{bmatrix} -1 \\ 1 \end{bmatrix} + c_3 \begin{bmatrix} 2 \\ 3 \end{bmatrix} = \begin{bmatrix} 0 \\ 0 \end{bmatrix},$$

then

$$c_1 - c_2 + 2c_3 = 0$$
$$c_1 + c_2 + 3c_3 = 0.$$

Since we know that a homogeneous linear system with more variables than equations always has a nontrivial solution (Theorem 1.1), the three given vectors are linearly dependent and hence do not form a basis for $\mathbb{R}^2$. ●

There are other ways of characterizing bases besides saying they are linearly independent spanning sets. Theorem 2.7 describes one other way.

THEOREM 2.7   Suppose that $v_1, v_2, \ldots, v_n$ are vectors in a vector space V. Then $v_1, v_2, \ldots, v_n$ form a basis for V if and only if each vector in V is uniquely expressible as a linear combination of $v_1, v_2, \ldots, v_n$.

Proof   First suppose $v_1, v_2, \ldots, v_n$ form a basis for V. Let v be a vector in V. Since $v_1, v_2, \ldots, v_n$ span V, there are scalars $c_1, c_2, \ldots, c_n$ so that

$$v = c_1 v_1 + c_2 v_2 + \cdots + c_n v_n.$$

We must show this is unique. To do so, suppose we also have

$$v = d_1 v_1 + d_2 v_2 + \cdots + d_n v_n$$

where $d_1, d_2, \ldots, d_n$ are scalars. Subtracting these two expressions for v, we obtain

$$v - v = 0 = (c_1 - d_1)v_1 + (c_2 - d_2)v_2 + \cdots + (c_n - d_n)v_n.$$

Now because $v_1, v_2, \ldots, v_n$ are linearly independent, we have

$$c_1 - d_1 = 0, \qquad c_2 - d_2 = 0, \qquad \ldots, \qquad c_n - d_n = 0$$

or

$$c_1 = d_1, \qquad c_2 = d_2, \qquad \ldots, \qquad c_n = d_n.$$

Hence we have the desired uniqueness of the linear combination.

To prove the converse, first note that if every vector in V is uniquely expressible as a linear combination of $v_1, v_2, \ldots, v_n$, we immediately have that $v_1, v_2, \ldots, v_n$ span V. Suppose

$$c_1 v_1 + c_2 v_2 + \cdots + c_n v_n = 0.$$

Since the trivial linear combination of $v_1, v_2, \ldots, v_n$ is the zero vector, the uniqueness property gives us

$$c_1 = 0, \qquad c_2 = 0, \qquad \ldots, \qquad c_n = 0.$$

Hence $v_1, v_2, \ldots, v_n$ are linearly independent, which completes our proof.   ●

As a matter of convenience, we shall sometimes denote bases by lowercase Greek letters in this text. If α is a basis for a vector space V consisting of the vectors $v_1, v_2, \ldots, v_n$ and v is a vector in V, when we write v uniquely as

$$v = c_1 v_1 + c_2 v_2 + \cdots + c_n v_n,$$

the scalars $c_1, c_2, \ldots, c_n$ are called the **coordinates of v relative to the basis** α of V. The column vector

$$[v]_\alpha = \begin{bmatrix} c_1 \\ c_2 \\ \vdots \\ c_n \end{bmatrix}$$

is called the **coordinate vector of v relative to the basis** α. To illustrate, suppose α is the standard basis

$$e_1 = \begin{bmatrix} 1 \\ 0 \end{bmatrix}, \qquad e_2 = \begin{bmatrix} 0 \\ 1 \end{bmatrix}$$

for $\mathbb{R}^2$. Since for any vector

$$v = \begin{bmatrix} a \\ b \end{bmatrix},$$

we have

$$v = \begin{bmatrix} a \\ b \end{bmatrix} = ae_1 + be_2,$$

we then get

$$[v]_\alpha = \begin{bmatrix} a \\ b \end{bmatrix}.$$

In other words, the coordinates and coordinate vectors relative to the standard basis of $\mathbb{R}^2$ are just the usual coordinates and column vectors. More generally, the same thing occurs when we use the standard basis for $\mathbb{R}^n$. If we use the standard basis for $M_{m \times n}(\mathbb{R})$, the coordinates of a matrix relative to it are the entries of the matrix. If we use the standard basis for P_n, the coordinates of a polynomial in P_n relative to it are the coefficients of the polynomial. If we do not use standard bases, we have to work harder to determine the coordinates relative to the basis. Consider Examples 12 and 13.

EXAMPLE 12   Find the coordinate vector of

$$v = \begin{bmatrix} 1 \\ 3 \\ 7 \end{bmatrix}$$

relative to the basis β for $\mathbb{R}^3$ in Example 8 consisting of

$$\begin{bmatrix} 1 \\ 0 \\ 1 \end{bmatrix}, \begin{bmatrix} 1 \\ 1 \\ 1 \end{bmatrix}, \begin{bmatrix} -1 \\ 1 \\ 1 \end{bmatrix}.$$

Solution   We need to find c_1, c_2, c_3 so that

$$c_1 \begin{bmatrix} 1 \\ 0 \\ 1 \end{bmatrix} + c_2 \begin{bmatrix} 1 \\ 1 \\ 1 \end{bmatrix} + c_3 \begin{bmatrix} -1 \\ 1 \\ 1 \end{bmatrix} = \begin{bmatrix} 1 \\ 3 \\ 7 \end{bmatrix}.$$

Reducing the augmented matrix for the resulting system of equations,

$$\left[\begin{array}{ccc:c} 1 & 1 & -1 & 1 \\ 0 & 1 & 1 & 3 \\ 1 & 1 & 1 & 7 \end{array} \right] \rightarrow \left[\begin{array}{ccc:c} 1 & 1 & -1 & 1 \\ 0 & 1 & 1 & 3 \\ 0 & 0 & 2 & 6 \end{array} \right],$$

we see

$$c_3 = 3, \quad c_2 = 0, \quad c_1 = 4.$$

Hence the coordinate vector relative to β is

$$[v]_\beta = \begin{bmatrix} 4 \\ 0 \\ 3 \end{bmatrix}.$$

●

EXAMPLE 13   Find the coordinate vector of $v = x + 1$ relative to the basis γ for P_2 in Example 9 consisting of $x^2 + x - 3, x - 5, 3$.

Solution   We need to find c_1, c_2, c_3 so that

$$c_1(x^2 + x - 3) + c_2(x - 5) + 3c_3 = x + 1.$$

We then have the system

$$c_1 = 0$$
$$c_1 + c_2 = 1$$
$$-3c_1 - 5c_2 + 3c_3 = 1,$$

which has the solution

$$c_1 = 0, \quad c_2 = 1, \quad c_3 = 2.$$

Hence

$$[v]_\gamma = \begin{bmatrix} 0 \\ 1 \\ 2 \end{bmatrix}.$$

●

We can reverse the procedure of translating vectors into coordinate vectors as the following example illustrates.

EXAMPLE 14   Find v in P_2 if

$$[v]_\gamma = \begin{bmatrix} 1 \\ 2 \\ -1 \end{bmatrix}$$

where γ is the basis for P_2 in Examples 9 and 13 consisting of $x^2 + x - 3, x - 5, 3$.

Solution   We have

$$v = 1 \cdot (x^2 + x - 3) + 2(x - 5) + (-1) \cdot 3 = x^2 + 3x - 16.$$

●

Using coordinates relative to a basis gives us a way of translating elements in a vector space into vectors with numbers. One place where this is useful is when working on a computer: Translate vectors into coordinate vectors, then do the desired work on the computer in terms of coordinate vectors, and finally translate the computer's results in coordinate vectors back into the original type of vectors. We will further develop these number translation ideas when we come to the study of linear transformations in Chapter 5.

EXERCISES 2.3

In each of Exercises 1–10, determine whether the given vectors are linearly dependent or linearly independent.

1. $\begin{bmatrix} 1 \\ -1 \end{bmatrix}, \begin{bmatrix} 2 \\ 1 \end{bmatrix}$ **2.** $\begin{bmatrix} 4 \\ -1 \end{bmatrix}, \begin{bmatrix} -8 \\ 2 \end{bmatrix}$

3. $\begin{bmatrix} 6 \\ -4 \\ 2 \end{bmatrix}, \begin{bmatrix} -9 \\ 6 \\ -3 \end{bmatrix}$ **4.** $\begin{bmatrix} 1 \\ 3 \\ -1 \end{bmatrix}, \begin{bmatrix} 4 \\ 1 \\ 1 \end{bmatrix}$

5. $\begin{bmatrix} 1 \\ -1 \\ 1 \end{bmatrix}, \begin{bmatrix} 1 \\ 2 \\ -1 \end{bmatrix}, \begin{bmatrix} 1 \\ 1 \\ 0 \end{bmatrix}$

6. $\begin{bmatrix} 0 \\ 4 \\ -1 \end{bmatrix}, \begin{bmatrix} 1 \\ 5 \\ -3 \end{bmatrix}, \begin{bmatrix} 1 \\ -3 \\ -1 \end{bmatrix}$

7. $\begin{bmatrix} 1 & 0 \\ -1 & 1 \end{bmatrix}, \begin{bmatrix} 0 & 1 \\ 1 & 0 \end{bmatrix}, \begin{bmatrix} 1 & 1 \\ 1 & 1 \end{bmatrix}$

8. $\begin{bmatrix} 1 & -1 \\ 2 & 1 \end{bmatrix}, \begin{bmatrix} 1 & 0 \\ 0 & -1 \end{bmatrix},$
$\begin{bmatrix} 1 & -2 \\ 0 & 1 \end{bmatrix}, \begin{bmatrix} 3 & -3 \\ 2 & 1 \end{bmatrix}$

9. $x^2 + x + 2, x^2 + 2x + 1, 2x^2 + 5x + 1$

10. $x^3 - 1, x^2 - 1, x - 1, 1$

11. Show that $\begin{bmatrix} 1 \\ -1 \end{bmatrix}, \begin{bmatrix} 2 \\ 3 \end{bmatrix}$ form a basis for $\mathbb{R}^2$.

12. Show that $\begin{bmatrix} 1 \\ 2 \end{bmatrix}, \begin{bmatrix} 5 \\ 4 \end{bmatrix}$ form a basis for $\mathbb{R}^2$.

13. Show that $\begin{bmatrix} 1 \\ 3 \\ -1 \end{bmatrix}, \begin{bmatrix} 0 \\ -1 \\ 2 \end{bmatrix}, \begin{bmatrix} 2 \\ 1 \\ 3 \end{bmatrix}$ form a basis for $\mathbb{R}^3$.

14. Show that $\begin{bmatrix} 2 \\ -1 \\ 0 \end{bmatrix}, \begin{bmatrix} 1 \\ 3 \\ -1 \end{bmatrix}, \begin{bmatrix} 1 \\ -4 \\ -1 \end{bmatrix}$ form a basis for $\mathbb{R}^3$.

15. Show that $\begin{bmatrix} 1 & -1 \\ 0 & 1 \end{bmatrix}, \begin{bmatrix} 1 & 2 \\ 1 & 1 \end{bmatrix},$
$\begin{bmatrix} 0 & 1 \\ -1 & 0 \end{bmatrix}, \begin{bmatrix} 1 & 2 \\ 1 & 2 \end{bmatrix}$ form a basis for $M_{2\times2}(\mathbb{R})$.

16. Show that $\begin{bmatrix} 1 & 0 & 1 \\ -1 & 0 & 1 \end{bmatrix}, \begin{bmatrix} 0 & 1 & -1 \\ -1 & 2 & 0 \end{bmatrix},$
$\begin{bmatrix} 1 & 1 & 1 \\ 0 & 2 & 1 \end{bmatrix}, \begin{bmatrix} 1 & 2 & 1 \\ 0 & 1 & 2 \end{bmatrix}, \begin{bmatrix} 1 & 2 & 1 \\ 0 & -1 & 2 \end{bmatrix},$
$\begin{bmatrix} 2 & 3 & 0 \\ 0 & 2 & 3 \end{bmatrix}$ form a basis for $M_{2\times3}(\mathbb{R})$.

17. Show that $x^2 + x + 1, x^2 - x + 1, x^2 - 1$ form a basis for P_2.

18. Show that $x^3 + x, x^2 - x, x + 1, x^3 + 1$ form a basis for P_3.

19. Show that $\begin{bmatrix} 1 \\ 2 \\ 1 \end{bmatrix}, \begin{bmatrix} -1 \\ 0 \\ 1 \end{bmatrix}$ do not form a basis for $\mathbb{R}^3$.

20. Show that $x^2 - 3x, x + 7$ do not form a basis for P_2.

21. Show that $x + 1, x + 2, x + 3$ do not form a basis for P_1.

22. Show that $\begin{bmatrix} -1 \\ 2 \\ 1 \end{bmatrix}, \begin{bmatrix} 1 \\ 1 \\ 0 \end{bmatrix}, \begin{bmatrix} 0 \\ 1 \\ 1 \end{bmatrix}, \begin{bmatrix} 1 \\ 1 \\ -1 \end{bmatrix}$ do not form a basis for $\mathbb{R}^3$.

23. If α is the basis in Exercise 13, find:

a) $[v]_\alpha$ for $v = \begin{bmatrix} 1 \\ 2 \\ -1 \end{bmatrix}$,

b) v if $[v]_\alpha = \begin{bmatrix} 1 \\ 2 \\ -1 \end{bmatrix}$.

24. If β is the basis in Exercise 14, find:

a) $[v]_\beta$ if $v = \begin{bmatrix} -1 \\ 1 \\ 0 \end{bmatrix}$,

b) v if $[v]_\beta = \begin{bmatrix} -1 \\ 1 \\ 0 \end{bmatrix}$.

25. If β is the basis in Exercise 17, find:

a) $[v]_\beta$ if $v = 2x^2 + 3x$,

b) v if $[v]_\beta = \begin{bmatrix} 1 \\ 2 \\ 3 \end{bmatrix}$.

26. If γ is the basis of Exercise 18, find:

a) $[v]_\gamma$ if $v = x^3 + x^2 + x + 1$,

b) v if $[v]_\gamma = \begin{bmatrix} -2 \\ 2 \\ 0 \\ 1 \end{bmatrix}$.

27. Show that any set of vectors that contains the zero vector is a linearly dependent set of vectors.

28. Show that if $v_1, v_2, \ldots, v_n$ is a linearly independent set of vectors, then any subset of these vectors is also linearly independent.

29. Suppose v_1 and v_2 are nonzero vectors in $\mathbb{R}^3$ and L is a line in $\mathbb{R}^3$ parallel to v_1. What are necessary and sufficient conditions in terms of the line L for v_1 and v_2 to be linearly independent?

30. Let v_1 and v_2 be vectors in $\mathbb{R}^3$ that are linearly independent.

a) If all initial points of vectors are placed at the same point in $\mathbb{R}^3$, what geometric object do the linear combinations of v_1 and v_2 determine?

b) If v_3 is a third vector in $\mathbb{R}^3$, what are necessary and sufficient conditions for v_1, v_2, v_3 to be

linearly independent in terms of this geometric object?

Use the system of linear equation solving capabilities of Maple or another appropriate software package in Exercises 31 and 32.

31. a) Show that the vectors

$$\begin{bmatrix} 13 \\ 15 \\ 11 \\ -1 \\ 53 \\ 16 \end{bmatrix}, \begin{bmatrix} 1 \\ 22 \\ 12 \\ -9 \\ 18 \\ 77 \end{bmatrix}, \begin{bmatrix} -5 \\ -14 \\ 13 \\ 10 \\ 3 \\ 1 \end{bmatrix}, \begin{bmatrix} 8 \\ 81 \\ 14 \\ 101 \\ 11 \\ 15 \end{bmatrix},$$

$$\begin{bmatrix} 3 \\ 3 \\ 15 \\ 2 \\ 99 \\ -68 \end{bmatrix}, \begin{bmatrix} 7 \\ 3 \\ 16 \\ 88 \\ -49 \\ 1 \end{bmatrix}$$

form a basis for $\mathbb{R}^6$.

b) Find the coordinates of

$$\begin{bmatrix} 29 \\ 4 \\ -9 \\ 13 \\ 71 \\ -51 \end{bmatrix}$$

relative to the basis in part (a).

32. a) Show that

$$x^6 - x^5 + 3x + 1, \ x^5 + x^4 + x^2 + x + 1,$$
$$-x^6 + x^4 + x^3 + 3x^2 + x + 1,$$
$$2x^6 - x^4 - 2x^3 - 5x^2 - 1,$$
$$3x^6 + x^5 + 2x^4 + x^3 - x^2 - x - 1,$$
$$x^6 + 3x^5 + 2x^4 - 3x^3 + 2x^2 - 2x - 1,$$
$$2x^6 - x^5 + 2x^3 + x^2 - 3x + 1$$

form a basis for P_6.

b) Find the coordinates of $7x^6 + 6x^5 - 5x^4 - 4x^3 - 3x^2 + 2x - 1$ relative to the basis in part (a).

33. We saw in the solution to Example 11 that the three vectors given in this example are linearly dependent in $\mathbb{R}^2$. Show that this can be generalized to the following: If $v_1, v_2, \ldots, v_m$ are vectors in $\mathbb{R}^n$ and if $m > n$, then $v_1, v_2, \ldots, v_m$ are linearly dependent.

34. Let $v_1, v_2, \ldots, v_n$ be vectors in $\mathbb{R}^n$. Show that $v_1, v_2, \ldots, v_n$ form a basis for $\mathbb{R}^n$ if and only if the matrix $[v_1 \ v_2 \ \ldots \ v_n]$ is nonsingular.

35. Use the result of Exercise 34 and a suitable test for invertibility of a matrix in Maple or another appropriate software package to show that the vectors in Exercise 31(a) form a basis for $\mathbb{R}^6$.

2.4 DIMENSION; NULLSPACE, ROW SPACE, AND COLUMN SPACE

Looking back through the last section at all the examples of bases for $\mathbb{R}^n$ we did for various values of n or any exercises you did involving bases for these spaces, you might notice every basis had n elements. This is no accident. In fact, our first main objective in this section will be to show that once a vector space has a basis of n vectors, then every other basis also has n vectors. To reach this objective, we first prove the following lemma.

LEMMA 2.8 If $v_1, v_2, \ldots, v_n$ are a basis for a vector space V, then every set of vectors $w_1, w_2, \ldots, w_m$ in V where $m > n$ is linearly dependent.

Proof We must show that there is a nontrivial linear combination

$$c_1 w_1 + c_2 w_2 + \cdots + c_m w_m = 0. \tag{1}$$

To this end, let us first write each w_i as a linear combination of our basis vectors $v_1, v_2, \ldots, v_n$:

$$\begin{aligned} w_1 &= a_{11}v_1 + a_{21}v_2 + \cdots + a_{n1}v_n \\ w_2 &= a_{12}v_1 + a_{22}v_2 + \cdots + a_{n2}v_n \\ &\vdots \\ w_m &= a_{1m}v_1 + a_{2m}v_2 + \cdots + a_{nm}v_n. \end{aligned} \tag{2}$$

Substituting the results of Equations (2) into Equation (1), we have

$$c_1(a_{11}v_1 + a_{21}v_2 + \cdots + a_{n1}v_n) + c_2(a_{12}v_1 + a_{22}v_2 + \cdots + a_{n2}v_n)$$
$$+ \cdots + c_m(a_{1m}v_1 + a_{2m}v_2 + \cdots + a_{nm}v_n) =$$
$$(a_{11}c_1 + a_{12}c_2 + \cdots + a_{1m}c_m)v_1 + (a_{21}c_1 + a_{22}c_2 + \cdots + a_{2m}c_m)v_2$$
$$+ \cdots + (a_{n1}c_1 + a_{n2}c_2 + \cdots + a_{nm}c_m)v_n = 0.$$

Since $v_1, v_2, \ldots, v_n$ are linearly independent, the last equation tells us that we have the homogeneous system:

$$a_{11}c_1 + a_{12}c_2 + \cdots + a_{1m}c_m = 0$$
$$a_{21}c_1 + a_{22}c_2 + \cdots + a_{2m}c_m = 0$$
$$\vdots$$
$$a_{n1}c_1 + a_{n2}c_2 + \cdots + a_{nm}c_m = 0.$$

Since $m > n$, we know by Theorem 1.1 that this system has nontrivial solutions. Thus we have nontrivial linear combinations equal to the zero vector in Equation (1) and hence $w_1, w_2, \ldots, w_m$ are linearly dependent.   ●

Now we are ready to prove the result referred to at the beginning of this section, which we state as follows.

THEOREM 2.9   If $v_1, v_2, \ldots, v_n$ and $w_1, w_2, \ldots, w_m$ both form bases for a vector space V, then $n = m$.

Proof   Applying Lemma 2.8 with $v_1, v_2, \ldots, v_n$ as the basis, we must have $m \le n$ or else $w_1, w_2, \ldots, w_m$ would be linearly dependent. Interchanging the roles of $v_1, v_2, \ldots, v_n$ and $w_1, w_2, \ldots, w_m$, we obtain $n \le m$. Hence $n = m$.   ●

The number of vectors in a basis (which Theorem 2.9 tells us is always the same) is what we call the dimension of the vector space.

> **DEFINITION**   If a vector space V has a basis of n vectors, we say the **dimension** of V is n.

We denote the dimension of a vector space V by

$$\boxed{\dim(V).}$$

Thus, for example, since the standard basis $e_1, e_2, \ldots, e_n$ for $\mathbb{R}^n$ has n vectors,

$$\boxed{\dim(\mathbb{R}^n) = n.}$$

Since the standard basis for $M_{m \times n}(\mathbb{R})$ consists of the mn matrices E_{ij} with 1 in the ij-position and 0s elsewhere,

$$\boxed{\dim(M_{m \times n}(\mathbb{R})) = mn.}$$

Since the standard basis $x^n, \ldots, x, 1$ for P_n has $n + 1$ elements,

$$\boxed{\dim(P_n) = n + 1.}$$

Not every vector space V has a basis consisting of a finite number of vectors, but we can still introduce dimensions for such vector spaces. One such case occurs when V is the zero vector space consisting of only the zero vector. The zero vector space does not have a basis. In fact, the set consisting of the zero vector is the only spanning set for the zero vector space. But it is not a basis since, as you were asked to show in Exercise 27 in the previous section, no set containing the zero vector is linearly independent. For obvious reasons, if V is the zero vector space, we take the dimension of V to be 0 and write $\dim(V) = 0$. The zero vector space along with vector spaces that have bases with a finite number of vectors are called **finite dimensional** vector spaces. Vector spaces V that are not finite dimensional are called **infinite dimensional** vector spaces and we write $\dim(V) = \infty$. It can be proven that infinite dimensional vector spaces have bases with infinitely many vectors, but we will not attempt to prove this here.[7] The set of all polynomials P is an example of an infinite dimensional vector space. Indeed, the polynomials $1, x, x^2, x^3, \ldots$ form a basis for P. Many of the other function spaces we looked at in Sections 2.1 and 2.2, such as $F(a, b)$, $C(a, b)$, $D(a, b)$, and $C^\infty(a, b)$, are infinite dimensional, but we will not attempt to give bases for these vector spaces.

We next develop some facts that will be useful to us from time to time. We begin with the following lemma.

LEMMA 2.10   Let $v_1, v_2, \ldots, v_n$ and $w_1, w_2, \ldots, w_m$ be vectors in a vector space V. Then $\mathrm{Span}\{v_1, v_2, \ldots, v_n\} = \mathrm{Span}\{w_1, w_2, \ldots, w_m\}$ if and only if each v_i is a linear combination of $w_1, w_2, \ldots, w_m$ and each w_i is a linear combination of $v_1, v_2, \ldots, v_n$.

Proof   Suppose that $\mathrm{Span}\{v_1, v_2, \ldots, v_n\} = \mathrm{Span}\{w_1, w_2, \ldots, w_m\}$. Since each v_i is in $\mathrm{Span}\{v_1, v_2, \ldots, v_n\}$ (use 1 for the scalar on v_i and 0 for the scalar on all other v_j to write v_i as a linear combination of $v_1, v_2, \ldots, v_n$), v_i is in $\mathrm{Span}\{w_1, w_2, \ldots, w_m\}$. Hence v_i is a linear combination of $w_1, w_2, \ldots, w_m$. Likewise each w_i is a linear combination of $v_1, v_2, \ldots, v_n$.

To prove the converse, first note that if each v_i is a linear combination of $w_1, w_2, \ldots, w_m$, then each v_i is in $\mathrm{Span}\{w_1, w_2, \ldots, w_m\}$. Since subspaces are closed under addition and scalar multiplication, they are also closed under linear combinations. Hence any linear combination of $v_1, v_2, \ldots, v_n$ lies in $\mathrm{Span}\{w_1, w_2, \ldots, w_m\}$ giving us that $\mathrm{Span}\{v_1, v_2, \ldots, v_n\}$ is contained in $\mathrm{Span}\{w_1, w_2, \ldots, w_m\}$. Likewise we will have $\mathrm{Span}\{w_1, w_2, \ldots, w_m\}$ is contained in $\mathrm{Span}\{v_1, v_2, \ldots, v_n\}$ so that these two subspaces are equal as desired.   ●

The next lemma tells us that we can extend linearly independent sets of vectors to bases and reduce spanning sets to bases by eliminating vectors if necessary.

LEMMA 2.11   Suppose that V is a vector space of positive dimension n.

1. If $v_1, v_2, \ldots v_k$ are linearly independent vectors in V, then there exist vectors $v_{k+1}, \ldots, v_n$ so that $v_1, \ldots, v_k, v_{k+1}, \ldots, v_n$ form a basis for V.

2. If $v_1, v_2, \ldots, v_k$ span V, then there exists a subset of $v_1, v_2, \ldots, v_k$ that forms a basis of V.

---

[7] Typical proofs involve using a result called Zorn's Lemma, which you may well encounter if you continue your study of mathematics.

Proof

1. Notice that $k \leq n$ by Lemma 2.8. If $k < n$, $v_1, v_2, \ldots, v_k$ cannot span V; otherwise $\dim(V) = k$ not n. Thus there is a vector v_{k+1} not in $\text{Span}\{v_1, v_2, \ldots, v_k\}$. We must have $v_1, v_2, \ldots, v_k, v_{k+1}$ are linearly independent. To see this, suppose

$$c_1 v_1 + c_2 v_2 + \cdots + c_k v_k + c_{k+1} v_{k+1} = 0$$

where $c_1, c_2, \ldots, c_k, c_{k+1}$ are scalars. Were $c_{k+1} \neq 0$, we could solve this equation for v_{k+1} obtaining v_{k+1} is in $\text{Span}\{v_1, v_2, \ldots, v_k\}$, which we know is not the case. Now that $c_{k+1} = 0$ in this linear combination, the linear independence of $v_1, v_2, \ldots, v_k$ tells us we also have $c_1 = 0, c_2 = 0, \ldots, c_k = 0$. Hence we have the linear independence of $v_1, v_2, \ldots, v_{k+1}$. If $k + 1 < n$, repeat this procedure again. After $n - k$ steps we arrive at a set of n linearly independent vectors $v_1, v_2, \ldots, v_n$. These n vectors must span V; otherwise we could repeat our procedure again obtaining $n + 1$ linearly independent vectors $v_1, v_2, \ldots, v_n, v_{n+1}$, which is impossible by Lemma 2.8. Thus we have arrived at the desired basis.

2. If $v_1, v_2, \ldots, v_k$ are linearly independent, they then form the desired basis. If not, one of them must be a linear combination of the others. Relabeling if necessary to make the notation simpler, we may assume v_k is a linear combination of $v_1, v_2, \ldots, v_{k-1}$. Since each $v_1, v_2, \ldots, v_k$ is a linear combination of $v_1, v_2, \ldots, v_{k-1}$ and vice versa, it follows by Lemma 2.10 that $v_1, v_2, \ldots, v_{k-1}$ span V. If $v_1, v_2, \ldots, v_{k-1}$ are also linearly independent, we have our desired basis; if not, repeat the procedure we just did again. Since such steps cannot go on forever, we must obtain the desired type of basis after a finite number of steps. ●

At the beginning of this section we noted that the examples and exercises of the previous section suggested that once one basis has n elements, all other bases also have n elements. We then went on to prove this (Theorem 2.9). Another thing you might notice from the examples and exercises in the last section is that every time we had n vectors in a vector space of dimension n, having one of the properties of linear independence or spanning seemed to force the other property to hold too. The next theorem tells us this is indeed the case.

THEOREM 2.12   Suppose that V is a vector space of dimension n.

1. If the vectors $v_1, v_2, \ldots, v_n$ are linearly independent, then $v_1, v_2, \ldots, v_n$ form a basis for V.

2. If the vectors $v_1, v_2, \ldots, v_n$ span V, then $v_1, v_2, \ldots, v_n$ form a basis for V.

Proof

1. By part (1) of Lemma 2.11, we can extend $v_1, v_2, \ldots, v_n$ to a basis of V. But since every basis of V has n elements, $v_1, v_2, \ldots, v_n$ must already form a basis for V.

2. By part (2) of Lemma 2.11, we know some subset of the set of vectors $v_1, v_2, \ldots, v_n$ forms a basis of V. Again since every basis has n elements, this subset forming the basis must be the entire set of vectors $v_1, v_2, \ldots, v_n$. ●

EXAMPLE 1   Show that $x^2 - 1, x^2 + 1, x + 1$ form a basis for P_2.

Solution   Since $\dim(P_2) = 3$ and we are given three vectors in P_2, Theorem 2.12 tells us we can get by with showing either linear independence or spanning. Let us do linear independence. If

$$c_1(x^2 - 1) + c_2(x^2 + 1) + c_3(x + 1) = 0,$$

we have the system:

$$c_1 + c_2 = 0$$
$$c_3 = 0$$
$$-c_1 + c_2 + c_3 = 0.$$

It is easily seen that this system has only the trivial solution. Thus $x^2 - 1, x^2 + 1, x + 1$ are linearly independent and hence form a basis for P_2. ●

Of course, notice that Theorem 2.12 allows us to get by with checking for linear independence or spanning only when we already know the dimension of a finite dimensional vector space V and are given the same number of vectors as $\dim(V)$. If we do not know $\dim(V)$ for a finite dimensional vector space V, we must check both linear independence and spanning. Notice too that if we do know the dimension of a vector space V is n, no set of vectors with fewer or more than n elements could form a basis since all bases must have exactly n elements. For example, neither the set of two vectors $(1, 1, 1), (1, -1, 3)$ nor the set of four vectors $(1, 1, 1), (1, -1, 3), (0, 1, 1), (2, 1, -1)$ could form a basis for the three-dimensional space $\mathbb{R}^3$.

We conclude this section with a discussion of techniques for finding bases for three important subspaces associated with a matrix. Recall that the solutions to the homogeneous system $AX = 0$ where A is an $m \times n$ matrix form a subspace of $\mathbb{R}^n$ (Theorem 2.4). This vector space of solutions is called the **nullspace** or **kernel** of the matrix A and we shall denote it by

$$\boxed{NS(A).}$$

The manner in which we wrote our solutions to homogeneous systems in Chapter 1 leads us naturally to a basis for $NS(A)$. Consider the following example.

EXAMPLE 2   Find a basis for $NS(A)$ if

$$A = \begin{bmatrix} 1 & 2 & -1 & 3 & 0 \\ 1 & 1 & 0 & 4 & 1 \\ 1 & 4 & -3 & 1 & -2 \end{bmatrix}.$$

Solution   Reducing the augmented matrix for the system $AX = 0$ to reduced row-echelon form,

$$\left[\begin{array}{ccccc|c} 1 & 2 & -1 & 3 & 0 & 0 \\ 1 & 1 & 0 & 4 & 1 & 0 \\ 1 & 4 & -3 & 1 & -2 & 0 \end{array}\right] \rightarrow \left[\begin{array}{ccccc|c} 1 & 2 & -1 & 3 & 0 & 0 \\ 0 & -1 & 1 & 1 & 1 & 0 \\ 0 & 2 & -2 & -2 & -2 & 0 \end{array}\right]$$

$$\rightarrow \left[\begin{array}{ccccc|c} 1 & 0 & 1 & 5 & 2 & 0 \\ 0 & 1 & -1 & -1 & -1 & 0 \\ 0 & 0 & 0 & 0 & 0 & 0 \end{array}\right],$$

we see our solutions are

$$\begin{bmatrix} x_1 \\ x_2 \\ x_3 \\ x_4 \\ x_5 \end{bmatrix} = \begin{bmatrix} -x_3 - 5x_4 - 2x_5 \\ x_3 + x_4 + x_5 \\ x_3 \\ x_4 \\ x_5 \end{bmatrix}.$$

If we express this column vector as

$$x_3 \begin{bmatrix} -1 \\ 1 \\ 1 \\ 0 \\ 0 \end{bmatrix} + x_4 \begin{bmatrix} -5 \\ 1 \\ 0 \\ 1 \\ 0 \end{bmatrix} + x_5 \begin{bmatrix} -2 \\ 1 \\ 0 \\ 0 \\ 1 \end{bmatrix},$$

we immediately have that every solution is a linear combination of

$$\begin{bmatrix} -1 \\ 1 \\ 1 \\ 0 \\ 0 \end{bmatrix}, \begin{bmatrix} -5 \\ 1 \\ 0 \\ 1 \\ 0 \end{bmatrix}, \begin{bmatrix} -2 \\ 1 \\ 0 \\ 0 \\ 1 \end{bmatrix}$$

so that these three columns span $NS(A)$. They also are easily seen to be linearly independent. (The only such linear combination of them that is the zero column is the one with $x_3 = 0$, $x_4 = 0$, $x_5 = 0$.) Hence these three columns form a basis for $NS(A)$. ●

The subspace of $M_{1 \times n}(\mathbb{R})$ spanned by the rows of an $m \times n$ matrix A is called the **row space** of A and is denoted

$$\boxed{RS(A).}$$

For instance, if A is the matrix in Example 2, the row space of A is

$$\text{Span}\left\{\begin{bmatrix} 1 & 2 & -1 & 3 & 0 \end{bmatrix}, \begin{bmatrix} 1 & 1 & 0 & 4 & 1 \end{bmatrix}, \begin{bmatrix} 1 & 4 & -3 & 1 & -2 \end{bmatrix}\right\}.$$

Notice that if we perform an elementary row operation on a matrix A, the resulting matrix has its rows being linear combinations of the rows of A and vice versa since elementary row operations are reversible. Consequently, A and the resulting matrix have the same row space by Lemma 2.10. More generally this holds if we repeatedly do elementary row operations and hence we have the following theorem.

THEOREM 2.13   If A and B are row equivalent matrices, then

$$RS(A) = RS(B).$$

Because of Theorem 2.13, if B is the reduced row-echelon form of a matrix A, then B has the same row space as A. It is easy to obtain a basis for $RS(B)$ (which equals $RS(A)$) as the following example illustrates.

EXAMPLE 3   Find a basis for the row space of the matrix

$$A = \begin{bmatrix} 1 & 2 & -1 & 3 & 0 \\ 1 & 1 & 0 & 4 & 1 \\ 1 & 4 & -3 & 1 & -2 \end{bmatrix}$$

of Example 2.

Solution   From the solution to Example 2, we see the reduced row-echelon form of A is

$$B = \begin{bmatrix} 1 & 0 & 1 & 5 & 2 \\ 0 & 1 & -1 & -1 & -1 \\ 0 & 0 & 0 & 0 & 0 \end{bmatrix}.$$

The nonzero rows of B span $RS(B) = RS(A)$. They also are linearly independent. (If

$$c_1 \begin{bmatrix} 1 & 0 & 1 & 5 & 2 \end{bmatrix} + c_2 \begin{bmatrix} 0 & 1 & -1 & -1 & -1 \end{bmatrix} = \begin{bmatrix} 0 & 0 & 0 & 0 & 0 \end{bmatrix},$$

we see from the first two entries that $c_1 = 0$ and $c_2 = 0$.) Hence

$$\begin{bmatrix} 1 & 0 & 1 & 5 & 2 \end{bmatrix}, \begin{bmatrix} 0 & 1 & -1 & -1 & -1 \end{bmatrix}$$

form a basis for $RS(A)$.   ●

There is a relationship between the dimensions of $RS(A)$ and $NS(A)$ and the number of columns n of an $m \times n$ matrix A. Notice that $\dim(RS(A))$ is the number of nonzero rows of the reduced row-echelon form of A, which is the same as the number of nonfree variables in the solutions of the homogeneous system $AX = 0$. Also notice that $\dim(NS(A))$ is the number of free variables in the solutions of the homogeneous system $AX = 0$. Since the number of nonfree variables plus the number of free variables is the total number of variables in the system $AX = 0$, which is n, we have the following theorem.

THEOREM 2.14   If A is an $m \times n$ matrix,

$$\dim(RS(A)) + \dim(NS(A)) = n.$$

The final subspace associated with an $m \times n$ matrix A we shall introduce is the subspace of $\mathbb{R}^m$ spannned by the columns of A, which is called the **column space** of A and is denoted

$$\boxed{CS(A).}$$

For the matrix A in Examples 2 and 3, the column space is

$$\text{Span} \left\{ \begin{bmatrix} 1 \\ 1 \\ 1 \end{bmatrix}, \begin{bmatrix} 2 \\ 1 \\ 4 \end{bmatrix}, \begin{bmatrix} -1 \\ 0 \\ -3 \end{bmatrix}, \begin{bmatrix} 3 \\ 4 \\ 1 \end{bmatrix}, \begin{bmatrix} 0 \\ 1 \\ -2 \end{bmatrix} \right\}.$$

In the same manner as we use row operations to find bases for the row space of a matrix A, we could find a basis for the column space of A by using column operations to get the column reduced echelon form of A. The nonzero columns of the column reduced echelon form will form a basis for $CS(A)$. But if you feel more comfortable using row operations as we authors do, notice that a basis for $CS(A)$ can be found by reducing A^T to row-echelon form and then transposing the basis vectors of $RS(A^T)$ back to column vectors. This is the approach we take in the next example.

EXAMPLE 4   Find a basis for the column space of the matrix

$$A = \begin{bmatrix} 1 & 2 & -1 & 3 & 0 \\ 1 & 1 & 0 & 4 & 1 \\ 1 & 4 & -3 & 1 & -2 \end{bmatrix}$$

of Examples 2 and 3.

Solution   Reducing A^T to reduced row-echelon form,

$$A^T = \begin{bmatrix} 1 & 1 & 1 \\ 2 & 1 & 4 \\ -1 & 0 & -3 \\ 3 & 4 & 1 \\ 0 & 1 & -2 \end{bmatrix} \rightarrow \begin{bmatrix} 1 & 1 & 1 \\ 0 & -1 & 2 \\ 0 & 1 & -2 \\ 0 & 1 & -2 \\ 0 & 1 & -2 \end{bmatrix} \rightarrow \begin{bmatrix} 1 & 0 & 3 \\ 0 & 1 & -2 \\ 0 & 0 & 0 \\ 0 & 0 & 0 \\ 0 & 0 & 0 \end{bmatrix}$$

from which we see

$$\begin{bmatrix} 1 & 0 & 3 \end{bmatrix}^T = \begin{bmatrix} 1 \\ 0 \\ 3 \end{bmatrix}, \qquad \begin{bmatrix} 0 & 1 & -2 \end{bmatrix}^T = \begin{bmatrix} 0 \\ 1 \\ -2 \end{bmatrix}$$

form a basis for $CS(A)$.   ●

You might note that we actually do not have to go all the way to reduced row-echelon form to see a basis. For instance, it is easy to see that the first two rows in the second matrix of the solution to Example 4 would have to give us a basis so that we could equally well use

$$\begin{bmatrix} 1 \\ 1 \\ 1 \end{bmatrix}, \begin{bmatrix} 0 \\ -1 \\ 2 \end{bmatrix}$$

as a basis for $CS(A)$ in Example 4.

The dimensions of $RS(A)$ and $CS(A)$ are the same for the matrix A in Examples 3 and 4. (Both dimensions are 2.) In Corollary 5.5 of Chapter 5 we shall prove that this is always the case; that is, for any matrix A,

$$\boxed{\dim(RS(A)) = \dim(CS(A)).}$$

This common dimension is called the **rank** of the matrix A and is denoted rank(A). For instance, if A is the matrix in Examples 3 and 4, then rank$(A) = 2$. Observe that the rank of a matrix is the same as the number of nonzero rows (columns) in its reduced row (column)-echelon form.

Sometimes we have to find a basis for a subspace of $\mathbb{R}^n$ spanned by several vectors of $\mathbb{R}^n$. This is the same as finding a basis for the column space of a matrix as the final example of this section illustrates.

EXAMPLE 5 Find a basis for the subspace of $\mathbb{R}^3$ spanned by

$$\begin{bmatrix} 1 \\ -1 \\ 0 \end{bmatrix}, \begin{bmatrix} 1 \\ 1 \\ 1 \end{bmatrix}, \begin{bmatrix} 2 \\ 0 \\ 1 \end{bmatrix}.$$

Solution The subspace spanned by these three vectors is the same as the column space of the matrix

$$A = \begin{bmatrix} 1 & 1 & 2 \\ -1 & 1 & 0 \\ 0 & 1 & 1 \end{bmatrix}.$$

Reducing A^T,

$$A^T = \begin{bmatrix} 1 & -1 & 0 \\ 1 & 1 & 1 \\ 2 & 0 & 1 \end{bmatrix} \rightarrow \begin{bmatrix} 1 & -1 & 0 \\ 0 & 2 & 1 \\ 0 & 2 & 1 \end{bmatrix}.$$

It is apparent that

$$\begin{bmatrix} 1 \\ -1 \\ 0 \end{bmatrix}, \begin{bmatrix} 0 \\ 2 \\ 1 \end{bmatrix}$$

form a basis for $CS(A)$ and hence for the subspace spanned by the three given vectors. ●

EXERCISES 2.4

1. In each of parts (a)–(d), determine whether the given vectors form a basis for $\mathbb{R}^2$.

a) $\begin{bmatrix} 1 \\ 2 \end{bmatrix}$

b) $\begin{bmatrix} 1 \\ 1 \end{bmatrix}, \begin{bmatrix} 2 \\ -1 \end{bmatrix}$

c) $\begin{bmatrix} -12 \\ 16 \end{bmatrix}, \begin{bmatrix} 9 \\ -12 \end{bmatrix}$

d) $\begin{bmatrix} -2 \\ -7 \end{bmatrix}, \begin{bmatrix} 4 \\ 3 \end{bmatrix}, \begin{bmatrix} -5 \\ 8 \end{bmatrix}$

2. In each of parts (a)–(d), determine whether the given vectors form a basis for $\mathbb{R}^3$.

a) $\begin{bmatrix} 1 \\ 1 \\ 4 \end{bmatrix}, \begin{bmatrix} 2 \\ -1 \\ 0 \end{bmatrix}, \begin{bmatrix} 0 \\ -1 \\ 8 \end{bmatrix}$

b) $\begin{bmatrix} 3 \\ 2 \\ 1 \end{bmatrix}, \begin{bmatrix} -1 \\ -1 \\ 0 \end{bmatrix}, \begin{bmatrix} 3 \\ 1 \\ 2 \end{bmatrix}$

c) $\begin{bmatrix} 1 \\ 1 \\ 1 \end{bmatrix}, \begin{bmatrix} -3 \\ 2 \\ 5 \end{bmatrix}, \begin{bmatrix} 4 \\ 4 \\ 5 \end{bmatrix}, \begin{bmatrix} 0 \\ -3 \\ 1 \end{bmatrix}$

d) $\begin{bmatrix} -7 \\ -9 \\ 1 \end{bmatrix}, \begin{bmatrix} 1 \\ 2 \\ 2 \end{bmatrix}$

3. In each of parts (a)–(d), determine whether the given polynomials form a basis for P_2.

a) $x^2 + x - 1, 2x^2 - 3, x^2 + x + 2$

b) $5 - 4x^2, 3 - 2x$

c) $x^2 + x - 1, x^2 + x + 2, x^2 + x + 14$

d) $x^2 + x, x + 1, x^2 + 1, 1$

4. In each of parts (a)–(d), determine whether the given matrices form a basis for $M_{2 \times 2}(\mathbb{R})$.

a) $\begin{bmatrix} 1 & 1 \\ 0 & 1 \end{bmatrix}, \begin{bmatrix} 1 & 0 \\ 0 & -1 \end{bmatrix}, \begin{bmatrix} 1 & 1 \\ 0 & 0 \end{bmatrix},$ $\begin{bmatrix} 1 & 0 \\ 0 & 1 \end{bmatrix}$

b) $\begin{bmatrix} 1 & 0 \\ 0 & 1 \end{bmatrix}, \begin{bmatrix} 0 & 1 \\ 1 & 0 \end{bmatrix}, \begin{bmatrix} 1 & 0 \\ 0 & 0 \end{bmatrix}$

c) $\begin{bmatrix} 0 & 1 \\ 0 & 1 \end{bmatrix}, \begin{bmatrix} 1 & 0 \\ 0 & 1 \end{bmatrix}, \begin{bmatrix} 1 & 0 \\ 0 & 0 \end{bmatrix}, \begin{bmatrix} 0 & 1 \\ 1 & 0 \end{bmatrix}$

d) $\begin{bmatrix} 1 & 0 \\ 0 & 1 \end{bmatrix}, \begin{bmatrix} 1 & 1 \\ 0 & 1 \end{bmatrix}, \begin{bmatrix} 1 & 0 \\ 0 & -1 \end{bmatrix},$ $\begin{bmatrix} 1 & 1 \\ 0 & 0 \end{bmatrix}, \begin{bmatrix} 1 & 0 \\ 0 & 1 \end{bmatrix}$

Do the following for the matrices in Exercises 5–12:

a) Find a basis for the nullspace of the matrix.

b) Find a basis for the row space of the matrix.

c) Find a basis for the column space of the matrix.

d) Determine the rank of the matrix.

(Parts (a)–(c) do not have unique answers.)

5. $\begin{bmatrix} 1 & 1 \\ 1 & -2 \end{bmatrix}$

6. $\begin{bmatrix} -2 & 4 \\ 1 & -2 \end{bmatrix}$

7. $\begin{bmatrix} 1 & -1 & 1 \\ -1 & 1 & 0 \\ 1 & -1 & 2 \end{bmatrix}$

8. $\begin{bmatrix} 2 & -1 & 0 \\ 1 & 1 & -1 \\ 1 & 0 & 1 \end{bmatrix}$

9. $\begin{bmatrix} 1 & 1 & 0 & 3 \\ 1 & 1 & 1 & -2 \\ 3 & 3 & 2 & -1 \end{bmatrix}$

10. $\begin{bmatrix} 2 & -1 & 3 & 4 \\ 1 & 0 & -1 & 3 \end{bmatrix}$

11. $\begin{bmatrix} 1 & -1 & 2 & 1 \\ 2 & 1 & -1 & 0 \\ 1 & 2 & -3 & -1 \\ 0 & 3 & -5 & -2 \\ 4 & -1 & 3 & 2 \end{bmatrix}$

12. $\begin{bmatrix} 1 & -1 & -1 & 2 & 0 \\ -2 & 1 & 1 & -1 & 0 \\ 1 & 1 & -2 & 1 & 1 \end{bmatrix}$

In Exercises 13–15, find a basis for the subspace spanned by the given vectors. (These do not have unique answers.)

13. $\begin{bmatrix} 1 \\ -2 \\ 1 \end{bmatrix}$, $\begin{bmatrix} 0 \\ 1 \\ 3 \end{bmatrix}$, $\begin{bmatrix} 5 \\ -6 \\ -7 \end{bmatrix}$

14. $\begin{bmatrix} 0 \\ -2 \\ 1 \end{bmatrix}$, $\begin{bmatrix} 1 \\ -1 \\ 0 \end{bmatrix}$, $\begin{bmatrix} 2 \\ 0 \\ -1 \end{bmatrix}$, $\begin{bmatrix} 4 \\ -2 \\ -1 \end{bmatrix}$

15. $\begin{bmatrix} -2 \\ 1 \\ 3 \\ -4 \end{bmatrix}$, $\begin{bmatrix} 1 \\ -2 \\ 1 \\ -1 \end{bmatrix}$, $\begin{bmatrix} -1 \\ -4 \\ 9 \\ -11 \end{bmatrix}$

16. Without formally showing that there is a nontrivial linear combination of $x^4 - 2x$, $x^4 + x^3 - 1$, $x^3 + x + 3$, $x - x^2 - x^4$, $10x - 91$, $\pi x^4 + \sqrt{3}x^3 - 7x^2$ equal to the zero polynomial, we can conclude these polynomials are linearly dependent. Why is this?

17. Show that $\det(A) \neq 0$ for an $n \times n$ matrix A if and only if $\text{rank}(A) = n$.

18. Suppose that A is an $m \times n$ matrix. Show that if $m > n$, then the rows of A are linearly dependent.

19. Suppose that A is an $m \times n$ matrix. Show that if $m < n$, then the columns of A are linearly dependent.

20. Consider the linearly independent polynomials $p_1(x) = x^2 + x$, $p_2(x) = x + 1$. Find a polynomial $p_3(x)$ so that $p_1(x)$, $p_2(x)$, $p_3(x)$ form a basis for P_2 in the manner of the proof of part (1) of Lemma 2.11.

21. a) Show that the polynomials $p_1(x) = x^2 - 1$, $p_2(x) = x^2 + 1$, $p_3(x) = x - 1$, $p_4(x) = x + 1$ span P_2.

b) Find a subset of $p_1(x)$, $p_2(x)$, $p_3(x)$, $p_4(x)$ that forms a basis for P_2 in the manner of the proof of part (2) of Lemma 2.11.

22. If the initial points of vectors in $\mathbb{R}^3$ are placed at the origin, what geometric object is a subspace of $\mathbb{R}^3$ of dimension one? What geometric object is a subspace of $\mathbb{R}^3$ of dimension two?

23. Explain why the reduced row-echelon form of a matrix is unique. Is the reduced column-echelon form unique? Why or why not?

24. Use the system of linear equation solving capabilities of Maple or another appropriate software package to show that the matrices

$\begin{bmatrix} 1 & 0 \\ -3 & -33 \\ 11 & -2 \end{bmatrix}$, $\begin{bmatrix} 1 & -3 \\ 1 & 13 \\ 16 & -2 \end{bmatrix}$,

$\begin{bmatrix} 1 & 17 \\ -3 & 46 \\ 32 & -9 \end{bmatrix}$, $\begin{bmatrix} -1 & 91 \\ 6 & -98 \\ -1 & 1 \end{bmatrix}$,

$\begin{bmatrix} -2 & -1 \\ 21 & 0 \\ 0 & 0 \end{bmatrix}$, $\begin{bmatrix} 3 & 2 \\ 5 & -5 \\ -4 & 0 \end{bmatrix}$

form a basis for $M_{3\times 2}(\mathbb{R})$.

25. Let A be the matrix

$$A = \begin{bmatrix} -2 & 3 & -1 & 4 & -2 \\ 3 & -1 & 2 & 6 & 8 \\ 1 & 7 & -5 & 1 & 4 \\ 0 & 12 & -5 & 15 & 8 \end{bmatrix}.$$

a) In Maple, the command *nullspace* (or equivalently, *kernel*) can be used to find a basis for the nullspace of a matrix. (The basis vectors will be given as row vectors instead of column vectors.) Use this command or a corresponding command in an appropriate software package to find a basis for the nullspace of the matrix A.

b) Bases for the row space of a matrix can be found with Maple by using the *gausselim* or *gaussjord* (or equivalently, *rref*) commands. Such bases may also be found using the *rowspace* and *rowspan* commands. Find bases for row space of the matrix A using each of these four Maple commands or corresponding commands in another appropriate software package. If you are using Maple, compare your results obtained with these four different commands.

c) Bases for the column space of a matrix can be found with Maple by using the *gausselim* or *gaussjord* commands on the transpose of the matrix. Such bases may also be found by using the *colspace* and *colspan* commands. Find bases for the column space of the matrix A using each of these four methods in Maple or corresponding commands in another appropriate software package. If you are using Maple, compare your results obtained with these four different approaches.

26. The *basis* command of Maple may be used to find a basis for the subspace spanned by a finite number of row vectors in $M_{1 \times n}(\mathbb{R})$. Use this command or the corresponding command in an appropriate software package to find a basis for the subspace of $M_{1 \times 5}(\mathbb{R})$ spanned by the vectors

$$\begin{bmatrix} -2 & 3 & -1 & 4 & -2 \end{bmatrix}, \begin{bmatrix} 3 & -1 & 2 & 6 & 8 \end{bmatrix},$$

$$\begin{bmatrix} 1 & 7 & -5 & 1 & 4 \end{bmatrix}, \begin{bmatrix} 0 & 12 & -5 & 15 & 8 \end{bmatrix}.$$

Compare the result you obtain here with your results in Exercise 25(b).

27. Suppose that $v_1, v_2, \ldots, v_k$ are linearly independent vectors in $\mathbb{R}^n$. Let v_{k+1} be another vector in $\mathbb{R}^n$. How could the *gausselim* or *gaussjord* commands of Maple or corresponding commands in another appropriate software package be used to determine if v_{k+1} is in $\text{Span}\{v_1, v_2, \ldots, v_k\}$?

28. Use the result of Exercise 27 to extend the set consisting of the vectors

$$\begin{bmatrix} 1 \\ 2 \\ 3 \\ -1 \end{bmatrix}, \begin{bmatrix} 2 \\ 1 \\ -1 \\ 0 \end{bmatrix},$$

to a basis of $\mathbb{R}^4$ in the manner of the proof of part (1) of Lemma 2.11.

2.5 WRONSKIANS

In our work with linear differential equations in Chapter 4 we will have instances where we will have to determine whether a set of functions forms a linearly independent set. For instance, problems such as the following will arise: Are the functions given by e^x, $\cos x$, $\sin x$ linearly independent? Consider a linear combination of these functions that equals the zero function:

$$c_1 e^x + c_2 \cos x + c_3 \sin x = 0.$$

The only such linear combination is the trivial one. One way to see this is to choose three values of x such as $x = 0, \pi/2, \pi$. Substituting these values of x into our linear combination, we have the system:

$$c_1 + c_2 = 0$$
$$e^{\pi/2} c_1 + c_3 = 0$$
$$e^{\pi} c_1 - c_2 = 0.$$

Adding the first and third equations, we obtain $(1 + e^{\pi}) c_1 = 0$ from which we get $c_1 = 0$. It then follows that $c_2 = 0$ and $c_3 = 0$. Hence e^x, $\cos x$, and $\sin x$ are linearly independent. But there is another way to arrive at a system of equations involving derivatives that is often more convenient to use for showing linear independence.

Suppose that we have n functions $f_1, f_2, \ldots, f_n$ each of which have $(n-1)$st derivatives on an open interval (a, b); that is, assume $f_1, f_2, \ldots, f_n$ all lie in $D^{n-1}(a, b)$.[8] Consider a linear combination of these functions that is equal to the zero function:

$$c_1 f_1(x) + c_2 f_2(x) + \cdots + c_n f_n(x) = 0.$$

Considering this equation along with the first $n-1$ derivatives of each side of it, we arrive at the following system:

$$
\begin{aligned}
c_1 f_1(x) + c_2 f_2(x) + \cdots + c_n f_n(x) &= 0 \\
c_1 f_1'(x) + c_2 f_2'(x) + \cdots + c_n f_n'(x) &= 0 \\
c_1 f_1''(x) + c_2 f_2''(x) + \cdots + c_n f_n''(x) &= 0 \\
&\vdots \\
c_1 f_1^{(n-1)}(x) + c_2 f_2^{(n-1)}(x) + \cdots + c_n f_n^{(n-1)}(x) &= 0.
\end{aligned}
\tag{1}
$$

If there is some x in (a, b) for which this system has only the trivial solution, then $f_1, f_2, \ldots, f_n$ will be linearly independent. Having such an x is the same as having an x in (a, b) for which the the matrix

$$
\begin{bmatrix}
f_1(x) & f_2(x) & \cdots & f_n(x) \\
f_1'(x) & f_2'(x) & \cdots & f_n'(x) \\
f_1''(x) & f_2''(x) & \cdots & f_n''(x) \\
\vdots & \vdots & \vdots & \vdots \\
f_1^{(n-1)}(x) & f_2^{(n-1)}(x) & \cdots & f_n^{(n-1)}(x)
\end{bmatrix}
$$

is nonsingular. A convenient way of seeing if there is an x in (a, b) where this matrix is nonsingular is by looking at its determinant, which is called the **Wronskian**[9] of the functions $f_1, f_2, \ldots, f_n$. We will denote the Wronskian of $f_1, f_2, \ldots, f_n$ by $w(f_1(x), f_2(x), \ldots, f_n(x))$:

$$
w(f_1(x), f_2(x), \ldots, f_n(x)) =
\begin{vmatrix}
f_1(x) & f_2(x) & \cdots & f_n(x) \\
f_1'(x) & f_2'(x) & \cdots & f_n'(x) \\
f_1''(x) & f_2''(x) & \cdots & f_n''(x) \\
\vdots & \vdots & \vdots & \vdots \\
f_1^{(n-1)}(x) & f_2^{(n-1)}(x) & \cdots & f_n^{(n-1)}(x)
\end{vmatrix}.
$$

Since a square matrix is nonsingular if and only if its determinant is nonzero (Theorem 1.21), we have the following theorem.

---

[8] In fact, what we are about to do does not require the interval to be open; it will work on other types of intervals (such as closed ones) as well.

[9] The Wronskian is named in honor of the Polish-French mathematician Jósef Maria Hoëné-Wronski (1776–1853).

THEOREM 2.15   Suppose that $f_1, f_2, \ldots, f_n$ are functions in $D^{n-1}(a, b)$. If the Wronskian $w(f_1(x),$ $f_2(x), \ldots, f_n(x))$ of $f_1, f_2, \ldots, f_n$ is nonzero for some x in (a, b), then $f_1, f_2, \ldots, f_n$ are linearly independent elements of $D^{n-1}(a, b)$.

Let us use the Wronskian to show $e^x, \cos x, \sin x$ are linearly independent (with $(-\infty, \infty)$ for the interval).

EXAMPLE 1   Determine if $e^x, \cos x, \sin x$ are linearly independent.

Solution   We have

$$w(e^x, \cos x, \sin x) = \begin{vmatrix} e^x & \cos x & \sin x \\ e^x & -\sin x & \cos x \\ e^x & -\cos x & -\sin x \end{vmatrix}$$
$$= e^x(\sin^2 x + \cos^2 x) - \cos x(-e^x \sin x - e^x \cos x)$$
$$+ \sin x(-e^x \cos x + e^x \sin x)$$
$$= 2e^x.$$

Since $2e^x$ is not zero for some x (in fact, it is not zero for every x), we have that $e^x, \cos x, \sin x$ are linearly independent by Theorem 2.15.   ●

Be careful not to read too much into Theorem 2.15. It only tells us that if the Wronskian is nonzero for some x, then the functions are linearly independent. It does not tell us that the converse is true; that is, it does not tell us that linearly independent functions have their Wronskian being nonzero for some x (or equivalently, that if the Wronskian is zero for all x, then the functions are linearly dependent). In fact, the converse of Theorem 2.15 does not hold in general. Here is an example illustrating this.

EXAMPLE 2   Show that the Wronskian of the functions f and g where $f(x) = x^2$ and $g(x) = x|x|$ is zero for every x and that f and g are linearly independent on $(-\infty, \infty)$.

Solution   To calculate the Wronskian of f and g, we need their derivatives. This is easy for f. To get the derivative of g notice that we can also express $g(x)$ as

$$g(x) = \begin{cases} x^2 & \text{if } x \geq 0 \\ -x^2 & \text{if } x < 0 \end{cases}.$$

The graph of g appears in Figure 2.3.
We have

$$g'(x) = \begin{cases} 2x & \text{if } x > 0 \\ -2x & \text{if } x < 0 \end{cases}.$$

At $x = 0$,

$$g'(0) = \lim_{h \to 0} \frac{g(0+h) - g(0)}{h} = \lim_{h \to 0} \frac{h|h|}{h} = \lim_{h \to 0} |h| = 0.$$

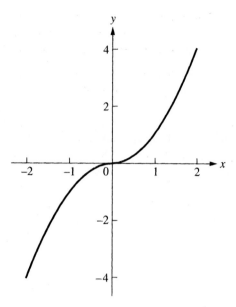

Figure 2.3

Thus we can say

$$g'(x) = \begin{cases} 2x & \text{if } x \geq 0 \\ -2x & \text{if } x < 0 \end{cases}.$$

If $x \geq 0$,

$$w(f(x), g(x)) = \begin{vmatrix} x^2 & x^2 \\ 2x & 2x \end{vmatrix} = 0.$$

If $x < 0$,

$$w(f(x), g(x)) = \begin{vmatrix} x^2 & -x^2 \\ 2x & -2x \end{vmatrix} = 0.$$

Hence we have that the Wronskian of f and g is zero for every x.

To see that f and g are linearly independent, suppose that

$$c_1 f(x) + c_2 g(x) = 0$$

where c_1 and c_2 are scalars. Substituting $x = 1$ and $x = -1$ into this equation, we arrive at the system

$$c_1 + c_2 = 0$$
$$c_1 - c_2 = 0,$$

which has only the trivial solution. Hence f and g are linearly independent as desired. ●

It is possible to obtain a converse of Theorem 2.15 provided we put some additional restrictions on the types of functions being considered. In Chapter 4 we will see a case in which this occurs.

EXERCISES 2.5

In each of Exercises 1–9, show that the given functions are linearly independent on $(-\infty, \infty)$.

1. e^{3x}, e^{-2x}

2. $\cos 5x, \sin 5x$

3. $e^{2x} \cos x, e^{2x} \sin x$

4. e^{-x}, xe^{-x}

5. $x^2 - 1, x^2 + 1, x + 1$

6. e^x, e^{2x}, e^{3x}

7. $e^{4x}, xe^{4x}, x^2 e^{4x}$

8. $e^x, e^x \cos x, e^x \sin x$

9. $x^3, |x|^3$

10. Show that $1/x, x$ are linearly independent on $(0, \infty)$.

11. Show that $x + 1, x - 1, x$ are linearly dependent on $(-\infty, \infty)$.

12. Show that $\sin^2 x, \cos^2 x, \cos 2x$ are linearly dependent on $(-\infty, \infty)$.

13. Show that the two functions in Example 2 are linearly dependent on $[0, \infty)$.

14. Suppose that $a < c < d < b$ and that $f_1, f_2, \ldots, f_n$ are functions in $F(a, b)$.

a) If $f_1, f_2, \ldots, f_n$ are linearly independent on (a, b), are $f_1, f_2, \ldots, f_n$ necessarily linearly independent on (c, d)? Why or why not?

b) If $f_1, f_2, \ldots, f_n$ are linearly independent on (c, d), are $f_1, f_2, \ldots, f_n$ necessarily linearly independent on (a, b)? Why or why not?

15. a) Find the Wronskian of $1, x, x^2, \ldots, x^{n-1}$.

b) Show that

$$w(g(x)f_1(x), g(x)f_2(x), \ldots, g(x)f_n(x)) = [g(x)]^n w(f_1(x), f_2(x), \ldots, f_n(x)).$$

c) Show that $e^{rx}, xe^{rx}, x^2 e^{rx}, \ldots, x^{n-1} e^{rx}$ are linearly independent on $(-\infty, \infty)$.

16. Use the *wronskian* and *det* commands in Maple or appropriate commands in another software package to find the Wronskian of the functions $e^{3x}, xe^{3x}, e^{3x} \cos 2x, e^{3x} \sin 2x, x \cos x, x \sin x$. Are these functions linearly independent?

First Order Ordinary Differential Equations

In Chapters 1 and 2 you were introduced to linear algebra. A field of mathematics that has a deep relationship with linear algebra is the field of differential equations, as you will see in Chapters 4, 6, and 7. A differential equation is an equation that involves a function, its variables, and its derivatives. Such equations arise in modeling many physical situations. Here is one you probably have encountered in your calculus courses:

> *It is well documented that most radioactive materials decay at a rate proportional to the amount present. What is an equation describing this activity?*

Suppose we use $Q(t)$ to denote the amount of radioactive material present at time t. The fact that "the materials decay at a rate proportional to the amount present" means that the rate $Q'(t)$ is a constant (called the **constant of proportionality** or the **constant of variation**) multiple of the amount $Q(t)$. Moreover, since $Q'(t)$ is negative because $Q(t)$ is decreasing, the constant of proportionality must be negative. Thus we can model this activity by the equation

$$Q'(t) = -kQ(t)$$

where $k > 0$ is a constant, which is a differential equation since it involves the derivative $Q'(t)$.

As we proceed through this and later chapters, we will see several other examples of phenomena modeled by differential equations. However, we will merely scratch the surface on applications of differential equations in this text. There are widespread applications of differential equations throughout the sciences, engineering, and finance and economics that must necessarily be left for later courses. In this text we will study some of the mathematical aspects of differential equations and learn how to model some

111

applications using differential equations. We begin this chapter with a general discussion of differential equations in the first section. Later sections deal with techniques for finding and approximating solutions of differential equations and some applications of them.

3.1 INTRODUCTION TO DIFFERENTIAL EQUATIONS

As mentioned in the opening of this chapter, a **differential equation** is an equation that involves a function, its variables, and its derivatives. For instance,

$$y' = x, \tag{1}$$

$$y' = x^2 + y^2, \tag{2}$$

$$y^{(4)}(x) - 3y^{(3)}(x) + xy''(x) = 0, \tag{3}$$

and

$$u_t = uu_y + u_x y \tag{4}$$

are examples of differential equations. If the function depends on only one variable, we call the differential equation an **ordinary** differential equation; those with more than one variable are called **partial** differential equations. Equations (1), (2), and (3) are ordinary differential equations, and Equation (4) is an example of a partial differential equation. In this book we will only consider ordinary differential equations. The **order** of a differential equation is the order of the highest derivative appearing in the equation. Equations (1) and (2) are first order, and Equation (3) is fourth order. Sometimes we will write differential equations in terms of differentials instead of derivatives. When doing so, we will say that the differential equation is in differential form. For instance, the differential equation

$$2xy + (x^2 + 3y^2) \frac{dy}{dx} = 0$$

is

$$2xy \, dx + (x^2 + 3y^2) \, dy = 0$$

in differential form.

A **solution** of a differential equation is a function that satisfies the equation. For example, $y = x^2/2$ is a solution to $y' = x$. We leave it for you to check that any function of the form $Q(t) = Ce^{-kt}$ where C is a constant is a solution of the differential equation $Q'(t) = -kQ(t)$ modeling radioactive decay given in the introduction to this chapter. As a third illustration, you can easily verify that $y = e^{-x}$ and $y = e^{3x}$ are solutions to $y'' - 2y' - 3y = 0$.

A type of differential equation that you already know how to solve is one of the form

$$y' = f(x).$$

Indeed, from calculus you know that every solution is given by

$$y = \int f(x)\, dx + C.^1$$

As an illustration, if

$$y' = x,$$

all the solutions to this differential equation have the form

$$y = \int x\, dx + C = \frac{x^2}{2} + C.$$

A general form of a function giving us all the solutions to a given differential equation is called the **general solution** of the differential equation. Thus, for instance, the general solution of a differential equation of the form $y' = f(x)$ is $y = \int f(x)\, dx + C$. In particular, the general solution of $y' = x$ is $y = x^2/2 + C$. Using techniques presented in the next section, we will be able to obtain that the general solution of $Q'(t) = -kQ(t)$ is $Q(t) = Ce^{-kt}$ where C is a constant. Using techniques presented in Chapter 4, we shall see that the general solution of $y'' - 2y' - 3y = 0$ is $y = c_1 e^{-x} + c_2 e^{3x}$ where c_1 and c_2 are constants.

A first order differential equation along with a condition specifying a value y_0 of y at a particular value x_0 of x is called a first order **initial value problem.** We will indicate such an initial value problem by writing the differential equation along with the condition in a manner illustrated by the following three examples:

$$y' = x, \qquad y(1) = 2;$$

$$Q'(t) = -kQ(t), \qquad Q(0) = 1;$$

$$2xy\, dx + (x^2 + 3y^2)\, dy = 0, \qquad y(1) = 1.$$

While there are infinitely many solutions given by the general solution $y = x^2/2 + C$ to the differential equation $y' = x$, there is only one solution also satisfying the initial condition $y(1) = 2$. Indeed, substituting $x = 1$ into the general solution and equating with $y(1) = 2$ gives us

$$y(1) = \frac{1}{2} + C = 2.$$

Solving for C we get $C = 3/2$ and hence $y = x^2/2 + 3/2$ is the one and only one solution to the initial value problem $y' = x, y(1) = 2$.

As another illustration, consider the initial value problem

$$Q'(t) = -kQ(t), \qquad Q(0) = 1.$$

---

[1] In this book we use the symbol $\int f(x)\, dx$ to indicate one antiderivative of $f(x)$. For example, $\int x\, dx = x^2/2$.)

We have noted that functions of the form $Q(t) = Ce^{-kt}$ are solutions to this differential equation. The initial condition gives us $Q(0) = C = 1$. Hence $Q(t) = e^{-kt}$ is one solution satisfying this initial value problem. Suppose there is another solution, $w(t)$. Then

$$\frac{d}{dt}\left[\frac{Q(t)}{w(t)}\right] = \frac{Q'(t)w(t) - Q(t)w'(t)}{[w(t)]^2} = \frac{-ke^{-kt}w(t) - e^{-kt}(-kw(t))}{[w(t)]^2} = 0,$$

since $w'(t) = -kw(t)$. Therefore, we know that $Q(t)/w(t)$ is a constant. Since $Q(0)/w(0) = 1$, the constant must be 1 and $Q(t) = w(t)$. Once again, we see that imposing an initial condition results in one and only one solution.

The fact that both of the initial value problems $y' = x$, $y(1) = 2$ and $Q'(t) = -kQ(t)$, $Q(0) = 1$ had unique solutions illustrates the following theorem whose proof will be sketched in Section 3.8.

THEOREM 3.1   Let $a, b > 0$ and suppose f and $\partial f/\partial y$ are continuous on the rectangle $|x - x_0| < a$ and $|y - y_0| < b$. Then there exists an $h > 0$ so that the initial value problem

$$y' = f(x, y), \qquad y(x_0) = y_0$$

has one and only one solution y for $|x - x_0| \le h$.

The existence part of Theorem 3.1 does not tell us how to produce the solution to an initial value problem. In succeeding sections we will develop some techniques for finding general solutions to certain types of differential equations that are then used to produce the solution to an initial value problem involving the differential equation. Even if we do not know a technique for solving a differential equation, however, we can still obtain graphical information about its solutions.

To illustrate, consider the differential equation $y' = 2y$. The differential equation determines the slope of a solution at a point (x, y) to be $y'(x) = 2y(x)$. For example, at the point $(0, 1)$, the slope is $y'(0) = 2y(0) = 2$; at the point $(2, 3)$, the slope is $y'(2) = 6$. Software packages such as Maple can be used to display the slopes at various points. This display is called a **direction field.**

To use Maple (see the accompanying *Technology Resource Manual* for doing these with Mathematica or MATLAB) to obtain direction fields, you must first load Maple's *DEtools* package by typing and entering

```
with(DEtools);
```

at the command prompt. To generate the direction field to $y'(x) = 2y(x)$ for $-1 \le x \le 1$ and $-3 \le y \le 3$, we type and enter

```
dfieldplot(diff(y(x),x)=2*y(x),y(x),x=-1..1,y=-3..3);
```

at the prompt command.[2] This direction field appears in Figure 3.1(a).

Graphs of solutions to initial value problems are called **phase portraits.** To graph the phase portrait for the solution to the initial value problem $y' = 2y$, $y(0) = 1$ for $-1 \le x \le 1$ using Maple (see the accompanying *Technology Resource Manual* for

---

[2] In Maple `diff(y(x),x)` is used for dy/dx or y'. Notice that you must explicitly indicate that y is a function of x.

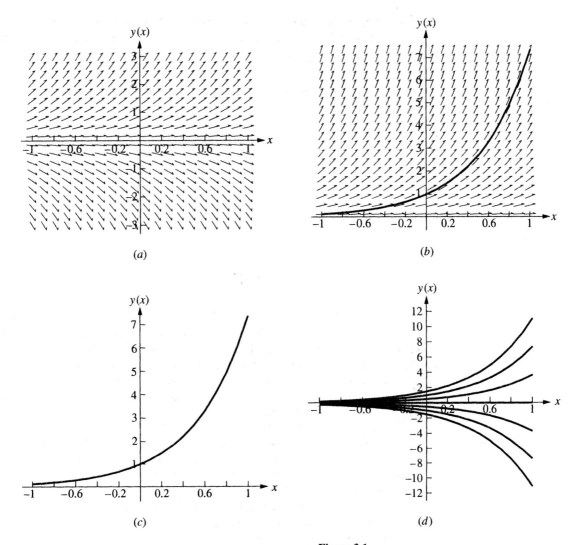

Figure 3.1

doing these with Mathematica or MATLAB), we can type and enter:

```
phaseportrait(diff(y(x),x)=2*y(x),y(x),x=-1..1,[[y(0)=1]]);
```

This phase portrait appears in Figure 3.1(b). We can also generate Figure 3.1(b) by using the command *DEplot* in place of *phaseportrait*. To leave slope vectors off the phase portrait add *,arrows=none* as follows:

```
phaseportrait(diff(y(x),x)=2*y(x),y(x),x=-1..1,
              [[y(0)=1]],arrows=none);
```

This graph appears in Figure 3.1(c). One way of graphing the phase portrait for solutions satisfying the initial conditions $y(0) = k/2$, $k = -3, -2, -1, 0, 1, 2, 3$ without slope vectors is by typing and entering:

```
phaseportrait(diff(y(x),x)=2*y(x),y(x),x=-1..1,
    [seq([y(0)=k/2],k=-3..3)], arrows=none);
```

This graph appears in Figure 3.1(d).

Direction fields and phase portraits can be useful in many circumstances. Direction fields illustrate the rates of change of the solutions to a differential equation giving us a sense of the behavior of the solutions. Phase portraits give us an approximate graph of the solution to an initial value problem. Knowledge of such graphs can be used, for example, to make predictions. As another example, suppose we have some data and wonder if a certain initial value problem models this data. By plotting this data and graphing the phase portrait to this initial value problem on the same coordinate system, we can we see how well the initial value problem fits our data.

An **equilibrium solution** of an ordinary differential equation is a constant solution. These constant solutions are horizontal lines that divide the xy-plane into regions. If $f(x, y)$ satisfies the conditions of Theorem 3.1, then no other solutions of $y' = f(x, y)$ can cross these horizontal lines. Otherwise, if another solution crosses an equilibrium solution $y = c$ at $x = x_0$, then both solutions satisfy the condition $y(x_0) = c$ violating the uniqueness of a solution in Theorem 3.1. Thus a nonequilibrium solution lies entirely within one of the regions determined by the equilibrium solutions. In the following examples we use equilibrium solutions along with the signs of the first and second derivatives to help us anticipate the phase portraits in these regions.

For the differential equations in Examples 1–4 do the following.

1. Determine the equilibrium solutions.

2. On each region determined by the equilibrium solutions, determine when the graph of the solution is increasing, decreasing, and concave up and down.

3. Use Maple (or another appropriate software package) to sketch a phase portrait that illustrates the results of parts (1) and (2).

EXAMPLE 1 $y' = y - y^2$

Solution   We can rewrite this equation as

$$y' = y(1 - y)$$

and see that $y = 0$ and $y = 1$ are the equilibrium solutions. In the region $y > 1$ we see that $y' < 0$, hence y is decreasing in this region. To determine the concavity on this region we look at the second derivative,

$$y'' = y' - 2yy' = y'(1 - 2y) = y(1 - y)(1 - 2y).$$

If $y > 1$, we see that $y'' > 0$, so the solutions are concave up in this region. In the region $0 < y < 1$ we see that $y' > 0$, hence y is increasing in this region. In this region we see that $y'' > 0$ if $y < 1/2$ and $y'' < 0$ for $y > 1/2$. Hence in the region $0 < y < 1$, solutions are concave up for $y < 1/2$ and concave down for $y > 1/2$. In the region $y < 0$, we see that $y' < 0$ and $y'' < 0$ so that solutions are decreasing and concave down in this region. In Figure 3.2 we have used Maple to generate a phase portrait making sure to use initial conditions that give the equilibrium solutions and solutions in each of the regions by using the initial conditions $y(0) = k/2$, $k = -1, 0, 1, 2, 3$. This graph was obtained by typing and entering

```
phaseportrait(diff(y(x),x)=y(x)-(y(x))^2,y(x),x=-2..2,
    [seq([y(0)=k/2],k=-1..3)],y=-2..2,arrows=none);
```

at the command prompt. Notice how we restricted y so that $-2 \le y \le 2$. Initially, we tried to get Maple to draw these phase portraits without including this restriction on y but got an error message. After some experimentation, we found we could fix the problem by introducing this restriction. ●

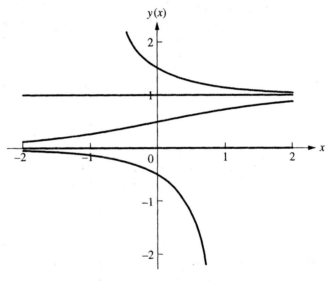

Figure 3.2

EXAMPLE 2 $y' = xy$

Solution   We see that $y = 0$ is the only equilibrium solution. In the region $y > 0$, we see that y is increasing when $x > 0$ and y is decreasing when $x < 0$. Since

$$y'' = xy' + y = x^2 y + y = y(x^2 + 1)$$

we see that $y'' > 0$ if $y > 0$. Hence solutions are concave up when $y > 0$. In the region $y < 0$, we see that y is decreasing when $x > 0$ and y is increasing when $x < 0$ and that $y'' < 0$; hence solutions are concave down on this region. These features are illustrated by the phase portraits in Figure 3.3 obtained by typing and entering

```
phaseportrait(diff(y(x),x)=x*y(x),y(x),x=-3..3,
    [seq([y(0)=k/2],k=-2..2)],y=-3..3,arrows=none);
```

at the command prompt. Maple did give us these phase portraits without including the restriction $-3 \le y \le 3$. However, the five curves were closely bunched about the origin since a scale containing large values of $|y|$ was used. We then decided to "zoom in" by restricting the size of $|y|$ obtaining a picture better distinguishing the five solutions close to the origin. ●

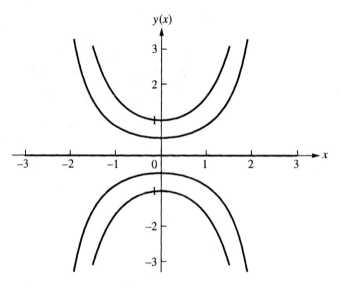

Figure 3.3

EXAMPLE 3 $y' = x^2 + y^2$

Solution If $y(x) = C$ is an equilibrium solution, then $y'(x) = 0 = x^2 + C^2$ for all x, which is impossible. Therefore, there are no equilibrium solutions. Since $y' > 0$ except at $(0, 0)$ y is always increasing. The second derivative is

$$y'' = 2x + 2yy' = 2x + 2y(x^2 + y^2) = 2(x + y(x^2 + y^2)).$$

We see that $y'' > 0$ if $x > 0$ and $y > 0$, and hence solutions are concave up in the first quadrant. Similarly, $y'' < 0$ in the third quadrant, and hence solutions are concave down in this quadrant. In the second and fourth quadrants, the concavity is determined by whether or not $x + y(x^2 + y^2)$ is positive or negative. Figure 3.4 illustrates these properties. These phase portraits were obtained by typing and entering

```
phaseportrait (diff (y(x),x),x)=x^2 +(y(x))^2,y(x),x=-
0.8..0.8, [seq (y(0)=k/2],k=-3..3)],y=-2..2,arrows=none);
```

at the command prompt. As in Example 1, we found ourselves having to include a restriction on y in order to avoid an error message. ●

EXAMPLE 4 $y' = x \sin y$

Solution The equilibrium solutions are $k\pi$ for $k = 0, \pm1, \pm2, \dots$. Phase portraits appear in Figure 3.5. No restrictions were placed on y to obtain these with Maple. Can you determine the command we used to obtain them? ●

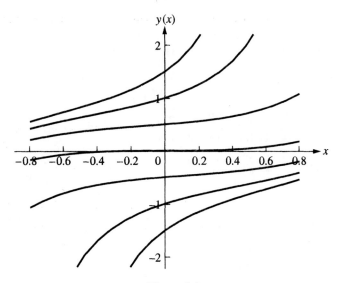

Figure 3.4

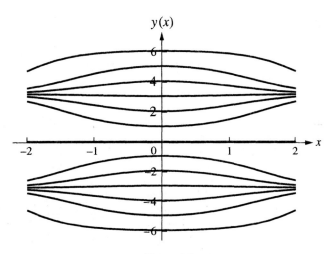

Figure 3.5

EXERCISES 3.1

In Exercises 1–4, determine the order of the differential equation.

1. $x^2 - 3y + 4(y'')^3 = 0$

2. $y'' - 4y' + 5xy''' = 2$

3. $x^2yy'' - 2x + xy' = 4y$

4. $y'' = 2y' \sin x - 3xy = 5$

In Exercises 5–8, determine whether the given function is a solution of the differential equation. If this function is a solution, determine whether it satisfies the indicated initial condition.

5. $y = e^{2x} - 2$, $y' = 2y + 4$, $y(0) = -2$

6. $y = x^2$, $y' = xy^2 - 2x$, $y(1) = 1$

7. $y = e^x$, $y'' - 2y' + y = 0$, $y(0) = -1$

8. $y = 3\cos 2x$, $y'' + 4y = 0$, $y(0) = 3$

In Exercises 9–12, determine the slopes of the tangent lines to the solutions of the differential equation at the given points. Also use Maple (or another appropriate software package) to graph a direction field for the differential equation.

9. $y' = 2xy$; $(1, 0)$, $(2, 4)$, $(-2, -2)$

10. $y' = y \cos x$; $(0, 2)$, $(3, 2)$, $(-3, -5)$

11. $y' = xy^2 + 2x$; $(1, -1)$, $(4, 2)$, $(-4, 2)$

12. $y' = xe^x y - 2 \sin x$; $(0, -3)$, $(\pi/2, 0)$, $(-\pi/4, -1)$

For the differential equations in Exercises 13–18, do the following:

1. Determine the equilibrium solutions.

2. On each region determined by the equilibrium solutions, determine when the graph of the solution is increasing, decreasing, and concave up and down.

3. Use Maple (or another appropriate software package) to sketch a phase portrait that illustrates the results of parts (1) and (2).

13. $y' = 2y + 4$

14. $y' = xy + 4x$

15. $y' = y^2 - 4y$

16. $y' = y \cos x$

17. $y' = y^3 - 9y$

18. $y' = xy - xy^3$

19. Show that both $\sqrt{1 - e^{-2x}}$ and $-\sqrt{1 - e^{-2x}}$ satisfy the initial value problem $y' = -y + y^{-1}$, $y(0) = 0$. Explain why this is not a contradiction to Theorem 3.1.

20. Show that for every real number c,

$$\phi(x) = (x + c\sqrt{x})^2$$

satisfies the initial value problem

$$y' = y/x + y^{1/2}, \; y(0) = 0.$$

Explain why this is not a contradiction to Theorem 3.1.

The *dsolve* command of Maple is used to find the general solution to a differential equation or the solution to an initial value problem with Maple. Use Maple (or another appropriate software package) to find the general solution of the differential equation or the solution to the initial value problem in Exercises 21–26.

21. $y' = 2y$ **22.** $y' - 2xy = x^3$

23. $2xy \, dx + (x^2 - y^2) \, dy = 0$

24. $(x^2 - y^2) \, dx + xy \, dy = 0$

25. $y' = xy^2 + 2x$, $y(1) = -1$

26. $2y \sin(xy) \, dx + (2x \sin(xy) + 3y^2) \, dy = 0$, $y(0) = 1$

3.2 SEPARABLE DIFFERENTIAL EQUATIONS

An ordinary differential equation that can be written in the form

$$\boxed{y' = M(x)N(y)}$$

or

$$\boxed{\frac{dy}{dx} = M(x)N(y)}$$

is called a **separable** differential equation. A separable differential equation may be solved by using integration. We first arrange the equation so that all the expressions involving y are on the left and all the expressions involving x are on the right:

$$\frac{dy}{N(y)} = M(x) \, dx.$$

Now, if we integrate each side,

$$\int \frac{dy}{N(y)} = \int M(x) \, dx,$$

we obtain an equation in x and y that implicitly gives us the solutions to the differential equation. Following are three examples illustrating this process.

EXAMPLE 1   Solve

$$y' = ay$$

where a is a constant.

Solution   Replacing y' by $\dfrac{dy}{dx}$, our equation becomes

$$\frac{dy}{dx} = ay.$$

Dividing by y and multiplying by dx, we have

$$\frac{dy}{y} = a\, dx.$$

Integrating each side,

$$\int \frac{dy}{y} = \int a\, dx,$$

gives us

$$\ln |y| = ax + C.$$

We will usually leave our answers as an equation in x and y, but let us solve for y in this example. Applying the exponent function to each side, we have

$$e^{\ln |y|} = e^{ax+C}.$$

This gives us

$$|y| = e^{ax} e^{C}.$$

Hence

$$y = \pm e^{C} e^{ax} = k e^{ax}$$

where k is the constant $\pm e^{C}$.   ●

Recall that in the last section you verified that the functions given by $Q(t) = Ce^{-kt}$ are solutions to the differential equation $Q'(t) = -kQ(t)$ describing radioactive decay. In Example 1, we solved this differential equation with x in place of t, y in place of $Q(t)$, and a in place of $-k$ obtaining $Q(t) = Ce^{-kt}$ when k is replaced by C, a is replaced by $-k$, and x is replaced by t in the answer to Example 1.

EXAMPLE 2   Solve

$$y' = xy - 4x.$$

Solution   We factor out x on the right-hand side and get

$$\frac{dy}{dx} = x(y-4)$$

from which we obtain

$$\frac{dy}{y-4} = x \, dx.$$

After integrating each side, we find

$$\ln |y-4| = \frac{x^2}{2} + C. \qquad \bullet$$

EXAMPLE 3   Show the differential equation

$$x^2 y \, dx + (y^2 - 1) \, dy = 0$$

is separable and solve it.

Solution   We can rewrite this equation as

$$(y^2 - 1) \, dy = -x^2 y \, dx$$

and then as

$$\frac{y^2 - 1}{y} \, dy = -x^2 \, dx.$$

Hence it is separable. Integrating,

$$\int \left(y - \frac{1}{y} \right) dy = \int -x^2 \, dx,$$

we obtain that our solutions are given by the equation

$$\frac{y^2}{2} - \ln |y| = -\frac{x^3}{3} + C. \qquad \bullet$$

It is not convenient to solve the equation for y in our answer to Example 3 to obtain the solutions as functions of x. Phase portraits of the solutions in Example 3 for particular values of C can be easily plotted using Maple. One way to do this is to use the *phaseportrait* command as in the last section. But there is another way when we have an equation describing the solutions as we do here—we simply have Maple graph this equation. To do so, we first have to load Maple's graphing package by typing and entering

```
with(plots);
```

at the command prompt. Next we type and enter

```
implicitplot(y^ 2/2-ln(abs(y)) + x^ 3/3=0,x=-5..5,y=-5..5);
```

and we get the graph shown in Figure 3.6 of the solution for $C = 0$. If you wish, plot some solutions for other values of C yourself.

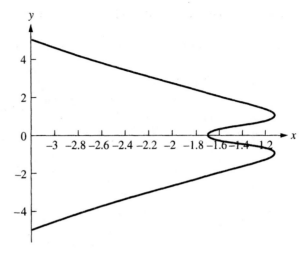

Figure 3.6

In the final example of this section, we solve an initial value problem.

EXAMPLE 4   Solve

$$y' = y \cos x - xy, \qquad y(0) = 1.$$

Solution   We change this equation to

$$\frac{dy}{y} = (\cos x - x)\, dx$$

and integrate obtaining

$$\ln |y| = \sin x - \frac{x^2}{2} + C.$$

Since our initial condition requires y to be positive, we drop the absolute value on y and have

$$\ln y = \sin x - \frac{x^2}{2} + C.$$

Substituting in the initial condition $x = 0$ and $y = 1$ gives us

$$\ln 1 = \sin 0 - 0 + C,$$

from which we see $C = 0$. The solution is given by the equation

$$\ln y = \sin x - \frac{x^2}{2}$$

●

EXERCISES 3.2

In Exercises 1–4, determine if the differential equation is separable.

1. $y' = x \cos y + 2xy^2$

2. $3xy^2 \, dx - 6 \sin y \, dy = 0$

3. $\dfrac{dy}{dx} = x^2 + 2xy$ **4.** $\sin xy' = xy + 4x$

Solve the differential equations in Exercises 5–12.

5. $y' = 2x^2y - 4y$ **6.** $e^x y \, dx + 4y^3 \, dy = 0$

7. $(x^2 + 1) \dfrac{dy}{dx} = xy^2 + x$

8. $\sec xy' - xy/(y + 4) = 0$

9. $e^x(y^2 - 4y) \, dx + 4 \, dy = 0$

10. $\dfrac{dy}{dx} = \dfrac{x^2y^3 + 3xy^3}{2x^2y}$ **11.** $t \dfrac{dy}{dt} + \dfrac{dy}{dt} = te^y$

12. $(r^2 + 1) \cos \theta \dfrac{dr}{d\theta} = r \sin \theta$

Solve the initial value problems in Exercises 13–16.

13. $y' + x^2/y = 0, \ y(0) = 1$

14. $3 \dfrac{dy}{dx} = 2xy - y, \ y(2) = 1$

15. $\dfrac{dy}{dx} = \dfrac{4xy}{y^2 + 4}, \ y(1) = 1$

16. $ye^x \, dy - \sec y \, dx = 0, \ y(0) = \pi$

17. a) Solve the equation in the answer to Exercise 13 for y.

   b) Use the result of part (a) to determine the values of x for which the solution to the initial value problem in Exercise 13 is valid.

18. a) Solve the equation in the answer to Exercise 14 for y.

   b) Use the result of part (a) to determine the values of x for which the solution to the initial value problem in Exercise 14 is valid.

19. Use Maple (or another appropriate software package) to graph the solution in Exercise 13.

20. Use Maple (or another appropriate software package) to graph the solution in Exercise 14.

21. a) Show that the equation $y' = (y + x + 1)^2$ is not separable.

   b) Show that the substitution $v = y + x + 1$ converts the differential equation in part (a) into a separable one.

   c) Solve the differential equation in part (a).

3.3   EXACT DIFFERENTIAL EQUATIONS

In the last section we learned how to solve separable differential equations. In this section we will consider a type of equation whose solutions arise from implicit differentiation. If we use the Chain Rule to differentiate the equation

$$F(x, y) = C$$

implicitly with respect to x,

$$\frac{d}{dx} F(x, y) = \frac{\partial F}{\partial x} + \frac{\partial F}{\partial y} \frac{dy}{dx} = 0. \tag{1}$$

Thus, if we have an equation of the form in (1), we know that the solutions are given by the equation $F(x, y) = C$. In differential form, the differential equation in (1) is

$$\frac{\partial F}{\partial x} dx + \frac{\partial F}{\partial y} dy = 0. \tag{2}$$

Notice that the left-hand side of Equation (2) is the total differential of F. Differential equations of this form are called **exact** differential equations. For instance, the differential equation

$$2x \cos y \, dx - (x^2 \sin y + 2y) \, dy = 0$$

is exact since its left-hand side is the total differential of

$$F(x, y) = x^2 \cos y - y^2.$$

The solutions to this differential equation are given by

$$x^2 \cos y - y^2 = C.$$

The differential equation in (2) is one of the form

$$M(x, y) \, dx + N(x, y) \, dy = 0. \tag{3}$$

We are going to study how we can tell if a differential equation of the form in (3) is exact. If Equation (3) is exact, then there is a function F of x and y so that

$$F_x(x, y) = M(x, y) \quad \text{and} \quad F_y(x, y) = N(x, y).$$

If $F_x(x, y)$, $F_y(x, y)$, $F_{xy}(x, y)$, and $F_{yx}(x, y)$ are all continuous, we have that the mixed partials $F_{xy}(x, y)$ and $F_{yx}(x, y)$ are the same. But having these mixed partials equal is the same as having

$$M_y(x, y) = N_x(x, y). \tag{4}$$

That is, if Equation (3) is exact, then Equation (4) is true. Amazingly enough, the converse of this also holds so that we have the following theorem.

THEOREM 3.2   Suppose the functions M, N, M_y, and N_x are continuous on a rectangular region Ω. Then

$$\boxed{M(x, y) \, dx + N(x, y) \, dy = 0}$$

is an exact differential equation on Ω if and only if

$$\boxed{M_y(x, y) = N_x(x, y)}$$

on Ω.

We are going to sketch the proof of the converse part of Theorem 3.2. Before doing so, we point out that the rectangular region is needed in Theorem 3.2 to guarantee the existence of antiderivatives.[3] We will assume that antiderivatives exist whenever we need them. A rigorous proof of the converse part of Theorem 3.2 that includes the construction of the necessary antiderivatives is left for other courses.

To prove the converse part, we must construct a function F so that

$$F_x(x, y) = M(x, y), \qquad F_y(x, y) = N(x, y).$$

---

[3] In fact, Theorem 3.2 holds under the weaker assumption that the region is simply connected. Roughly speaking, a simply connected region is one that is a single piece and contains no holes.

To construct F, we first integrate $F_x(x, y) = M(x, y)$ with respect to x and set

$$F(x, y) = \int M(x, y)\, dx + h(y). \qquad (5)$$

The function $h(y)$ is the constant of integration, which may involve y since y is treated as a constant in this integral. This gives us (up to a yet to be determined $h(y)$) a function F so that $F_x(x, y) = M(x, y)$. Now we need to determine $h(y)$ so that $F_y(x, y) = N(x, y)$ for the function $F(x, y)$ constructed in Equation (5). This means we must have

$$\frac{\partial F}{\partial y} = \frac{\partial}{\partial y} \int M(x, y)\, dx + h'(y) = N(x, y).$$

A theorem from calculus allows us to interchange the order of the partial derivative and integral in this equation. Doing so, we get

$$F_y(x, y) = \int M_y(x, y)\, dx + h'(y) = N(x, y).$$

Solving this equation for $h'(y)$, we have

$$h'(y) = N(x, y) - \int M_y(x, y)\, dx. \qquad (6)$$

If we show that $N(x, y) - \int M_y(x, y)\, dx$ depends only on y, then we can integrate the right-hand side of Equation (6) to get $h(y)$ and hence complete the construction of $F(x, y)$. This is done by showing that the derivative of the right-hand side of Equation (6) with respect to x is zero. Doing this differentiation, we get

$$\frac{\partial}{\partial x}(N(x, y) - \int M_y(x, y)\, dx) = N_x(x, y) - M_y(x, y),$$

which is equal to zero by our hypothesis.   ●

We now look at some examples where we use the equality of the partial derivatives as stated in Theorem 3.2 to test for exactness of the differential equations. Once we find them to be exact, we use the construction in our proof of the converse of Theorem 3.2 to find the solutions to the differential equations.

EXAMPLE 1   Solve the differential equation

$$y' = \frac{e^x \cos y - 2xy}{e^x \sin y + x^2}.$$

Solution   It is easily seen that this equation is not separable. Let us see if it is exact. We rewrite it in differential form obtaining

$$(e^x \cos y - 2xy)\, dx + (-e^x \sin y - x^2)\, dy = 0.$$

To test if this equation is exact we let

$$M(x, y) = (e^x \cos y - 2xy) \text{ and } N(x, y) = (-e^x \sin y - x^2).$$

We see that

$$\frac{\partial}{\partial y} M = -e^x \sin y - 2x \quad \text{and} \quad \frac{\partial}{\partial x} N = -e^x \sin y - 2x,$$

and therefore this equation is exact. We then set

$$F(x, y) = \int M(x, y) \, dx = \int (e^x \cos y - 2xy) \, dx = e^x \cos y - x^2 y + h(y),$$

where $h(y)$ is to be determined. Since

$$\frac{\partial}{\partial y} F = -e^x \sin y - x^2 + h'(y),$$

has to be the same as N, we have

$$-e^x \sin y - x^2 + h'(y) = -e^x \sin y - x^2.$$

Solving for $h'(y)$ gives $h'(y) = 0$. Hence we can use any constant for $h(y)$. Using $h(y) = 0$ gives

$$F(x, y) = e^x \cos y - x^2 y.$$

Therefore, solutions are given by

$$e^x \cos y - x^2 y = C$$

where C is an arbitrary constant. ●

The proof of the converse part of Theorem 3.2 may also be done by integrating with respect to y first. We leave the proof of this approach as an exercise (Exercise 18). The next example illustrates how this method looks in practice.

EXAMPLE 2 Solve the initial value problem

$$y' = \frac{2xy - \sin x - 3y^2 e^x}{6ye^x - x^2}, \qquad y(0) = 2.$$

Solution   Writing the equation in differential form gives us

$$(\sin x + 3y^2 e^x - 2xy) \, dx + (6ye^x - x^2) \, dy = 0.$$

Checking for exactness,

$$\frac{\partial}{\partial y} (\sin x + 3y^2 e^x - 2xy) = 6ye^x - 2x = \frac{\partial}{\partial x} (6ye^x - x^2)$$

and we see that our differential equation is exact. We now let

$$F(x, y) = \int (6ye^x - x^2) \, dy = 3y^2 e^x - x^2 y + g(x).$$

Since

$$\frac{\partial}{\partial x} F = 3y^2 e^x - 2xy + g'(x) = M(x, y)$$
$$= \sin x + 3y^2 e^x - 2xy,$$

we see

$$g'(x) = \sin x,$$

and hence we may use

$$g(x) = -\cos x.$$

We now have

$$F(x, y) = 3y^2 e^x - x^2 y - \cos x.$$

The solutions to the differential equation are then given by

$$3y^2 e^x - x^2 y - \cos x = C.$$

Letting $x = 0$, $y = 2$, we find $C = 11$ so that the solution to the initial value problem is given by

$$3y^2 e^x - x^2 y - \cos x = 11.$$

The graph of the solution appears in Figure 3.7.   ●

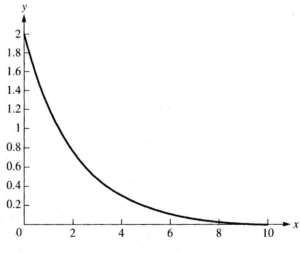

Figure 3.7

We conclude this section with an illustration of why exactness is necessary for finding the solutions to equations of the form $M(x, y) \, dx + N(x, y) \, dy = 0$ using our methods for constructing $F(x, y)$. Consider the differential equation

$$(3xy + y^2) \, dx + (x^2 + xy) \, dy = 0.$$

Notice that

$$\frac{\partial}{\partial y} M = 3x + 2y \quad \text{and} \quad \frac{\partial}{\partial x} N = 2x + y$$

so that this equation is not exact. Suppose we try to solve this equation as we did in Example 1. We have

$$F(x, y) = \int M(x, y)\, dx = \frac{3}{2}x^2 y + xy^2 + h(y).$$

Next,

$$\frac{\partial}{\partial y} F = \frac{3}{2}x^2 + 2xy + h'(y).$$

However, there is no function $h(y)$ that would make this equal to $N = x^2 + xy$, for were this the case,

$$\frac{3}{2}x^2 + 2xy + h'(y) = x^2 + xy,$$

which implies

$$h'(y) = -\frac{1}{2}x^2 - xy.$$

But this is impossible since the right-hand side depends on x and is not a function of y alone. We will see how to solve an equation such as this in Section 3.5.

EXERCISES 3.3

In Exercises 1–10, determine if the differential equation is exact. If it is exact, find its solution.

1. $(3x^2 - 4y^2)\, dx - (8xy - 12y^3)\, dy = 0$

2. $(3xy + 4y^2)\, dx + (5x^2 y + 2x^2)\, dy = 0$

3. $(2xy + ye^x)\, dx + (x^2 + e^x)\, dy = 0$

4. $(2xe^{xy} + x^2 ye^{xy} - 2)\, dx + x^3 e^{xy}\, dy = 0$

5. $(2x \cos y - x^2)\, dx + x^2 \sin y\, dy = 0$

6. $(y \cos x + 3e^x \cos y)\, dx + (\sin x - 3e^x \sin y)\, dy = 0$

7. $y' = \dfrac{1 - 2xy}{x^2}$ **8.** $y' = \dfrac{x^2 - 2xy + 1}{x^2 - y^2}$

9. $\dfrac{ds}{dt} = \dfrac{e^t - 2t \cos s}{e^s - t^2 \sin s}$ **10.** $\dfrac{dr}{d\theta} = \dfrac{\theta - r \cos \theta}{r + \sin \theta}$

In Exercises 11–14, find the solution of the initial value problem.

11. $2xy\, dx + (x^2 + 3y^2)\, dy = 0$, $y(1) = 1$

12. $y' = \dfrac{2xe^y - 3x^2 y}{x^3 - x^2 e^y}$, $y(1) = 0$

13. $2y \sin(xy)\, dx + (2x \sin(xy) + 3y^2)\, dy = 0$, $y(0) = 1$

14. $\left(1 - \dfrac{x}{x^2 + y^2}\right) dy + \dfrac{y}{x^2 + y^2}\, dx = 0$, $y(0) = 1$

15. Use Maple (or another appropriate software package) to graph the solution in Exercise 11.

16. Use Maple (or another appropriate software package) to graph the solution in Exercise 12.

17. Show a separable differential equation is exact.

18. Show the converse of Theorem 3.2 can be proved by integrating with respect to y first.

19. Determine conditions on a, b, c, and d so that the differential equation

$$y' = \frac{ax + by}{cx + dy}$$

is exact and, for a differential equation satisfying these conditions, solve the differential equation.

20. A function F of two variables x and y is said to be harmonic if $F_{xx}(x, y) + F_{yy}(x, y) = 0$. Show that the differential equation F is harmonic if and only if the differential equation

$$F_y(x, y)\, dx - F_x(x, y)\, dy = 0$$

is exact.

3.4   LINEAR DIFFERENTIAL EQUATIONS

An ordinary differential equation that can be written in the form

$$\boxed{q_1(x)y' + q_0(x)y = g(x)}$$

is called a **linear** first order differential equation. If $g(x) = 0$, the differential equation is called a **homogeneous** linear differential equation. We shall assume that the functions q_1, q_0, and g are continuous on some open interval (a, b) throughout this section. We will also assume that $q_1(x) \neq 0$ for all x in this interval (a, b). Doing so allows us to divide this differential equation by $q_1(x)$. This puts the differential equation in the form

$$\boxed{y' + p(x)y = q(x)}$$

where $p(x) = q_0(x)/q_1(x)$ and $q(x) = g(x)/q_1(x)$, which is the form we will work with as we solve first order linear differential equations in this section. These assumptions on q_1, q_0, and g also will ensure that the hypothesis of Theorem 3.1 is satisfied and that the initial value problem

$$q_1(x)y' + q_0(x)y = g(x), \qquad y(x_0) = y_0$$

with x_0 in the interval (a, b) has a unique solution on (a, b).

In the next chapter we will study higher order linear differential equations. In some ways linear differential equations are the simplest type of differential equations. The theory of linear differential equations is very rich, and we will see in the next chapter that linear differential equations have an intimate relationship with linear algebra. Indeed, this relationship with linear algebra will be used to understand and solve linear differential equations.

We now develop a technique that can be used to solve any first order linear differential equation that has been written in the form

$$y' + p(x)y = q(x)$$

(up to a couple of integrals). The technique hinges on the fact that we can put linear differential equations in a form where we can integrate both sides of the equation with respect to x.

If we differentiate uy where u and y are functions of x, we get

$$\frac{d}{dx}(uy) = uy' + u'y.$$

Multiplying the linear differential equation

$$y' + p(x)y = q(x)$$

by u gives us

$$uy' + up(x)y = uq(x).$$

Notice that the left-hand side of this equation is similar to the derivative of the product uy. As a matter of fact, they will be equal if

$$u' = up(x).$$

This is a separable differential equation in u. Solving it:

$$\frac{du}{dx} = up(x)$$

$$\frac{du}{u} = p(x)\, dx$$

$$\ln u = \int p(x)\, dx$$

$$\boxed{u = e^{\int p(x)\, dx}.}$$

With this u our equation takes on the form

$$\boxed{\frac{d}{dx}(uy) = uq(x).}$$

Integrating this equation with respect to x, we obtain

$$\boxed{uy = \int uq(x)\, dx + C.}$$

Solving for y,

$$\boxed{y = u^{-1}\int uq(x)\, dx + Cu^{-1}.}$$

We call the function u an **integrating factor** because it allows us to solve the differential equation by integrating.

Using our formulas for u and y, we can write the solutions to a linear differential equation as

$$\boxed{y(x) = e^{-\int p(x)\, dx}\int e^{\int p(x)\, dx}q(x)\, dx + Ce^{-\int p(x)\, dx}.}$$

In general, we will not use this formula. Instead, we will determine the integrating factor u, multiply the differential equation by u, and then integrate. We illustrate the procedure with some examples.

EXAMPLE 1   Solve the differential equation

$$y' = 2y + x.$$

Solution   Expressing this differential equation in the form

$$y' - 2y = x,$$

we see $p(x) = -2$. Therefore, an integrating factor is

$$u = e^{\int -2\,dx} = e^{-2x}.$$

Multiplying by this integrating factor gives us

$$e^{-2x}y' - 2e^{-2x}y = xe^{-2x}.$$

Since the left-hand side of this equation is

$$\frac{d}{dx}(uy) = \frac{d}{dx}(e^{-2x}y) = e^{-2x}y' - 2e^{-2x}y,$$

we have

$$\frac{d}{dx}(e^{-2x}y) = xe^{-2x}.$$

Thus

$$e^{-2x}y = \int xe^{-2x}\,dx + C.$$

Using integration by parts on the right-hand side we have

$$e^{-2x}y = -\frac{1}{2}xe^{-2x} - \frac{1}{4}e^{-2x} + C.$$

Multiplying through by $u^{-1} = e^{2x}$ we see that

$$y = -\frac{1}{2}x - \frac{1}{4} + Ce^{2x}.$$

●

EXAMPLE 2   Solve the differential equation

$$y' - \frac{1}{x}y = 0$$

for $x > 0$.

Solution   An integrating factor is

$$u = e^{\int -\frac{1}{x}\,dx} = e^{-\ln x} = \frac{1}{x}.$$

Multiplying through by this integrating factor gives us

$$\frac{1}{x}y' - \frac{1}{x^2}y = 0.$$

Therefore

$$\frac{d}{dx}\left(\frac{1}{x}y\right) = 0$$

so that

$$\frac{1}{x}y = C \quad \text{or} \quad y = Cx. \qquad \bullet$$

You might notice that the differential equation in Example 2 can also be solved as a separable equation.

EXAMPLE 3   Solve the initial value problem

$$y' = xy - x, \qquad y(1) = 2.$$

Solution   We see that this equation is both separable and linear, but we will solve it as a linear equation. Writing the equation as

$$y' - xy = -x,$$

$$u = e^{\int -x \, dx} = e^{-x^2/2}.$$

Multiplying by the integrating factor u gives us

$$\frac{d}{dx}(e^{-x^2/2}y) = -xe^{-x^2/2}.$$

Thus

$$e^{-x^2/2}y = e^{-x^2/2} + C$$

and

$$y = 1 + Ce^{x^2/2}.$$

Substituting in the initial condition, we find $C = e^{-1/2}$. (Try it.) The solution is then

$$y = 1 + e^{-1/2}e^{x^2/2}. \qquad \bullet$$

EXAMPLE 4   Solve the initial value problem

$$xy' + 2y = 3x, \qquad y(1) = 0.$$

Solution   Dividing by x gives us the differential equation

$$y' + \frac{2}{x}y = 3.$$

An integrating factor is

$$e^{\int \frac{2}{x} \, dx} = x^2$$

and multiplying by it gives us

$$\frac{d}{dx}(x^2 y) = 3x^2.$$

Integrating and solving for y, we obtain

$$y = x + Cx^{-2}.$$

Using the initial condition, we find $C = -1$ and consequently the solution to the initial value problem is

$$y = x - x^{-2}.$$

●

EXAMPLE 5   Solve the initial value problem

$$y' \cos x + xy \cos x = \cos x, \qquad y(0) = 1.$$

Solution   Upon dividing by $\cos x$, we obtain

$$y' + xy = 1.$$

We find that an integrating factor is

$$e^{\int x\, dx} = e^{x^2/2}.$$

Multiplying the differential equation by it gives us

$$\frac{d}{dx}(e^{x^2/2} y) = e^{x^2/2}.$$

Since there is no closed form for an antiderivative of $e^{x^2/2}$, we will use an antiderivative of the form $F(x) = \int_a^x f(t)\, dt$ for $e^{x^2/2}$:

$$\int_0^x e^{t^2/2}\, dt.$$

We now have

$$\frac{d}{dx}(e^{x^2/2} y) = \frac{d}{dx}\int_0^x e^{t^2/2}\, dt,$$

so that

$$e^{x^2/2} y = \int_0^x e^{t^2/2}\, dt + C.$$

Using the initial condition $y(0) = 1$ gives us

$$y(0) = \int_0^0 e^{t^2/2}\, dt + C = 0 + C = C = 1.$$

Therefore we have

$$e^{x^2/2}y = \int_0^x e^{t^2/2}\, dt + 1,$$

or

$$y = e^{-x^2/2}\int_0^x e^{t^2/2}\, dt + e^{-x^2/2}.$$ ●

We can use Maple to graph the solution to Example 5 by typing and entering

```
plot(exp(-x^2/2)*int(exp(t^2/2),t=0..x)+exp(-x^2/2),x=0..5);
```

This graph appears in Figure 3.8.

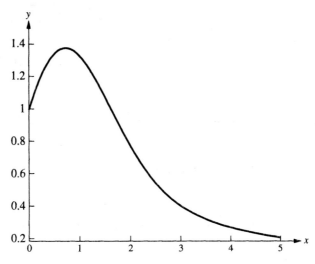

Figure 3.8

EXERCISES 3.4

In Exercises 1–12, determine the integrating factor and solve the differential equation.

1. $y' + y/x^2 = 0$

2. $y' = y/x - 2, x > 0$

3. $y' - 2xy = x$

4. $y' = 4y + 2x$

5. $y' = 1 + \dfrac{y}{1+2x}, x > 0$

6. $y' = x^2 - \dfrac{2xy}{1+x^2}$

7. $xy' + y = x^2, x > 0$

8. $xy' + y = x^2, x < 0$

9. $(1+xy)\, dx - x^2\, dy = 0, x < 0$

10. $(1+xy)\, dx - x^2\, dy = 0, x > 0$

11. $\dfrac{dy}{dt} + e^t y = e^t$

12. $\dfrac{dr}{d\theta} = r\tan\theta + \sin\theta, 0 < \theta < \pi/2$

In Exercises 13–18, solve the initial value problems.

13. $y' + 4y = 2, y(1) = 2$

14. $2xy' + y = 1, y(4) = 0$

15. $y' + \dfrac{y}{x+1} = 2, y(0) = 2$

16. $y' + 2xy = 1, y(0) = -1$

17. $y' = \cos 2x - y/x, y(\pi/2) = 0$

18. $x(x+1)y' = 2 + y, y(1) = 0$

19. Use Maple (or another appropriate software package) to graph the solution in Exercise 13.

20. Use Maple (or another appropriate software package) to graph the solution in Exercise 14.

21. Show that the solution to the linear first order initial value problem

$$y' + p(x)y = q(x), \qquad y(x_0) = y_0$$

is given by

$$y(x) = e^{-\int_{x_0}^{x} p(t)\, dt} \int_{x_0}^{x} u(t)q(t)\, dt + y_0 e^{-\int_{x_0}^{x} p(t)\, dt}$$

where $u(t) = e^{\int_{x_0}^{t} p(s)\, ds}$.

22. Use the result of Exercise 21 to do Exercise 13.

3.5   MORE TECHNIQUES FOR SOLVING FIRST ORDER DIFFERENTIAL EQUATIONS

In the previous three sections we considered three types of first order differential equations: separable equations, exact equations, and linear equations. If a first order equation is not of one of these types, there are methods that sometimes can be used to convert the equation to one of these three types, which then gives us a way of solving the differential equation. In this section we will give a sampling of some of these techniques.

The first approach we consider involves attempting to convert a differential equation

$$\boxed{r(x, y)\, dx + s(x, y)\, dy = 0} \tag{1}$$

that is not exact into an exact one. The idea is to try to find a function I of x and y so that when we multiply our differential equation in (1) by $I(x, y)$ obtaining

$$\boxed{I(x, y)r(x, y)\, dx + I(x, y)s(x, y)\, dy = 0,} \tag{2}$$

we have an exact differential equation.

In the previous section we multiplied the linear differential equation

$$y' + p(x)y = q(x)$$

by $u = e^{\int p(x)\, dx}$ because it made the left-hand side of the linear differential equation $d/dx(uy)$. Since we could now integrate this derivative, we called u an integrating factor. In the same spirit, we will refer to $I(x, y)$ as an **integrating factor** for the differential equation in (1) if it converts Equation (1) into Equation (2) with a left-hand side that is a total differential. (Actually, multiplying the linear equation $y' + p(x)y = q(x)$ by u does make the differential equation into an exact one—see Exercise 7. Thus our old meaning of integrating factor is a special case of its new meaning.)

If we let

$$M(x, y) = I(x, y)r(x, y)$$

and

$$N(x, y) = I(x, y)s(x, y)$$

in Theorem 3.2, for exactness we must have

$$\frac{\partial}{\partial y} I(x, y)r(x, y) = \frac{\partial}{\partial x} I(x, y)s(x, y). \tag{3}$$

Numerous techniques have been devised in attempts to find integrating factors that satisfy the above condition. One ploy is to try to find an integrating factor of the form

$$I(x, y) = x^m y^n$$

where m and n are constants. The following example illustrates this approach.

EXAMPLE 1   Solve the differential equation

$$(x^2 y + y^2)\, dx + (x^3 + 2xy)\, dy = 0.$$

Solution   It is easily seen that this equation is not exact. Let us see if we can find an integrating factor of the form $I(x, y) = x^m y^n$. Multiplying the differential equation by this, we have the differential equation

$$x^m y^n [(x^2 y + y^2)\, dx + (x^3 + 2xy)\, dy] = 0$$

or

$$(x^{m+2} y^{n+1} + x^m y^{n+2})\, dx + (x^{m+3} y^n + 2x^{m+1} y^{n+1})\, dy = 0. \qquad \textbf{(4)}$$

With

$$M(x, y) = x^{m+2} y^{n+1} + x^m y^{n+2} \quad \text{and} \quad N(x, y) = x^{m+3} y^n + 2x^{m+1} y^{n+1},$$

Equation (3) tells us that we must have

$$M_y(x, y) = (n + 1)x^{m+2} y^n + (n + 2)x^m y^{n+1}$$

the same as

$$N_x(x, y) = (m + 3)x^{m+2} y^n + 2(m + 1)x^m y^{n+1}$$

for Equation (4) to be exact. At this point we make the observation that $M_y(x, y)$ and $N_x(x, y)$ will be the same if the coefficients of the $x^{m+2} y^n$ terms are the same and the coefficients of the $x^m y^{n+1}$ terms are the same in these two expressions. This gives us the system of linear equations

$$n + 1 = m + 3$$
$$n + 2 = 2(m + 1),$$

which you can easily solve finding

$$m = 2 \quad \text{and} \quad n = 4.$$

Substituting these values of m and n into Equation (4), we have the exact equation

$$(x^4 y^5 + x^2 y^6)\, dx + (x^5 y^4 + 2x^3 y^5)\, dy = 0.$$

Let us now solve this exact equation. We have

$$F(x, y) = \int (x^4 y^5 + x^2 y^6) \, dx + h(y)$$

$$= \frac{x^5 y^5}{5} + \frac{x^3 y^6}{3} + h(y)$$

and

$$F_y(x, y) = x^5 y^4 + 2x^3 y^5 + h'(y) = x^5 y^4 + 2x^3 y^5$$

so that

$$h'(y) = 0$$

and

$$h(y) = 0.$$

Our solutions are then given by

$$\frac{x^5 y^5}{5} + \frac{x^3 y^6}{3} = C.$$

●

Do not expect the technique of Example 1 to always work. Indeed, if you were to try to apply it to the equation

$$(x^2 + y^2 + 1) \, dx + (xy + y) \, dy = 0,$$

you would find that the resulting system of equations in n and m obtained by comparing coefficients as we did in Example 1 has no solution. (Try it.) There is, however, a way of finding an integrating factor for this equation. It involves looking for one that is a function of x only; that is, we look for an integrating factor of the form

$$\boxed{I(x, y) = f(x).}$$

Indeed, Equation (3) tells us the differential equation is exact if and only if

$$\frac{\partial}{\partial y}(f(x)r(x, y)) = \frac{\partial}{\partial x}(f(x)s(x, y))$$

or

$$f(x)r_y(x, y) = f(x)s_x(x, y) + f'(x)s(x, y).$$

Solving this equation for $f'(x)$, we have

$$f'(x) = \frac{r_y(x, y) - s_x(x, y)}{s(x, y)} f(x)$$

or, using u for $f(x)$,

$$\boxed{\frac{du}{dx} = \frac{r_y(x, y) - s_x(x, y)}{s(x, y)} u.} \tag{5}$$

If the expression

$$\boxed{\frac{r_y(x, y) - s_x(x, y)}{s(x, y)}} \tag{6}$$

is a function of x only, Equation (5) is a separable equation that can be solved for $u = f(x)$. In other words, we have just arrived at the following procedure for finding an integrating factor that is a function of x:

1. Calculate the expression in (6).
2. If the expression in (6) is a function of x only, solve the separable differential equation in (5) to obtain an integrating factor $u = f(x)$.
3. Use this integrating factor to solve the differential equation.

Of course, our procedure does not apply if the expression in (6) involves y. A similar approach can be used to attempt to obtain integrating factors that are functions of y only. See Exercise 8 for the details.

Let us use the procedure we have just obtained to solve a differential equation.

EXAMPLE 2 Solve the differential equation

$$(x^2 + y^2 + 1) \, dx + (xy + y) \, dy = 0.$$

Solution With

$$r(x, y) = x^2 + y^2 + 1 \quad \text{and} \quad s(x, y) = xy + y,$$

we have

$$\frac{r_y(x, y) - s_x(x, y)}{s(x, y)} = \frac{2y - y}{xy + y} = \frac{1}{x + 1}$$

is a function of x only. Now we solve the separable equation

$$\frac{du}{dx} = \frac{1}{x + 1} \cdot u$$

for the integrating factor u. This gives

$$\frac{du}{u} = \frac{1}{x + 1} \, dx$$

$$\ln u = \ln(x + 1)$$

or

$$u = x + 1.$$

(We leave off the constant of integration since we only have to find one solution for u to obtain an integrating factor.) Multiplying the differential equation in the statement of this example by $u = x + 1$ we have the exact equation

$$(x^3 + xy^2 + x + x^2 + y^2 + 1)\, dx + (x^2 y + 2xy + y)\, dy = 0.$$

Solving this exact equation, we find that our solutions are given by the equation

$$\frac{x^4}{4} + \frac{x^2 y^2}{2} + \frac{x^2}{2} + \frac{x^3}{3} + xy^2 + x + \frac{y^2}{2} = C. \qquad\bullet$$

We next consider a type of equation that can be converted to a separable equation. A first order equation of the form

$$\boxed{M(x, y)\, dx + N(x, y)\, dy = 0}$$

is said to have **homogeneous coefficients**[4] if

$$\boxed{\frac{M(ax, ay)}{N(ax, ay)} = \frac{M(x, y)}{N(x, y)}} \qquad (7)$$

for all values of a for which these fractional expressions are defined. The differential equation

$$(x^2 + y^2)\, dx + xy\, dy = 0$$

is an example of an equation with homogeneous coefficients since

$$\frac{(ax)^2 + (ay)^2}{(ax)(ay)} = \frac{x^2 + y^2}{xy}.$$

Suppose the equation

$$M(x, y)\, dx + N(x, y)\, dy = 0$$

has homogeneous coefficients. To solve it, we first write it in the form

$$\boxed{\frac{dy}{dx} = -\frac{M(x, y)}{N(x, y)}.} \qquad (8)$$

Using x for a in Equation (7), we may express $M(x, y)/N(x, y)$ as

$$\frac{M(x, y)}{N(x, y)} = \frac{M(x \cdot 1, x(y/x))}{N(x \cdot 1, x(y/x))} = \frac{M(1, y/x)}{N(1, y/x)}, \qquad (9)$$

---

[4] The word homogeneous is used in many different senses throughout mathematics. Here it has a different meaning than it does with respect to the linear equations of Chapter 1 or the homogeneous linear differential equations that were mentioned in Section 3.4.

and hence our differential equation has the form

$$\frac{dy}{dx} = -\frac{M(1, y/x)}{N(1, y/x)}.$$

In other words, our differential equation can be expressed as one where we have dy/dx as a function of y/x:

$$\frac{dy}{dx} = f\left(\frac{y}{x}\right).$$ (10)

Let us set

$$\frac{y}{x} = u.$$ (11)

Since $y = ux$,

$$\frac{dy}{dx} = u + x\,\frac{du}{dx}.$$ (12)

Substituting the results of Equations (11) and (12) into Equation (10), we have

$$u + x\,\frac{du}{dx} = f(u).$$

This is a separable equation in u and x which we can solve for u and then use the fact that $y = ux$ to find our solution for y. The following example illustrates the procedure we have just obtained.

EXAMPLE 3   Solve the differential equation

$$(x^2 - y^2)\,dx + xy\,dy = 0.$$

Solution   Writing our equation in the form of Equation (8), we have

$$\frac{dy}{dx} = -\frac{x^2 - y^2}{xy} = \frac{y^2 - x^2}{xy}.$$

We could apply the steps in Equation (9) to write the right-hand side of this equation as a function of y/x, but notice that we can more easily get this by first doing term-by-term division by xy on the right-hand side obtaining

$$\frac{dy}{dx} = \frac{y}{x} - \frac{x}{y} = \frac{y}{x} - \frac{1}{\frac{y}{x}}.$$

Making the substitutions from Equations (11) and (12), we arrive at the differential equation

$$u + x \frac{du}{dx} = u - \frac{1}{u}.$$

Let us now solve this separable equation for u:

$$x \frac{du}{dx} = -\frac{1}{u}$$

$$u \, du = -\frac{dx}{x}$$

$$\frac{u^2}{2} = -\ln |x| + C$$

$$u^2 = \ln x^{-2} + 2C.$$

Since $u = y/x$, the solutions are given by

$$\frac{y^2}{x^2} = \ln x^{-2} + K$$

where K is a constant.   ●

The differential equation in Example 3 also may be solved by using the techniques of Examples 1 or 2. Try doing it these two other ways. Among the three methods for solving it, which do you find the easiest to use?

For our final technique, we consider a type of equation that can be converted to a linear differential equation. An equation of the form

$$\boxed{y' + p(x)y = q(x)y^n}$$

is called a **Bernoulli** equation (named after Jakoub Bernoulli, 1654–1705). If $n = 0, 1$ the equation is linear. For other values of n, this equation can be converted to a linear differential equation by making the substitution

$$\boxed{v = y^{1-n}.} \tag{13}$$

To see why, first note that since $y = v^{1/(1-n)}$, we have

$$y' = \frac{1}{1-n} v^{n/(1-n)} v'.$$

Substituting this into the Bernoulli equation gives us

$$\frac{1}{1-n} v^{n/(1-n)} v' + p(x)v^{1/(1-n)} = q(x)v^{n/(1-n)}.$$

Dividing by $v^{n/(1-n)}/(1-n)$, we have the linear differential equation

$$\boxed{v' + (1-n)p(x)v = (1-n)q(x).} \tag{14}$$

EXAMPLE 4   Solve the differential equation

$$y' + xy = xy^2.$$

Solution   Making the substitution in Equation (13), which is

$$v = y^{-1}$$

here, our differential equation becomes

$$v' - xv = -x$$

by Equation (14). Solving this linear equation, we find

$$v = 1 + Ce^{x^2/2}$$

and hence

$$y = v^{-1} = \frac{1}{1 + Ce^{x^2/2}}.$$ ●

EXERCISES 3.5

In each of Exercises 1–6, find integrating factors for the differential equation and then solve it.

1. $(x^2 + y^2)\,dx - xy\,dy = 0$

2. $(2y^2 - xy)\,dx + (xy - 2x^2)\,dy = 0$

3. $(x - 3x^2y)\dfrac{dy}{dx} = xy^2 + y$

4. $(x^2 + y^2)y' = 2xy$

5. $(2x^2 + 3y^2 - 2)\,dx - 2xy\,dy = 0$

6. $(3x + 2e^y)\,dx + xe^y\,dy = 0$

7. Show that multiplying the differential equation

$$y' + p(x)y = q(x) \text{ by } e^{\int p(x)\,dx}$$

converts the differential equation into an exact one.

8. Show that if

$$\frac{s_x(x, y) - r_y(x, y)}{r(x, y)}$$

is a function of y only, then a solution to the separable differential equation

$$\frac{du}{dy} = \frac{s_x(x, y) - r_y(x, y)}{r(x, y)}u$$

is an integrating factor for the differential equation

$$r(x, y)\,dx + s(x, y)\,dy = 0.$$

Use the result of Exercise 8 to find an integrating factor and solve the differential equations in Exercises 9 and 10.

9. $(xy + x)\,dx + (x^2 + y^2 - 1)\,dy = 0$

10. $y\,dx + (2x - ye^y)\,dy = 0$

In each of Exercises 11–14, show that the differential equation has homogeneous coefficients and then solve it.

11. $(x^2 - y^2)\,dx + xy\,dy = 0$

12. $(x^2 + y^2)\,dx + xy\,dy = 0$

13. $(y + xe^{y/x})\,dx - x\,dy = 0$

14. $\left(x\cos\dfrac{y}{x} - y\sin\dfrac{y}{x}\right)dx + x\sin\dfrac{y}{x}\,dy = 0$

15. Show that if the differential equation

$$M(x, y)\,dx + N(x, y)\,dy = 0$$

has homogeneous coefficients, then

$$I(x, y) = \frac{1}{xM(x, y) + yN(x, y)}$$

is an integrating factor for the differential equation.

16. Use the result of Exercise 15 to solve the differential equation in Exercise 11.

Solve the differential equations in Exercises 17 and 18.

17. $xy' + y = y^2$ **18.** $y' = 2y - y^3$

3.6   MODELING WITH DIFFERENTIAL EQUATIONS

In this section we will look at several different phenomena that can be described by differential equations. We begin with some that arise from exponential functions.

No doubt you have seen a number of places where exponential functions are used to model phenomena. As an illustration, suppose a type of bacteria divide every hour and a petri dish intially contains $N(0)$ of these bacteria. Assuming there are no factors restricting the continued growth of the bacteria, after 1 hour the petri dish will contain $N(0) \cdot 2$ bacteria, after 2 hours it will contain $N(0) \cdot 2^2$ bacteria, after 3 hours it will contain $N(0) \cdot 2^3$ bacteria, and after t hours it will contain

$$N = N(0)2^t = N(0)e^{t \ln 2}$$

bacteria.

A similar analysis may be used to get a formula modeling the amount of a radioactive substance in terms of its half-life.[5] To illustrate, suppose that we have a sample containing radioactive carbon-14, which has a half-life of about 5730 years. By considering the amounts of radioactive carbon-14 after 5730 years, $2 \cdot 5730$ years, $3 \cdot 5730$ years, and so on, you should be able to convince yourself that the quantity Q of radioactive carbon-14 in the sample after t years is given approximately by

$$Q = Q(0) \left(\frac{1}{2} \right)^{t/5730} = Q(0)e^{-(\ln 2)t/5730}$$

where $Q(0)$ is the initial quantity of radioactive carbon-14 present in the sample.

As a final elementary illustration of an exponential model, consider compounded interest. If $A(0)$ dollars are invested at 5% interest compounded yearly, the formula for the value of the investment after t years is

$$A = A(0)(1 + 0.05)^t = A(0)e^{t \ln 1.05}.$$

(Why?) If interest is compounded quarterly instead of yearly, the formula becomes

$$A = A(0) \left(1 + \frac{0.05}{4} \right)^{4t} = A(0)e^{4t \ln 1.0125}.$$

(Why?) If interest is compounded daily (ignoring leap years), the formula becomes

$$A = A(0) \left(1 + \frac{0.05}{365} \right)^{365t} = A(0)e^{365t \ln(1+0.05/365)}.$$

(Why?) In general, if $A(0)$ dollars is invested at r% interest compounded k times per year, the formula for the value of the investment after t years is

$$A = A(0) \left(1 + \frac{r}{100k} \right)^{kt} = A(0)e^{kt \ln(1+r/(100k))}.$$

---

[5] The **half-life** of a radioactive substance is the amount of time it takes for the substance to decay to half of its original amount.

If we let k approach infinity in this formula,

$$\lim_{k \to \infty} A(0)e^{kt \ln(1+r/(100k))},$$

an application of l'Hôpital's rule gives us the formula for **continuously compounded interest**:

$$A = A(0)e^{rt/100}. \tag{1}$$

The population, radioactive decay, and compound interest examples we have just considered are all modeled by functions of the form

$$\boxed{y = Ce^{kt}} \tag{2}$$

where $C > 0$ and k are constants. In these models, we say we have **exponential growth** if $k > 0$ and **exponential decay** if $k < 0$. The derivative of the function in Equation (2) is $y' = kCe^{kt}$ or

$$\boxed{y' = ky.} \tag{3}$$

Of course, now you know that if you solve the differential equation in (3) you find its solutions are given by Equation (2). In effect, we have found ourselves on a two-way street. At the beginning of this section we applied common sense to obtain the exponential models of the form in Equation (2) from which we get differential equations of the form in Equation (3). We could equally well start with the differential equations in the form of Equation (3) and solve them to obtain the exponential models in Equation (2).

You may well feel that the approach we used at the beginning of this section to arrive at exponential models is more natural than the differential equations approach and wonder why we bother with the differential equation approach. One reason for taking the differential equation point of view is that there are many instances where we have to adjust the rate of change of y in order to model the problem at hand. Consider the following example.

EXAMPLE 1   Suppose that an investment program pays 6% interest a year compounded continuously. If an investor initially invests $1000 and then contributes $500 a month at a continuous rate in this account, how much will have accumulated in her account after 30 years? Also, how much of this will be interest?

Solution   Were it not for the contributions, Equation (1) gives us that this investor would have

$$A = A(0)e^{rt/100} = 1000e^{0.06t}$$

in her account after t years and the annual rate of growth of her investment would be governed by the differential equation

$$A' = (0.06)1000e^{0.06t} \quad \text{or} \quad A' = 0.06A.$$

In this problem, we adjust the growth rate by noting that her continous contributions of $500 per month increase her investment's growth rate by $12 \cdot \$500 = \$6000$ per year. Hence we can model the amount A of her investment after t years by the differential equation

$$A' = 0.06A + 6000,$$

with initial condition

$$A(0) = 1000.$$

This differential equation is linear. Solving it, we find

$$A = Ce^{0.06t} - \frac{6000}{0.06} = Ce^{0.06t} - 100,000.$$

From the initial condition we get

$$C = 101,000$$

so that

$$A = 101,000e^{0.06t} - 100,000.$$

The graph of the solution to the initial value problem in this example appears in Figure 3.9. The amount in her account after 30 years is

$$A(30) \approx \$511,014.39$$

(rounded to the nearest cent). After 30 years she will have contributed $1000 + 30 \cdot \$6000 = \$181, 000$, and hence she will have received $330,104.39 in interest (rounded to the nearest cent). ●

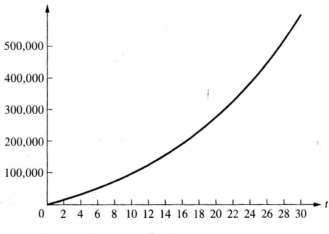

Figure 3.9

Our next model is a type of problem called a **mixture problem.**

EXAMPLE 2   Ten pounds of salt is dissolved in a 200-gallon tank containing 100 gallons of water. A saltwater solution containing 1 lb salt/gal is then poured into the tank at a rate of 2 gal/min. The tank is continuously well-stirred and drained at a rate of 1 gal/min. How much salt is in the tank after a half hour? How much salt is in the tank when the tank overfills?

Solution   Figure 3.10 illustrates this problem.

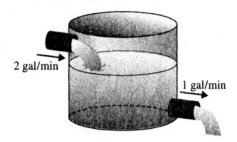

2 gal/min

1 gal/min

Figure 3.10

Note that the volume of salt water in the tank is increasing at a rate of 1 gal/min. (Why?) Therefore, the volume of salt water in the tank after t min is $(100 + t)$ gal. (Why?) The amount of salt in the tank is also changing. To solve this problem, we first determine the differential equation describing the rate of change of the salt.

The rate of pounds of salt entering the tank is given by

$$1 \frac{\text{lb}}{\text{gal}} \cdot 2 \frac{\text{gal}}{\text{min}} = 2 \frac{\text{lb}}{\text{min}}.$$

The amount of salt leaving the tank is determined in a similar manner. Denote the number of pounds of salt in the tank at time t by $S(t)$. Then

$$\frac{S(t)}{100 + t} \frac{\text{lb}}{\text{gal}}$$

is the concentration of salt in the tank and is also the concentration of salt that will leave the tank. Multiplying this concentration by 1 gal/min gives us

$$\frac{S(t)}{100 + t} \frac{\text{lb}}{\text{gal}} \cdot 1 \frac{\text{gal}}{\text{min}} = \frac{S(t)}{100 + t} \frac{\text{lb}}{\text{min}}$$

for the rate of pounds of salt leaving the tank every minute. Since the rate of change of salt in the tank is given by the rate of salt entering the tank minus the rate of salt leaving the tank,

$$S'(t) = 2 - \frac{S(t)}{100 + t}.$$

This is a linear differential equation. Its general solution is

$$S(t) = (100 + t) + \frac{C}{100 + t}.$$

Using the initial condition $S(0) = 10$ gives us $C = -9000$ so that

$$S(t) = 100 + t - \frac{9000}{100 + t}.$$

The graph of this solution to our initial value problem appears in Figure 3.11. After a half hour, $S(30) = 130 - 900/13$ lb salt are in the tank. Noting that the tank overfills at $t = 100$ min, $S(100) = 155$ lb salt are in the tank when it overfills. ●

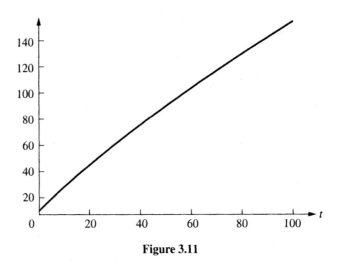

Figure 3.11

We next look at a cooling problem.

EXAMPLE 3   Newton's law of cooling states that a body of uniform composition when placed in an environment with a constant temperature will approach the temperature of the environment at a rate that is proportional to the difference in temperature of the body and the environment. A turkey has been in a refrigerator for several days and has a uniform temperature of 40° F. An oven is preheated to 325° F. The turkey is placed in the oven for 20 minutes and then taken out and its temperature is found to be 60° F. Approximately how long does the turkey have to stay in the oven to have a temperature of 185° F?

Solution   Let θ be the temperature of the turkey at time t and assume the turkey has uniform composition. By Newton's law of cooling

$$\theta' = k(325 - \theta).$$

We can solve this equation either as a separable or linear differential equation. Doing either, we find the solutions are

$$\theta = Ce^{-kt} + 325.$$

Using $\theta(0) = 40$, we find

$$C = -285.$$

Then, using $\theta(20) = 60$, we get

$$k = \frac{\ln(57/53)}{20}.$$

Therefore,

$$\theta = -285e^{-\frac{\ln(57/53)}{20}t} + 325.$$

The graph of this function appears in Figure 3.12. To finish the problem, we need to find t so that

$$\theta = -285e^{-\frac{\ln(57/53)}{20}t} + 325 = 185.$$

Solving for t we find

$$t = -20\frac{\ln(28/57)}{\ln(57/53)} \approx 195.397 \text{ minutes},$$

or about 3.26 hours.   ●

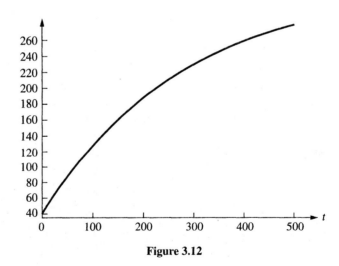

Figure 3.12

The final two examples of this section involve the motion of an object.

EXAMPLE 4   Newton's law of momentum says that the force acting on a body is equal to its mass times acceleration. An object falling in the earth's atmosphere is subject to the force of gravity and the friction of air. Assuming that the force due to friction of air is proportional to the velocity of the object, determine the speed at which an object dropped from rest will hit the ground.

Solution   Let v be the velocity of the object. Then dv/dt is the acceleration of the object. If we let M be the mass of the object and g represent the acceleration due to gravity, then

Newton's law of momentum gives us

$$\text{Force} = M\,\frac{dv}{dt} = Mg - kv$$

where $k > 0$ is the friction constant. (The constant k is positive because friction opposes the motion.) This equation is both separable and linear. Solving it either way gives

$$v = ce^{-\frac{k}{M}t} + \frac{Mg}{k}.$$

Using the initial condition $v(0) = 0$, we find that

$$v = \frac{Mg}{k}(1 - e^{-\frac{k}{M}t}). \qquad\qquad \bullet$$

EXAMPLE 5   Newton's law of gravity states that if the radius of a planet is R and if x is the distance of an object of mass M in space from the planet, then the weight of the object is given by

$$w = \frac{MgR^2}{(R+x)^2}$$

where g is the acceleration due to the gravity of the planet. Suppose an object of mass M is projected upward from the earth's surface with initial velocity v_0. Determine the velocity and the maximum distance from the earth of the object.

Solution   The differential equation (ignoring air friction) describing the velocity of the object is

$$M\,\frac{dv}{dt} = -w = -\frac{MgR^2}{(R+x)^2}$$

where the minus sign indicates that the weight is opposing the upward motion of the object. From the Chain Rule we know that

$$\frac{dv}{dt} = \frac{dv}{dx}\frac{dx}{dt} = v\frac{dv}{dx}.$$

Using this our differential equation takes on the form

$$v\,\frac{dv}{dx} = -\frac{gR^2}{(R+x)^2}.$$

This is a separable equation whose solutions are given by

$$\frac{v^2}{2} = \frac{gR^2}{R+x} + C.$$

Since $x = 0$ when $t = 0$ we have

$$\frac{v_0^2}{2} = \frac{gR^2}{R} + C,$$

which gives us $C = v_0^2/2 - gR$. Consequently,

$$v = \pm \left(v_0^2 - 2gR + \frac{2gR^2}{R + x} \right)^{1/2}.$$

As the object is rising, v must be positive and hence

$$v = \left(v_0^2 - 2gR + \frac{2gR^2}{R + x} \right)^{1/2}.$$

Since the object reaches its maximum distance from the earth when $v = 0$, we obtain

$$x = \frac{v_0^2 R}{2gR - v_0^2}$$

as the maximum distance from the earth achieved by the object.   ●

You will need to use the results of Examples 4 and 5 in Exercises 13–16. A few of the many other applications involving first order differential equations not covered in the examples of this section can be found in Exercises 17–22.

EXERCISES 3.6

1. After his company goes public, an entrepreneur retires at the age of 32 with a retirement portfolio worth $2 million. If his portfolio grows at a rate of 10% per year compounded continuously and he withdraws $250,000 per year at a continuous rate to live on, how old will he be when he runs out of money?

2. A company offers a retirement plan for its employees. If the retirement plan has a yearly interest rate of 10% compounded continuously and an employee continuously invests in this plan at a rate of $100 per month, how much money will the employee have in this plan after 30 years?

3. An amoeba population in a jar of water starts at 1000 amoeba and doubles in an hour. After this time, amoeba are removed continuously from the jar at a rate of 100 amoeba per hour. Assuming continuous growth, what will the amoeba population in the jar be after 3 hours?

4. Suppose that a culture initially contains 1 million bacteria that increases to 1.5 million in a half hour. Further suppose after this time that 0.5 million bacteria are added to this culture every hour at a continuous rate. Assuming continuous growth, how many bacteria will be in the culture after 5 hours?

5. Cesium 137 has a half-life of approximately 30 years. If cesium 137 is entering a cesium-free pond at a rate of 1 lb/year, approximately how many pounds of cesium 137 will be in the pond after 30 years?

6. A drum contains 30 g of radioactive cobalt 60 suspended in liquid. Because the drum has a small leak, 0.1 g of radioactive cobalt 60 escapes from the drum each year. If the half-life of radioactive cobalt 60 is about 5.3 years, about how many grams of radioactive cobalt 60 are left in the drum after 10 years?

7. A 100-gal tank contains 40 gal of an alcohol-water solution 2 gal of which is alcohol. A solution containing 0.5 gal alcohol/gal runs into the tank at a rate of 3 gal/min and the well-stirred mixture leaves the tank at a rate of 2 gal/min.

a) How many gallons of alcohol are in the tank after 10 minutes?

b) How many gallons of alcohol are in the tank when it overflows?

8. A 500-gal tank contains 200 gal fresh water. Salt water containing 0.5 lb/gal enters the tank at 4 gal/min. The well-stirred solution leaves the tank at the same rate.

a) How much salt is in the tank after 10 minutes?

b) Suppose after 10 minutes the salt water entering the tank is stopped and replaced by pure water entering at 4 gal/min. If the well-stirred mixture now leaves at 3 gal/min, how much salt is in the tank after another 10 minutes?

9. A 10,000-cu-ft room contains 20% carbon dioxide. Pure oxygen will be pumped into the room at a rate of 5 cu ft/min. The well-mixed air escapes from the room at a rate of 5 cu ft/min.

a) How long will it take to reduce the carbon-dioxide level to 5%?

b) How long will it take to reduce the carbon-dioxide level to 5% if smokers are putting in carbon dioxide at a rate of 1 cu ft/min?

10. A pond with a stream flowing in and a stream flowing out maintains a nearly constant volume of 1,000,000 gal. The flow rate is about 2 gal/min. An industrial plant located upstream from the pond has been allowing pollutants to flow into the pond for some time. Water samples show that about 10% of the pond water consists of pollutants. Biologists have determined that for the species depending on the pond water the pollution level has to be less than 1%. What is the rate that the plant can discharge pollutants if this level is to be reached within 5 years?

11. A can of frozen juice is taken from a freezer at 20° F into a room with a temperature of 70° F. After 35 min, the temperature of the juice is 28° F. How long will it take for the temperature of the juice to be 32° F?

12. A cup of liquid is put in a microwave oven and heated to 180° F. The cup is then taken out and put in a room of constant temperature 70° F. After 2 min the liquid has cooled to 160° F. What will the temperature of the liquid be at 10 minutes?

13. A dead body is discovered by an investigator. The temperature of the body is taken and found to be 90° F. The investigator records the temperature an hour later and finds the temperature to be 85° F. If the air temperature is 75° F, estimate the time of death. (Assume that normal body temperature is 98.6° F.)

14. A dead body is discovered by an investigator. At the time of discovery, the body's temperature is 85° F. The air temperature is 75° F. The body is then put into a refrigerated system with a temperature of 40° F for 2 hours. At the end of these 2 hours the temperature of the body is 65° F. Assuming the con-

stant of proportionality depends only on the body, estimate how long the person has been dead.

15. A body of mass 5 kg is dropped from a height of 200 m. Assuming the gravitational effect is constant over the 200 m and air resistance is proportional to the velocity of the falling object with proportionality constant 10 kg/sec, determine the velocity of the object after 2 seconds. (Use 10 m/sec^2 for the acceleration of gravity.)

16. Suppose that the body in Exercise 15 is projected downward with a velocity of 1 m/sec instead of being dropped. Again determine the velocity of the body after 2 seconds.

17. A body of mass 10 kg is projected from a height 5 m above the surface of the earth with velocity 20 m/sec. Ignoring friction, find the maximum height obtained by the object.

18. Determine the initial velocity an object projected vertically at the surface of the earth must have to escape the earth's gravitational pull.

19. A geometrical problem that arises in physics and engineering is to find the curves that intersect a given family of curves orthogonally. (That is, the tangent lines are perpendicular at the intersection points.) These curves are called the **orthogonal trajectories** of the given family of curves. For instance, consider the family of circles $x^2 + y^2 = a^2$ where a is a constant. From the geometry of circles, we can see that the orthogonal trajectories of this family of circles consist of the lines through the origin. Another way of obtaining this that works more generally for other families of curves involves using differential equations.

a) Show that the orthogonal trajectories of the family of circles $x^2 + y^2 = a^2$ satisfy the differential equation

$$\frac{dy}{dx} = \frac{y}{x}.$$

(*Hint:* Recall that the slopes of perpendicular lines are the negative reciprocals of each other.)

b) Solve the differential equation in part (a) to determine the orthogonal trajectories of the family of circles $x^2 + y^2 = a^2$.

20. Find the orthogonal trajectories of the family of hyperbolas $xy = k$ where k is a constant.

21. As populations increase in size, the growth rate of the population tends to slow as the environment inhibits the growth rate. The logistic growth curve is one means of trying to model this. This curve is given by a differential equation of the form

$$\frac{dP}{dt} = \frac{kP(C - P)}{C}$$

where k is a constant, P is the population, and C is the maximum population the environment can sustain, called the carrying capacity of the environment. Notice that if P is small relative to C, the differential equation for logistic growth is approximately the same as the one $dP/dt = kP$ modeling exponential growth since $(C - P)/C$ is close to one. As P approaches C, $(C - P)/C$ is close to zero so that the growth rate in logistic growth is close to zero. Find the logistic growth curve of a population P if the carrying capacity of the environment is 1 trillion and the initial population is 10 billion.

22. As an epidemic spreads through a population, the rate at which people are infected is often proportional to the product of the number of people infected and the number of people who are not infected.

    a) If C denotes the total population, P denotes the number of infected people, and P_0 denotes the initial number of infected people, write down an initial value problem modeling such an epidemic.

    b) Solve the initial value problem in part (a).

23. In thermodynamics, the equation relating pressure P, volume V, number of moles N, and temperature T is given by $PV = NRT$ where R is Avogadro's constant.

    a) Show that $P\,dV + V\,dP = N R\,dT$.

    b) Using the adiabatic condition $dU = -P\,dV$ and the energy equation $dU = Nc_v\,dT$ where U is the total energy and c_v is the molar heat capacity, show the differential equation in part (a) can be expressed as
    $$P\,dV + V\,dP = -(R/c_v)P\,dV.$$

    c) Solve the differential equation in part (b) to obtain the equation for the ideal gas process relating P and V.

24. Financial analysts use the following differential equation to model the interest rate process of a market economy

$$d(f(r)) = (\theta(t) - a(r))\,dt + \sigma(t)\,dz$$

where dz models the randomness of the economy.

    a) The Ho-Lee process uses $f(r) = r$, $a(r) = 0$, and $\theta(t) = R(t) + \sigma(t)$. Solve the Ho-Lee equation if $\sigma(t) = 0$ and $R(t) = 0.05t$.

    b) Another process has $f(r) = \ln r$ and $a(r) = \ln r$. Solve this equation assuming $\theta(t) = R(t) + \sigma(t)$, $\sigma(t) = 0$, and $R(t) = 0.05t$.
    (Solutions for $\sigma \neq 0$ are called stochastic solutions and are beyond the scope of this text.)

3.7 REDUCTION OF ORDER

In this section we will consider some higher order differential equations that can be solved by reducing them to first order differential equations. One case where we can do this is when we have a second order differential equation that does not contain y. Such an equation can be reduced to a first order equation by substituting

$$\boxed{v = y'}$$

as the first two examples of this section illustrate.

EXAMPLE 1 Solve

$$y'' + 2y' = x.$$

Solution   Letting $v = y'$ we have

$$v' + 2v = x.$$

This is a linear equation. Solving this equation (you will find that you have to use integration by parts), we obtain

$$v = C_1 e^{-2x} + \frac{1}{2}x - \frac{1}{4}.$$

Since $y' = v$, integrating v gives us

$$y = -\frac{1}{2}C_1 e^{-2x} + \frac{1}{4}x^2 - \frac{1}{4}x + C_2.$$ ●

We have seen that the general solution to a first order differential equation contains a constant. Notice that the general solution to the second order differential equation in Example 1 has two constants. This pattern continues; that is, the general solution to an nth order differential equation will have n constants.

EXAMPLE 2   Solve the initial value problem

$$y'' = xy' + 2x; \qquad y(0) = 0, \qquad y'(0) = 1.$$

Solution   Substituting $v = y'$ gives us the initial value problem

$$v' = xv + 2x, \qquad v(0) = 1.$$

The differential equation is a linear equation with solution

$$v = -2 + Ce^{x^2/2}.$$

Using initial condition $v(0) = 1$, we have

$$-2 + C = 1,$$
$$C = 3,$$

and hence

$$v = -2 + 3e^{x^2/2}.$$

Now we integrate v to obtain y. Unfortunately, $\int e^{x^2/2}\,dx$ does not have a closed form. Let us use $\int_0^x e^{t^2/2}\,dt$ as an antiderivative of $e^{x^2/2}$. Doing so, we have

$$y = -2x + 3\int_0^x e^{t^2/2}\,dt + C.$$

Using the initial condition $y(0) = 0$, we find

$$C = 0.$$

Therefore, the solution is

$$y = -2x + 3 \int_0^x e^{t^2/2} \, dt. \qquad \bullet$$

The approach of Examples 1 and 2 can be extended to third order differential equations that do not have y' and y. In fact, it can be extended to nth order differential equations not having $y^{(n-2)}$, $y^{(n-3)}$, ..., y. (How would we do this?)

The technique we used in Examples 1 and 2 worked because the equation did not contain y. There is another technique we can use to reduce the order of second order equations that have y but not x. We do this by again letting

$$\boxed{v = y'.}$$

Instead of using v' for y'', however, we use the Chain Rule to write y'' as

$$\boxed{y'' = \frac{dv}{dx} = \frac{dv}{dy}\frac{dy}{dx} = v\frac{dv}{dy}.}$$

The next two examples illustrate how these two substitutions are used.

EXAMPLE 3   Solve the differential equation $yy'' = (y')^2$.

Solution   Substituting

$$y' = v \quad \text{and} \quad y'' = v\frac{dv}{dy},$$

our equation becomes

$$yv\frac{dv}{dy} = v^2,$$

which is a separable equation. Upon rewriting this equation as

$$\frac{dv}{v} = \frac{dy}{y}$$

and integrating, we find

$$\ln|v| = \ln|y| + C.$$

Now we need to solve this first order differential equation. To do so, let us first solve for v:

$$e^{\ln|v|} = e^{\ln|y|+C}$$
$$|v| = e^C|y|$$
$$v = \pm e^C y.$$

Relabeling the constant $\pm e^C$ as C_1, we have

$$v = C_1 y \quad \text{or} \quad y' = C_1 y.$$

We can solve this last equation as either a separable or a linear one. Doing it either way, we find its solutions have the form

$$y = C_2 e^{C_1 x}. \qquad \bullet$$

EXAMPLE 4   Solve the differential equation $y'' + y = 0$.

Solution   Making the substitutions $v = y'$ and $y'' = v \dfrac{dv}{dy}$ in this equation, we obtain

$$v \frac{dv}{dy} + y = 0.$$

This is a separable differential equation whose solutions are given by

$$\frac{v^2}{2} = -\frac{y^2}{2} + C.$$

Since $y' = v$, we have

$$y'^2 + y^2 = C_1^2$$

where $C_1^2 = 2C$. (The choice of C_1 was made in hindsight to make the form of our answer nicer.) Solving this equation for $y' = dy/dx$ gives us the separable differential equation

$$\frac{dy}{dx} = \pm (C_1^2 - y^2)^{1/2}$$

or

$$\frac{dy}{(C_1^2 - y^2)^{1/2}} = \pm dx.$$

Integrating we get

$$\sin^{-1}\left(\frac{y}{C_1}\right) = \pm x + C,$$

so that

$$y = C_1 \sin(\pm x + C).$$

Using the fact that $\sin(\pm \theta) = \pm \sin(\theta)$, we express the solutions in the form

$$y = C_1 \sin(x + C_2). \qquad \bullet$$

We will see a much easier way of finding the solution to the differential equation in Example 4 in the next chapter.

EXERCISES 3.7

In Exercises 1–8, solve the differential equations.

1. $y'' - 9y' = x$ **2.** $y'' + y' = 1$

3. $y'' - (y')^3 = 0$ **4.** $y'' + xy'/(1+x) = 0$

5. $yy'' - 4(y')^2 = 0$ **6.** $y'' - e^y y' = 0$

7. $y'' + 4y = 0$ **8.** $y'' - 25y = 0$

In Exercises 9–15, solve the initial value problems.

9. $y''y' = 1$; $y(0) = 0$, $y'(0) = 1$

10. $y'' = x(y')^2$; $y(0) = 0$, $y'(0) = 1$

11. $x^2 y'' + 2xy' = 1$; $y(1) = 0$, $y'(1) = 1$

12. $y'' - y = 0$; $y(0) = 1$, $y'(0) = 1$

13. $y'' - 3y^2 = 0$; $y(0) = 2$, $y'(0) = 4$

14. $y^3 y'' = 9$; $y(0) = 1$, $y'(0) = 4$

15. $xy''' = y''$; $y(1) = 1$, $y'(1) = 2$, $y''(1) = 3$

16. If y is the height of the object in Example 4 of Section 3.6, the differential equation could also be written as

$$\frac{d^2 y}{dt^2} = Mg - k\frac{dy}{dt}.$$

Determine the height y of the object at a time t if it is dropped from a height y_0.

17. The initial value problem in Example 5 of Section 3.6 could also be written as

$$M\frac{d^2 x}{dt^2} = -\frac{MgR^2}{(R+x)^2}; \quad v(0) = v_0, \quad x(0) = x_0$$

where x_0 is the initial distance of the object from the planet. Solve this initial value problem. You may leave your answer in terms of an integral.

18. The motion of a simple pendulum of length l is described by the initial value problem

$$l\frac{d^2\theta}{dt^2} + g\sin\theta = 0; \quad \theta'(0) = 0, \quad \theta(0) = \theta_0$$

where θ is the angle the pendulum makes with the vertical, θ_0 is the initial angle the pendulum makes with the vertical, and g is the acceleration of gravity. Solve this initial value problem. You may leave your answer in terms of an integral.

3.8   THE THEORY OF FIRST ORDER DIFFERENTIAL EQUATIONS

In this section we will consider the existence and uniqueness of solutions to the initial value problem for first order differential equations. In particular, we will consider the proof of the following theorem restated from Section 3.1.

THEOREM 3.3   Let $a, b > 0$ and suppose f and $\partial f/\partial y$ are continuous on the rectangle R given by $|x - x_0| < a$ and $|y - y_0| < b$. Then there exists an $h > 0$ so that the initial value problem

$$y' = f(x, y), \qquad y(x_0) = y_0$$

has one and only one solution $y = y(x)$ for $|x - x_0| \le h$.

There are many technical issues that have to be dealt with in proving this theorem that must be left for a course in advanced calculus or an advanced differential equations course with advanced calculus as a prerequisite. We will point out these technical issues as we outline a proof of the theorem. One consequence of f and f_y being continuous on R that we will use throughout this section and the exercises is that there is a number $K > 0$ and a number $M > 0$ so that $|f(x, y)| \le K$ and $|f_y(x, y)| \le M$ for all (x, y) in R. (We will reserve K and M for the bounds of f and f_y on R, respectively, throughout this section.) The proof of the uniqueness part of Theorem 3.3 is the easier part and we begin with it.

To prove the uniqueness part, we make use of one of the fundamental theorems of integral calculus that tells us if the function F is continuous on an open interval containing x_0 and x, then an antiderivative of F on this open interval is

$$G(x) = \int_{x_0}^{x} F(t)\, dt.$$

Notice that

$$G(x_0) = \int_{x_0}^{x_0} F(t)\, dt = 0$$

and hence $G(x)$ is a solution to the initial value problem $y' = F(x)$, $y(x_0) = 0$. Suppose $y(x)$ is a solution to the initial value problem

$$\boxed{y' = f(x, y), \qquad y(x_0) = y_0.}$$

If we let

$$\boxed{F(x) = f(x, y(x)),}$$

then $y(x)$ satisfies the initial value problem

$$\boxed{y'(x) = F(x), \qquad y(x_0) = y_0,}$$

and it follows that $y(x)$ is given by

$$\boxed{y(x) = y_0 + \int_{x_0}^{x} F(t)\, dt = y_0 + \int_{x_0}^{x} f(t, y(t))\, dt.} \qquad \textbf{(1)}$$

We will call Equation (1) an integral equation. A solution to it is a solution to the initial value problem and vice versa. In particular, if the integral equation has a unique solution, then so does the initial value problem.

We now prove the uniqueness of the solution to the initial value problem by proving the uniqueness of the solution to the integral equation. Suppose that $u(x)$ and $v(x)$ are solutions to the integral equation

$$y(x) = y_0 + \int_{x_0}^{x} f(t, y(t))\, dt.$$

That is,

$$u(x) = y_0 + \int_{x_0}^{x} f(t, u(t))\, dt$$

and

$$v(x) = y_0 + \int_{x_0}^{x} f(t, v(t))\, dt.$$

Subtracting these equations gives us

$$u(x) - v(x) = \int_{x_0}^{x} (f(t, u(t)) - f(t, v(t)))\, dt.$$

Thus

$$|u(x) - v(x)| = \left| \int_{x_0}^{x} (f(t, u(t)) - f(t, v(t)))dt \right| \tag{2}$$

$$\leq \int_{x_0}^{x} |f(t, u(t)) - f(t, v(t))|dt.$$

Using the Mean Value Theorem, for each t we can express $f(t, u(t)) - f(t, v(t))$ as

$$f(t, u(t)) - f(t, v(t)) = f_y(t, y^*(t))(u(t) - v(t))$$

for some $y^*(t)$ between $v(t)$ and $u(t)$. Substituting this, the inequality in Equation (2) becomes

$$|u(x) - v(x)| \leq \int_{x_0}^{x} |f_y(t, y^*(t))(u(t) - v(t))|\, dt$$

$$= \int_{x_0}^{x} |f_y(t, y^*(t))|\, |(u(t) - v(t))|\, dt.$$

Since $|f_y(t, y)| \leq M$ on R, we have

$$|u(x) - v(x)| \leq \int_{x_0}^{x} M|u(t) - v(t)|dt \leq M \int_{x_0}^{x} |u(t) - v(t)|\, dt$$

on this rectangle.

We now let

$$w(x) = |u(x) - v(x)|.$$

We have that

$$w(x_0) = |u(x_0) - v(x_0)| = 0,$$

$$w(x) \geq 0,$$

and

$$w(x) \leq M \int_{x_0}^{x} w(t)\, dt.$$

Setting

$$W(x) = \int_{x_0}^{x} w(t)\, dt,$$

$$w(x) = W'(x) \leq MW(x)$$

or

$$W'(x) - MW(x) \le 0.$$

This looks like a first order linear differential equation only with an inequality instead of an equality. Let us multiply by the integrating factor

$$e^{\int -M \, dx} = e^{-Mx}$$

to obtain

$$e^{-Mx}(W'(x) - MW(x)) = (W(x)e^{-Mx})' \le 0.$$

Replacing x by t, integrating each side of this inequality from x_0 to x, and using the fact that

$$W(x_0) = \int_{x_0}^{x_0} w(t) \, dt = 0$$

gives us

$$\int_{x_0}^{x} (W(t)e^{-Mt})' \, dt = W(t)e^{-Mt}\Big|_{x_0}^{x} = W(x)e^{-Mx} \le 0.$$

Since $e^{-Mx} \ge 0$, we have on the one hand that

$$W(x) \le 0$$

for $|x - x_0| < a$. On the other hand,

$$W(x) = \int_{x_0}^{x} w(t) \, dt = \int_{x_0}^{x} |u(t) - v(t)| \, dt \ge 0$$

for $|x - x_0| < a$. Therefore,

$$W(x) = 0$$

and

$$w(x) = W'(x) = 0$$

for $|x - x_0| < a$. Since

$$w(x) = |u(x) - v(x)|,$$

we must have that

$$u(x) = v(x)$$

for $|x - x_0| < a$, completing our proof of the uniqueness part.
    We now turn our attention to the existence part. Notice that if

$$y' = f(x, y), \qquad y(x_0) = y_0,$$

then

$$y'(x_0) = f(x_0, y_0).$$

Consequently, the initial value problem determines the first degree Taylor polynomial

$$y_0 + y'(x_0)(x - x_0) = y_0 + f(x_0, y_0)(x - x_0)$$

about x_0 of the solution $y(x)$. Applying the Chain Rule to the differential equation $y' = f(x, y)$ to find y'', we have

$$y'' = \frac{d}{dx} y' = f_x(x, y) \frac{dx}{dx} + f_y(x, y) \frac{dy}{dx}$$

$$= f_x(x, y) + f_y(x, y) y'.$$

This allows us to find $y''(x_0)$ as

$$y''(x_0) = f_x(x_0, y_0) + f_y(x_0, y_0) y'(x_0)$$

from which we can obtain the second degree Taylor polynomial about x_0 for the solution $y(x)$. This process can be continued to obtain the Taylor series of the solution $y(x)$. Indeed, this is an approach to producing solutions that we will study in Chapter 8. This procedure would show the existence of a solution (in Taylor series form) provided we fill in some gaps. One of these gaps is: Do the required higher order derivatives needed to produce the Taylor series exist? Another gap, assuming we can produce the Taylor series, is: Does the series converge to a solution? Answers to these questions are beyond the scope of this text.

Another approach to proving the existence of solutions that does not require differentiating $f(x, y(x))$ involves a method called **Picard iteration,**[6] or the **method of successive approximation.** As we did in the proof of the uniqueness part, we consider the integral equation

$$y(x) = y_0 + \int_{x_0}^{x} f(t, y(t)) \, dt,$$

which is equivalent to the initial value problem. The idea is to create a sequence of functions that converge to the solution of the integral equation. (This is like creating the sequence of Taylor polynomials that converge to the Taylor series.)

To start the sequence of functions, we define the 0th Picard iterate to be the constant function

$$\boxed{p_0(x) = y_0.}$$

By this definition p_0 agrees with the solution to the integral equation at x_0:

$$p_0(x_0) = y(x_0) = y_0.$$

---

[6] Named for the French mathematician Émile Picard (1856–1941).

We now define a sequence of functions that also agree with y at x_0 by setting

$$p_{k+1}(x) = y_0 + \int_{x_0}^{x} f(t, p_k(t)) \, dt \text{ for } k = 0, 1, 2, \ldots.$$

Thus, for example,

$$p_1(x) = y_0 + \int_{x_0}^{x} f(t, p_0(t)) \, dt = y_0 + \int_{x_0}^{x} f(t, y_0) \, dt$$

and

$$p_2(x) = y_0 + \int_{x_0}^{x} f(t, p_1(t)) \, dt.$$

We call $p_k(x)$ the kth Picard iterate. It is easily checked that

$$p_k(x_0) = y_0$$

for each $k = 1, 2, 3 \ldots$.

To give you a feel for this iterative process, let us apply it to some initial value problems.

EXAMPLE 1   Determine the Picard iterates for the following initial value problem.

$$y' = y, \qquad y(0) = 1$$

Solution   We first note that

$$f(x, y) = y$$

and that the equivalent integral equation to this problem is

$$y(x) = 1 + \int_{0}^{x} f(t, y(t)) \, dt = 1 + \int_{0}^{x} y(t) \, dt.$$

We now develop the Picard iterates. The 0th Picard iterate is

$$p_0(x) = y(0) = 1.$$

The first Picard iterate is

$$p_1(x) = y_0 + \int_{x_0}^{x} f(t, p_0(t)) \, dt = 1 + \int_{0}^{x} f(t, 1) \, dt = 1 + \int_{0}^{x} 1 \, dt = 1 + x.$$

The second Picard iterate is

$$p_2(x) = y_0 + \int_{x_0}^{x} f(t, p_1(t)) \, dt = 1 + \int_{0}^{x} f(t, 1 + t) \, dt = 1 + \int_{0}^{x} (1 + t) \, dt$$

$$= 1 + x + \frac{1}{2}x^2.$$

The third Picard iterate is

$$p_3(x) = y_0 + \int_{x_0}^x f(t, p_2(t)) \, dt = 1 + \int_0^x f\left(t, 1 + t + \frac{1}{2}t^2\right) dt$$

$$= 1 + \int_0^x \left(1 + t + \frac{1}{2}t^2\right) dt = 1 + x + \frac{1}{2}x^2 + \frac{1}{3!}x^3.$$

From the pattern developing here, we see that the kth Picard iterate is

$$p_k(x) = 1 + x + \frac{1}{2}x^2 + \frac{1}{3!}x^3 + \cdots + \frac{1}{k!}x^k = \sum_{n=0}^k \frac{1}{n!}x^n.$$

Notice that $p_k(x)$ is the kth Maclaurin polynomial for e^x and that e^x is the solution to this initial value problem. The Picard iterates converge to the solution to the initial value problem in this example. ●

EXAMPLE 2   Determine the Picard iterates for the following initial value problem.

$$y' = 1 + y^2, \qquad y(0) = 0$$

Solution   We have

$$f(x, y) = 1 + y^2$$

and the equivalent integral equation is

$$y(x) = 1 + \int_0^x f(t, y(t)) \, dt = 1 + \int_0^x (1 + y(t)^2) \, dt.$$

The 0th Picard iterate is

$$p_0(x) = y(0) = 0.$$

The first Picard iterate is

$$p_1(x) = 0 + \int_0^x f(t, 0) \, dt = \int_0^x 1 \, dt = x.$$

The second Picard iterate is

$$p_2(x) = 0 + \int_0^x f(t, t) \, dt = \int_0^x (1 + t^2) \, dt = x + \frac{1}{3}x^3.$$

The third Picard iterate is

$$p_3(x) = 0 + \int_0^x f\left(t, t + \frac{1}{3}t^3\right) dt = \int_0^x \left(1 + \left(t + \frac{1}{3}t^3\right)^2\right) dt$$

$$= x + \frac{1}{3}x^3 + \frac{2}{15}x^5 + \frac{1}{63}x^7.$$

The fourth Picard iterate is

$$p_4(x) = 0 + \int_0^x f\left(t, t + \frac{1}{3}t^3 + \frac{2}{15}t^5 + \frac{1}{63}t^7\right) dt$$

$$= \int_0^x \left(1 + \left(t + \frac{1}{3}t^3 + \frac{2}{15}t^5 + \frac{1}{63}t^7\right)^2\right) dt$$

$$= x + \frac{1}{3}x^3 + \frac{2}{15}x^5 + \frac{17}{315}x^7 + \frac{38}{2835}x^9 + \frac{134}{51975}x^{11} + \frac{4}{12285}x^{13} + \frac{1}{59535}x^{15}.$$

The pattern developing here is not as apparent as it was in Example 1. However, it can be verified that the terms of degree less than or equal to k of $p_k(x)$ form the kth degree Maclaurin polynomial for $\tan x$ and that $\tan x$ is the solution to the initial value problem. Here again the Picard iterates converge to the solution of the initial value problem. ●

EXAMPLE 3   Determine the Picard iterates for the following initial value problem.

$$y' = x^2 + y^2, \qquad y(-1) = 0$$

Solution   We have

$$f(x, y) = x^2 + y^2$$

and the equivalent integral equation is

$$y(x) = 0 + \int_{-1}^x f(t, y(t)) \, dt = 0 + \int_{-1}^x t^2 + y(t)^2 \, dt.$$

The 0th Picard iterate is

$$p_0(x) = y(-1) = 0.$$

The first Picard iterate is

$$p_1(x) = 0 + \int_{-1}^x f(t, 0) \, dt = \int_{-1}^x t^2 \, dt = \frac{1}{3} + \frac{1}{3}x^3.$$

The second Picard iterate is

$$p_2(x) = 0 + \int_{-1}^x f\left(t, \frac{1}{3} + \frac{1}{3}t^3\right) dt$$

$$= \int_{-1}^x \left(t^2 + \left(\frac{1}{3} + \frac{1}{3}t^3\right)^2\right) dt = \frac{17}{42} + \frac{1}{9}x + \frac{1}{3}x^3 + \frac{1}{18}x^4 + \frac{1}{63}x^7.$$

Even if we continue, no pattern for $p_k(x)$ is discernible as in Examples 1 and 2. However, as we shall argue next, Picard iterates such as these do converge to the solution to the initial value problem. Exercises 11–13 illustrate how to modify an initial value problem such as this so that the Picard iterates generate the Taylor polynomials about x_0 of the original initial value problem. ●

The questions that have to be addressed regarding the Picard iterates to obtain that they converge to the solution of the initial value problem in Theorem 3.1 are:

1. Are the Picard iterates defined for each k?

2. Do the Picard iterates converge?

3. If the Picard iterates converge, do they converge to the solution of the initial value problem?

In answer to question 1, for the Picard iterates to be defined the integrals defining them must exist. This is why Theorem 3.1 requires that f and $\partial f / \partial y$ be continuous. In Examples 1–3 f is not only continuous, but infinitely differentiable. However, continuity of f and f_y is all we need to guarantee the existence of the integral. The number h in Theorem 3.1 is needed to guarantee that $(x, p_k(x))$ always lies in the rectangle R where f and f_y are continuous. Exercise 15 addresses this issue. Assuming that the Picard iterates are defined for each k, we move on to question 2.

Note that

$$p_1(x) = p_0(x) + (p_1(x) - p_0(x)),$$

$$p_2(x) = p_0(x) + (p_1(x) - p_0(x)) + (p_2(x) - p_1(x))$$

$$= p_0(x) + \sum_{n=0}^{1} (p_{n+1}(x) - p_n(x)),$$

and that in general

$$p_{k+1}(x) = p_0(x) + (p_1(x) - p_0(x)) + \cdots + (p_{k+1}(x) - p_k(x))$$

$$= p_0(x) + \sum_{n=0}^{k} (p_{n+1}(x) - p_n(x)).$$

Showing that the sequence of functions $p_k(x)$ converges is the same as showing that the series

$$p_0(x) + \sum_{n=0}^{\infty} (p_{n+1}(x) - p_n(x))$$

converges. In fact, it will be shown that this series converges absolutely for $|x| \le h$ for an $h > 0$.

Consider the partial sum

$$|p_0(x)| + \sum_{n=0}^{k} |p_{n+1}(x) - p_n(x)|.$$

Substituting in the integral definition of $p_n(x)$ gives us

$$|p_0(x)| + \sum_{n=0}^{k} |p_{n+1}(x) - p_n(x)| = |y_0| + \left| y_0 + \int_{x_0}^{x} f(t, p_0(t)) \, dt - y_0 \right| + \cdots$$

$$+ \left| y_0 + \int_{x_0}^{x} f(t, p_k(t)) \, dt - \left(y_0 + \int_{x_0}^{x} f(t, p_{k-1}(t)) \, dt \right) \right|$$

$$= |y_0| + \left| \int_{x_0}^{x} f(t, p_0(t)) \, dt \right| + \sum_{n=1}^{k} \left| \int_{x_0}^{x} (f(t, p_n(t)) - f(t, p_{n-1}(t)) \, dt \right|$$

$$\leq |y_0| + \int_{x_0}^{x} |f(t, p_0(t))| \, dt + \sum_{n=1}^{k} \int_{x_0}^{x} |(f(t, p_n(t)) - f(t, p_{n-1}(t))| \, dt.$$

Using a Mean Value Theorem approach similar to the one we used in the proof of the uniqueness part, we can obtain (the details of this are left as Exercise 16)

$$|p_0(x)| + \sum_{n=0}^{k} |p_{n+1}(x) - p_n(x)| \leq |y_0| + K|x - x_0| + \sum_{n=1}^{k} M \int_{x_0}^{x} |p_n(t) - p_{n-1}(t)| \, dt. \quad (3)$$

Another inequality we need whose proof we leave as Exercise 17 is

$$|p_n(t) - p_{n-1}(t)| \leq \frac{M K^{n-1}}{(n-1)!} |t|^{n-1} \quad (4)$$

for $n \geq 2$. Substituting the result of (4) into (3) gives us

$$|p_0(x)| + \sum_{n=0}^{k} |p_{n+1}(x) - p_n(x)| \leq |y_0| + K|x - x_0| + \sum_{n=1}^{k} M \int_{x_0}^{x} \frac{M K^{n-1}}{(n-1)!} |t|^{n-1} \, dt$$

$$\leq |y_0| + K|x| + \sum_{n=1}^{k} \frac{M^2 K^{n-1}}{n!} |x - x_0|^n$$

for $|x - x_0| \leq h$. We leave it as Exercise 18 for you to show that the series

$$|y_0| + K|x - x_0| + \sum_{n=1}^{\infty} \frac{M^2 K^{n-1}}{n!} |x - x_0|^n$$

converges. Since

$$|p_0(x)| + \sum_{n=0}^{k} |p_{n+1}(x) - p_n(x)| \leq |y_0| + K|x - x_0| + \sum_{n=1}^{\infty} \frac{M^2 K^{n-1}}{n!} |x - x_0|^n,$$

the sequence of partial sums $|p_0(x)| + \sum_{n=0}^{k} |p_{n+1}(x) - p_n(x)|$ is a nondecreasing bounded sequence and hence converges. Thus the series

$$|p_0(x)| + \sum_{n=0}^{\infty} |p_{n+1}(x) - p_n(x)|$$

converges, which completes the proof that the sequence of functions $p_k(x)$ converges.

Now that we have question 2 behind us, we come to question 3. We need to verify that

$$p(x) = \lim_{k \to \infty} p_k(x)$$

is a solution to the initial value problem. This is done by showing that $p(x)$ is a solution to the equivalent integral equation

$$y = y_0 + \int_{x_0}^{x} f(t, p(t)) \, dt. \tag{5}$$

We get this by taking the limit of

$$p_{k+1}(x) = y_0 + \int_{x_0}^{x} f(t, p_k(t)) \, dt$$

as k approaches infinity. Beginning to do this, we have

$$p(x) = \lim_{k \to \infty} p_{k+1}(x) = \lim_{k \to \infty} \left(y_0 + \int_{x_0}^{x} f(t, p_k(t)) \, dt \right)$$

$$= y_0 + \lim_{k \to \infty} \int_{x_0}^{x} f(t, p_k(t)) \, dt.$$

At this point, we move the limit inside the integral[7] obtaining

$$p(x) = y_0 + \int_{x_0}^{x} \lim_{k \to \infty} f(t, p_k(t)) \, dt.$$

Finally, because of the continuity of f, we can move the limit inside f and get

$$p(x) = y_0 + \int_{x_0}^{x} f(t, \lim_{k \to \infty} p_k(t)) \, dt$$

$$= y_0 + \int_{x_0}^{x} f(t, p(t)) \, dt$$

showing that $p(x)$ is a solution of the integral equation in (5).

EXERCISES 3.8

In Exercises 1–2, give the equivalent integral equation to the initial value problem.

1. $y' = x + 2y$, $y(0) = 0$

2. $y' = e^y + \sin x$, $y(1) = -1$

In Exercises 3–6, solve the initial value problem, find the first four Picard iterates, and compare these iterates to the first four Maclaurin polynomials of the solution to the initial value problem.

3. $y' = 2y$, $y(0) = 1$ **4.** $y' = -y$, $y(0) = 2$

5. $y' = 2y^2$, $y(0) = 1$ **6.** $y' = xy$, $y(0) = -1$

In Exercises 7–8, find the first two Picard iterates for the initial value problem.

7. $y' = x^2 - y^2$, $y(1) = 0$

8. $y' = x + y^2$, $y(-1) = 1$

In Exercises 9–10, generate the Picard iterates until you can no longer compute the integrals in a closed form. These exercises illustrate that Picard iterates must sometimes be expressed in integral form.

9. $y' = \cos y$, $y(0) = 0$ **10.** $y' = e^y$, $y(0) = 0$

---

[7] If you take an advanced calculus course, you will learn that interchanging an integral and limit as we are doing here requires a stronger type of convergence called uniform convergence. It is possible to show that the convergence of the sequence of functions $p_k(x)$ is uniform, but the details of this are left for more advanced courses.

11. Show that if $u(x)$ is a solution to the initial value problem $u' = f(x + x_0, u)$, $u(0) = y_0$, then $y(x) = u(x - x_0)$ is a solution to the initial value problem $y' = f(x, y)$, $y(x_0) = y_0$.

In Exercises 12–13, use the approach of Exercise 11 to generate the first three Picard iterates for u and then use $y(x) = u(x - x_0)$ to obtain the first three Picard iterates for y. Also verify that the first three terms of the third Picard iterate for y is the third degree Taylor polynomial about x_0 of the solution to the initial value problem at x_0.

12. $y' = x^2 + y^2$, $y(-1) = 0$

13. $y' = x - y^2$, $y(1) = 1$

14. Consider the sequence of functions $p_n(x) = nxe^{-nx^2}$ show that

$$\lim_{n \to \infty} \int_0^1 p_n(t) \, dt \neq \int_0^1 \lim_{n \to \infty} p_n(t) \, dt.$$

This illustrates that limits and integrals cannot be interchanged in general.

15. This exercise outlines the proof of the fact that the Picard iterates are defined. Show that if $(x, p_n(x))$ lies in the rectangle $|x - x_0| < a$ and $|y - y_0| < b$, then $|p'_n(x)| = |f(x, p_{n-1}(x))| \leq K$. Use this to

show that if h is the minimum of b/K and a, then the Picard iterates are defined for $|x - x_0| \leq h$.

16. a) Show that if $|x - x_0| \leq h$, then
$$|p_1(x) - p_0(x)| \leq K|x - x_0| \leq Kh.$$

b) Using the Mean Value Theorem and part (a), show that
$$|p_0(x)| + \sum_{n=0}^{k} |p_{n+1}(x) - p_n(x)|$$
$$\leq |y_0| + K|x - x_0|$$
$$+ \sum_{n=1}^{k} M \int_{x_0}^{x} |p_n(t) - p_{n-1}(t)| \, dt.$$

17. Using induction on the Picard iterates show that
$$|p_n(t) - p_{n-1}(t)| \leq \frac{MK^{n-1}}{(n-1)!} |t|^{n-1}.$$

18. Show that $|y_0| + K|x - x_0| + \sum_{n=1}^{\infty} \frac{M^2 K^{n-1}}{n!} |x - x_0|^n$ converges for all x.

19. We chose $p_0(x) = y_0$ for the Picard iteration. Show that for any choice of $p_0(x)$ that is continuous for $|x - x_0| < a$ with $|p_0(x)| < b$ for $|x - x_0| < a$ the Picard iterates converge to the unique solution of the initial value problem.

3.9   NUMERICAL SOLUTIONS OF ORDINARY DIFFERENTIAL EQUATIONS

In Section 3.8 we proved that the initial value problem

$$y' = f(x, y), \qquad y(x_0) = y_0$$

has a unique solution under the conditions of Theorem 3.1. In Sections 3.2–3.6 we developed some methods for solving initial value problems with first order differential equations. However, many first order differential equation initial value problems cannot be solved by any of these methods. Nevertheless, it is often possible to obtain approximations to the solutions to initial value problems. These approximations are called **numerical solutions.**

In this section we will show you three methods for determining numerical solutions to initial value problems. In a numerical analysis class one studies methods to determine how close the numerical solution is to the solution. We will give only an introduction to numerical solutions. A more in-depth study is left for courses in numerical analysis.

Each of our three methods will be presented in a subsection. Our first subsection deals with a technique known as Euler's method.

3.9.1 Euler's Method

If $y = \phi(x)$ is the unique solution to our initial value problem

$$y' = f(x, y), \qquad y(x_0) = y_0,$$

then the slope of the tangent line of the solution at x_0 is

$$\phi'(x_0) = f(x_0, y_0).$$

Therefore, an equation of the tangent line to the solution of our initial value problem at x_0 is

$$y = y_0 + f(x_0, y_0)(x - x_0).$$

If x is close to x_0, then the tangent line is close to the solution $y = \phi(x)$ for x near x_0. Therefore, if x_1 is close to x_0, then

$$y_1 = y_0 + f(x_0, y_0)(x_1 - x_0)$$

is a good approximation of $\phi(x_1)$. Now we repeat the process with x_0 replaced by x_1. An equation for the tangent line to the solution of our initial value problem at x_1 is

$$y = \phi(x_1) + f(x_1, \phi(x_1))(x - x_1).$$

We now replace $\phi(x_1)$ by y_1 obtaining

$$y_2 = y_1 + f(x_1, y_1)(x - x_1).$$

Continuing this process, we obtain a set of points

$$(x_0, y_0), (x_1, y_1), (x_2, y_2), \ldots, (x_N, y_N)$$

that (we hope) are close to the points

$$(x_0, y_0), (x_1, \phi(x_1)), (x_2, \phi(x_2)), \ldots, (x_N, \phi(x_N))$$

of the exact solution.

This method is Euler's method. We formalize it as follows: Given the x-values

$$x_0, x_1, x_2, \ldots, x_N,$$

the Euler method approximation to the solution at the points $x_0, x_1, x_2, \ldots, x_N$ of the initial value problem

$$y' = f(x, y), \qquad y(x_0) = y_0$$

is given by

$$y_{n+1} = y_n + f(x_n, y_n)(x_{n+1} - x_n), \qquad n = 0, 1, 2, \ldots, N - 1. \tag{1}$$

We apply the Euler method in the following example.

EXAMPLE 1   Use the Euler method approximation on

$$y' = y + x, \qquad y(1) = 0$$

at the x-values

$$1.1, 1.2, 1.4, 1.5.$$

Solution   Substituting $x_1 = 1.1$, $x_2 = 1.2$, $x_3 = 1.4$, and $x_4 = 1.5$ into Equation (1), we have

$$y_1 = 0 + f(1, 0)(1.1 - 1) = 0.1$$
$$y_2 = 0.1 + f(1.1, 0.1)(1.2 - 1.1) = 0.22$$
$$y_3 = 0.22 + f(1.2, 0.22)(1.4 - 1.2) = 0.504$$
$$y_4 = 0.504 + f(1.4, 0.504)(1.5 - 1.4) = 0.6944.$$ ●

Since the differential equation in Example 1 is linear, we have the exact solution

$$y = \phi(x) = 2e^{x-1} - x - 1$$

to the initial value problem. The values of ϕ at $x_1 = 1.1$, $x_2 = 1.2$, $x_3 = 1.4$, and $x_4 = 1.5$ are

$$\phi(1.1) = 2e^{0.1} - 2.1 \approx 0.11034$$
$$\phi(1.2) \approx 0.24281$$
$$\phi(1.4) \approx 0.58365$$
$$\phi(1.5) \approx 0.79744.$$

Notice how these compare with the approximations we obtained in Example 1.

If we connect the points

$$(x_0, y_0), (x_1, y_1), (x_2, y_2), \ldots, (x_N, y_N)$$

obtained from Euler's method with line segments, we get a graph that is close to the graph of the solution to the intial value problem $y = \phi(x)$. Figure 3.13 illustrates this for the initial value problem and points in Example 1. The filled in circles in Figure 3.13 represent the points of the numerical solution.

In Example 1 notice that the error in the approximation increases as n increases. There are ways to control the errors in the approximations given by Euler's method, but these are left for a numerical analysis course.

3.9.2   Taylor Methods

In calculus you saw how Taylor polynomials can be used to approximate functions and you saw how to bound the error of the approximation. In this subsection we see how to modify these ideas for an initial value problem. (Another approach for using Taylor polynomials to approximate solutions appears in Chapter 8.)

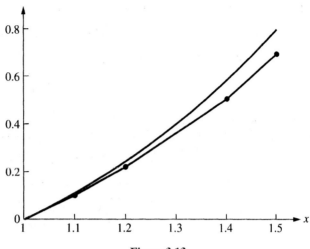

Figure 3.13

Recall that Taylor's theorem tells us that if a function p is $n + 1$ times continuously differentiable on an interval containing a, then

$$p(x) = p(a) + p'(a)(x - a) + \frac{p''(a)}{2}(x - a)^2 + \frac{p'''(a)}{3!}(x - a)^3$$

$$+ \cdots + \frac{p^{(n)}(a)}{n!}(x - a)^n + \frac{p^{(n+1)}(c)}{(n + 1)!}(x - a)^{n+1}$$

for some number c between x and a. The polynomial

$$T_n(x) = p(a) + p'(a)(x - a) + \frac{p''(a)}{2}(x - a)^2 + \frac{p'''(a)}{3!}(x - a)^3 + \cdots + \frac{p^{(n)}(a)}{n!}(x - a)^n \tag{2}$$

is called the nth degree **Taylor polynomial** for p at a. This polynomial, $T_n(x)$, and its first n derivatives at a agree with $p(x)$ and its first n derivatives at a. That is,

$$T_n(a) = p(a), \qquad T_n'(a) = p'(a), \qquad T_n''(a) = p''(a), \qquad \ldots, \qquad T_n^{(n)}(a) = p^{(n)}(a).$$

The expression

$$\frac{p^{(n+1)}(c)}{(n + 1)!}(x - a)^{n+1}$$

is called the Lagrange form of the remainder and is the error in approximating $p(x)$ by $T_n(x)$. Determining techniques for minimizing this error is an important topic in the study of numerical solutions to initial value problems.

As in Euler's method, suppose we wish to obtain approximations at the x-values

$$x_0, x_1, x_2, \ldots, x_N$$

to the solution $y = \phi(x)$. In the Taylor method approach we begin by substituting y for p in Equation (2), which gives us

$$
T_n(x) = y(a) + y'(a)(x - a) + \frac{y''(a)}{2}(x - a)^2 + \frac{y'''(a)}{3!}(x - a)^3
$$
$$
+ \cdots + \frac{y^{(n)}(a)}{n!}(x - a)^n. \tag{3}
$$

We then substitute x_0 for a and x_1 for x in Equation (3), obtaining

$$
T_n(x_1) = y(x_0) + y'(x_0)(x_1 - x_0) + \frac{y''(x_0)}{2}(x_1 - x_0)^2 + \frac{y'''(x_0)}{3!}(x_1 - x_0)^3
$$
$$
+ \cdots + \frac{y^{(n)}(x_0)}{n!}(x_1 - x_0)^n.
$$

We will use $T_n(x_1)$ as our approximation to $\phi(x_1)$. Substituting x_1 for a and x_2 for x in Equation (3), we obtain

$$
T_n(x_2) = y(x_1) + y'(x_1)(x_2 - x_1) + \frac{y''(x_1)}{2}(x_2 - x_1)^2 + \frac{y'''(x_1)}{3!}(x_2 - x_1)^3
$$
$$
+ \cdots + \frac{y^{(n)}(x_1)}{n!}(x_2 - x_1)^n
$$

as our approximation for $\phi(x_2)$. Continuing this process, we obtain

$$
T_n(x_{k+1}) = y(x_k) + y'(x_k)(x_{k+1} - x_k) + \frac{1}{2}y''(x_k)(x_{k+1} - x_k)^2
$$
$$
+ \cdots + \frac{1}{n!}y^{(k)}(x_k)(x_{k+1} - x_k)^n
$$

as an approximation for $\phi(x_{k+1})$.

The next example illustrates Taylor's method for $n = 2$.

EXAMPLE 2   Find the second order Taylor approximation to

$$
y' = y + x, \qquad y(1) = 0
$$

at the x-values

$$
1, 1.1, 1.2, 1.4, 1.5.
$$

Solution   For the second order Taylor approximation we need the second derivative of the solution.

This is

$$y'' = \frac{d}{dx}y' = \frac{d}{dx}(y+x) = y' + 1.$$

Since

$$T_2(x) = y(x_0) + y'(x_0)(x - x_0) + \frac{1}{2}y''(x_0)(x - x_0)^2,$$

we have that

$$T_2(x_1) = y(1) + y'(1)(x_1 - 1) + \frac{1}{2}y''(1)(x_1 - 1)^2$$

$$= y(1) + y'(1)(1.1 - 1) + \frac{1}{2}y''(1)(1.1 - 1)^2$$

$$= 0 + 0.1 + \frac{1}{2}(2)(0.1)^2 = 0.11.$$

We now develop the second order Taylor polynomial at $(x_1, T_2(x_1)) = (1.1, 0.11)$. Since

$$T_2(x) = y(x_1) + y'(x_1)(x - x_1) + \frac{1}{2}y''(x_1)(x - x_1)^2,$$

we have that

$$T_2(x_2) = y(1.1) + y'(1.1)(x_2 - 1.1) + \frac{1}{2}y''(1.1)(x_2 - 1.1)^2$$

$$= y(1.1) + y'(1.1)(1.2 - 1.1) + \frac{1}{2}y''(1.1)(1.2 - 1.1)^2$$

$$\approx 0.11 + 1.21(0.1) + \frac{1}{2}(2.21)(0.1)^2 \approx 0.24205.$$

(We have used the approximation symbol since each of the values we substituted for $y(1.1)$, $y'(1.1)$, and $y''(1.1)$ are approximations.) Continuing, we have

$$T_2(x) = y(x_2) + y'(x_2)(x - x_2) + \frac{1}{2}y''(x_2)(x - x_2)^2,$$

so that

$$T_2(x_3) = y(1.2) + y'(1.2)(x_3 - 1.2) + \frac{1}{2}y''(1.2)(x_3 - 1.2)^2$$

$$= y(1.2) + y'(1.2)(1.4 - 1.2) + \frac{1}{2}y''(1.2)(1.4 - 1.1)^2$$

$$\approx 0.24205 + 1.44205(0.2) + \frac{1}{2}(2.44205)(0.2)^2 \approx 0.579301$$

and

$$T_2(x) = y(x_3) + y'(x_3)(x - x_3) + \frac{1}{2}y''(x_3)(x - x_3)^2,$$

so that

$$T_2(x_4) = y(1.4) + y'(1.4)(x_4 - 1.4) + \frac{1}{2}y''(1.4)(x_4 - 1.4)^2$$

$$= y(1.4) + y'(1.4)(1.5 - 1.4) + \frac{1}{2}y''(1.4)(1.5 - 1.4)^2$$

$$\approx 0.579301 + 1.979301(0.1) + \frac{1}{2}(2.979301)(0.1)^2$$

$$\approx 0.7921276. \qquad\qquad \bullet$$

Comparing with Example 1, notice that the second order Taylor polynomial gives a better approximation of $\phi(x)$ at each of the x-values than Euler's method.

In Example 3 we use the fourth order Taylor method.

EXAMPLE 3   Give the fourth order Taylor method approximation to

$$y' = x^2 + xy, \qquad y(0) = 1$$

at the x-values

$$0, 0.1, 0.2, 0.3, 0.4.$$

Solution   We need y'', y''', and $y^{(4)}$. Upon differentiating we obtain

$$y'' = \frac{d}{dx}y' = \frac{d}{dx}(x^2 + xy) = 2x + y + xy',$$

$$y''' = \frac{d}{dx}y'' = \frac{d}{dx}(2x + y + xy') = 2 + 2y' + xy'',$$

$$y^{(4)} = \frac{d}{dx}y''' = \frac{d}{dx}(2 + 2y' + xy'') = 3y'' + xy'''.$$

Using the formula for the nth order Taylor polynomial with $n = 4$ gives us

$$y_{n+1} = y_n + (x_n^2 + x_n y_n)(x_{n+1} - x_n) + \frac{1}{2}(2x_n + y_n + x_n y'(x_n, y_n))(x_{n+1} - x_n)^2$$

$$+ \frac{1}{3!}(2 + 2y'(x_n, y_n) + x_n y''(x_n, y_n))(x_{n+1} - x_n)^3$$

$$+ \frac{1}{4!}(3y''(x_n, y_n) + x_n y'''(x_n, y_n))(x_{n+1} - x_n)^4.$$

Beginning by substituting in $x_0 = 0$ and $y_0 = 1$ gives us:

$$y_1 \approx 1.005345833$$
$$y_2 \approx 1.022887920$$
$$y_3 \approx 1.055189331$$
$$y_4 \approx 1.053149. \qquad\qquad \bullet$$

In practice, one usually chooses $x_1, x_2, \ldots x_n$ so that they are equally spaced. That is,

$$x_1 = x_0 + h$$
$$x_2 = x_1 + h = x_0 + 2h$$
$$\vdots$$
$$x_{N+1} = x_N + h = x_0 + Nh,$$

for some fixed number h. For example, $h = 0.1$ in Example 3. In Exercise 9 we ask you to obtain formulas for the Euler and Taylor methods in this special case.

3.9.3   Runge-Kutta Methods

In 1895 the German mathematician Carle Runge (1856–1927) used Taylor's formula to develop other techniques for generating numerical solutions to initial value problems. In 1901 the German mathematician M. W. Kutta (1867–1944) improved on Runge's works. The techniques that these two individuals developed are now known as Runge-Kutta methods and are very popular for approximating solutions to initial value problems. We only present their fourth order method and will not attempt to explain how the formula is developed. We leave the development of this formula as well as the other Runge-Kutta methods for a numerical analysis course.

The fourth order Runge-Kutta method for the initial value problem

$$\boxed{y' = f(x, y), \qquad y(x_0) = y_0}$$

is a weighted average of values of $f(x, y)$. We present the fourth order Runge-Kutta method only for equally spaced x-values.

$$\boxed{x_0, x_0 + h, x_0 + 2h \ldots x_0 + Nh}$$

We set

$$\boxed{\begin{aligned}
a_n &= f(x_n, y_n) \\
b_n &= f\left(x_n + \frac{h}{2}, y_n + \frac{h}{2}a_n\right) \\
c_n &= f\left(x_n + \frac{h}{2}, y_n + \frac{h}{2}b_n\right) \\
d_n &= f\left(x_n + h, y_n + hc_n\right).
\end{aligned}}$$

The approximation for $\phi(x_{n+1})$ is then given by

$$y_{n+1} = y_n + \frac{h}{6}(a_n + 2b_n + 2c_n + d_n).$$

The term

$$\frac{1}{6}(a_n + 2b_n + 2c_n + d_n)$$

may be interpreted as a weighted average slope.

In the exercises we ask you to compare the numerical solutions of the Euler method, fourth order Taylor method, and fourth order Runge-Kutta method for some initial value problems.

EXERCISES 3.9

In Exercises 1–4, calculate the Euler method approximations to the solution of the initial value problem at the given x-values. Compare your results to the exact solution at these x-values.

1. $y' = 2y - x$; $y(0) = 1$, $x = 0.1, 0.2, 0.4, 0.5$

2. $y' = xy + 2$; $y(0) = 0$, $x = 0.1, 0.2, 0.4, 0.5$

3. $y' = x^2 y^2$; $y(1) = 1$, $x = 1.2, 1.4, 1.6, 1.8$

4. $y' = y + y^2$; $y(1) = -1$, $x = 1.2, 1.4, 1.6, 1.8$

In Exercises 5–8, use the Taylor method of the stated order n to obtain numerical solutions to the indicated exercise of this section. Compare the results from this method with the method of this exercise.

5. Exercise 1, $n = 2$ **6.** Exercise 2, $n = 2$

7. Exercise 3, $n = 3$ **8.** Exercise 4, $n = 3$

9. Derive formulas in the Euler and Taylor methods for
$x_{n+1} = x_n + h = x_0 + nh, n = 0, 1, 2, \ldots, N.$

In Exercises 10–13, use the fourth order Runge-Kutta method with $h = 0.1$ and $N = 5$ to obtain numerical solutions. In Exercise 10, compare with the results of Exercises 2 and 6, and in Exercise 11, compare with the results of Exercises 1 and 5.

10. $y' = xy + 2$, $y(0) = 0$

11. $y' = 2y - x$, $y(0) = 1$

12. $y' = x^2 - y^2$, $y(0) = 0$

13. $y' = x^2 + y^2$, $y(0) = 0$

In Exercises 14–19, use a software package such as Maple to help with the calculations.

14. Consider the initial value problem $y' = x \tan y$, $y(0) = 1$. Use Euler's method with the following values for (h, N): $(0.1, 10)$, $(0.05, 20)$, $(0.025, 40)$, $(0.0125, 80)$. Solve this initial value problem and compare these approximating values with those of the actual solution. What conclusions can you make regarding the accuracy of the Euler method for this initial value problem? What happens if you increase N?

15. Consider the initial value problem $y' = x^2 y$, $y(0) = 1$. Use Euler's method with the following values for (h, N): $(0.1, 10)$, $(0.05, 20)$, $(0.025, 40)$, $(0.0125, 80)$. Solve this initial value problem and compare these approximating values with those of the actual solution. What conclusions can you make regarding the accuracy of the Euler method for this initial value problem? What happens if you increase N?

16. Compare the fourth order Taylor and fourth order Runge-Kutta methods for $y' = x^2 y + 4x$, $y(0) = 0$ with $h = 0.1$ and $N = 40$. Solve this initial value problem and compare these approximating values with those of the actual solution. Which approximations are more accurate?

17. Compare the fourth order Taylor and fourth order Runge-Kutta methods for $y' = xy - 6$, $y(0) = 0$ with $h = 0.1$ and $N = 40$. Solve this initial value problem and compare these approximating values with those of the actual solution. Which approximations are more accurate?

18. A closed form solution cannot be obtained for the initial value problem $y' = x^2 + y^2$, $y(0) = 0$. Use the fourth order Taylor and fourth order Runge-Kutta methods on the interval $0 < x < 1$ first with $h = 0.2$ and then with $h = 0.1$ to obtain numerical solutions to this initial value problem. Then use Maple (or an appropriate software package) to plot the points obtained with these numerical solutions and to graph the phase portrait of this initial value problem. Which method appears to give a more accurate answer?

19. Do Exercise 18 for the initial value problem $y' = \cos y + \sin x$, $y(0) = 0$, which also does not have a closed form solution.

In Exercises 20–23, use the *dsolve* command in Maple with the *numeric option* to determine a numerical solution at $x = 0.5$ with Maple's default method. Then use Maple's *odeplot* command to graph Maple's numerical solution over the interval $[0, 1]$. (Or use another appropriate software package.) Finally, use Maple to graph the phase portrait of the initial value problem over this interval and compare this graph with the graph of the numerical solution.

20. $y' = x^2 + y^2$, $y(0) = 0$

21. $y' = \cos y + \sin x$, $y(0) = 0$

22. $y' = \sin y + e^{-x}$, $y(1) = 0$

23. $y' = x^4 + y^2$, $y(1) = -1$

Linear Differential Equations

In the previous chapter we focused most of our attention on techniques for solving first order differential equations. Among the types of equations considered were the first order linear differential equations

$$q_1(x)y' + q_0(x)y = g(x).$$

In this chapter we are going to study higher order linear differential equations. As we do so, you will see how many of the vector space concepts from Chapter 2 come into play in this study. Higher order linear differential equations arise in many applications. Once you have learned techniques for solving these differential equations, we will turn our attention to some of the applications involving them, such as mass-spring systems and electrical circuits in the last section of this chapter. To get started, we first discuss the theory of these equations.

4.1 THE THEORY OF HIGHER ORDER LINEAR DIFFERENTIAL EQUATIONS

An **nth order linear differential equation** is a differential equation that can be written in the form

$$q_n(x)y^{(n)} + q_{n-1}(x)y^{(n-1)} + \cdots + q_1(x)y' + q_0(x)y = g(x). \tag{1}$$

Throughout this chapter, we will assume that the functions $q_n, q_{n-1}, \ldots, q_0$ and g are continuous on an interval (a, b). We shall further assume that $q_n(x)$ is not zero for any x in (a, b). (Recall that we included such assumptions for the first order linear differential equations $q_1(x)y' + q_0(x)y = g(x)$ in Chapter 3.) Note that if $y = f(x)$ is a solution to the differential equation (1) on (a, b), then f certainly must have an nth derivative on

(a, b) and consequently is a function in the vector space of functions with nth derivatives on (a, b), $D^n(a, b)$. Actually, we can say a bit more. Since

$$q_n(x)f^{(n)}(x) + q_{n-1}(x)f^{(n-1)}(x) + \cdots + q_1(x)f'(x) + q_0(x)f(x) = g(x),$$

solving for $f^{(n)}(x)$ leads to

$$f^{(n)}(x) = \frac{g(x)}{q_n(x)} - \frac{q_{n-1}(x)f^{(n-1)}(x)}{q_n(x)} - \cdots - \frac{q_0(x)f(x)}{q_n(x)},$$

showing that $f^{(n)}$ is continuous on (a, b). Hence the solutions of an nth order linear differential equation on (a, b) will be elements of the vector space of functions with continuous nth derivatives on (a, b), $C^n(a, b)$.

If $g(x) = 0$ for all x in (a, b) in Equation (1), then we call the nth order linear differential equation **homogeneous**; otherwise we call it **nonhomogeneous**. If g is nonzero, we refer to

$$\boxed{q_n(x)y^{(n)} + q_{n-1}(x)y^{(n-1)} + \cdots + q_1(x)y' + q_0(x)y = 0}$$

as the **corresponding homogeneous equation** of

$$q_n(x)y^{(n)}(x) + q_{n-1}(x)y^{(n-1)}(x) + \cdots + q_1(x)y'(x) + q_0(x)y(x) = g(x).$$

To illustrate our terminology, consider the following differential equations.

$$x^2y'' - x(1+x)y' + y = x^2 e^{2x} \tag{2}$$

$$3y''' + 2y'' - 2y' + 4y = 0 \tag{3}$$

$$y^{(4)} + x^2y'' - 2e^x y' + \cos xy = \sin x \tag{4}$$

$$y'''y + x^2y'' - 2xy' + y = 0 \tag{5}$$

$$x^2y'' - 2x(y')^2 + y = x \tag{6}$$

$$y^{(4)} + x^2\cos y'' - 2xy' + y = \sin x \tag{7}$$

Equation (2) is a second order nonhomogeneous linear differential equation. In it, $q_2(x) = x^2$, $q_1(x) = -x(1+x)$, $q_0(x) = 1$, and $g(x) = x^2 e^{2x}$. The corresponding homogeneous equation is

$$x^2y'' - x(1+x)y' + y = 0.$$

Equation (3) is a third order homogeneous linear differential equation. Equation (4) is a fourth order nonhomogeneous nonlinear differential equation. Equation (5) is nonlinear due to the $y'''y$ term, Equation (6) is nonlinear due to the term involving the square of y', and Equation (7) is nonlinear due to the term involving $\cos y''$.

By an **initial value problem** with an nth order linear differential equation, we mean an nth order linear differential equation along with initial conditions on $y^{(n-1)}$, $y^{(n-2)}, \ldots, y', y$ at a value x_0 in (a, b). We will write our initial value problems in the form

$$q_n(x)y^{(n)} + q_{n-1}(x)y^{(n-1)} + \cdots + q_1(x)y' + q_0(x)y = g(x);$$
$$y(x_0) = k_0, \quad y'(x_0) = k_1, \quad \ldots, \quad y^{(n-1)}(x_0) = k_{n-1}.$$

Theorem 3.1 of the last chapter told us that first order initial value problems have unique solutions (provided the function f in the differential equation $y' = f(x, y)$ is sufficiently well-behaved). Similar theorems hold for higher order differential equations. The next theorem, whose proof will be omitted, is the version of this for nth order linear differential equations.

THEOREM 4.1   Suppose $q_n, q_{n-1}, \ldots, q_1, q_0$, and g are continuous on an interval (a, b) containing x_0 and suppose $q_n(x) \neq 0$ for all x in (a, b). Then the initial value problem

$$q_n(x)y^{(n)} + q_{n-1}(x)y^{(n-1)} + \cdots + q_1(x)y' + q_0(x)y = g(x);$$
$$y(x_0) = k_0, \quad y'(x_0) = k_1, \quad \ldots, \quad y^{(n-1)}(x_0) = k_{n-1}$$

where $k_0, k_1, \ldots, k_{n-1}$ are constants has one and only one solution on the interval (a, b).

Using Theorem 4.1, we can prove the following crucial theorem about homogeneous linear differential equations.

THEOREM 4.2   The solutions to an nth order homogeneous linear differential equation

$$q_n(x)y^{(n)} + q_{n-1}(x)y^{n-1} + \cdots + q_0(x)y = 0$$

on an interval (a, b) where $q_n, q_{n-1}, \ldots, q_0$ are continuous on (a, b) and $q_n(x)$ is nonzero for all x in (a, b) form a vector space of dimension n.

Proof   We have already noted that the solutions are functions in $C^n(a, b)$. To show that they form a vector space, we shall show that they are a subspace of the vector space $C^n(a, b)$. Suppose that $f_1(x)$ and $f_2(x)$ are solutions of the homogeneous linear differential equation. Since

$$q_n(x)(f_1(x) + f_2(x))^{(n)} + q_{n-1}(x)(f_1(x) + f_2(x))^{(n-1)} + \cdots + q_0(x)(f_1(x) + f_2(x))$$
$$= q_n(x)f_1^{(n)}(x) + q_{n-1}(x)f_1^{(n-1)}(x) + \cdots + q_0(x)f_1(x)$$
$$+ q_n(x)f_2^{(n)}(x) + q_{n-1}(x)f_2^{(n-1)}(x) + \cdots + q_0(x)f_2(x)$$
$$= 0 + 0 = 0,$$

$f_1(x) + f_2(x)$ is a solution and the solutions are then closed under addition. For any scalar c,

$$q_n(x)(cf_1(x))^{(n)} + q_{n-1}(x)(cf_1(x))^{(n-1)} + \cdots + q_0(x)(cf_1(x))$$
$$= c(q_n(x)f_1^{(n)}(x) + q_{n-1}(x)f_1^{(n-1)}(x) + \cdots + q_0(x)f_1(x)) = c \cdot 0 = 0$$

and hence $cf_1(x)$ is a solution giving us the solutions are closed under scalar multiplication. Thus the solutions do form a subspace of $C^n(a, b)$.

To complete this proof, we show that the solutions have a basis of n functions. Choose a value x_0 in (a, b). By Theorem 4.1, there is a solution $y_1(x)$ satisfying the initial value problem

$$q_n(x)y^{(n)} + q_{n-1}(x)y^{n-1} + \cdots + q_0(x)y = 0;$$

$$y_1(x_0) = 1, \quad y_1'(x_0) = 0, \quad y_1''(x_0) = 0, \quad \ldots, \quad y_1^{(n-1)}(x_0) = 0,$$

a solution $y_2(x)$ satisfying the initial value problem

$$q_n(x)y^{(n)} + q_{n-1}(x)y^{n-1} + \cdots + q_0(x)y = 0;$$

$$y_2(x_0) = 0, \quad y_2'(x_0) = 1, \quad y_2''(x_0) = 0, \quad \ldots, \quad y_2^{(n-1)}(x_0) = 0,$$

$$\vdots$$

a solution $y_n(x)$ satisfying the initial value problem

$$q_n(x)y^{(n)} + q_{n-1}(x)y^{n-1} + \cdots + q_0(x)y = 0;$$

$$y_n(x_0) = 0, \quad y_n'(x_0) = 0, \quad y_n''(x_0) = 0, \quad \ldots, \quad y_n^{(n-1)}(x_0) = 1.$$

We show that these n solutions $y_1(x), y_2(x), \ldots, y_n(x)$ form a basis for our vector space of solutions. They are linearly independent since their Wronskian at x_0 is

$$w(y_1(x_0), y_2(x_0), \cdots, y_n(x_0)) = \begin{vmatrix} 1 & 0 & \cdots & 0 \\ 0 & 1 & \cdots & 0 \\ \vdots & \vdots & & \vdots \\ 0 & 0 & \cdots & 1 \end{vmatrix} \neq 0.$$

To see that $y_1(x), y_2(x), \ldots, y_n(x)$ span the vector space of solutions, suppose that $f(x)$ is a solution of the homogeneous linear differential equation. Let us denote the values of $f(x), f'(x), \ldots, f^{(n-1)}(x)$ at x_0 by $c_1, c_2, \ldots, c_n$:

$$f(x_0) = c_1, \quad f'(x_0) = c_2, \quad \ldots, \quad f^{(n-1)}(x_0) = c_n.$$

Now let

$$F(x) = c_1 y_1(x) + c_2 y_2(x) + \cdots + c_n y_n(x).$$

Then

$$F(x_0) = c_1 y_1(x_0) + c_2 y_2(x_0) + \cdots + c_n y_n(x_0) = c_1,$$
$$F'(x_0) = c_1 y_1'(x_0) + c_2 y_2'(x_0) + \cdots + c_n y_n'(x_0) = c_2,$$

$$\vdots$$

$$F^{(n-1)}(x_0) = c_1 y_1^{(n-1)}(x_0) + c_2 y_2^{(n-1)}(x_0) + \cdots + c_n y_n^{(n-1)}(x_0) = c_n.$$

Thus $F(x)$ is a solution to the same initial value problem as $f(x)$. Since such solutions are unique by Theorem 4.1,

$$f(x) = F(x) = c_1 y_1(x) + c_2 y_2(x) + \cdots + c_n y_n(x)$$

and hence $y_1(x), y_2(x), \ldots, y_n(x)$ do span the vector space of solutions. ●

The last theorem indicates that we should look for a set of solutions to the homogeneous equation that forms a basis for the vector space of all solutions to the homogeneous equation. A set of n linearly independent functions $y_1, \ldots, y_n$ each of which is a solution of an nth order homogeneous linear differential equation is called a **fundamental set of solutions** for the homogeneous equation. Recalling part (1) of Theorem 2.12, which tells us that n linearly independent vectors in an n-dimensional vector space form a basis for the vector space, it follows that a fundamental set of solutions is a basis for the solutions to the homogeneous equation. If $y_1, \ldots, y_n$ is a fundamental set of solutions to an nth order homogeneous linear differential equation, then the general solution to the homogeneous equation is

$$y_H = c_1 y_1 + \cdots + c_n y_n.$$

The next theorem shows how we use the general solution to the corresponding homogeneous equation to obtain the general solution to a nonhomogeneous equation.

THEOREM 4.3   Suppose that $y_1, \ldots, y_n$ forms a fundamental set of solutions to the homogeneous linear differential equation

$$q_n(x)y^{(n)} + q_{n-1}(x)y^{(n-1)} + \cdots + q_1(x)y' + q_0(x)y = 0$$

and that y_P is a solution to the nonhomogeneous linear differential equation

$$q_n(x)y'' + q_{n-1}(x)y^{(n-1)} + \cdots + q_1(x)y' + q_0(x)y = g(x).$$

Then the general solution to this nonhomogeneous linear differential equation is

$$y = y_H + y_P = c_1 y_1 + \cdots + c_n y_n + y_P$$

where $c_1, \ldots, c_n$ are constants.

Proof   We first show that any function given in the form $c_1 y_1 + \cdots + c_n y_n + y_P$ is a solution to the nonhomogeneous equation. Substituting this into the left-hand side of the differential equation, we have

$$q_n(x)(c_1 y_1 + \cdots + c_n y_n + y_P)^{(n)} + q_{n-1}(x)(c_1 y_1 + \cdots + c_n y_n + y_P)^{(n-1)}$$
$$+ \cdots + q_0(x)(c_1 y_1 + \cdots + c_n y_n + y_P)$$
$$= q_n(x)(c_1 y_1 + \cdots + c_n y_n)^{(n)} + q_{n-1}(x)(c_1 y_1 + \cdots + c_n y_n)^{(n-1)}$$
$$+ \cdots + q_0(x)(c_1 y_1 + \cdots + c_n y_n) + q_n(x)y_P^{(n)} + q_{n-1}y_P^{(n-1)} + \cdots + q_0(x)y_P$$
$$= 0 + g(x) = g(x)$$

and hence $c_1 y_1 + \cdots + c_n y_n + y_P$ is indeed a solution to the nonhomogeneous solution.

To see that the general solution is $c_1 y_1 + \cdots + c_n y_n + y_P$, we must show every solution of the nonhomogeneous equation has this form for some constants $c_1, \ldots, c_n$.

Suppose y is a solution of the nonhomogeneous equation. Since

$$q_n(x)(y - y_P)^{(n)} + q_{n-1}(x)(y - y_P)^{(n-1)} + \cdots + q_0(x)(y - y_P)$$
$$= q_n(x)y^{(n)} + q_{n-1}(x)y^{(n-1)} + \cdots + q_0(x)y - (q_n(x)y_P^{(n)} + q_{n-1}(x)y_P^{(n-1)}$$
$$+ \cdots + q_0(x)y_P)$$
$$= g(x) - g(x) = 0,$$

$y - y_P$ is a solution to the homogeneous equation. Thus there are scalars $c_1, \ldots, c_n$ so that

$$y - y_P = c_1 y_1 + \cdots + c_n y_n$$

or

$$y = c_1 y_1 + \cdots + c_n y_n + y_P$$

as needed.   ●

We call the function y_P of Theorem 4.3 a **particular solution** to the nonhomogeneous differential equation. This theorem tells us that we may find the general solution to the nonhomogeneous equation by first determining the general solution y_H to the corresponding homogeneous solution, then determining a particular solution y_P to the nonhomogeneous equation, and finally adding y_H and y_P together.

We now look at some examples illustrating our theorems.

EXAMPLE 1   Consider the differential equation

$$y'' - y = x.$$

(a) Determine the largest interval for which a unique solution to the initial value problem

$$y'' - y = x; \qquad y(0) = 1, \qquad y'(0) = 0$$

is guaranteed.

(b) Give the corresponding homogeneous equation, show that $y_1 = e^x$, $y_2 = e^{-x}$ form a fundamental set of solutions to the homogeneous equation and give the general solution to this homogeneous equation.

(c) Solve the initial value problem $y'' - y = 0$; $y(0) = 1$, $y'(0) = 0$.

(d) Show $y_P = -x$ is a particular solution and give the general solution to the nonhomogeneous equation.

(e) Solve the initial value problem $y'' - y = x$; $y(0) = 1$, $y'(0) = 0$.

Solution

(a) Since $q_2(x) = 1$, $q_1(x) = 0$, $q_0(x) = -1$, and $g(x) = x$ are continuous for all x and $q_2(x) \neq 0$ for all x, a unique solution to the initial value problem is guaranteed for all x or, to put it another way, on the interval $(-\infty, \infty)$.

(b) The corresponding homogeneous equation is

$$y'' - y = 0.$$

Substituting in $y_1 = e^x$ and $y_2 = e^{-x}$ shows they are solutions of this homogeneous equation. Since the Wronskian of y_1 and y_2 is

$$w(e^x, e^{-x}) = \begin{vmatrix} e^x & e^{-x} \\ e^x & -e^{-x} \end{vmatrix} = -e^x e^{-x} - e^x e^{-x} = -2 \neq 0,$$

y_1, y_2 form a fundamental set of solutions to this second order homogeneous linear differential equation. Hence, the general solution to the homogeneous equation is given by

$$y_H = c_1 y_1 + c_2 y_2 = c_1 e^x + c_2 e^{-x}.$$

(c) Since

$$y_H = c_1 e^x + c_2 e^{-x} \quad \text{and} \quad y'_H = c_1 e^x - c_2 e^{-x},$$

in order for this y_H to solve the homogeneous initial value problem c_1 and c_2 must satisfy

$$y(0) = c_1 + c_2 = 1$$
$$y'(0) = c_1 - c_2 = 0.$$

Solving this system gives us $c_1 = 1/2$, $c_2 = 1/2$. Therefore, the unique solution to the homogeneous initial value problem is given by

$$y_H = \frac{1}{2}e^x + \frac{1}{2}e^{-x} = \frac{1}{2}(e^x + e^{-x}).$$

(d) Substituting $y_P = -x$ into the nonhomogeneous equation

$$y'' - y = x,$$

we see it satisfies this equation. The general solution to the nonhomogeneous equation is

$$y = c_1 y_1 + c_2 y_2 + y_P = c_1 e^x + c_2 e^{-x} - x.$$

(e) Since

$$y = c_1 e^x + c_2 e^{-x} - x \quad \text{and} \quad y' = c_1 e^x - c_2 e^{-x} - 1,$$

c_1 and c_2 must satisfy

$$y(0) = c_1 + c_2 = 1$$
$$y'(0) = c_1 - c_2 - 1 = 0.$$

Solving this system gives us $c_1 = 1$, $c_2 = 0$. Therefore, the unique solution to the nonhomogeneous initial value problem is

$$y = e^x - x.$$

EXAMPLE 2   Show $1, \cos 2x, \sin 2x$ form a fundamental set of solutions for

$$y''' + 4y' = 0.$$

Also, give the general solution to this homogeneous equation.

Solution   Substituting each of $1, \cos 2x, \sin 2x$ into this homogeneous equation, we find each is a solution of the differential equation. Their Wronskian is

$$w(1, \cos 2x, \sin 2x) = \begin{vmatrix} 1 & \cos 2x & \sin 2x \\ 0 & -2\sin 2x & 2\cos 2x \\ 0 & -4\cos 2x & -4\sin 2x \end{vmatrix}$$

$$= 8.$$

Since this is not zero for any x, we have that $1, \cos 2x, \sin 2x$ is a fundamental set of solutions and

$$y_H = c_1 + c_2 \cos 2x + c_3 \sin 2x$$

is the general solution to this third order homogeneous differential equation.   ●

EXAMPLE 3   Determine $g(x)$ so that $y_P = x^4$ is a particular solution to

$$y''' + 4y' = g(x).$$

For this $g(x)$, solve the initial value problem for the initial conditions $y(1) = 0$, $y'(1) = -1$, $y''(1) = 0$.

Solution   Substituting in $y_P = x^4$ leads to

$$y''' + 4y' = 24x + 16x^3.$$

Therefore,

$$g(x) = 24x + 16x^3.$$

Using the results of Example 2, the general solution of

$$y''' + 4y' = 24x + 16x^3$$

is given by

$$y = c_1 + c_2 \cos 2x + c_3 \sin 2x + x^4.$$

Also

$$y' = -2c_2 \sin 2x + 2c_3 \cos 2x + 4x^3$$

and

$$y'' = -4c_2 \cos 2x - 4c_3 \sin 2x + 12x^2.$$

Using the initial conditions, we obtain the system of equations:

$$y(1) = c_1 + c_2 \cos 2 + c_3 \sin 2 + 1 = 0$$
$$y'(1) = -2c_2 \sin 2 + 2c_3 \cos 2 + 4 = -1$$
$$y''(1) = -4c_2 \cos 2 - 4c_3 \sin 2 + 12 = 0.$$

Solving this system of equations gives us

$$c_1 = -4, \qquad c_2 = \frac{6 \cos 2 + 5 \sin 2}{2}, \qquad c_3 = -\frac{5 \cos 2 - 6 \sin 2}{2}.$$

Therefore, the unique solution to this initial value problem is

$$y = -4 + \frac{6 \cos 2 + 5 \sin 2}{2} \cos 2x - \frac{5 \cos 2 - 6 \sin 2}{2} \sin 2x + x^4. \qquad \bullet$$

We conclude this section with some important results regarding the Wronskian of solutions to nth order homogeneous linear differential equations. In Section 2.5 we saw that the Wronskian of a set of functions could be zero and the set of functions is still linearly independent. The following theorem shows this cannot be true for solutions to an nth order homogeneous linear differential equation.

THEOREM 4.4   Let $y_1, \ldots, y_n$ be solutions to the differential equation

$$q_n(x)y^{(n)} + q_{n-1}(x)y^{(n-1)} + \cdots + q_1(x)y' + q_0(x)y = 0$$

on an interval (a, b). If $w(y_1(x_0), \ldots, y_n(x_0)) = 0$ for any x_0 in (a, b), then $y_1, \ldots, y_n$ are linearly dependent on (a, b).

Proof   We prove this for $n = 2$ and leave the proof for general n as an exercise (Exercise 19). If

$$w(y_1(x_0), y_2(x_0)) = \begin{vmatrix} y_1(x_0) & y_2(x_0) \\ y_1'(x_0) & y_2'(x_0) \end{vmatrix} = 0,$$

then the system of equations

$$c_1 y_1(x_0) + c_2 y_2(x_0) = 0$$
$$c_1 y_1'(x_0) + c_2 y_2'(x_0) = 0$$

has infinitely many solutions. Let $c_1 = A$, $c_2 = B$ be one of the solutions to this system and then let

$$u(x) = A y_1(x) + B y_2(x).$$

We have

$$u(x_0) = A y_1(x_0) + B y_2(x_0) = 0, \qquad u'(x_0) = A y_1'(x_0) + B y_2'(x_0) = 0.$$

Thus u is a solution to the initial value problem

$$q_2(x)y'' + q_1(x)y' + q_0(x)y = 0; \qquad y(x_0) = 0, \qquad y'(x_0) = 0.$$

But so is $y(x) = 0$. Hence, by uniqueness

$$u(x) = Ay_1(x) + By_2(x) = 0,$$

and y_1, y_2 are linearly dependent. ●

As a consequence of Theorem 4.4 we have the following result, which we will use in Section 4.4.

COROLLARY 4.5   Let $y_1, \ldots, y_n$ be a fundamental set of solutions to the differential equation

$$q_n(x)y^{(n)} + q_{n-1}(x)y^{(n-1)} + \cdots + q_1(x)y' + q_0(x)y = 0$$

on an interval (a, b). Then $w(y_1, \ldots, y_n)$ is never zero on (a, b).

Proof   If $w(y_1(x_0), \ldots, y_n(x_0)) = 0$ for some x_0 in (a, b), then $y_1, \ldots, y_n$ are linearly dependent by Theorem 4.4. ●

EXERCISES 4.1

In Exercises 1–4, determine if the differential equation is linear or not. If it is linear, give the order of the differential equation.

1. $e^x y''' + (2 \sin x)y'' - 4x^2 y' - 5y = 3$

2. $e^y y'' - 4y' + 5xy = 0$

3. $5x \sin x = 3y^{(4)} - 4xy''' + \frac{5}{x}y'' - 2y' + x^3 y$

4. $y'' = 2y' \sin x - 3xy$

In Exercises 5–8, give the largest interval for which Theorem 4.1 guarantees a unique solution to the initial value problem.

5. $y'' - 2xy' + 4y = 5x$; $y(0) = 1$, $y'(0) = 0$

6. $x(x^2 - 1)y'' - 2xy' + 4y = 0$; $y(2) = 0$, $y'(2) = -1$

7. $e^x \ln xy''' - xy'' + 2y' - 5xy = 2$; $y(1) = 0$, $y'(1) = 0$, $y''(1) = -1$

8. $4y^{(4)} - 3y''' + 2y'' + y' - 6y = 7x^2 - 3x + 4$; $y(-1) = 0$, $y'(-1) = 2$, $y''(-1) = 0$, $y'''(-1) = -\pi$

In Exercises 9–12, show the given set of functions forms a fundamental set of solutions to the homogeneous differential equation, give the general solution to the homogeneous differential equation, and then solve the initial value problems for the given initial conditions.

9. $\{e^{-x}, e^{2x}\}$, $y'' - y' - 2y = 0$; $y(0) = 1$, $y'(0) = 0$

10. $\{1/x, x\}$, $x^2 y'' + xy' - y = 0$; $y(1) = 0$, $y'(1) = -1$

11. $\{e^x, e^{2x}, e^{-3x}\}$, $y''' - 7y' + 6y = 0$; $y(0) = 1$, $y'(0) = 0$, $y''(0) = 0$

12. $\{x, x^2, 1/x\}$, $x^3 y''' + x^2 y'' - 2xy' + 2y = 0$; $y(-1) = 0$, $y'(-1) = 0$, $y''(-1) = 0$

In Exercises 13–16, show y_P is a solution of the nonhomogeneous differential equation, give the general solution to the nonhomogeneous equation by using the fact that Exercises 9–12 contain the respective corresponding homogeneous equation, and then solve the initial value problems for the given initial conditions.

13. $y_P = -2$, $y'' - y' - 2y = 4$; $y(0) = 1$, $y'(0) = 0$

14. $y_P = -x/4 + (1/2)x \ln x$, $x^2 y'' + xy' - y = x$; $y(1) = 0$, $y'(1) = -1$

15. $y_P = e^{-x}$, $y''' - 7y' + 6y = 12e^{-x}$; $y(0) = 1$, $y'(0) = 0$, $y''(0) = 0$

16. $y_P = x^4/15$, $x^3 y''' + x^2 y'' - 2xy' + 2y = 2x^4$; $y(-1) = 0$, $y'(-1) = 0$, $y''(-1) = 0$

In Exercises 17 and 18, show the given set of functions forms a fundamental set of solutions to the corresponding homogeneous equation. Then determine $g(x)$ so that y_P is a solution to the given differential equation. Finally, for this $g(x)$, determine the solution to the initial value problem for the given initial conditions.

17. $\{x, x^2\}$, $(1/2)x^2 y'' - xy' + y = g(x)$, $y_P = e^x$; $y(1) = 1$, $y'(1) = 0$

18. $\{e^{x/3}, e^{-2x}, e^{(5/2)x}\}$, $6y''' - 5y'' - 29y' + 10y = g(x)$, $y_P = x^2$; $y(0) = 0$, $y'(0) = 1$, $y''(0) = 0$

19. Prove Theorem 4.4 for an arbitrary positive integer n.

20. Prove the converse of Corollary 4.5.

21. Can x, x^2 form a fundamental set of solutions to a second order homogeneous linear differential equation on $(-\infty, \infty)$? Why or why not?

22. Can $x^3, x^2|x|$ form a fundamental set of solutions to a second order homogeneous linear differential equation on some interval? Why or why not?

23. Let y_1, y_2, y_3 be solutions to a homogeneous linear differential equation. What can you say about the

three solutions if the Wronskian at $x = 1$ is

$$w(y_1(1), y_2(1), y_3(1)) = \begin{vmatrix} 1 & 1 & 1 \\ 1 & 2 & 3 \\ 1 & -1 & 1 \end{vmatrix} ?$$

Use the *dsolve* command in Maple (or the appropriate command in another suitable software package) to find the general solution to the differential equations in Exercises 24–27. Determine y_H and y_P from the output.

24. $2y'' + 3y' - 2y = \cos 2x$

25. $x^2 y'' + 4xy' + 2y = xe^x$

26. $3y''' + 2y'' - 61y' + 20y = 2x - 5$

27. $x^3 y''' + 2x^2 y'' - xy' + y = x^3 + 5x$

4.2   HOMOGENEOUS CONSTANT COEFFICIENT LINEAR DIFFERENTIAL EQUATIONS

A linear differential equation that can be written in the form

$$a_n y^{(n)} + a_{n-1} y^{(n-1)} + \cdots + a_1 y' + a_0 y = g(x)$$

where $a_0, a_1, \ldots, a_n$ are constant real numbers is called a **constant coefficient** nth order linear differential equation. In this section we shall see how to find the general solution of the homogeneous equation

$$a_n y^n + a_{n-1} y^{(n-1)} + \cdots + a_1 y' + a_0 y = 0.$$

After we master this homogeneous case, we will turn our attention to finding particular solutions in the nonhomogeneous case in the next two sections.

To solve constant coefficient linear homogeneous differential equations, we will seek solutions of the form $y = e^{\lambda x}$. Since

$$\frac{d}{dx}(e^{\lambda x}) = \lambda e^{\lambda x}, \qquad \frac{d^2}{dx^2}(e^{\lambda x}) = \lambda^2 e^{\lambda x}, \qquad \ldots, \qquad \frac{d^n}{dx^n}(e^{\lambda x}) = \lambda^n e^{\lambda x},$$

substituting $y = e^{\lambda x}$ in the constant coefficient linear homogeneous differential equation

$$a_n y^{(n)} + a_{n-1} y^{(n-1)} + \cdots + a_1 y' + a_0 y = 0,$$

gives us

$$a_n \lambda^n e^{\lambda x} + a_{n-1} \lambda^{n-1} e^{\lambda x} + \cdots + a_1 \lambda e^{\lambda x} + a_0 e^{\lambda x} = 0. \tag{1}$$

Factoring out $e^{\lambda x}$, we obtain

$$(a_n \lambda^n + a_{n-1} \lambda^{n-1} + \cdots + a_1 \lambda + a_0) e^{\lambda x} = 0.$$

If $p(\lambda)$ is the polynomial of degree n defined by

$$p(\lambda) = a_n\lambda^n + a_{n-1}\lambda^{n-1} + \cdots + a_1\lambda + a_0,$$

then we see that we can rewrite Equation (1) as

$$p(\lambda)e^{\lambda x} = (a_n\lambda^n + a_{n-1}\lambda^{n-1} + \cdots + a_1\lambda + a_0)e^{\lambda x} = 0.$$

Since $e^{\lambda x}$ is never zero, in order for $p(\lambda)e^{\lambda x} = 0$ we must have $p(\lambda) = 0$. That is, the nth order constant coefficient linear homogeneous differential equation has solutions given by $e^{\lambda x}$ for λ satisfying $p(\lambda) = 0$. Therefore, to solve the nth order constant coefficient linear homogeneous differential equation we must determine the roots of $p(\lambda)$. We call $p(\lambda)$ the **characteristic polynomial** or **auxiliary polynomial** for the nth order constant coefficient linear homogeneous differential equation and call the equation $p(\lambda) = 0$ the **characteristic equation** or **auxiliary equation.** The nature of the roots of the characteristic polynomial $p(\lambda)$ will determine how we get a fundamental set of solutions. A polynomial of degree n with real coefficients has (counting multiplicities) n roots all of which may be real or some (including all) of which may be imaginary. We are going to break our treatment into three subsections. In the first subsection, we consider the case in which all of the roots of $p(\lambda)$ are real and distinct. In the second subsection, we consider the case in which all of the roots of $p(\lambda)$ are real but some of them are repeated. In the final subsection, we consider the case where we have imaginary roots of $p(\lambda)$.

4.2.1   Characteristic Equations with Real Distinct Roots

In this subsection we consider differential equations

$$a_n y^n + a_{n-1}y^{(n-1)} + \cdots + a_1 y' + a_0 y = 0$$

that have a characteristic polynomial with n distinct real roots. The following examples illustrate how we determine the general solution to this type of differential equation.

EXAMPLE 1   Determine the general solution to the following differential equation.

$$y'' + 4y' + 3y = 0$$

Solution   The characteristic equation is

$$\lambda^2 + 4\lambda + 3 = 0.$$

Factoring, we have

$$(\lambda + 3)(\lambda + 1) = 0,$$

which has roots $\lambda = -3, -1$. Thus we have solutions

$$e^{-3x}, e^{-x}.$$

The Wronskian of these two functions is

$$\begin{vmatrix} e^{-3x} & e^{-x} \\ -3e^{-3x} & -e^{-x} \end{vmatrix} = 2e^{-4x} \neq 0.$$

Hence e^{-3x}, e^{-x} are linearly independent and form a fundamental set of solutions for this second order equation. The general solution is then

$$y = c_1 e^{-3x} + c_2 e^{-x}.$$ ●

EXAMPLE 2   Determine the general solution to the following differential equation.

$$y''' - 4y' = 0$$

Solution   The characteristic equation is

$$\lambda^3 - 4\lambda = 0.$$

This factors into

$$\lambda(\lambda + 2)(\lambda - 2) = 0.$$

The roots are

$$0, -2, 2.$$

The corresponding solutions to the differential equation are

$$e^{0x} = 1, e^{-2x}, e^{2x}.$$

The Wronskian of these three functions is

$$\begin{vmatrix} 1 & e^{-2x} & e^{2x} \\ 0 & -2e^{-2x} & 2e^{2x} \\ 0 & 4e^{-2x} & 4e^{2x} \end{vmatrix} = -16 \neq 0.$$

Hence these solutions are linearly independent and form a fundamental set of solutions for this third order differential equation. The general solution is then

$$y = c_1 + c_2 e^{2x} + c_3 e^{-2x}.$$ ●

EXAMPLE 3   Determine the general solution to the following differential equation.

$$y''' - 5y'' + 6y' - 2y = 0$$

Solution   The characteristic equation is

$$\lambda^3 - 5\lambda^2 + 6\lambda - 2 = 0.$$

Possible rational roots of this characteristic polynomial are ± 1 and ± 2.[1] We begin to substitute the values $1, -1, 2, -2$ into our characteristic polynomial and find 1 is a root. This tells us $\lambda - 1$ is a factor of our characteristic polynomial. The other factor can be found by dividing $\lambda - 1$ into our characteristic polynomial (using either long or synthetic division). Doing so, we find the quotient (which is the other factor; the remainder is necessarily zero) is $\lambda^2 - 4\lambda + 2$ and hence our characteristic equation becomes

$$(\lambda - 1)(\lambda^2 - 4\lambda + 2) = 0.$$

Using the quadratic formula, we find the roots of the quadratic factor $\lambda^2 - 4\lambda + 2$ are

$$2 + \sqrt{2}, 2 - \sqrt{2}$$

so that all the roots of the characteristic polynomial are

$$1, 2 + \sqrt{2}, 2 - \sqrt{2}.$$

You can also find these roots using the *solve* command in Maple or the appropriate command in a suitable software package. Typing and entering

```
solve(lambda^3-5 * lambda ^ 2+6 * lambda-2=0, lambda)
```

Maple also gives us the roots $1, 2 + \sqrt{2}, 2 - \sqrt{2}$. We now have that

$$e^x, e^{(2+\sqrt{2})x}, e^{(2-\sqrt{2})x}$$

are solutions to this differential equation. Computing the Wronskian of these three functions, we find it to be nonzero. (Try it.) Hence these three functions are linearly independent and form a fundamental set of solutions for this third order equation. The general solution is then

$$y = c_1 e^x + c_2 e^{(2+\sqrt{2})x} + c_3 e^{(2-\sqrt{2})x}. \qquad \bullet$$

In Examples 1–3 you will notice that the functions corresponding to the distinct real roots of the characteristic polynomial have nonzero Wronskian and hence are linearly independent. We leave it as an exercise (Exercise 35) to prove this is always the case. Thus we have the following theorem.

THEOREM 4.6 If $r_1, \ldots, r_k$ are distinct real roots for the characteristic polynomial of

$$a_n y^{(n)} + a_{n-1} y^{(n-1)} + \cdots + a_1 y' + a_0 y = 0,$$

then $e^{r_1 x}, \ldots, e^{r_k x}$ are linearly independent solutions of this differential equation.

---

[1] A well-known result about polynomials over the integers called the Rational Roots Theorem states that if

$$P(x) = a_n x^n + a_{n-1} x^{n-1} + \cdots + a_1 x + a_0 \qquad (a_n \neq 0)$$

is a polynomial where each a_i is an integer and if p/q is a rational root of $P(x)$ in reduced form where p and q are integers, then p divides a_0 and q divides a_n. Thus a rational root p/q of $\lambda^3 - 5\lambda^2 + 6\lambda - 2 = 0$ must have p dividing 2 and q dividing 1 giving us ± 1 and ± 2 as the only possible rational roots.

If the number of distinct real roots is equal to the order of the linear constant co-efficient differential equation (that is, $k = n$ in Theorem 4.6), then the solutions corresponding to these roots must form a fundamental set of solutions which we record as the following corollary.

COROLLARY 4.7   If $r_1, \ldots, r_n$ are distinct real roots for the characteristic polynomial of

$$a_n y^{(n)} + a_{n-1} y^{(n-1)} + \cdots + a_1 y' + a_0 y = 0,$$

then $e^{r_1 x}, \ldots, e^{r_n x}$ form a fundamental set of solutions for this differential equation.

As a final example of this subsection, we do an initial value problem.

EXAMPLE 4   Solve the initial value problem

$$y''' + 3y'' + 2y' = 0; \qquad y(0) = 1, \qquad y'(0) = 0, \qquad y''(0) = -1.$$

Solution   The characteristic equation is

$$\lambda^3 + 3\lambda^2 + 2\lambda = 0.$$

Factoring,

$$\lambda(\lambda + 2)(\lambda + 1) = 0.$$

By Corollary 4.7, it follows that the general solution to the differential equation is

$$y = c_1 + c_2 e^{-2x} + c_3 e^{-x}.$$

Using the initial conditions, we get the system of equations

$$y(0) = c_1 + c_2 + c_3 = 1$$
$$y'(0) = -2c_2 - c_3 = 0$$
$$y''(0) = 4c_2 + c_3 = -1$$

whose solution is $c_1 = 1/2, c_2 = -1/2, c_3 = 1$. Therefore, the solution to the initial value problem is

$$y = \frac{1}{2} - \frac{1}{2}e^{-2x} + e^{-x}. \qquad \bullet$$

4.2.2   Characteristic Equations with Real Repeated Roots

If the characteristic polynomial has repeated roots, our method for obtaining solutions to a constant coefficient homogeneous linear differential equation does not produce enough solutions to form a fundamental set of solutions. To illustrate, consider

$$y'' - 2y' + y = 0.$$

Its characteristic equation is

$$\lambda^2 - 2\lambda + 1 = (\lambda - 1)^2 = 0$$

from which we get the solution

$$y_1 = e^x.$$

But we need another solution y_2 linearly independent of this one to get a fundamental set of solutions to this differential equation. How can we obtain such a y_2? One way is to use a reduction of order technique that involves seeking a solution of the form $y_2 = u(x)y_1 = ue^x.$[2] Substituting $y_2 = ue^x$ into $y'' - 2y' + y = 0$ gives us

$$y'' - 2y' + y = (ue^x)'' - 2(ue^x)' + ue^x$$
$$= e^x(u'' + 2u' + u) - 2e^x(u' + u) + e^x u = e^x u'' = 0.$$

Thus if $u'' = 0$, $y_2 = ue^x$ gives us a second solution. Since the functions with zero second derivative are linear functions, any u of the form $u = mx + b$ will work. We cannot use one where $m = 0$, however, since the resulting solution be^x will not produce a solution linearly independent of e^x. Let us use the one with $m = 1$ and $b = 0$ giving $u = x$ and thus xe^x as a second solution. Since

$$w(e^x, xe^x) = \begin{vmatrix} e^x & xe^x \\ e^x & xe^x + e^x \end{vmatrix} = e^x(x + 1)e^x - e^x xe^x = e^{2x} \neq 0,$$

these two solutions are linearly independent and the general solution of

$$y'' - 2y' + y = 0$$

is

$$y = c_1 e^x + c_2 xe^x.$$

The second solution $y_2 = xe^x$ that was produced for the differential equation $y'' - 2y' + y = 0$ is x times the solution e^x that we obtained from the root of multiplicity 2,[3] $\lambda = 1$. This works in general. That is, if r is a root of the characteristic polynomial with multiplicity 2, then

$$e^{rx}, xe^{rx}$$

---

[2] See Exercises 31–34 for other applications of this technique. On the surface, it may not appear to be clear why this is a reduction of order technique. Exercise 48 illustrates why seeking a solution of the form $y_2 = uy_1$ results in an equation that can be solved for u using one of our reduction of order techniques from Section 3.7.

[3] The **multiplicity** of a root $x = r$ of a polynomial $p(x)$ is the largest power of $x - r$ that is a factor of $p(x)$. To illustrate, consider

$$p(x) = (x - 3)(x + 2)^3(x^2 + 4)^2(x^2 + 2x - 1)^4,$$

which has roots $x = 3$, $x = -2$, $x = \pm 2i$, and $x = -1 \pm \sqrt{2}$. Since this polynomial factors as

$$p(x) = (x - 3)(x + 2)^3[(x - 2i)(x + 2i)]^2[(x + 1 - \sqrt{2})(x + 1 + \sqrt{2})]^4$$
$$= (x - 3)(x + 2)^3(x - 2i)^2(x + 2i)^2(x + 1 - \sqrt{2})^4(x + 1 + \sqrt{2})^4,$$

3 is a root of multiplicity 1, -2 is a root of multiplicity 3, $2i$ and $-2i$ are roots of multiplicity 2, and $-1 + \sqrt{2}$ and $-1 - \sqrt{2}$ are roots of multiplicity 4.

are solutions. Further, this extends to roots of higher multiplicity. For example, if r is a root of the characteristic polynomial with multiplicity 3, then

$$e^{rx}, xe^{rx}, x^2 e^{rx}$$

are solutions and so on.

THEOREM 4.8   If $\lambda = r$ is a root of multiplicity m of the characteristic polynomial of the homogeneous linear differential equation

$$a_n y^{(n)} + a_{n-1} y^{(n-1)} + \cdots + a_1 y' + y = 0,$$

then

$$\boxed{e^{rx}, xe^{rx}, x^2 e^{rx}, \dots, x^{m-1} e^{rx}}$$

are solutions to this differential equation.

We will postpone the proof of this theorem until Section 5.2. From part (c) of Exercise 15 of Section 2.5, we have that the solutions given in Theorem 4.8 are linearly independent.

EXAMPLE 5   Determine the general solution to the following differential equation.

$$y'' + 4y' + 4y = 0$$

Solution   The characteristic equation is

$$\lambda^2 + 4\lambda + 4 = (\lambda + 2)^2 = 0.$$

The general solution is

$$y = c_1 e^{-2x} + c_2 x e^{-2x}.$$

●

EXAMPLE 6   Determine the general solution to the following differential equation.

$$y''' - 9y'' + 27y' - 27y = 0$$

Solution   The characteristic equation is

$$\lambda^3 - 9\lambda^2 + 27\lambda - 27 = (\lambda - 3)^3 = 0.$$

The general solution is

$$y = c_1 e^{3x} + c_2 x e^{3x} + c_3 x^2 e^{3x}.$$

●

EXAMPLE 7   Determine the general solution to the following differential equation.

$$y^{(4)} - 7y''' + 2y'' + 64y' - 96y = 0$$

Solution   The roots of the characteristic equation can be found by using the rational root approach as in Example 3 or by using a software package like Maple. They are 2, -3, and 4 where 4 is a root of multiplicity 2. The general solution is

$$y = c_1 e^{2x} + c_2 e^{-3x} + c_3 e^{4x} + c_4 x e^{4x}.$$

●

Here is an example of an initial value problem with repeated real roots.

EXAMPLE 8   Solve the initial value problem

$$y''' - 16y'' + 64y' = 0; \qquad y(0) = 0, \qquad y'(0) = 0, \qquad y''(0) = 2.$$

Solution   The roots of the characteristic equation are easily found to be 0 and 8 where 8 is a root of multiplicity 2. The general solution is

$$y = c_1 + c_2 e^{8x} + c_3 x e^{8x}.$$

We have

$$y'(x) = 8c_2 e^{8x} + c_3(8xe^{8x} + e^{8x})$$
$$y''(x) = 64c_2 e^{8x} + c_3(8e^{8x} + 64xe^{8x} + 8e^{8x}).$$

Using the initial conditions gives us

$$y(0) = c_1 + c_2 = 0$$
$$y'(0) = 8c_2 + c_3 = 0$$
$$y''(0) = 64c_2 + 16c_3 = 2.$$

The solution to this system of equations is $c_1 = 1/32, c_2 = -1/32, c_3 = 1/4$. Therefore, the solution to the initial value problem is

$$y = \frac{1}{32} - \frac{1}{32}e^{8x} + \frac{1}{4}xe^{8x}.$$

●

We now can determine the general solution to a constant coefficient linear differential equation whose characteristic equation has only real roots.

4.2.3   Characteristic Equations with Complex Roots

In order to say that we have completely studied how to find solutions of constant coefficient homogeneous linear differential equations, we must fill in one gap: What do we do if the characteristic polynomial has complex roots? For instance, if our differential equation is

$$y'' + 4y = 0,$$

the characteristic equation is

$$\lambda^2 + 4 = 0,$$

which has imaginary roots

$$\lambda = \pm\sqrt{-4} = \pm 2i.$$

We are now going to show you how to use such imaginary roots to produce real-valued functions that are solutions to the differential equation. To get this, we are first going to show you how we raise e to an imaginary exponent. You will then see that this gives us complex-valued solutions e^{rx} when r is an imaginary number, which will lead to two real-valued functions that are solutions.

Let us review some basic aspects of the complex numbers. Recall that if a, b are real numbers, then $z = a + bi$ is a complex number where $i^2 = -1$. We call a the **real part** of z and write $Re(z) = a$; b is called the **imaginary part** of z and we write $Im(z) = b$. The conjugate of z, denoted $\bar{z}$, is $\bar{z} = a - bi$. For example, if $z = 2 + 3i$, then $Re(z) = 2$, $Im(z) = 3$, and $\bar{z} = \overline{2 + 3i} = 2 - 3i$.

We also note that since $i^2 = -1$, we have

$$i^3 = -i, \qquad i^4 = (i^2)^2 = 1, \qquad i^5 = i, \dots$$

Consequently, i raised to an even power is ± 1 and i raised to an odd power is $\pm i$.

From calculus we know that e^t is equal to its Maclaurin series when t is a real number. That is,

$$e^t = \sum_{k=0}^{\infty} \frac{t^k}{k!} = 1 + t + \frac{t^2}{2} + \frac{t^3}{3!} + \frac{t^4}{4!} + \cdots$$

We use this same series to define e^t when t is the imaginary number $t = i\theta$:[4]

$$e^{i\theta} = \sum_{k=0}^{\infty} \frac{(i\theta)^k}{k!} = 1 + i\theta + \frac{(i\theta)^2}{2} + \frac{(i\theta)^3}{3!} + \frac{(i\theta)^4}{4!} + \cdots$$

Rearranging this series, we obtain

$$e^{i\theta} = 1 + i\theta + \frac{i^2\theta^2}{2} + \frac{i^3\theta^3}{3!} + \frac{i^4\theta^4}{4!} + \cdots$$
$$= \left(1 - \frac{\theta^2}{2} + \frac{\theta^4}{4!} - \cdots\right) + i\left(\theta - \frac{\theta^3}{3!} + \frac{\theta^5}{5!} - \cdots\right).$$

Notice that $Re(e^{i\theta})$ is the Maclaurin series for $\cos\theta$ and that $Im(e^{i\theta})$ is the Maclaurin series for $\sin\theta$. This gives us an important identity discovered by Euler called **Euler's formula**:

$$\boxed{e^{i\theta} = \cos\theta + i\sin\theta.}$$

To illustrate, we have that

$$e^{2ix} = \cos 2x + i\sin 2x$$

---

[4] We will work formally and ignore questions about the convergence of the series we work with here. All of this can be carefully justified. These details are left for a course in complex analysis.

and

$$e^{-2ix} = \cos(-2x) + i\sin(-2x) = \cos 2x - i\sin 2x.$$

In general, we have that

$$e^{-i\theta} = \cos(-\theta) + i\sin(-\theta) = \cos\theta - i\sin\theta.$$

That is, $e^{-i\theta}$ is the conjugate of $e^{i\theta}$.

We extend the definition of $e^{i\theta}$ to e^{a+bi} by setting

$$e^{a+bi} = e^a e^{ib} = e^a(\cos b + i\sin b) = e^a\cos b + ie^a\sin b.$$

This extension is necessary because we will need to use it for $e^{(a+bi)x}$ when $a + bi$ is a root of the characteristic polynomial. In this case we get

$$\boxed{e^{(a+bi)x} = e^{ax+ibx} = e^{ax}\cos bx + ie^{ax}\sin bx.}$$

For example,

$$e^{(2+i)x} = e^{2x}\cos x + ie^{2x}\sin x.$$

Notice that $e^{(a+bi)x}$ is a complex-valued function of the form

$$w(x) = u(x) + iv(x)$$

where u and v are real-valued functions. The derivative of such a function is

$$w'(x) = u'(x) + iv'(x).$$

We leave it as an exercise (Exercise 36) to show that

$$\frac{d}{dx}(e^{(a+bi)x}) = (a + bi)e^{(a+bi)x}.$$

Because of this, it follows that if $\lambda = a + bi$ is a root of the characteristic polynomial of

$$a_n y^{(n)} + a_{n-1}y^{(n-1)} + \cdots + a_1 y' + a_0 y = 0,$$

then $e^{(a+bi)x}$ is a complex-valued solution of this differential equation.

The next theorem tells us how to obtain real-valued solutions from the complex-valued ones.

THEOREM 4.9   If $w(x) = u(x) + iv(x)$ is a solution to

$$q_n(x)y^{(n)} + q_{n-1}(x)y^{(n-1)} + \cdots + q_1(x)y' + q_0(x)y = 0,$$

then $u(x)$ and $v(x)$ are real-valued solutions of this differential equation.

Proof   Substitute $w(x)$ into the differential equation and separate into the real and imaginary parts. We leave the details as Exercise 37.   ●

The following corollary is used in solving the constant coefficient homogeneous linear differential equation whose characteristic polynomial has complex roots.

COROLLARY 4.10   If $a + bi$ is a root of a characteristic equation for a constant coefficient linear homogeneous differential equation, then $e^{ax} \cos bx$ and $e^{ax} \sin bx$ are two linearly independent solutions to the differential equation.

Proof   The fact that they are solutions follows from Theorem 4.9. Calculating the Wronskian of $e^{ax} \cos bx$ and $e^{ax} \sin bx$ shows they are linearly independent. (We leave this as Exercise 38.)   ●

To illustrate how to use Corollary 4.10, consider again the differential equation

$$y'' + 4y = 0.$$

We know that the roots of its characteristic equation are $\pm 2i$. Using the root $2i$, Corollary 4.10 tells us that $y_1 = \cos 2x$ and $y_2 = \sin 2x$ form a fundamental set of solutions to this second order differential equation. Its general solution is then given by

$$y = c_1 \cos 2x + c_2 \sin 2x.$$

If we use the conjugate root $-2i$, we have $y_1 = \cos 2x$ and $y_2 = -\sin 2x$. However, their linear combinations would give the same general solution as $y_1 = \cos 2x$ and $y_2 = \sin 2x$.

It is always the case that if r is a complex root of a polynomial with real coefficients, then its conjugate, $\bar{r}$, is also a root of the polynomial. Thus complex roots of a characteristic polynomial come in conjugate pairs. If $a + bi$ is one of these roots producing the solutions $e^{ax} \cos bx$ and $e^{ax} \sin bx$, we will always have the solutions $e^{ax} \cos bx$ and $-e^{ax} \sin bx$ produced by the conjugate root $a - bi$. However, these solutions produced by the conjugate $a - bi$ are just linear combinations of the solutions produced by $a + bi$ and hence we do not need them when forming the general solution.

We now look at some more examples involving complex roots.

EXAMPLE 9   Determine the general solution to the following differential equation.

$$2y'' + 8y' + 26y = 0$$

Solution   The characteristic equation is $2\lambda^2 + 8\lambda + 26 = 0$ and its roots using the quadratic formula are found to be

$$\lambda = \frac{-8 \pm \sqrt{64 - 4(2)(26)}}{4} = -2 \pm 3i.$$

Thus the general solution is

$$y = c_1 e^{-2x} \cos 3x + c_2 e^{-2x} \sin 3x = e^{-2x}(c_1 \cos 3x + c_2 \sin 3x).$$ ●

EXAMPLE 10   Determine the general solution to the following differential equation.

$$y''' + y'' - y' + 15y = 0$$

Solution   The characteristic equation is $\lambda^3 + \lambda^2 - \lambda + 15 = 0$. Using the Rational Root Theorem or a software package like Maple, the roots are found to be $-3, 1 + 2i, 1 - 2i$. (If you use Maple, you will see it uses I for i.) The general solution is, therefore,

$$y = c_1 e^{-3x} + e^x (c_2 \cos 2x + c_2 \sin 2x).$$ ●

The next example is an initial value problem.

EXAMPLE 11   Solve the initial value problem

$$y'' + 2y' + 4y = 0; \qquad y(0) = 0, \qquad y'(0) = 1.$$

Solution   The roots of the characteristic equation are $-1 + i\sqrt{3}, -1 - i\sqrt{3}$. The general solution is

$$y = e^{-x}(c_1 \cos \sqrt{3}x + c_2 \sin \sqrt{3}x).$$

The initial conditions give us the system of equations

$$c_1 = 0$$
$$-c_1 + \sqrt{3}c_2 = 1$$

from which we obtain

$$y = \frac{\sqrt{3}}{3} e^{-x} \sin \sqrt{3}x$$

as the solution to the initial value problem.   ●

If the characteristic polynomial has a repeated real root $\lambda = r$ of multiplicity m, we know we can produce additional solutions by multiplying e^{rx} by $x, x^2, \ldots, x^{m-1}$ (Theorem 4.8). The same procedure will work for complex roots: If $\lambda = a + bi$ is a complex root of the characteristic polynomial of multiplicity m, it follows from our differentiation rule $w'(x) = u'(x) + iv'(x)$ for a complex-valued function $w(x) = u(x) + iv(x)$ that multiplying

$$e^{(a+bi)x} = e^{ax} \cos bx + i e^{ax} \sin bx$$

by $x, x^2, \ldots, x^{m-1}$ produces additional complex-valued solutions. These in turn produce the additional real-valued solutions

$$xe^{ax} \cos bx, xe^{ax} \sin bx, x^2 e^{ax} \cos bx, x^2 e^{ax} \sin bx, \ldots, x^{m-1} e^{ax} \cos bx, x^{m-1} e^{ax} \sin bx.$$

We use this fact in the next example.

EXAMPLE 12   Determine the general solution of

$$y^{(4)} + 2y'' + y = 0.$$

Solution   The characteristic equation is

$$r^4 + 2r^2 + 1 = 0.$$

Factoring, we have

$$(r^2 + 1)^2 = 0$$

and hence i and $-i$ are both roots of multiplicity 2. Using i, we obtain the solutions

$$\cos x, \sin x, x \cos x, x \sin x.$$

Calculating the Wronskian of these four functions, we find them to be linearly independent (try it) and consequently the general solution is

$$c_1 \cos x + c_2 \sin x + c_3 x \cos x + c_4 x \sin x. \qquad \bullet$$

We now can determine the general solution for any constant coefficient homogeneous linear differential equation provided we can determine all the roots of its characteristic polynomial. The final example of this section makes us use all of the techniques we have developed here.

EXAMPLE 13   The characteristic polynomial of a constant coefficient homogeneous linear differential equation is

$$\lambda(\lambda - 3)^4(\lambda + 4)(\lambda - 8)(\lambda^2 - 2\lambda + 5)(\lambda^2 + 4\lambda + 13)^2.$$

Give the general solution to the differential equation.

Solution   The roots of the characteristic equation are $0, 3, -4, 8, 1 + 2i, 1 - 2i, -2 - 3i, -2 + 3i$. The root 3 has multiplicity 4 while the complex roots $-2 - 3i, -2 + 3i$ have multiplicity 2. This leads to the general solution

$$y = c_1 + c_2 e^{3x} + c_3 x e^{3x} + c_4 x^2 e^{3x} + c_5 x^3 e^{3x}$$
$$+ c_6 e^{-4x} + c_7 e^{8x} + e^x(c_8 \cos 2x + c_9 \sin 2x) \qquad \bullet$$
$$+ e^{-2x}(c_{10} \cos 3x + c_{11} \sin 3x + c_{12} x \cos 3x + c_{13} x \sin 3x).$$

EXERCISES 4.2

In Exercises 1–6, the characteristic polynomial of the differential equation has distinct real roots. Determine the general solution of the differential equation.

1. $y'' + 8y' = 0$ **2.** $y'' - y' - 6y = 0$

3. $y'' + 4y' + y = 0$ **4.** $2y'' - 5y' - y = 0$

5. $y''' + 4y'' - y' - 4y = 0$

6. $y''' - 4y'' + 3y' + 2y = 0$

In Exercises 7–10, the characteristic polynomial of the differential equation has real repeated roots. Determine the general solution of the differential equation.

7. $y'' + 6y' + 9y = 0$

8. $y''' - 5y'' + 8y' - 4y = 0$

9. $3y''' + 18y'' + 36y' + 24y = 0$

10. $y^{(4)} + 2y''' + y'' = 0$

In Exercises 11–14, the characteristic polynomial of the differential equation has complex roots. Determine the general solution of the differential equation.

11. $4y'' + 16y = 0$ **12.** $2y'' - 8y' + 14y = 0$

13. $y''' - 2y' + 4y = 0$

14. $y^{(4)} - 12y''' + 56y'' + 120y' + 100 = 0$

In Exercises 15–26, find the general solution of the differential equation in part (a) and the solution to the initial value problem in part (b) for the differential equation in part (a).

15. a) $y'' - y = 0$
    b) $y(1) = 0, y'(1) = -1$

16. a) $y'' + y = 0$
    b) $y(\pi) = -1, y'(\pi) = 1$

17. a) $y'' + 4y' + 8y = 0$
    b) $y(0) = 0, y'(0) = -1$

18. a) $y'' + 4y' + 4y = 0$
    b) $y(0) = 0, y'(0) = -1$

19. a) $4y'' + 4y' + y = 0$
    b) $y(0) = 0, y'(0) = -1$

20. a) $4y'' + 5y' + y = 0$
    b) $y(0) = 0, y'(0) = -1$

21. a) $y''' + 5y'' = 0$
    b) $y(-1) = 1, y'(-1) = 0, y''(-1) = 2$

22. a) $y''' - 5y'' = 0$
    b) $y(-1) = 1, y'(-1) = 0, y''(-1) = 2$

23. a) $y''' - y'' - 6y' = 0$
    b) $y(0) = 1, y'(0) = 0, y''(0) = 2$

24. a) $y''' - 4y'' + 5y' = 0$
    b) $y(1) = 1, y'(1) = 0, y''(1) = 2$

25. a) $y''' - 5y'' + 3y' + 9y = 0$
    b) $y(0) = 0, y'(0) = -1, y''(0) = 1$

26. a) $y^{(4)} + 2y''' - 2y'' + 8y = 0$
    b) $y(0) = 1, y'(0) = 0, y''(0) = 0, y'''(0) = 0$

In Exercises 27–30, give the general solution to the homogeneous linear differential equation having the given characteristic polynomial.

27. $p(\lambda) = (\lambda - 4)^3(\lambda^2 + 4)^2$
28. $p(\lambda) = (\lambda - 3)^4(\lambda^2 + \lambda - 6)(\lambda^2 - 4\lambda + 29)$
29. $p(\lambda) = (2\lambda + 4)(\lambda^2 - 9)^2(\lambda^2 + 9)^3$
30. $p(\lambda) = \lambda^3(\lambda + 4)^2(2\lambda^2 + 4\lambda + 4)^2(2\lambda^2 - 2\lambda + 5)$

In Exercises 31–34, show y_1 is a solution to the homogeneous equation. Then use the reduction of order technique introduced in subsection 4.2.2 to determine a second linearly independent solution.

31. $x^2y'' - xy' = 0, y_1 = x^2$
32. $x^2y'' - 6y = 0, y_1 = x^3$

33. $(1 + x)y'' + xy' - y = 0, y_1 = e^{-x}$

34. $x^2y'' - 2xy' + 2y = 0, y_1 = x^2$

35. Prove Theorem 4.6. (*Suggestion:* Use the result of Exercise 7(d) of Section 1.7.)

36. Show $\dfrac{d}{dx}(e^{(a+bi)x}) = (a + bi)e^{(a+bi)x}$.

37. Prove Theorem 4.9.

38. Show that $e^{ax}\cos bx$ and $e^{ax}\sin bx$ are linearly independent.

39. For the initial value problems in part (b) of Exercises 15–26, determine the limits as $x \to \infty$. Also, use Maple (or an appropriate software package) to plot the solutions to these initial value problems to illustrate these limits.

In many cases it is important to determine what a solution to a differential equation tends to (that is, determine its limit as $x \to \infty$). Use the results from Exercise 39 to help with Exercises 40–42.

40. Suppose the roots $r_1, \ldots, r_n$ of the characteristic equation are real and distinct. What is (are) the condition(s) on $r_1, \ldots, r_n$ so that the limit is 0?

41. Suppose some of the real roots $r_1, \ldots, r_n$ of the characteristic equation are not distinct. What is (are) the condition(s) on $r_1, \ldots, r_n$ so that the limit is 0?

42. Suppose some of the roots $r_1, \ldots, r_n$ of the characteristic equation are complex. What is (are) the condition(s) on $r_1, \ldots, r_n$ so that the limit is 0?

43. There are nonconstant coefficient linear differential equations that can be converted to constant coefficient linear differential equations through a substitution. **Euler type** equations, which have the form

$$x^2y'' + Axy' + By = 0, \qquad x > 0,$$

for real constants A and B, are one such class of equations.

a) Show that if $z = \ln x$, then

$$y' = \frac{dy}{dx} = \frac{1}{x}\frac{dy}{dz}$$

and

$$y'' = \frac{d}{dx}y' = \frac{1}{x^2}\frac{d^2y}{dz^2} - \frac{1}{x^2}\frac{dy}{dz}.$$

b) Substitute y' and y'' from part (a) into the Euler type equation to get the constant coefficient linear differential equation

$$\frac{d^2y}{dz^2} + (A-1)\frac{dy}{dz} + By = 0$$

In Exercises 44–47, use Exercise 43 to transform the given Euler type differential equation into a constant coefficient linear differential equation. Then determine the general solution to the given Euler type differential equation.

44. $x^2y'' + 2xy' - 2y = 0$ **45.** $x^2y'' - 3xy' - 5y = 0$

46. $x^2y'' + 4xy' + 2y = 0$ **47.** $x^2y'' - xy' + y = 0$

48. Suppose that y_1 is a solution to the differential equation

$$q_2(x)y'' + q_1(x)y' + q_0(x)y = 0.$$

Show that substituting $y = uy_1$ into this differential equation results in the differential equation

$$y_1q_2(x)u'' + (2y_1'q_2(x) + y_1q_1(x))u' = 0.$$

Observe that this latter second order equation does not contain u and hence can be solved by the first reduction of order technique in Section 3.7.

4.3   THE METHOD OF UNDETERMINED COEFFICIENTS

In the last section we learned how to solve the constant coefficient homogeneous equation

$$a_ny^n + a_{n-1}y^{(n-1)} + \cdots + a_1y' + a_0y = 0.$$

In this and the next section we are going to consider nonhomogeneous equations. Recall from Theorem 4.3 that we can obtain the general solution to a nonhomogeneous equation by adding together the general solution y_H to its corresponding homogeneous equation and a particular solution y_P of the nonhomogeneous equation. Now that the constant coefficient homogeneous case has been studied, we focus our attention on finding a particular solution to the nonhomogeneous constant coefficient equation. In this section we consider a method for finding particular solutions called the **method of undetermined coefficients.** The basic idea behind this method is trial and error using common sense. We illustrate it with some examples.

EXAMPLE 1   Determine the general solution to the following nonhomogeneous equation.

$$y'' + 2y' - 3y = 4e^{2x}$$

Solution   We first solve the corresponding homogeneous equation

$$y'' + 2y' - 3y = 0.$$

Its characteristic equation is

$$\lambda^2 + 2\lambda - 3 = (\lambda - 1)(\lambda + 3) = 0.$$

Consequently, it has general solution

$$y_H = c_1e^x + c_2e^{-3x}.$$

Now, let us try to find a particular solution to the nonhomogeneous equation. We must find a function y_P so that when we substitute y_P into the left-hand side of the differential equation, we get the right-hand side, $4e^{2x}$. Would it not make sense (from what we know about the derivatives of e^{2x}) to look for a particular solution of the same form—that is, a particular solution of the form

$$y_P = Ae^{2x}$$

where A is a constant? Let us see if we can find such a solution. Since

$$y_P' = 2Ae^{2x} \quad \text{and} \quad y_P'' = 4Ae^{2x},$$

we obtain

$$4Ae^{2x} + 4Ae^{2x} - 3Ae^{2x} = 4e^{2x}$$

or

$$5Ae^{2x} = 4e^{2x}$$

upon substituting $y_P = Ae^{2x}$ into the differential equation. Hence we obtain a solution if

$$5A = 4 \quad \text{or} \quad A = \frac{4}{5};$$

that is,

$$y_P = \frac{4}{5}e^{2x}$$

is a particular solution. The general solution to the nonhomogeneous equation in this example is then

$$y = c_1 e^x + c_2 e^{-3x} + \frac{4}{5}e^{2x}. \qquad \bullet$$

EXAMPLE 2   Determine the general solution to the following nonhomogeneous equation.

$$y'' + 2y' - 3y = x^2$$

Solution   From Example 1 we know that the general solution to the corresponding homogeneous equation is

$$y_H = c_1 e^{-3x} + c_2 e^x.$$

Based on Example 1, you might be tempted to try $y_P = Ax^2$ as a particular solution. However, if we substitute this into the differential equation, we obtain

$$2A + 4Ax - 3Ax^2 = x^2.$$

Notice that we have three equations for A:

$$2A = 0,$$
$$4A = 0,$$
$$-3A = 1.$$

Obviously, there is no real number A that satisfies these three equations. What caused this not to work is that we introduced an x-term and a constant term on the left-hand side of the differential equation when differentiating y_P. If we add an x-term and a constant term to y_P we can get this approach to work. That is, let us instead look for a particular solution of the form

$$y_P = Ax^2 + Bx + C.$$

Substituting this y_P into the differential equation gives us

$$y_P'' + 2y_P' - 3y_P = 2A + 2(2Ax + B) - 3(Ax^2 + Bx + C)$$
$$= -3Ax^2 + (4A - 3B)x + 2A + 2B - 3C = x^2.$$

Now we obtain the system

$$2A + 2B - 3C = 0$$
$$4A - 3B = 0$$
$$-3A = 1,$$

which has solution

$$A = -\frac{1}{3}, \qquad B = -\frac{4}{9}, \qquad C = -\frac{14}{27}.$$

Therefore, the general solution is

$$y = c_1 e^{-3x} + c_2 e^x - \frac{1}{3}x^2 - \frac{4}{9}x - \frac{14}{27}. \qquad\qquad \bullet$$

Notice how we included terms involving the derivatives of x^2 in y_P in Example 2. To put it another way, we sought a particular solution y_P that was a linear combination of the type of function on the right-hand side of the equation and its derivatives. In Example 1, the derivatives of e^{2x} are expressions of this same form so that we did not have to include additional terms.

The name method of undetermined coefficients for the approach of Examples 1 and 2 derives from the fact that the scalars we seek in these linear combinations are coefficients that we must determine. Let us do two more examples.

EXAMPLE 3   Find the general solution to the equation.

$$y'' + 2y' - 3y = 3 \sin x$$

Solution   Again from Example 1 we know that the general solution to the corresponding homogeneous equation is

$$y_H = c_1 e^{-3x} + c_2 e^x.$$

Because the derivative of $\sin x$ is $\cos x$, we try the linear combination $y_P = A \cos x + B \sin x$. Substituting this into the differential equation gives us

$$y_P'' + 2y_P' - 3y_P = (-A \cos x - B \sin x) + 2(-A \sin x + B \cos x)$$
$$- 3(A \cos x + B \sin x)$$
$$= (-4A + 2B) \cos x + (-2A - 4B) \sin x = 3 \sin x.$$

In order for $A \cos x + B \sin x$ to be a particular solution, we see that A and B must satisfy the system of equations

$$-4A + 2B = 0$$
$$-2A - 4B = 3.$$

This system has solution

$$A = -\frac{3}{10}, \qquad B = -\frac{3}{5}.$$

The general solution to the differential equation is then

$$y = c_1 e^{-3x} + c_2 e^x - \frac{3}{10} \cos x - \frac{3}{5} \sin x. \qquad \bullet$$

EXAMPLE 4   Find the general solution to the equation.

$$y'' + 2y' - 3y = -2e^x \cos x$$

Solution   Proceeding as in Example 3, we try to find a particular solution of the form $y_P = Ae^x \cos x + Be^x \sin x$. (Why?) Here,

$$y_P' = Ae^x \cos x - Ae^x \sin x + Be^x \sin x + Be^x \cos x$$
$$= (A + B)e^x \cos x + (B - A)e^x \sin x,$$
$$y_P'' = (A + B)e^x \cos x - (A + B)e^x \sin x + (B - A)e^x \sin x + (B - A)e^x \cos x$$
$$= 2Be^x \cos x - 2Ae^x \sin x.$$

Substituting this into the differential equation gives us

$$y_P'' + 2y_P' - 3y_P = 2Be^x \cos x - 2Ae^x \sin x$$
$$+ 2((A + B)e^x \cos x + (B - A)e^x \sin x)$$
$$- 3(Ae^x \cos x + Be^x \sin x)$$
$$= (4B - A)e^x \cos x + (-4A - B)e^x \sin x = -2e^x \cos x.$$

In order for $Ae^x \cos x + Be^x \sin x$ to be a particular solution, we must have

$$-A + 4B = -2$$
$$-4A - B = 0.$$

The solution to this system is

$$A = \frac{2}{17}, \qquad B = -\frac{8}{17}.$$

Therefore, the general solution is

$$y = c_1 e^{-3x} + c_2 e^x + \frac{2}{17} e^x \cos x - \frac{8}{17} e^x \sin x.$$ ●

Notice that in Examples 1–4 none of the terms in the linear combination we tried for a particular solution appeared in the general solution to the corresponding homogeneous equation. The next example illustrates that this procedure will not work if this does not hold and shows how we modify the procedure to obtain a particular solution.

EXAMPLE 5   Determine the general solution to

$$y'' + 2y' - 3y = -2e^x.$$

Solution   As in Example 1, we try $y_P = Ae^x$. However, substituting in Ae^x gives us

$$y_P'' + 2y_P' - 3y_P = Ae^x + 2Ae^x - 3Ae^x = 0 \neq -2e^x,$$

for any A. Hence, we cannot have a particular solution of this form. We could have seen this in advance since Ae^x is a solution to the homogeneous equation. To get a particular solution, let us try using the same idea of multiplying by x we used for constant coefficient linear homogeneous differential equations when we had repeated roots. If we substitute $y_P = Axe^x$ into the differential equation, we get

$$y_P'' + 2y_P' - 3y_P = A(2 + x)e^x + 2A(1 + x)e^x - 3Axe^x = 4Ae^x = -2e^x.$$

Now we see that Axe^x is a particular solution if $A = -1/2$. The general solution is then

$$y = c_1 e^{-3x} + c_2 e^x - \frac{1}{2} xe^x.$$ ●

The next example is another example where we have to modify our original approach.

EXAMPLE 6   Find the general solution to

$$y'' + y = x \cos x.$$

Solution   The roots of the characteristic equation for the corresponding homogeneous equation

$$y'' + y = 0$$

are $\pm i$. The general solution to this homogeneous equation is

$$y_H = c_1 \cos x + c_2 \sin x.$$

The derivatives of $x \cos x$ involve terms with itself, $x \sin x$, $\cos x$, and $\sin x$. We do not need to include terms with $\cos x$ and $\sin x$ since these functions are solutions to the corresponding homogeneous equation. So let us seek a particular solution of the form

$$y_P = Ax \cos x + Bx \sin x.$$

This leads to

$$y_P'' + y_P = -2A \sin x - Ax \cos x + 2B \cos x - Bx \sin x + Ax \cos x + Bx \sin x$$
$$= 2B \cos x - 2A \sin x = x \cos x.$$

Comparing coefficients of $\cos x$, $\sin x$, and $x \cos x$, we have the system

$$2B = 0$$
$$-2A = 0$$
$$0 = 1,$$

which has no solution. If we were to multiply by x as in Example 5 and use

$$y_P = Ax^2 \cos x + Bx^2 \sin x,$$

we would find that this also does not work. If you do so, you will see the failure is caused by the same problem we encountered in Example 2: $x \cos x$ and $x \sin x$ terms arise in the derivatives of y_P. Therefore, we try

$$y_P = Ax^2 \cos x + Bx \cos x + Cx^2 \sin x + Dx \sin x.$$

Substituting in this y_P gives us

$$y_P'' + y_P = 4Cx \cos x - 4Ax \sin x + (2A + 2D) \cos x + (-2B + 2C) \sin x = x \cos x.$$

This leads to the system of equations

$$4C = 1$$
$$4A = 0$$
$$2A + 2D = 0$$
$$-2B + 2C = 0,$$

which has solution $C = 1/4$, $A = 0$, $D = 0$, $B = 1/4$. The general solution to the differential equation is then

$$y = c_1 \cos x + c_2 \sin x + \frac{1}{4}x \cos x + \frac{1}{4}x^2 \sin x. \qquad \bullet$$

Examples 1–6 show that if $g(x)$ is a linear combination of terms involving products of polynomials, e^{ax}, $\cos bx$, and/or $\sin bx$, then we should try to find a particular solution y_P that is a linear combination of these terms and their first and higher order derivatives. Furthermore, Examples 5 and 6 show that we should always determine the homogeneous solution before attempting to find a particular solution to see if any of the terms comprising the general homogeneous solution appear in the nonhomogeneous expression. If they do, we must make adjustments. The following theorem describes the forms of the terms we use in y_P.

THEOREM 4.11   Suppose that $\lambda = r$ is a root of multiplicity m of the characteristic polynomial $p(\lambda)$ of the homogeneous linear differential equation

$$a_n y^{(n)} + \cdots + a_1 y' + a_0 y = 0.$$

Let k be a nonnegative integer and A and B be constants.

1. If r is real, then the linear differential equation

$$\boxed{a_n y^{(n)} + \cdots + a_1 y' + a_0 y = Ax^k e^{rx}}$$

has a particular solution of the form

$$\boxed{x^m (A_k x^k + \cdots + A_1 x + A_0) e^{rx}.}$$

2. If $r = a + bi$ is imaginary, then the linear differential equation

$$\boxed{a_n y^{(n)} + \cdots + a_1 y' + a_0 y = Ax^k e^{ax} \cos bx + Bx^k e^{ax} \sin bx}$$

has a particular solution of the form

$$\boxed{\begin{aligned} &x^m (A_k x^k + \cdots + A_1 x + A_0) e^{ax} \cos bx \\ &+ x^m (B_k x^k + \cdots + B_1 x + B_0) e^{ax} \sin bx. \end{aligned}}$$

The proof of this theorem will be discussed in Section 5.2. While this theorem is stated for a root r of the characteristic polynomial $p(\lambda)$, its conclusions hold even if r is not a root of $p(\lambda)$ by considering the multiplicity of r to be $m = 0$. Thus this theorem guarantees us that we can always find a particular solution to a nonhomogeneous linear differential equation

$$a_n y^{(n)} + \cdots + a_1 y' + a_0 y = g(x)$$

by the method of undetermined coefficients provided $g(x)$ is a linear combination of functions of the form $x^k e^{rx}$, $x^k e^{ax} \cos bx$, or $x^k e^{ax} \sin bx$. (Note that this includes polynomials as terms in $g(x)$ since polynomials are linear combinations of terms of the form $x^k e^{rx}$ with $r = 0$.) We use Theorem 4.11 in the following examples.

EXAMPLE 7   Solve the initial value problem

$$y''' + 4y' = 2x + 3 \sin 2x - 3x^2 e^{2x}; \qquad y(0) = 1, \qquad y'(0) = 0, \qquad y''(0) = -1.$$

Solution   The roots of the characteristic equation for the corresponding homogeneous equation are $0, \pm 2i$. Therefore, the general solution to the corresponding homogeneous equation is

$$y_H = c_1 + c_2 \cos 2x + c_3 \sin 2x.$$

Using Theorem 4.11, for the $2x$ term we incorporate $x(Ax + B)$ into y_P (why?), for the $3 \sin 2x$ term we incorporate $Cx \cos 2x + Dx \sin 2x$ into y_P (why?), and for the $-3x^2e^{2x}$ term we incorporate $(Ex^2 + Fx + G)e^{2x}$ into y_P (why?). Substituting

$$y_P = x(Ax + B) + Cx \cos 2x + Dx \sin 2x + (Ex^2 + Fx + G)e^{2x}$$

into the differential equation gives us

$$y_P''' + 4y_P' = 4(2Ax + B) + (-8C \cos 2x - 8D \sin 2x) + 16Ex^2e^{2x}$$
$$+ (16F + 32E)xe^{2x} + (12E + 16F + 16G)e^{2x}$$
$$= 2x + 3 \sin 2x - 3x^2e^{2x}.$$

Solving the system of equations obtained by equating coefficients, we find

$$A = \frac{1}{4}, \quad B = 0, \quad C = 0, \quad D = -\frac{3}{8}, \quad E = -\frac{3}{16}, \quad F = \frac{3}{8}, \quad G = -\frac{15}{64}.$$

The general solution is then given by

$$y = c_1 + c_2 \cos 2x + c_3 \sin 2x + \frac{1}{4}x^2 - \frac{3}{8}x \sin 2x - \frac{3}{16}x^2e^{2x} + \frac{3}{8}xe^{2x} - \frac{15}{64}e^{2x}.$$

Using the initial conditions, we have

$$y(0) = c_1 + c_2 - \frac{15}{64} = 1$$

$$y'(0) = 2c_3 - \frac{3}{32} = 0$$

$$y''(0) = -4c_2 - \frac{13}{16} = -1.$$

Solving this system, the solution to the initial value problem is

$$y = \frac{19}{16} + \frac{3}{64} \cos 2x + \frac{3}{64} \sin 2x + \frac{1}{4}x^2 - \frac{3}{8}x \sin 2x - \frac{3}{16}x^2e^{2x} + \frac{3}{8}xe^{2x} - \frac{15}{64}e^{2x}. \quad \bullet$$

EXAMPLE 8   Give the form of the particular solution to

$$y^{(4)} - 8y''' + 33y'' - 68y' + 52y = 3xe^{2x} - 4x^3 + 12e^{2x} \sin 3x.$$

Do not determine the coefficients.

Solution   Using the Rational Root Theorem or a software package like Maple, we find the roots of the characteristic equation are $2, 2 + 3i, 2 - 3i$ where 2 is a root of multiplicity 2. Using Theorem 4.11, we see that a particular solution has the form

$$y_P = x^2(Ax + B)e^{2x} + Cx^3 + Dx^2 + Ex + F + Gxe^{2x} \cos 3x + Hxe^{2x} \sin 3x. \quad \bullet$$

EXERCISES 4.3

In Exercises 1–16, determine the general solution of the given differential equation.

1. $y'' - y' - 6y = 3e^{2x}$ **2.** $y'' + 2y' + 2y = 4$

3. $y'' + 25y = x^2 + 25$ **4.** $2y'' + 3y' + y = \cos x$

5. $y'' - 4y' + 13y = 4e^x \sin 3x$

6. $y'' + y' - 4y = x \sin 3x$

7. $y'' + y' = e^x + \cos 2x - x$

8. $y''' - 3y'' + 3y' - y' = e^{2x} + e^{3x}$

9. $4y'' + 16y = 3\cos 2x$

10. $y'' + 36y = -4x \sin 6x$

11. $y'' + 8y' = 2x^2 - 7x + 3$

12. $y'' - 6y' + 9y = xe^{3x}$

13. $y'' - 4y' + 13y = 4e^{2x} \sin 3x$

14. $3y'' + 6y' - 24y = 3x^2 - 5e^{2x} - 6\sin 4x$

15. $y''' - 3y'' - 4y' = x - e^{4x}$

16. $y''' + 4y'' - y' - 4y = 2 - 3\cos 3x + e^x$

In Exercises 17–22, solve the initial value problem.

17. $y'' - 2y' - 8y = x$; $y(0) = -2$, $y'(0) = 2$

18. $y'' + 4y' + 4y = 2x - \sin 3x$; $y(0) = 0$, $y'(0) = -1$

19. $y'' - y = 5e^x$; $y(1) = 0$, $y'(1) = -1$

20. $y'' + y = 2\cos x - 3\sin x$; $y(\pi) = 1$, $y'(\pi) = 1$

21. $y''' - 4y'' + 5y' = 7 - 2\cos x$; $y(0) = 1$, $y'(0) = 0$, $y''(0) = 2$

22. $y''' - 5y'' + 3y' + 9y = x^2 - e^{-x}$; $y(0) = 0$, $y'(0) = -1$, $y''(0) = 1$

In Exercises 23–26, determine a form for y_P of the given differential equation. (Do not determine the coefficients.)

23. $y'' + y = 2x^2 + 2x \cos x$

24. $y''' + 6y'' - 32y = 3xe^{2x} - 7e^{-4x}$

25. $y''' + y'' + 9y' + 9y = 4e^{-x} + 3x \sin 3x - 6\cos 3x$

26. $y'''' + 8y'' + 16y = 4x - 2e^{4x} - \cos 4x + 2x \sin 2x$

In Exercises 27–30, use the *dsolve* command in Maple to solve the differential equation in the indicated exercise of this section. Compare Maple's answers to yours. (Or use another appropriate software package.)

27. Exercise 1   **28.** Exercise 7

29. Exercise 13   **30.** Exercise 15

4.4   THE METHOD OF VARIATION OF PARAMETERS

In the last section we learned how to determine a particular solution to the nonhomogeneous constant coefficient equation

$$a_n y^n + a_{n-1}y^{(n-1)} + \cdots + a_1 y' + a_0 y = g(x)$$

when $g(x)$ is a linear combination of products of polynomials, e^{ax}, $\cos bx$, and/or $\sin bx$ using the method of undetermined coefficients. This method worked since the derivatives of functions of this form have this same form. In this section we will see another method for determining a particular solution to linear nonhomogeneous differential equations for any continuous function g. This method, called **variation of parameters,** was discovered by the famous French mathematician Joseph-Louis Lagrange (1736–1813). As with the method of undetermined coefficients, this method also depends on having a fundamental set of solutions to the corresponding homogeneous equation.

Indeed, we sometimes do need another method since the method of undetermined coefficients does not apply. To illustrate, consider

$$y'' + 4y = \tan 2x.$$

We cannot use the method of undetermined coefficients to find a particular solution to this differential equation. (Why is this the case? *Hint:* What are the first and higher order derivatives of $\tan 2x$?)

We now develop the method of variation of parameters for the general second order nonhomogeneous linear differential equation

$$q_2(x)y'' + q_1(x)y' + q_0(x)y = g(x).$$

Suppose that we have determined that y_1, y_2 form a fundamental set of solutions for the corresponding homogeneous equation giving us the general solution

$$y_H = c_1 y_1 + c_2 y_2$$

to the homogeneous differential equation. We vary the parameters c_1, c_2 in this general homogeneous solution to functions $u_1(x)$ and $u_2(x)$ and try to find a particular solution of the form

$$y_P = u_1(x)y_1 + u_2(x)y_2$$

to the nonhomogeneous differential equation. Differentiating gives us

$$y_P' = u_1 y_1' + u_1' y_1 + u_2 y_2' + u_2' y_2.$$

Imposing the condition

$$u_1' y_1 + u_2' y_2 = 0$$

will simplify our work and reduce y_P' to

$$y_P' = u_1 y_1' + u_2 y_2'.$$

The second derivative y_P'' is now

$$y_P'' = u_1 y_1'' + u_1' y_1' + u_2 y_2'' + u_2' y_2'.$$

Substituting y_P, y_P', y_P'' into the differential equation, the left-hand side becomes

$$
\begin{aligned}
q_2(x)y'' + q_1(x)y' + q_0(x)y &= q_2(x)(u_1 y_1'' + u_1' y_1' + u_2 y_2'' + u_2' y_2') \\
&\quad + q_1(x)(u_1 y_1' + u_2 y_2') + q_0(x)(u_1 y_1 + u_2 y_2) \\
&= u_1(q_2(x)y_1'' + q_1(x)y_1' + q_0(x)y_1) \\
&\quad + u_2(q_2(x)y_2'' + q_1(x)y_2' + q_0(x)y_2) \\
&\quad + q_2(x)(u_1' y_1' + u_2' y_2') \\
&= u_1 \cdot 0 + u_2 \cdot 0 + q_2(x)(u_1' y_1' + u_2' y_2')
\end{aligned}
$$

since y_1 and y_2 are solutions to the homogeneous equation. Thus, for y_P to be a solution,

$$q_2(x)(u_1' y_1' + u_2' y_2') = g(x).$$

Using this equation and our earlier equation $u_1' y_1 + u_2' y_2 = 0$, we have the system of equations

$$u_1' y_1 + u_2' y_2 = 0$$

$$u_1' y_1' + u_2' y_2' = \frac{g(x)}{q_2(x)}$$

in the unknowns u_1' and u_2'. In matrix form, this is

$$\begin{bmatrix} y_1 & y_2 \\ y_1' & y_2' \end{bmatrix} \begin{bmatrix} u_1' \\ u_2' \end{bmatrix} = \begin{bmatrix} 0 \\ g(x)/q_2(x) \end{bmatrix}.$$

Let us use Cramer's rule to solve this system for u_1' and u_2'. Notice that the determinant of the coefficient matrix of this system is

$$\begin{vmatrix} y_1 & y_2 \\ y_1' & y_2' \end{vmatrix},$$

which is the Wronskian $w(y_1, y_2)$. Thus

$$u_1' = \frac{\begin{vmatrix} 0 & y_2 \\ g(x)/q_2(x) & y_2' \end{vmatrix}}{w(y_1, y_2)}$$

and

$$u_2' = \frac{\begin{vmatrix} y_1 & 0 \\ y_1' & g(x)/q_2(x) \end{vmatrix}}{w(y_1, y_2)}.$$

Integrating u_1' and u_2' leads to the particular solution

$$y_P = u_1 y_1 + u_2 y_2 = y_1 \int \frac{\begin{vmatrix} 0 & y_2 \\ g(x)/q_2(x) & y_2' \end{vmatrix}}{w(y_1, y_2)} dx$$

$$+ y_2 \int \frac{\begin{vmatrix} y_1 & 0 \\ y_1' & g(x)/q_2(x) \end{vmatrix}}{w(y_1, y_2)} dx.$$

For notational purposes let

$$\boxed{w = w(y_1, y_2),}$$

$$\boxed{w_1 = \begin{vmatrix} 0 & y_2 \\ g(x)/q_2(x) & y_2' \end{vmatrix},}$$

and

$$w_2 = \begin{vmatrix} y_1 & 0 \\ y_1' & g(x)/q_2(x) \end{vmatrix}.$$

Our particular solution can then be easily remembered by the formulas

$$y_P = u_1 y_1 + u_2 y_2$$

and

$$u_1 = \int \frac{w_1}{w}\, dx, \qquad u_2 = \int \frac{w_2}{w}\, dx.$$

EXAMPLE 1   Determine the general solution to

$$y'' + 4y = \tan 2x.$$

Solution   The roots of the characteristic equation are $\pm 2i$. We use

$$y_1 = \cos 2x, \qquad y_2 = \sin 2x$$

for a fundamental set of solutions. We have

$$w = \begin{vmatrix} \cos 2x & \sin 2x \\ -2\sin 2x & 2\cos 2x \end{vmatrix} = 2,$$

$$w_1 = \begin{vmatrix} 0 & \sin 2x \\ \tan 2x & 2\cos 2x \end{vmatrix} = -\tan 2x \sin 2x,$$

and

$$w_2 = \begin{vmatrix} \cos 2x & 0 \\ -2\sin 2x & \tan 2x \end{vmatrix} = \cos 2x \tan 2x = \sin 2x.$$

Using the formulas for u_1 and u_2,

$$u_1 = \int \frac{w_1}{w}\, dx = \int \frac{-\tan 2x \sin 2x}{2} = -\frac{1}{2}\int \frac{\sin^2 2x}{\cos 2x} = \frac{1}{2}\int (\cos 2x - \sec 2x)\, dx$$

$$= \frac{1}{4}\sin 2x - \frac{1}{4}\ln|\sec 2x + \tan 2x|$$

and

$$u_2 = \int \frac{w_2}{w}\, dx = \int \frac{\sin 2x}{2}\, dx = -\frac{1}{4}\cos 2x.$$

Therefore, a particular solution is

$$y_P = \left(\frac{1}{4} \sin 2x - \frac{1}{4} \ln |\sec 2x + \tan 2x| \right) \cos 2x - \frac{1}{4} \cos 2x \sin 2x$$

$$= -\frac{1}{4} \cos 2x \ln |\sec 2x + \tan 2x|.$$

The general solution is then

$$y = c_1 \cos 2x + c_2 \sin 2x - \frac{1}{4} \cos 2x \ln |\sec 2x + \tan 2x|. \qquad \bullet$$

The procedure we used for constructing a particular solution in the second order case can be extended to higher order linear differential equations. (See Exercise 16.) Doing so gives us the following result.

THEOREM 4.12   Suppose that $q_n, q_{n-1}, \ldots, q_0, g$ are continuous on an interval (a, b) and $q_n(x) \neq 0$ for all x in (a, b). Further suppose that $y_1, y_2, \ldots, y_n$ form a fundamental set of solutions of the homogeneous differential equation

$$q_n(x)y^{(n)} + q_{n-1}(x)y^{(n-1)} + \cdots + q_0(x)y = 0.$$

Let w denote the Wronskian of $y_1, y_2, \ldots, y_n$, B denote the $n \times 1$ column vector

$$\begin{bmatrix} 0 \\ \vdots \\ 0 \\ g(x)/q_n(x) \end{bmatrix},$$

and w_i denote the determinant obtained from w by replacing the ith column of w by B for $i = 1, 2, \ldots, n$. Then

$$y_P = u_1 y_1 + u_2 y_2 + \cdots + u_n y_n$$

where

$$u_i = \int \frac{w_i}{w} \, dx$$

is a particular solution to the differential equation

$$q_n(x)y^{(n)} + q_{n-1}(x)y^{(n-1)} + \cdots + q_0(x)y = g(x).$$

Let us use Theorem 4.12 to solve a third order nonhomogeneous equation.

EXAMPLE 2   Determine the general solution to the differential equation

$$2y''' - 4y'' - 22y' + 24y = 2e^{4x}.$$

Solution   Solving the characteristic equation, we find its roots are -3, 1, 4.   We let

$y_1 = e^{-3x}$, $y_2 = e^x$, and $y_3 = e^{4x}$ be our fundamental set of solutions. Then

$$
w = \begin{vmatrix} e^{-3x} & e^x & e^{4x} \\ -3e^{-3x} & e^x & 4e^{4x} \\ 9e^{-3x} & e^x & 16e^{4x} \end{vmatrix} = 84e^{2x},
$$

$$
w_1 = \begin{vmatrix} 0 & e^x & e^{4x} \\ 0 & e^x & 4e^{4x} \\ e^{4x} & e^x & 16e^{4x} \end{vmatrix} = 3e^{9x},
$$

$$
w_2 = \begin{vmatrix} e^{-3x} & 0 & e^{4x} \\ -3e^{-3x} & 0 & 4e^{4x} \\ 9e^{-3x} & e^{4x} & 16e^{4x} \end{vmatrix} = -7e^{5x},
$$

and

$$
w_3 = \begin{vmatrix} e^{-3x} & e^x & 0 \\ -3e^{-3x} & e^x & 0 \\ 9e^{-3x} & e^x & e^{4x} \end{vmatrix} = 4e^{2x}.
$$

From this we obtain

$$
u_1 = \int \frac{w_1}{w}\, dx = \int \frac{3e^{9x}}{84e^{2x}}\, dx = \frac{1}{196} e^{7x},
$$

$$
u_2 = \int \frac{w_2}{w}\, dx = \int \frac{-7e^{5x}}{84e^{2x}}\, dx = -\frac{1}{36} e^{3x},
$$

and

$$
u_3 = \int \frac{w_3}{w}\, dx = \int \frac{4e^{2x}}{84e^{2x}}\, dx = \frac{1}{21} x,
$$

which leads to the particular solution

$$
y_P = u_1 e^{-3x} + u_2 e^x + u_3 e^{4x} = \frac{1}{196} e^{4x} - \frac{1}{36} e^{4x} + \frac{1}{21} x e^{4x}.
$$

The general solution is then

$$
y = c_1 e^{-3x} + c_2 e^x + c_3 e^{4x} + \frac{1}{21} x e^{4x}. \qquad \bullet
$$

By Corollary 4.5, the Wronskian w in Theorem 4.12 is never zero on the interval (a, b). Hence, once we know a fundamental set of solutions, the method of variation of parameters always gives us a particular solution (but it may be in terms of integrals that do not have closed forms) on the interval (a, b). Consequently, unlike the method of undetermined coefficients, which works only for certain types of $g(x)$, the method of variation of parameters works for any continuous $g(x)$. When applicable, however, the method of undetermined coefficients often is easier to use than variation of parameters. Exercise 1 of the following exercise set illustrates this.

EXERCISES 4.4

1. Determine a particular solution to

$$y'' - y = 3x^2 - 1$$

using

a) the method of undetermined coefficients and

b) the method of variation of parameters.

In Exercises 2–10, find the general solution using the method of variation of parameters to find a particular solution.

2. $y'' - y' - 6y = 4e^x$ **3.** $y'' + 6y' + 9y = 3e^{3x}$

4. $y'' + y = \sec x$ **5.** $y'' + 4y = 5 \csc 2x$

6. $y'' + 9y = \cot 3x$

7. $y'' - y' - 2y = e^x \cos x$

8. $2y'' - 4y' + 4y = e^x \tan x$

9. $y''' + 4y'' - y' - 4y = e^{-2x}$

10. $y''' + 4y'' - y' - 4y = xe^{-2x}$

The corresponding homogeneous differential equations for Exercises 11–14 are Exercises 31–34 in Section 4.2, respectively. Use the fundamental sets of solutions from those exercises and the method of variation of parameters to determine a particular solution to the differential equations in Exercises 11–14.

11. $x^2 y'' - xy' = x^3$ **12.** $x^2 y'' - 6y = 4 + x^2$

13. $(1 + x)y'' + xy' - y = (1 + x)^2 e^x$

14. $x^2 y'' - 2xy' + 2y = \ln x / x$

15. Show $1/x, x, x^2$ form a fundamental set of solutions for the corresponding homogeneous equation of

$$x^3 y''' + x^2 y'' - 2xy' + 2y = -5x^3, \qquad x > 0.$$

Then, using variation of parameters, find a particular solution to this nonhomogeneous equation.

16. a) Consider the third order nonhomogeneous linear differential equation

$$q_3(x)y''' + q_2(x)y'' + q_1(x)y' + q_0(x)y = g(x)$$

having y_1, y_2, y_3 as a fundamental set of solutions for the corresponding homogeneous linear differential equation. Extend the proof of Theorem 4.12 we did in the text for the case $n = 2$ to the case $n = 3$ as follows: Seek a particular solution of the form

$$y_P = u_1 y_1 + u_2 y_2 + u_3 y_3.$$

Impose the conditions

$$u_1' y_1 + u_2' y_2 + u_3' y_3 = 0 \qquad (1)$$

and

$$u_1' y_1' + u_2' y_2' + u_3' y_3' = 0. \qquad (2)$$

Substituting y_P into the differential equation, obtain that

$$u_1' y_1'' + u_2' y_2'' + u_3' y_3'' = \frac{g(x)}{q_3(x)}. \qquad (3)$$

Use Cramer's rule to solve the system of equations formed by Equations (1), (2), and (3) for u_1', u_2', u_3' and then do the final steps in obtaining Theorem 4.12 in this case.

b) Prove Theorem 4.12 in general by extending the process of part (a) from the third order case to nth order case.

4.5 SOME APPLICATIONS OF HIGHER ORDER DIFFERENTIAL EQUATIONS

In Section 3.6 we saw that differential equations can be used to describe stituations that change over time. In that section we modeled experiments using first order differential equations. In this section we will consider phenomena that can be modeled by second order differential equations. Of course, there are many phenomena that are described by differential equations that are of higher order than second. To keep with the spirit of this chapter, we will only consider linear differential equations. The reader should be aware of the fact that more and more disciplines and industries are using differential equations to model problems arising in their work. Therefore, applications of differential equations are increasing at an amazing rate. (No pun intended.)

Many second order differential equations arise from using Newton's law of motion: "The net force applied to an object equals the object's mass times its acceleration." If u represents the position of the object moving along a line, then

$u(t)$ is the object's position along the line at time t,

$v(t) = u'(t)$ is the object's velocity along the line at time t,

and

$a(t) = u''(t)$ is the object's acceleration along the line at time t.

Therefore, if m is the mass of the object and F is the force, then from Newton's law of motion we obtain

$$F = ma = mu'',$$

and we have an equation involving a second derivative. If we can write F in terms of u and v, then we will have a differential equation involving u, $v = u'$, and $a = u''$.

The first example we consider is a famous historical problem involving springs, which arises in many engineering applications. We consider an object of mass m attached to a spring. The spring can lie on a horizontal (line) track or be suspended vertically from a ceiling. (We will assume the motion of the object is always along the same line.) If you stretch or compress the spring and then let go, the spring will apply a force to the object and the object will move back and forth or up and down. The differential equation describing the horizontal motion will be equivalent to the differential equation describing the vertical motion so we develop only the equation for vertical motion and use this same equation for horizontal motion.

The British scientist Robert Hooke (1635–1703) discovered that if a spring is stretched a distance L, then the force F_s that the spring exerts is proportional to L (provided L is not too large). This is known as Hooke's law and, in equation form, is written

$$F_s = kL$$

where k is the constant of proportionality. Since k is a property of the spring, it is called the **spring constant.**

If an object of mass m is hung from a vertical spring with spring constant k, then the spring will be stretched a length L by the object, as illustrated in Figure 4.1. The force due

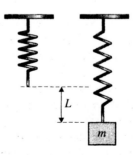

Figure 4.1

to gravity is given by mg where g is the acceleration due to gravity and acts downward. The spring force is $F_s = -kL$ by Hooke's law. (The minus sign is there because the spring force acts upward.) Since the system is in equilibrium, the acceleration of the object is 0. Summing the forces acting on the object and using Newton's law gives us

$$ma = 0 = F = mg - kL.$$

This equation tells us that we can determine the spring constant by hanging an object of **weight** $w = mg$ from the spring and dividing by the elongation L of the spring. That is,

$$k = \frac{mg}{L} = \frac{w}{L}.$$

When a mass-spring system is positioned so that the length of the spring is this distance L the spring is vertically stretched by the mass, we say the system is in **equilibrium.**

Now, suppose that the mass attached to the spring is moved u units from equilibrium, where we take u to be positive when the spring is stretched and u to be negative when the spring is compressed. (See Figure 4.2.) Then

$$mu''(t) = F(t)$$

where F is the sum of all the forces acting on the object. We now determine these forces.

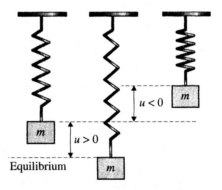

Figure 4.2

The force due to gravity is the weight of the mass $w = mg$. The force exerted by the spring is $-k(u + L)$. (The minus sign is there because the spring tries to pull the object up when $u > 0$ and the spring tries to push the object down when $u < 0$.) There is also force due to friction. The force due to friction has been studied extensively and there is experimental evidence showing that it is proportional to the velocity of the object. Since the frictional force opposes the motion, it has the opposite sign of the velocity. Therefore, we use $-fv = -fu'$ for the force due to friction, where f is a positive constant called the **friction constant.** If another force (such as a person pushing on the mass, for example) is applied to the object, we call such a force an **external force.** We will let $h(t)$ represent the external force. We now have that

$$mu''(t) = F(t) = mg - k(u(t) + L) - fu'(t) + h(t)$$
$$= mg - ku(t) - kL - fu'(t) + h(t)$$
$$= -ku(t) - fu'(t) + h(t)$$

since $mg = kL$. Rewriting this as

$$mu'' + fu' + ku = h(t),$$

we have a constant coefficient linear second order differential equation that can be solved by the techniques of this chapter. If the initial position $u(0)$ and the initial velocity $u'(0)$ are given, we have an initial value problem

$$\boxed{mu'' + fu' + ku = h(t); \quad u(0) = u_0, \quad u'(0) = u_1,}$$

which we know has a unique solution. Therefore, if we are given or can find m, f, k, h, u_0, and u_1, we can determine the solution giving the motion of the object.

The characteristic equation for this differential equation is

$$m\lambda^2 + f\lambda + k = 0$$

and its roots are

$$\lambda = \frac{-f \pm \sqrt{f^2 - 4km}}{2m}.$$

Since $f^2 - 4km$ can be positive, zero, or negative, we can have real, imaginary, or complex roots depending on the size of the friction constant f. In particular, we see that if there is no friction (that is, if $f = 0$), then the roots are imaginary. If there is friction (that is, if $f > 0$), then $Re(\lambda) < 0$. If there is no friction the mass-spring system is called **undamped,** and if there is friction the system is called **damped.**

We now look at examples that include each of these possibilities. First we consider an undamped system.

EXAMPLE 1   An object weighing 4 lb stretches a spring 6 in. Suppose the mass-spring system is on a horizontal track and that the mass is kept off the track by a cushion of air from a compressor. (In this case, the friction is virtually zero and will be ignored.) Determine the motion of the mass if no external force is applied and the object is pulled 2 in. from equilibrium and then released.

Solution   We determine m, k, and f first. Since $w = mg$,

$$m = \frac{w}{g} = \frac{4}{32} = \frac{1}{8}$$

using $g = 32$ ft/sec^2 for the acceleration of gravity. To be consistent with units, we use $L = 1/2$ ft, which gives us

$$k = \frac{w}{L} = \frac{4}{0.5} = 8.$$

We are assuming no friction, so

$$f = 0.$$

Since there is no external force, $h(t) = 0$. Because the object is moved to an *initial* position 2 in. beyond equilibrium, we have

$$u(0) = u_0 = \frac{1}{6}.$$

The fact that it is released from rest means

$$u'(0) = u_1 = 0.$$

The initial value problem describing the motion of the object attached to the spring is then

$$\frac{1}{8}u'' + 8u = 0; \qquad u(0) = \frac{1}{6}, \qquad u'(0) = 0.$$

The roots of the characteristic equation to this differential equation are $\pm 8i$. Therefore,

$$u(t) = c_1 \cos 8t + c_2 \sin 8t.$$

Since

$$u'(t) = -8c_1 \sin 8t + 8c_2 \cos 8t,$$

the initial conditions lead to

$$u(0) = c_1 = \frac{1}{6}$$
$$u'(0) = 8c_2 = 0.$$

The solution to the initial value problem is then

$$u = \frac{1}{6} \cos 8t.$$

We graph this solution using Maple in Figure 4.3. Notice that the period of the motion is

$$\frac{2\pi}{8} = \frac{\pi}{4}$$

and the amplitude is

$$\frac{1}{6}.$$

This means the object oscillates back and forth along the track covering a distance of 2/3 ft every $\pi/4$ sec. The value 8 is called the **frequency** of the motion.   ●

The next example illustrates the difference in a damped system from the undamped system in Example 1.

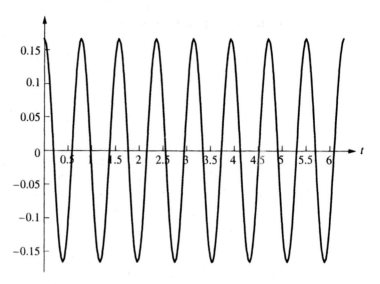

Figure 4.3

EXAMPLE 2   An object of mass 2 kg is attached to a vertical spring and stretches the spring 25 cm. The spring is hung in oil that offers a resistance to the motion of 8 kg/sec. Determine the motion of the mass if no external force is applied and the object is given an initial velocity of 25 cm/sec from its equilibrium position.

Solution   We immediately have that

$$m = 2 \quad \text{and} \quad f = 8.$$

Using $g = 10$ m/sec^2 gives us

$$k = \frac{mg}{l} = \frac{2(10)}{0.25} = 80.$$

The initial value problem describing the motion of the mass is then

$$2u'' + 8u' + 80u = 0; \quad u(0) = 0, \quad u'(0) = 0.25.$$

The roots of the characteristic equation for this differential equation are $-2 \pm 6i$. The general solution is given by

$$u(t) = e^{-2t}(c_1 \cos 6t + c_2 \sin 6t).$$

Using the initial conditions, we obtain

$$u(0) = c_1 = 0$$

and

$$u'(0) = 6c_2 = 0.25.$$

Therefore, the position of the object is given by

$$u = \frac{1}{24}e^{-2t}\sin 6t.$$

Since

$$-\frac{1}{24}e^{-2t} \le u \le \frac{1}{24}e^{-2t},$$

the solution oscillates with decaying amplitude $(1/24)e^{-2t}$ caused by the friction (in fact, $\lim_{t\to\infty} u = 0$). This is essentially the idea used in designing shock absorbers for vehicles. The value $2\pi/6 = \pi/3$ is often called the **quasi-period.** A graph of the solution using Maple depicting the motion and these properties is shown in Figure 4.4. ●

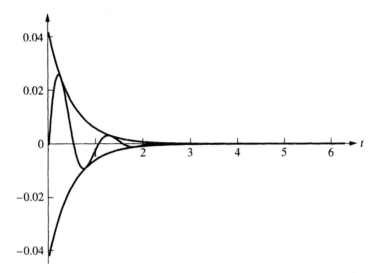

Figure 4.4

EXAMPLE 3   An object of mass 25 g is attached to a vertical spring and stretches the spring 25 cm. A 250-g mass is then attached to the spring and the spring is hung in an oil that offers a resistance to the motion of 1 kg/sec. Determine the motion of the mass if no external force is applied and the object is given an initial velocity of 5 cm/sec after being pushed up 10 cm from equilibrium.

Solution   We first find that

$$k = \frac{0.025(10)}{0.25} = 1.$$

Since $m = 0.25$, $f = 1$, and $k = 1$, the initial value problem describing the motion of the mass is

$$0.25u'' + u' + u = 0; \qquad u(0) = -0.10, \qquad u'(0) = 0.05.$$

The characteristic equation of the differential equation has 2 as a repeated root. The general solution is given by

$$u(t) = c_1 e^{-2t} + c_2 t e^{-2t}.$$

Using the initial conditions, we obtain

$$u(0) = c_1 = -0.10,$$
$$u'(0) = 0.2 + c_2 = 0.05,$$

which leads to

$$u = -0.1 e^{-2t} - 0.15 t e^{-2t}.$$

From the solution we see that $\lim_{t \to \infty} u(t) = 0$. This solution does not oscillate. This motion is often called **overdamped.** Notice how the graph of the solution in Figure 4.5 obtained using Maple verifies these properties.   ●

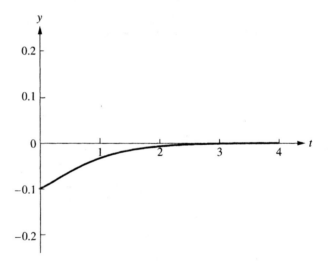

Figure 4.5

EXAMPLE 4   Do Example 1 if an external oscillating force, $h(t) = 4 \sin 8t$ lb, is applied to the object.

Solution   The initial value problem is

$$\frac{1}{8} u'' + 8u = 4 \sin 8t; \qquad u(0) = \frac{1}{6}, \qquad u'(0) = 0.$$

Since (from Example 1) the general solution to the homogeneous equation is

$$u(t) = c_1 \cos 8t + c_2 \sin 8t,$$

the method of undetermined coefficientstells us to look for a particular solution of the form

$$u_P = At \cos 8t + Bt \sin 8t.$$

Substituting u_P into the nonhomogeneous differential equation and solving for the coefficients, we find that

$$A = -2, \quad B = 0.$$

The general solution of the nonhomogeneous differential equation is then

$$u = c_1 \cos 8t + c_2 \sin 8t - 2t \cos 8t.$$

Finally, using the initial conditions, we find the solution for the position is

$$u = \frac{1}{6} \cos 8t + \frac{1}{4} \sin 8t - 2t \cos 8t.$$

As $t \to \infty$ the term $-2t \cos 8t$ grows without bound. Notice the effect of this in our graph of the solution in Figure 4.6. This solution indicates that the spring may be damaged as time becomes large. It is well known in physics and engineering that this will happen if a mass-spring system is oscillated by a forcing term with the same frequency as the system. This phenomenon is known as **resonance.**   ●

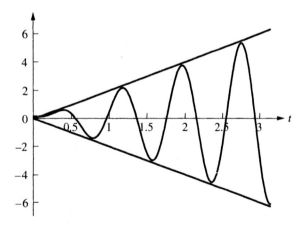

Figure 4.6

EXAMPLE 5   Do Example 2 if an external oscillating force is applied to the object given by $h(t) = 2 \cos 4t$ newtons.[5]

Solution   The initial value problem is

$$2u'' + 8u' + 80u = 2 \cos 4t; \quad u(0) = 0, \quad u'(0) = 0.25.$$

_____

[5] A newton is a unit of measure standing for 1 kg m/sec^2.

Since (from Example 2) the homogeneous equation has general solution

$$u(t) = e^{-2t}(c_1 \cos 6t + c_2 \sin 6t),$$

we try to find a particular solution of the form

$$u_P = A \cos 4t + B \sin 4t$$

for the nonhomogeneous problem. Substituting into the nonhomogeneous equation, we find that

$$A = \frac{3}{104}, \qquad B = \frac{1}{52}.$$

This gives us the general solution

$$u = e^{-2t}(c_1 \cos 6t + c_2 \sin 6t) + \frac{3}{104} \cos 4t + \frac{1}{52} \sin 4t.$$

Using the initial conditions, we obtain the position of the object to be

$$u = e^{-2t}\left(-\frac{3}{104} \cos 6t + \frac{1}{52} \sin 6t\right) + \frac{3}{104} \cos 4t + \frac{1}{52} \sin 4t.$$

Note that the term $e^{-2t}(-\frac{3}{104} \cos 6t + \frac{1}{52} \sin 6t)$ approaches 0 as $t \to \infty$. Therefore, u approaches $\frac{3}{104} \cos 4t + \frac{1}{52} \sin 4t$ as t becomes large. In other words, for large values of t the motion of the object is about the same as an object oscillating with motion given by $\frac{3}{104} \cos 4t + \frac{1}{52} \sin 4t$. The value 4 is called the **transient** frequency. Observe this motion in the graph displayed in Figure 4.7. ●

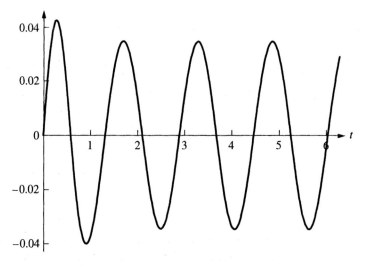

Figure 4.7

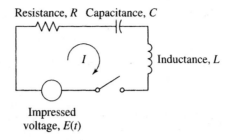

Figure 4.8

Another area where second order linear constant coefficient differential equations arise is in the flow of electric current in a circuit, as in Figure 4.8. The resistance, R, is measured in ohms; the capacitance, C, is measured in farads; and the inductance, L, is measured in henrys. We assume that all of these are constant values. The applied voltage, $E(t)$, is measured in volts and can change over time. We let I be the current in the circuit. If Q is the charge (measured in coulombs) on the capacitor, then

$$\frac{dQ}{dt} = I.$$

The flow of the current in the circuit is governed by Kirchoff's second law, which states: "In a closed circuit, the applied voltage is equal to the sum of the voltage drops across the circuit." Voltage drops are determined as follows:

The voltage drop across the resistor is IR.

The voltage drop across the capacitor is Q/C.

The voltage drop across the inductor is $L\dfrac{dI}{dt}$.

Combining these with Kirchoff's law, we obtain

$$LI' + RI + \frac{Q}{C} = E(t).$$

Differentiating this equation and using the fact that

$$\frac{dQ}{dt} = I,$$

gives us

$$LI'' + RI' + \frac{1}{C}I = E'(t),$$

which is a second order differential equation in I. If we also are given the initial charge and initial current or the initial current and the initial current's derivative, we have an initial value problem. Exercises 17–18 ask you to solve such initial value problems.

EXERCISES 4.5

In Exercises 1–8, use 32 ft/sec^2 or 10 m/sec^2 for the acceleration of gravity, g, as appropriate. Also, use Maple or another suitable software package to graph your solution and observe the behavior of the motion.

1. A 4-kg mass is attached to a vertically hanging spring. The object stretches the spring 10 cm. Assume air resistance on the object is negligible and the mass is pulled down an additional 50 cm and released from rest. Determine the motion of this mass. What are the frequency and amplitude of this motion?

2. Do Exercise 1 if the object is further given an initial velocity of 2 cm/sec.

3. A 2-lb object stretches a spring 6 in. The mass-spring system is placed in equilibrium on a horizontal track that has a cushion of air eliminating friction. Determine the motion of this object if the object is given an initial velocity of 2 ft/sec. What are the frequency and amplitude of this motion?

4. Do Exercise 3 if the spring is first stretched 2 in. from equilibrium and then given an initial velocity of 2 ft/sec from its resting position.

5. An 8-lb object stretches a spring 4 ft. The mass-spring system is placed on a horizontal track that has a friction constant of 1 lb-sec/ft. The object is pulled 6 in. from equilibrium and released. Determine the motion of the object.

6. A 5-kg mass stretches a spring 25 cm. The mass-spring system is hung vertically in a tall tank filled with oil offering a resistance to the motion of 40 kg/sec. The object is pulled 25 cm from rest and given an initial velocity of 50 cm/sec. Determine the motion of the object.

7. A 2-kg mass stretches a spring 1 m. This mass is hung vertically on the spring and then a shock absorber is attached that exerts a resistance of 14 kg/sec to motion. Determine the motion of the mass if it is pulled down 3 m and then released.

8. A 3-lb object stretches a spring 4 in. The mass-spring system is hung vertically and air offers a resistance to the motion of the object of 12 lb-sec/ft. The object is pushed up 3 in. from rest and given an initial velocity of 6 in./sec. Determine the motion of the object.

In Exercises 9–15, an external force is applied to the mass-spring system in the indicated exercise of this section. Determine the motion of the object when this additional external force is applied. Also, use Maple or another suitable software package to graph your solution and observe the behavior of the motion.

9. The external force 3 newtons is applied to the mass-spring system in Exercise 1.

10. The external force $-3\cos 10t$ newtons is applied to the mass-spring system in Exercise 2.

11. The external force $2\cos 2t$ lb is applied to the mass-spring system in Exercise 5.

12. The external force $6\sin t$ lb is applied to the mass-spring system in Exercise 5.

13. The external force $6e^{-2t}$ newtons is applied to the mass-spring system in Exercise 7.

14. The external force $3e^{-t/10}$ newtons is applied to the mass-spring system in Exercise 7.

15. Consider the mass-spring initial value problem

$$mu'' + fu' + ku = 0; \qquad u(0) = u_0, \qquad u'(0) = u_1.$$

a) Determine the value of the friction constant f where the roots of the characteristic polynomial for the differential equation change from complex to real. This value of f is called the **critical damping value.**

b) Determine the conditions for m, f, and k so that the system is (i) undamped, (ii) damped and oscillating, (iii) critically damped, and (iv) overdamped.

c) Let $m = 1 = k$, $u_0 = 1 = u_1$. Choose values for f so that the system is (i) undamped, (ii) damped and oscillating, (iii) critically damped, and (iv) overdamped. Solve the initial value problems for each of these systems and graph the four solutions on one coordinate system using Maple or another suitable software package.

16. Determine the general solution to

$$mu'' + ku = \sin \omega t.$$

Determine the value of ω that induces resonance.

17. Determine the current in an *RLC* circuit that has $L = 0.5$, $R = 2$, $C = 1$, and $E(t) = 0$ with initial charge 1 and initial current 1.

18. Determine the current in an *RLC* circuit that has $L = 0.5$, $R = 0.2$, $C = 1$, and $E(t) = 3\sin(2t)$ with no initial charge or current.

19. Determine the conditions on L, R, C, and ω so that the voltage $h(t) = \sin \omega t$ produces resonance in an *RLC* circuit.

20. Do Exercise 16 of Section 3.7 using the methods of this chapter.

Linear Transformations and Eigenvalues and Eigenvectors

Many areas of mathematics and its applications have special types of functions that play an important role in these areas. For example, differentiable functions play an important role in differential calculus; continuous functions play an important role in integral calculus. Linear transformations are a special type of function, playing an important role in linear algebra and its applications. In the first three sections of this chapter, we will study the basic aspects of linear transformations. We then will go on to study important numbers called eigenvalues and important vectors called eigenvectors, which are associated with matrices in particular and linear transformations in general that arise in numerous applications. In Chapter 6 we will see one such application of these eigenvalues and eigenvectors to systems of linear differential equations. But this is not the only application of them. You will likely encounter other applications of eigenvalues and eigenvectors in future mathematics, engineering, science, statistics, or computer science courses.

5.1 LINEAR TRANSFORMATIONS

Before telling you what a linear transformation is, let us present some notation and terminology associated with functions (also called mappings) that we frequently shall use. We shall denote a function f from a set X to a set Y by writing

$$f : X \to Y$$

(read "f maps X to Y"). The set X is called the **domain** of f. The set Y goes by various names. Some call it the **image set** of f; others call it the **codomain** of f, which is the term we shall employ. The subset

$$\{f(x)|x \in X\}$$

of Y is called the **range** of f. To illustrate, if $f : \mathbb{R} \to \mathbb{R}$ by

$$f(x) = e^x,$$

both the domain and codomain of f are $\mathbb{R}$; the range of this function f is the set of positive real numbers. (In particular, notice that the range of a function can be a proper subset of its codomain.) If $g : \mathbb{R}^2 \to \mathbb{R}$ by

$$g\left[\begin{array}{c} x \\ y \end{array} \right] = x^2 - y^2,$$

the domain of g is $\mathbb{R}^2$ and its codomain is $\mathbb{R}$; here the range of the function g is $\mathbb{R}$, the same as its codomain (why?).

With those preliminaries out of the way, we now state what we mean by a linear transformation.

DEFINITION   If V and W are vector spaces, a function $T : V \to W$ is called a **linear transformation** if, for all vectors u and v in V and all scalars c, the following two properties are satisfied:

1. $T(u + v) = T(u) + T(v)$.
2. $T(cv) = cT(v)$.

A function T from one vector space to another satisfying property 1 of our definition is said to *preserve addition*; if the function satisfies property 2, the function is said to *preserve scalar multiplication*. Thus a linear transformation is a function from one vector space to another that preserves both addition and scalar multiplication. In the special case of a linear transformation $T : V \to V$ from a vector space V to itself, T is sometimes called a **linear operator.**

Let us look at some examples of functions from one vector space to another that are and are not linear transformations.

EXAMPLE 1   Determine whether the given function is a linear transformation: $T : \mathbb{R}^3 \to \mathbb{R}^2$ by

$$T\left[\begin{array}{c} x \\ y \\ z \end{array} \right] = \left[\begin{array}{c} x + y - z \\ x + 2y + z \end{array} \right].$$

Solution   We have

$$T\left(\begin{bmatrix} x_1 \\ y_1 \\ z_1 \end{bmatrix} + \begin{bmatrix} x_2 \\ y_2 \\ z_2 \end{bmatrix}\right) = T\begin{bmatrix} x_1 + x_2 \\ y_1 + y_2 \\ z_1 + z_2 \end{bmatrix}$$

$$= \begin{bmatrix} x_1 + x_2 + y_1 + y_2 - z_1 - z_2 \\ x_1 + x_2 + 2y_1 + 2y_2 + z_1 + z_2 \end{bmatrix}$$

$$= \begin{bmatrix} x_1 + y_1 - z_1 \\ x_1 + 2y_1 + z_1 \end{bmatrix} + \begin{bmatrix} x_2 + y_2 - z_2 \\ x_2 + 2y_2 + z_2 \end{bmatrix}$$

$$= T\begin{bmatrix} x_1 \\ y_1 \\ z_1 \end{bmatrix} + T\begin{bmatrix} x_2 \\ y_2 \\ z_2 \end{bmatrix}$$

and hence T preserves addition. If c is any scalar,

$$T\left(c\begin{bmatrix} x \\ y \\ z \end{bmatrix}\right) = T\begin{bmatrix} cx \\ cy \\ cz \end{bmatrix} = \begin{bmatrix} cx + cy - cz \\ cx + 2cy + cz \end{bmatrix}$$

$$= c\begin{bmatrix} x + y - z \\ x + 2y + z \end{bmatrix} = cT\begin{bmatrix} x \\ y \\ z \end{bmatrix}$$

and consequently T preserves scalar multiplication. Thus T is a linear transformation. ●

EXAMPLE 2   Determine whether the given function is a linear transformation: $T : \mathbb{R}^2 \to \mathbb{R}^2$ by

$$T\begin{bmatrix} x \\ y \end{bmatrix} = \begin{bmatrix} x^2 \\ x + y + 1 \end{bmatrix}.$$

Solution   On the one hand,

$$T\left(\begin{bmatrix} x_1 \\ y_1 \end{bmatrix} + \begin{bmatrix} x_2 \\ y_2 \end{bmatrix}\right) = T\begin{bmatrix} x_1 + x_2 \\ y_1 + y_2 \end{bmatrix} = \begin{bmatrix} (x_1 + x_2)^2 \\ x_1 + x_2 + y_1 + y_2 + 1 \end{bmatrix},$$

while on the other hand,

$$T\begin{bmatrix} x_1 \\ y_1 \end{bmatrix} + T\begin{bmatrix} x_2 \\ y_2 \end{bmatrix} = \begin{bmatrix} x_1^2 \\ x_1 + y_1 + 1 \end{bmatrix} + \begin{bmatrix} x_2^2 \\ x_2 + y_2 + 1 \end{bmatrix}$$

$$= \begin{bmatrix} x_1^2 + x_2^2 \\ x_1 + x_2 + y_1 + y_2 + 2 \end{bmatrix}$$

and hence we see

$$T\left(\left[\begin{array}{c} x_1 \\ y_1 \end{array}\right] + \left[\begin{array}{c} x_2 \\ y_2 \end{array}\right]\right) \quad \text{is not the same as} \quad T\left[\begin{array}{c} x_1 \\ y_1 \end{array}\right] + T\left[\begin{array}{c} x_2 \\ y_2 \end{array}\right]$$

for any two vectors

$$\left[\begin{array}{c} x_1 \\ y_1 \end{array}\right] \quad \text{and} \quad \left[\begin{array}{c} x_2 \\ y_2 \end{array}\right]$$

in $\mathbb{R}^2$. Since T does not preserve addition, T is not a linear transformation. It is also the case that this function does not preserve scalar multiplication. (Verify this.)   ●

EXAMPLE 3   Determine whether the given function is a linear transformation: $T : P_2 \to P_1$ by

$$T(ax^2 + bx + c) = (a + b)x + b - c.$$

Solution   Since

$$T((a_1x^2 + b_1x + c_1) + (a_2x^2 + b_2x + c_2))$$
$$= T((a_1 + a_2)x^2 + (b_1 + b_2)x + c_1 + c_2)$$
$$= (a_1 + a_2 + b_1 + b_2)x + b_1 + b_2 - (c_1 + c_2)$$

is the same as

$$T(a_1x^2 + b_1x + c_1) + T(a_2x^2 + b_2x + c_2) = (a_1 + b_1)x + b_1 - c_1 + (a_2 + b_2)x + b_2 - c_2,$$

T preserves addition. It also preserves scalar multiplication since

$$T(d(ax^2 + bx + c)) = T(adx^2 + bdx + cd) = (ad + bd)x + bd - cd$$

is the same as

$$dT(ax^2 + bx + c) = d((a + b)x + b - c).$$

Hence T is a linear transformation.   ●

One commonly occurring type of linear transformation involves matrix multiplication.

THEOREM 5.1   If A is an $m \times n$ matrix, the function $T : \mathbb{R}^n \to \mathbb{R}^m$ defined by

$$T(X) = AX$$

is a linear transformation.

Proof   If X_1 and X_2 are two column vectors in $\mathbb{R}^n$,

$$T(X_1 + X_2) = A(X_1 + X_2) = AX_1 + AX_2 = T(X_1) + T(X_2).$$

If c is a scalar,

$$T(cX) = A(cX) = c(AX) = cT(X).$$

Hence T is a linear transformation.   ●

We shall call the type of linear transformation $T(X) = AX$ of Theorem 5.1 a **matrix transformation.** The linear tranformation $T : \mathbb{R}^3 \to \mathbb{R}^2$ by

$$T\begin{bmatrix} x \\ y \\ z \end{bmatrix} = \begin{bmatrix} x + y - z \\ x + 2y + z \end{bmatrix}$$

of Example 1 is actually a matrix transformation. To see why, observe that if A is the matrix

$$A = \begin{bmatrix} 1 & 1 & -1 \\ 1 & 2 & 1 \end{bmatrix},$$

then

$$T\begin{bmatrix} x \\ y \\ z \end{bmatrix} = \begin{bmatrix} 1 & 1 & -1 \\ 1 & 2 & 1 \end{bmatrix}\begin{bmatrix} x \\ y \\ z \end{bmatrix}.$$

In Section 5.3 we shall see that every linear transformation from $\mathbb{R}^n$ to $\mathbb{R}^m$ is in fact a matrix transformation. Even more generally, by using coordinate vectors we will see that every linear transformation from one finite dimensional vector space to another is, in a sense, given by such a matrix multiplication.

Differentiation and integration give rise to many important linear transformations. In the case of differentiation, recall from Chapter 2 that $D(a, b)$ denotes the set of all differentiable functions on an open interval (a, b), which is a subspace of the set of all functions $F(a, b)$ defined on (a, b). Writing the derivative of a function f as Df (that is, $Df = f'$), we may regard D as a function from $D(a, b)$ to $F(a, b)$,

$$D : D(a, b) \to F(a, b).$$

This function D is called the **differential operator** on $D(a, b)$. Since we know from calculus that

$$D(f + g) = (f + g)' = f' + g' = Df + Dg \quad \text{and} \quad D(cf) = (cf)' = cf' = cDf$$

for any two differentiable functions f and g and any constant (scalar) c, it follows that D is a linear transformation from $D(a, b)$ to $F(a, b)$. In the next section we will look back at the last chapter and see that much of the work we did there can be viewed in terms of linear transformations involving this differential operator D.

To give an illustration of how integration arises in connection with linear transformations, consider the vector space of continuous functions on a closed interval $[a, b]$ for finite numbers a and b, which we have been denoting as $C[a, b]$. Let $\text{Int}(f)$ denote the definite integral of a function f in $C[a, b]$ over $[a, b]$:

$$\text{Int}(f) = \int_a^b f(x)\, dx.$$

Again from calculus we know that

$$\text{Int}(f+g) = \int_a^b (f(x)+g(x))\,dx = \int_a^b f(x)\,dx + \int_a^b g(x)\,dx = \text{Int}(f) + \text{Int}(g)$$

and

$$\text{Int}(cf) = \int_a^b cf(x)\,dx = c\int_a^b f(x)\,dx = c\text{Int}(f)$$

for any f and g in $C[a,b]$ and any constant c and hence $\text{Int} : C[a,b] \to \mathbb{R}$ is a linear transformation. In Chapter 7, we shall employ this type of linear transformation (only on an interval of the form $[0,\infty)$) in our study of Laplace transforms.

The following theorem gives some elementary properties of linear transformations.

THEOREM 5.2 Suppose $T : V \to W$ is a linear transformation.

1. $T(0) = 0$.
2. $T(-v) = -T(v)$ for any v in V.
3. $T(u-v) = T(u) - T(v)$ for any u and v in V.
4. $T(c_1v_1 + c_2v_2 + \cdots + c_kv_k) = c_1T(v_1) + c_2T(v_2) + \cdots + c_kT(v_k)$ for any scalars $c_1, c_2, \ldots, c_k$ and any vectors $v_1, v_2, \ldots, v_k$ in V.

Proof We prove the first two parts here and leave the proof of the third and fourth parts as an exercise (Exercise 14). We obtain part (1) by first observing

$$T(0) = T(0+0) = T(0) + T(0).$$

Subtracting $T(0)$ from each side of this equation, we have $0 = T(0)$ as required.

To see part (2), notice that

$$T(v + (-v)) = T(v) + T(-v).$$

Also,

$$T(v + (-v)) = T(0) = 0$$

by part (1). Thus we have $T(v) + T(-v) = 0$, which implies $T(-v) = -T(v)$. ●

Part (4) of Theorem 5.2 can be described in words by saying, "T of a linear combination of vectors is the linear combination of T of the vectors." A consequence of it is that once we know what a linear transformation $T : V \to W$ does to a basis of V, we can determine what T does to any other vector of v of V. The following example illustrates how we do this.

EXAMPLE 4   Suppose that $T : \mathbb{R}^3 \to \mathbb{R}^2$ is a linear transformation so that

$$
T \begin{bmatrix} 1 \\ 1 \\ 0 \end{bmatrix} = \begin{bmatrix} 2 \\ 3 \end{bmatrix}, \quad
T \begin{bmatrix} 0 \\ 1 \\ 1 \end{bmatrix} = \begin{bmatrix} 0 \\ 3 \end{bmatrix}, \quad
T \begin{bmatrix} 1 \\ 0 \\ 1 \end{bmatrix} = \begin{bmatrix} 0 \\ 2 \end{bmatrix}.
$$

Find:

(a) $T \begin{bmatrix} 1 \\ 3 \\ 0 \end{bmatrix}$,

(b) $T \begin{bmatrix} x \\ y \\ z \end{bmatrix}$.

Solution   Before solving this example, we point out that the vectors

$$
\begin{bmatrix} 1 \\ 1 \\ 0 \end{bmatrix}, \begin{bmatrix} 0 \\ 1 \\ 1 \end{bmatrix}, \begin{bmatrix} 1 \\ 0 \\ 1 \end{bmatrix}
$$

do form a basis of $\mathbb{R}^3$. (Verify this if you are in doubt.)

(a) We first write

$$
\begin{bmatrix} 1 \\ 3 \\ 0 \end{bmatrix}
$$

as a linear combination of our basis vectors:

$$
c_1 \begin{bmatrix} 1 \\ 1 \\ 0 \end{bmatrix} + c_2 \begin{bmatrix} 0 \\ 1 \\ 1 \end{bmatrix} + c_3 \begin{bmatrix} 1 \\ 0 \\ 1 \end{bmatrix} = \begin{bmatrix} 1 \\ 3 \\ 0 \end{bmatrix}.
$$

This gives us the system

$$
\begin{aligned}
c_1 + c_3 &= 1 \\
c_1 + c_2 &= 3 \\
c_2 + c_3 &= 0,
\end{aligned}
$$

which has solution (try it):

$$
c_1 = 2, \qquad c_2 = 1, \qquad c_3 = -1.
$$

Thus

$$T\begin{bmatrix} 1 \\ 3 \\ 0 \end{bmatrix} = T\left(2\begin{bmatrix} 1 \\ 1 \\ 0 \end{bmatrix} + \begin{bmatrix} 0 \\ 1 \\ 1 \end{bmatrix} - \begin{bmatrix} 1 \\ 0 \\ 1 \end{bmatrix} \right)$$

$$= 2T\begin{bmatrix} 1 \\ 1 \\ 0 \end{bmatrix} + T\begin{bmatrix} 0 \\ 1 \\ 1 \end{bmatrix} - T\begin{bmatrix} 1 \\ 0 \\ 1 \end{bmatrix}$$

$$= 2\begin{bmatrix} 2 \\ 3 \end{bmatrix} + \begin{bmatrix} 0 \\ 3 \end{bmatrix} - \begin{bmatrix} 0 \\ 2 \end{bmatrix} = \begin{bmatrix} 4 \\ 7 \end{bmatrix}.$$

(b) We proceed in the same manner as in part (a), only using

$$\begin{bmatrix} x \\ y \\ z \end{bmatrix} \quad \text{in place of} \quad \begin{bmatrix} 1 \\ 3 \\ 0 \end{bmatrix} :$$

$$c_1\begin{bmatrix} 1 \\ 1 \\ 0 \end{bmatrix} + c_2\begin{bmatrix} 0 \\ 1 \\ 1 \end{bmatrix} + c_3\begin{bmatrix} 1 \\ 0 \\ 1 \end{bmatrix} = \begin{bmatrix} x \\ y \\ z \end{bmatrix}.$$

$$c_1 + c_3 = x$$
$$c_1 + c_2 = y$$
$$c_2 + c_3 = z$$

$$c_1 = \frac{x}{2} + \frac{y}{2} - \frac{z}{2}, \quad c_2 = -\frac{x}{2} + \frac{y}{2} + \frac{z}{2}, \quad c_3 = \frac{x}{2} - \frac{y}{2} + \frac{z}{2}$$

$$T\begin{bmatrix} x \\ y \\ z \end{bmatrix} = c_1 T\begin{bmatrix} 1 \\ 1 \\ 0 \end{bmatrix} + c_2 T\begin{bmatrix} 0 \\ 1 \\ 1 \end{bmatrix} + c_3 T\begin{bmatrix} 1 \\ 0 \\ 1 \end{bmatrix}$$

$$= \left(\frac{x}{2} + \frac{y}{2} - \frac{z}{2}\right)\begin{bmatrix} 2 \\ 3 \end{bmatrix} + \left(-\frac{x}{2} + \frac{y}{2} + \frac{z}{2}\right)\begin{bmatrix} 0 \\ 3 \end{bmatrix}$$

$$+ \left(\frac{x}{2} - \frac{y}{2} + \frac{z}{2}\right)\begin{bmatrix} 0 \\ 2 \end{bmatrix}$$

$$= \begin{bmatrix} x + y - z \\ x + 2y + z \end{bmatrix}$$

●

From the answer to part (b) of Example 4, we can see that the linear transformation in this example is the same linear transformation as in Example 1. Example 4 illustrates a common way of specifying a linear transformation: Simply indicate how the linear transformation acts on a basis for the domain. Once its action is known on a basis, the action of the linear transformation on any other vector may be determined by writing the other vector as a linear combination of the basis vectors and applying part (4) of Theorem 5.2.

An important set associated with a linear transformation $T : V \rightarrow W$ is the **kernel** of T, denoted ker(T), which is the set of all vectors v in V so that $T(v) = 0$; in set notation,

$$\ker(T) = \{v \in V | T(v) = 0\}.$$

We encountered kernels of linear transformations in the case of matrix transformations in Section 2.4. Recall that the nullspace of an $m \times n$ matrix A, $NS(A)$, is the set of all vectors X in $\mathbb{R}^n$ so that

$$AX = 0.$$

Equivalently, $NS(A)$ is the subspace of $\mathbb{R}^n$ consisting of the solutions to the homogeneous system $AX = 0$. Notice that the kernel of the matrix transformation

$$T(X) = AX$$

is then exactly the same as the nullspace of A:

$$\ker(T) = NS(A).$$

EXAMPLE 5   Find a basis for the kernel of the linear transformation

$$T \begin{bmatrix} x \\ y \\ z \end{bmatrix} = \begin{bmatrix} 1 & 1 & -1 \\ 1 & 2 & 1 \end{bmatrix} \begin{bmatrix} x \\ y \\ z \end{bmatrix}$$

of Example 1.

Solution   Solving the homogeneous system $AX = 0$,

$$\left[\begin{array}{ccc|c} 1 & 1 & -1 & 0 \\ 1 & 2 & 1 & 0 \end{array} \right] \rightarrow \left[\begin{array}{ccc|c} 1 & 1 & -1 & 0 \\ 0 & 1 & 2 & 0 \end{array} \right] \rightarrow \left[\begin{array}{ccc|c} 1 & 0 & -3 & 0 \\ 0 & 1 & 2 & 0 \end{array} \right],$$

we see the solutions are given by

$$x = 3z, \qquad y = -2z$$

and hence the vector

$$\begin{bmatrix} 3 \\ -2 \\ 1 \end{bmatrix}$$

forms a basis for ker(T).   ●

Another space associated with a matrix A from Section 2.4 that relates to the matrix transformation $T(X) = AX$ is the column space of A, which we denoted as $CS(A)$. Recall that the column space of an $m \times n$ matrix $A = [a_{ij}]$ is the subspace of $\mathbb{R}^m$ spanned by the columns of A. Consider

$$
AX = \begin{bmatrix} a_{11} & a_{12} & \cdots & a_{1n} \\ a_{21} & a_{22} & \cdots & a_{2n} \\ \vdots & \vdots & & \vdots \\ a_{m1} & a_{m2} & \cdots & a_{mn} \end{bmatrix} \begin{bmatrix} x_1 \\ x_2 \\ \vdots \\ x_n \end{bmatrix}.
$$

Noting that this product is the same as

$$
x_1 \begin{bmatrix} a_{11} \\ a_{21} \\ \vdots \\ a_{m1} \end{bmatrix} + x_2 \begin{bmatrix} a_{12} \\ a_{22} \\ \vdots \\ a_{m2} \end{bmatrix} + \cdots + x_n \begin{bmatrix} a_{1n} \\ a_{2n} \\ \vdots \\ a_{mn} \end{bmatrix},
$$

we see that the set of vectors AX as X runs over the elements of $\mathbb{R}^n$, which is the range of the function T, consists of the linear combinations of the columns of A, which is $CS(A)$. We are going to denote the range of a linear transformation T by

$$
\boxed{\text{range}(T).}
$$

We could then restate what we have just discovered by saying if T is a matrix transformation

$$
T(X) = AX,
$$

then

$$
\text{range}(T) = CS(A).
$$

EXAMPLE 6   Find a basis for the range of the linear transformation

$$
T \begin{bmatrix} x \\ y \\ z \end{bmatrix} = \begin{bmatrix} 1 & 1 & -1 \\ 1 & 2 & 1 \end{bmatrix} \begin{bmatrix} x \\ y \\ z \end{bmatrix}
$$

of Example 1.

Solution   Finding a basis for the column space of the matrix

$$
A = \begin{bmatrix} 1 & 1 & -1 \\ 1 & 2 & 1 \end{bmatrix}
$$

as we did in Section 2.4,

$$A^T = \begin{bmatrix} 1 & 1 \\ 1 & 2 \\ -1 & 1 \end{bmatrix} \rightarrow \begin{bmatrix} 1 & 1 \\ 0 & 1 \\ 0 & 2 \end{bmatrix},$$

we see that the vectors

$$\begin{bmatrix} 1 \\ 1 \end{bmatrix}, \begin{bmatrix} 0 \\ 1 \end{bmatrix}$$

form a basis for $CS(A) = \text{range}(T)$. ●

In the special case of a matrix transformation $T(X) = AX$ where A is an $m \times n$ matrix, we know that $\ker(T) = NS(A)$ is a subspace of the domain $\mathbb{R}^n$ of T and $\text{range}(T) = CS(A)$ is a subspace of the codomain $\mathbb{R}^m$ of T. This is true for any linear transformation.

THEOREM 5.3 If $T : V \rightarrow W$ is a linear transformation, then $\ker(T)$ is a subspace of V and $\text{range}(T)$ is a subspace of W.

Proof We have $\ker(T)$ is a nonempty set since the zero vector of V is in $\ker(T)$ by part (1) of Theorem 5.2 and $\text{range}(T)$ is certainly a nonempty set (since V contains elements v, $\text{range}(T)$ contains elements $T(v)$!), so we need to check the closure properties. Let u and v be vectors in $\ker(T)$. We have

$$T(u + v) = T(u) + T(v) = 0 + 0 = 0$$

and hence $\ker(T)$ is closed under addition. If c is a scalar,

$$T(cv) = cT(v) = c \cdot 0 = 0$$

and $\ker(T)$ is closed under scalar multiplication. Thus $\ker(T)$ is a subspace of V.
    To see $\text{range}(T)$ has the required closure properties, consider two vectors in $\text{range}(T)$. It will be convenient to write the vectors in the form $T(u)$ and $T(v)$ where u and v are in V. Since

$$T(u) + T(v) = T(u + v),$$

we have $T(u) + T(v)$ is the image of a vector of V (namely, $u + v$) under T and hence is in $\text{range}(T)$. Thus $\text{range}(T)$ is closed under addition. In a similar manner, the fact that

$$cT(v) = T(cv)$$

gives us that $cT(v)$ lies in $\text{range}(T)$ and hence $\text{range}(T)$ is closed under scalar multiplication. ●

The next theorem gives us a relationship between the dimensions of $\ker(T)$, $\text{range}(T)$, and V when V is finite dimensional.

THEOREM 5.4   If $T : V \rightarrow W$ is a linear transformation where V is a finite dimensional vector space, then

$$\dim(\ker(T)) + \dim(\text{range}(T)) = \dim(V).$$

Proof   We prove this here in the case when $0 < \dim(\ker(T)) < \dim(V)$. Other cases that must be considered are $\dim(V) = 0$, $0 = \dim(\ker(T)) < \dim(V)$, and $0 < \dim(\ker(T)) = \dim(V)$, which we leave as exercises (Exercise 30).

Let us set $\dim(V) = n$ and $\dim(\ker(T)) = k$. Choose a basis $v_1, v_2, \ldots, v_k$ for $\ker(T)$. By Lemma 2.11 we can extend this set of vectors to a basis $v_1, v_2, \ldots, v_k, v_{k+1}, \ldots, v_n$ of V. Our proof hinges on the truth of the following claim.

Claim $T(v_{k+1}), \ldots, T(v_n)$ form a basis for range(T).

We first show these vectors span range(T). Consider a vector $T(v)$ in range(T). Expressing v as

$$v = c_1 v_1 + \cdots + c_k v_k + c_{k+1} v_{k+1} + \cdots + c_n v_n,$$

we have

$$\begin{aligned}
T(v) &= T(c_1 v_1 + \cdots + c_k v_k + c_{k+1} v_{k+1} + \cdots + c_n v_n) \\
&= c_1 T(v_1) + \cdots + c_k T(v_k) + c_{k+1} T(v_{k+1}) + \cdots + c_n T(v_n) \\
&= c_1 \cdot 0 + \cdots + c_k \cdot 0 + c_{k+1} T(v_{k+1}) + \cdots + c_n T(v_n) \\
&= c_{k+1} T(v_{k+1}) + \cdots + c_n T(v_n).
\end{aligned}$$

This gives us that $T(v_{k+1}), \ldots, T(v_n)$ span range(T). To see why they are linearly independent, suppose

$$c_{k+1} T(v_{k+1}) + \cdots + c_n T(v_n) = 0.$$

Then

$$T(c_{k+1} v_{k+1} + \cdots + c_n v_n) = 0,$$

which tells us that $c_{k+1} v_{k+1} + \cdots + c_n v_n$ lies in $\ker(T)$. We leave it as an exercise (Exercise 29) to show that this implies $c_{k+1} = 0, \ldots, c_n = 0$. Hence $T(v_{k+1}), \ldots, T(v_n)$ are linearly independent and we have the claim.

By our claim, we now have

$$\dim(\text{range}(T)) = n - k = \dim(V) - \dim(\ker(T))$$

or

$$\dim(\ker(T)) + \dim(\text{range}(T)) = \dim(V). \qquad \bullet$$

As a corollary to Theorem 5.4, we have a result we promised in Section 2.4.

COROLLARY 5.5   If A is a matrix,

$$\dim(RS(A)) = \dim(CS(A)).$$

Proof Suppose that A is an $m \times n$ matrix. From Theorem 2.14, we have

$$\dim(RS(A)) = n - \dim(NS(A)).$$

Letting T be the matrix transformation $T(X) = AX$ from $\mathbb{R}^n$ to $\mathbb{R}^m$, we obtain

$$\dim(RS(A)) = \dim(\mathbb{R}^n) - \dim(\ker(T)) = \dim(\text{range}(T)) = \dim(CS(A)). \quad \bullet$$

EXERCISES 5.1

Determine whether the given function is a linear transformation in Exercises 1–12.

1. $T : \mathbb{R}^2 \to \mathbb{R}^2$ by $T \begin{bmatrix} x \\ y \end{bmatrix} = \begin{bmatrix} 3x - 2y \\ 5x + 3y \end{bmatrix}$.

2. $T : \mathbb{R}^2 \to \mathbb{R}^3$ by $T \begin{bmatrix} x \\ y \end{bmatrix} = \begin{bmatrix} x + 1 \\ y + 2 \\ x - y \end{bmatrix}$.

3. $T : \mathbb{R}^3 \to \mathbb{R}^3$ by $T \begin{bmatrix} x \\ y \\ z \end{bmatrix} = \begin{bmatrix} x + y + z \\ z - y - x \\ xyz \end{bmatrix}$.

4. $T : \mathbb{R}^3 \to \mathbb{R}^2$ by $T \begin{bmatrix} x \\ y \\ z \end{bmatrix} = \begin{bmatrix} 2x - 2y + 5z \\ x + 2z \end{bmatrix}$.

5. $T : P_2 \to P_1$ by $T(ax^2 + bx + c) = 2ax + b$.

6. $T : P_1 \to P_2$ by $T(ax + b) = ax^2/2 + bx$.

7. $T : P_2 \to P_2$ by $T(ax^2 + bx + c) = a(x + 1)^2 + b(x + 1) + c$.

8. $T : P_2 \to P_2$ by $T(ax^2 + bx + c) = ax^2 + bx + c + 1$.

9. $T : F(-\infty, \infty) \to F(-\infty, \infty)$ by $T(f(x)) = f(x) - 2$.

10. $T : F(-\infty, \infty) \to F(-\infty, \infty)$ by $T(f(x)) = f(x - 2)$.

11. $T : M_{m \times n}(\mathbb{R}) \to M_{n \times m}(\mathbb{R})$ by $T(A) = A^T$.

12. $T : M_{n \times n}(\mathbb{R}) \to \mathbb{R}$ by $T(A) = \det(A)$.

13. Recall from Section 2.1 that the set of positive real numbers $\mathbb{R}^+$ is a vector space under the "addition" $x \oplus y = xy$ and the "scalar multiplication" $c \odot x = x^c$.

 a) Show that the natural logarithm is a linear transformation from $\mathbb{R}^+$ to $\mathbb{R}$.

 b) Show that the exponential function is a linear transformation from $\mathbb{R}$ to $\mathbb{R}^+$.

14. Prove the following parts of Theorem 5.2.

 a) Part (3)

 b) Part (4)

In Exercises 15–18, find a matrix A that expresses the linear transformation T in the form of a matrix transformation $T(X) = AX$.

15. $T \begin{bmatrix} x \\ y \end{bmatrix} = \begin{bmatrix} 3x + 2y \\ 2x + 3y \end{bmatrix}$

16. $T \begin{bmatrix} x \\ y \\ z \end{bmatrix} = \begin{bmatrix} 2x + y - z \\ -3x + y - 4z \\ 5x + 2y - 5z \end{bmatrix}$

17. $T \begin{bmatrix} x \\ y \\ z \end{bmatrix} = \begin{bmatrix} x - y + z \\ 2x + y + 2z \\ 3x + 3y + 3x \\ x + 2y + z \end{bmatrix}$

18. $T \begin{bmatrix} x_1 \\ x_2 \\ x_3 \\ x_4 \end{bmatrix} = \begin{bmatrix} x_1 - x_2 + 3x_3 - x_4 \\ 2x_1 + 3x_2 - x_3 - 2x_4 \\ 3x_1 + 7x_2 - 5x_3 - 3x_4 \end{bmatrix}$

19. Suppose $T : \mathbb{R}^3 \to \mathbb{R}^3$ is a linear transformation so that

$$T \begin{bmatrix} 1 \\ 0 \\ 1 \end{bmatrix} = \begin{bmatrix} 1 \\ 2 \\ 1 \end{bmatrix}, \quad T \begin{bmatrix} 1 \\ 1 \\ 0 \end{bmatrix} = \begin{bmatrix} -1 \\ 0 \\ 1 \end{bmatrix},$$

$$T \begin{bmatrix} 1 \\ 1 \\ 1 \end{bmatrix} = \begin{bmatrix} 0 \\ 1 \\ 1 \end{bmatrix}.$$

 a) Find $T \begin{bmatrix} 3 \\ 2 \\ 2 \end{bmatrix}$. **b)** Find $T \begin{bmatrix} x \\ y \\ z \end{bmatrix}$.

20. Suppose $T : \mathbb{R}^3 \to \mathbb{R}^4$ is a linear transformation so that

$$T \begin{bmatrix} 1 \\ -1 \\ 0 \end{bmatrix} = \begin{bmatrix} 1 \\ 0 \\ -1 \\ 0 \end{bmatrix},$$

$$T \begin{bmatrix} 1 \\ 0 \\ 1 \end{bmatrix} = \begin{bmatrix} 2 \\ 1 \\ 0 \\ 0 \end{bmatrix},$$

$$T \begin{bmatrix} 0 \\ 1 \\ -1 \end{bmatrix} = \begin{bmatrix} 1 \\ 0 \\ 0 \\ -1 \end{bmatrix}.$$

a) Find $T \begin{bmatrix} 2 \\ 1 \\ -4 \end{bmatrix}$.

b) Find $T \begin{bmatrix} x \\ y \\ z \end{bmatrix}$.

21. Suppose $T : P_1 \to P_1$ is a linear transformation so that

$$T(x+1) = 2x + 1, \qquad T(x-1) = 2x - 1.$$

a) Find $T(x)$.

b) Find $T(ax + b)$.

22. Suppose $T : \mathbb{R}^3 \to P_2$ is a linear tranformation so that

$$T \begin{bmatrix} 1 \\ 1 \\ 0 \end{bmatrix} = x^2 + x, \quad T \begin{bmatrix} 1 \\ -1 \\ 1 \end{bmatrix} = x^2 - x + 1,$$

$$T \begin{bmatrix} 0 \\ 1 \\ 1 \end{bmatrix} = x + 1.$$

a) Find $T \begin{bmatrix} 1 \\ 0 \\ 0 \end{bmatrix}$.

b) Find $T \begin{bmatrix} a \\ b \\ c \end{bmatrix}$.

In Exercises 23–26, find bases for the kernel and the range of the linear transformation in the indicated exercise.

23. Exercise 15   **24.** Exercise 16

25. Exercise 17   **26.** Exercise 18

27. Describe the vectors in the kernel of the differential operator $D : D(a, b) \to F(a, b)$.

28. Describe the vectors in the kernel of the linear transformation Int $: C[a, b] \to \mathbb{R}$ where Int$(f) = \int_a^b f(x)\,dx$.

29. In the verification of the claim of proof of Theorem 5.4, show that if $c_{k+1}v_{k+1} + \cdots + c_n v_n$ lies in ker(T), then $c_{k+1} = 0, \ldots, c_n = 0$.

30. Prove Theorem 5.4 in the case when:

a) $\dim(V) = 0$.

b) $0 = \dim(\ker(T)) < \dim(V)$.

c) $0 < \dim(\ker(T)) = \dim(V)$.

31. Suppose that $f : \mathbb{R} \to \mathbb{R}$ preserves scalar multiplication. Show that $f(x) = mx$ for some constant m. (*Hint:* View $f(x)$ as $f(x \cdot 1)$.)

32. Suppose that V is a vector space and k is a scalar. Show that $T : V \to V$ by $T(v) = kv$ is a linear transformation. Such a linear transformation is called a dilation if $k > 1$, a contraction if $0 < k < 1$, and a reflection if $k = -1$. Why would they be given these names? (*Suggestion:* Consider the results of such transformations in $\mathbb{R}^2$ or $\mathbb{R}^3$.)

33. Let $T : \mathbb{R}^2 \to \mathbb{R}^2$ be defined by letting

$$T \begin{bmatrix} x \\ y \end{bmatrix}$$

be the vector obtained by rotating

$$\begin{bmatrix} x \\ y \end{bmatrix}$$

counterclockwise through the angle α. Show that T is the matrix transformation

$$T \begin{bmatrix} x \\ y \end{bmatrix} = \begin{bmatrix} \cos \alpha & -\sin \alpha \\ \sin \alpha & \cos \alpha \end{bmatrix} \begin{bmatrix} x \\ y \end{bmatrix}.$$

Suggestion: If (r, θ) are polar coordinates for the point (x, y),

$$\begin{bmatrix} x \\ y \end{bmatrix} = \begin{bmatrix} r \cos \theta \\ r \sin \theta \end{bmatrix}.$$

Observe that

$$T \begin{bmatrix} x \\ y \end{bmatrix} = \begin{bmatrix} r\cos(\theta + \alpha) \\ r\sin(\theta + \alpha) \end{bmatrix}.$$

34. Let $T : \mathbb{R}^2 \to \mathbb{R}^2$ be defined by letting

$$T \begin{bmatrix} x \\ y \end{bmatrix}$$

be the vector obtained by rotating

$$\begin{bmatrix} x \\ y \end{bmatrix}$$

clockwise through the angle α. Show that T is the matrix transformation

$$T \begin{bmatrix} x \\ y \end{bmatrix} = \begin{bmatrix} \cos\alpha & \sin\alpha \\ -\sin\alpha & \cos\alpha \end{bmatrix} \begin{bmatrix} x \\ y \end{bmatrix}.$$

35. Prove the following fact, which is sometimes called the Fredholm alternative:[1] If V is an n-dimensional vector space and $T : V \to V$ is a linear transformation, then exactly one of the following holds:

1. For each vector v in V there is a vector u in V so that $T(u) = v$.[2]
2. $\dim(\ker(T)) > 0$.

36. Recall that a function $f : X \to Y$ is one-to-one if whenever $f(x_1) = f(x_2)$, $x_1 = x_2$. Show that a linear transformation $T : V \to W$ is one-to-one if and only if $\ker(T)$ consists of only the zero vector.

37. Suppose V is a vector space, X is a set, and there is a function $f : V \to X$ that is one-to-one and onto. Thus, every element in X is uniquely expressible as $f(v)$ for some vector v in V. This allows us to define an addition on X as

$$f(v) + f(u) = f(v + u)$$

where v and u are elements of V and a scalar multiplication on X by

$$cf(v) = f(cv)$$

where c is a real number. Show that X is a vector space under this addition and this scalar multiplication.

5.2   THE ALGEBRA OF LINEAR TRANSFORMATIONS; DIFFERENTIAL OPERATORS AND DIFFERENTIAL EQUATIONS

In the same manner as we add and form scalar multiples of real-valued functions defined on an interval, we can add and form scalar multiples of linear transformations from one vector space to another as follows: Suppose $T : V \to W$ and $S : V \to W$ are linear transformations. We let $T + S : V \to W$ be the function defined by

$$(T + S)(v) = T(v) + S(v)$$

and $cT : V \to W$ be the function defined by

$$(cT)(v) = cT(v)$$

where c is a scalar. To illustrate, if T and S are the linear transformations from $\mathbb{R}^2$ to itself defined by

$$T \begin{bmatrix} x \\ y \end{bmatrix} = \begin{bmatrix} x + y \\ x - y \end{bmatrix}$$

---

[1] Named for the mathematical physicist Ivar Fredholm, 1866–1927.

[2] This is the same as saying the function T is onto.

and

$$S\begin{bmatrix} x \\ y \end{bmatrix} = \begin{bmatrix} 2x - y \\ x + 2y \end{bmatrix},$$

then

$$(T + S)\begin{bmatrix} x \\ y \end{bmatrix} = T\begin{bmatrix} x \\ y \end{bmatrix} + S\begin{bmatrix} x \\ y \end{bmatrix} = \begin{bmatrix} x + y \\ x - y \end{bmatrix} + \begin{bmatrix} 2x - y \\ x + 2y \end{bmatrix}$$

$$= \begin{bmatrix} 3x \\ 2x + y \end{bmatrix}$$

and

$$(5T)\begin{bmatrix} x \\ y \end{bmatrix} = 5\begin{bmatrix} x + y \\ x - y \end{bmatrix} = \begin{bmatrix} 5x + 5y \\ 5x - 5y \end{bmatrix}.$$

Our addition and scalar multiplication of linear transformations give us linear transformations back again.

THEOREM 5.6   If $T : V \to W$ and $S : V \to W$ are linear transformations, then so are $T + S$ and cT where c is a scalar.

Proof   We verify this for $T + S$ and leave the proof for cT as an exercise (Exercise 15). Suppose that u and v are vectors in V and a is a scalar. We have

$$(T + S)(u + v) = T(u + v) + S(u + v) = T(u) + T(v) + S(u) + S(v)$$
$$= T(u) + S(u) + T(v) + S(v) = (T + S)(u) + (T + S)(v)$$

and

$$(T + S)(av) = T(av) + S(av) = aT(v) + aS(v) = a(T(v) + S(v)) = a(T + S)(v)$$

giving us $T + S$ is a linear transformation.   ●

As an immediate consequence of Theorem 5.6, we get that linear combinations of linear transformations from one vector space to another are linear transformations.

COROLLARY 5.7   If $T_1, T_2, \ldots, T_n : V \to W$ are linear transformations and $c_1, c_2, \ldots, c_n$ are scalars, then

$$c_1 T_1 + c_2 T_2 + \cdots + c_n T_n : V \to W$$

is a linear transformation.

If $f : X \to Y$ and $g : Y \to Z$ are functions, the composite of f and g is the function from X to Z, denoted $g \circ f$, given by

$$g \circ f(x) = g(f(x))$$

where x is an element of X. Composition of functions may be thought of as a type of multiplication of functions. Indeed, throughout mathematics you often will find that the circle indicating composition is omitted so that $g \circ f$ is written as simply gf. This is the convention we shall follow in this book. For the linear transformations

$$T \begin{bmatrix} x \\ y \end{bmatrix} = \begin{bmatrix} x+y \\ x-y \end{bmatrix}$$

and

$$S \begin{bmatrix} x \\ y \end{bmatrix} = \begin{bmatrix} 2x-y \\ x+2y \end{bmatrix},$$

the composite ST is

$$ST \begin{bmatrix} x \\ y \end{bmatrix} = S \left(T \begin{bmatrix} x \\ y \end{bmatrix} \right) = S \begin{bmatrix} x+y \\ x-y \end{bmatrix}$$

$$= \begin{bmatrix} 2(x+y)-(x-y) \\ x+y+2(x-y) \end{bmatrix} = \begin{bmatrix} x+3y \\ 3x-y \end{bmatrix}.$$

Similar to Theorem 5.6, we have composites of linear transformations give us linear transformations back again.

THEOREM 5.8   If $T : V \to W$ and $S : W \to U$ are linear transformations, then the composite $ST : V \to U$ is a linear transformation.

Proof   For any two vectors u and v of V and any scalar c, we have

$$ST(u+v) = S(T(u+v)) = S(T(u)+T(v)) = S(T(u)) + S(T(v))$$
$$= ST(u) + ST(v)$$

and

$$ST(cv) = S(T(cv)) = S(cT(v)) = cS(T(v)) = cST(V). \qquad \bullet$$

With matrices we have an addition (provided sizes are the same), a scalar multiplication, and a multiplication (provided the number of columns in the left-hand factor equals the number of rows in the right-hand factor). Now, with linear transformations, we also have an addition (provided the domains and codomains of both linear transformations are the same), a scalar multiplication, and a multiplication (provided the domain of the left-hand factor equals the codomain of the right-hand factor). Properties we have for these operations on matrices carry over to linear transformations (compare with Theorems 1.2 and 1.3).

THEOREM 5.9   Provided the indicated operations are defined, the following properties hold where R, S, and T are linear transformations and c and d are scalars.

    1. $S + T = T + S$
    2. $R + (S + T) = (R + S) + T$
    3. $c(dT) = (cd)T$

4. $c(S + T) = cS + cT$

5. $(c + d)T = cT + dT$

6. $R(ST) = (RS)T$

7. $R(S + T) = RS + RT$

8. $(R + S)T = RT + ST$

9. $c(ST) = (cS)T = S(cT)$

Proof   We will prove the associativity property in part (6) and leave the proofs of the remaining parts as exercises (Exercise 16). In fact, this associativity property holds for all functions, not just linear transformations. To prove it (which is the approach used to prove all of these properties), we take a vector v in the domain and verify that the functions on each side of the equation applied to v give us the same result: Applying the definition of the composite of functions, we have

$$R(ST)(v) = R(ST(v)) = R(S(T(v)))$$

and

$$(RS)T(v) = RS(T(v)) = R(S(T(v))).$$

Therefore, the functions $R(ST)$ and $(RS)T$ are the same.   ●

One other similarity between matrices and linear transformations we mention at this time involves exponents. Recall that if A is a square matrix and n is a positive integer, we have defined the nth power of A, A^n, as the product of n factors of A. Notice that we can do likewise for a linear transformation that has the same domain and codomain. That is, if $T : V \rightarrow V$ is a linear transformation and n is a positive integer, then T^n is the product (composite) of n factors of T.

We are now going to see how we can use the ideas we have developed involving linear transformations to obtain a particularly elegant approach to the study of linear differential equations. To simplify our presentation, let us assume that our functions have derivatives of all orders on an open interval (a, b); that is, we will use $C^\infty(a, b)$ as our vector space of functions. We will make use of two basic types of linear transformations. One type involves the differential operator $D : C^\infty(a, b) \rightarrow C^\infty(a, b)$ and its powers D^n, which amount to taking the nth derivative. The other type involves multiplying functions f in $C^\infty(a, b)$ by a fixed function g in $C^\infty(a, b)$: If g is a function in $C^\infty(a, b)$, define

$$T_{g(x)} : C^\infty(a, b) \rightarrow C^\infty(a, b)$$

by

$$T_{g(x)}(f(x)) = g(x)f(x).$$

For example, if $g(x) = x^2$,

$$T_{g(x)}(f(x)) = T_{x^2}(f(x)) = x^2 f(x).$$

We leave it to you to verify that $T_{g(x)}$ is a linear transformation (Exercise 17).

Now, consider a linear differential equation

$$q_n(x)y^{(n)} + q_{n-1}(x)y^{(n-1)} + \cdots + q_1(x)y' + q_0(x)y = g(x).$$

Notice that the left-hand side of this differential equation is obtained by applying the linear transformation

$$L = T_{q_n(x)}D^n + T_{q_{n-1}(x)}D^{n-1} + \cdots + T_{q_1(x)}D + T_{q_0(x)} \tag{1}$$

to y:

$$L(y) = (T_{q_n(x)}D^n + T_{q_{n-1}(x)}D^{n-1} + \cdots + T_{q_1(x)}D + T_{q_0(x)})(y) \tag{2}$$

To simplify our notation, we leave off the Ts and write the linear transformation in Equation (1) as

$$L = q_n(x)D^n + q_{n-1}(x)D^{n-1} + \cdots + q_1(x)D + q_0(x) \tag{3}$$

where it is to be understood that a factor $q_i(x)$ appearing in a term indicates multiplication by $q_i(x)$. In Equation (2), we leave off not only the Ts but also the parentheses about y, writing this expression as

$$Ly = (q_n(x)D^n + q_{n-1}(x)D^{n-1} + \cdots + q_1(x)D + q_0(x))y. \tag{4}$$

With this notation, the linear differential equation at the beginning of this paragraph becomes

$$Ly = g(x).$$

For a homogeneous linear differential equation, our differential equation takes on the form

$$Ly = 0$$

in this differential operator notation. Notice that finding the solutions to such a linear differential equation is the same as finding the vectors in the kernel of the linear transformation in Equation (3). In effect, the techniques you learned for solving homogeneous linear differential equations in the previous chapter amount to methods for finding a basis for this kernel.

In the special case where the homogeneous linear differential equation has constant coefficients,

$$a_n y^{(n)} + a_{n-1}y^{(n-1)} + \cdots + a_1 y' + a_0 y = 0,$$

our notation takes on the form

$$Ly = (a_n D^n + a_{n-1}D^{n-1} + \cdots + a_1 D + a_0)y = 0$$

and our associated linear transformation is a polynomial in D

$$p(D) = L = a_n D^n + a_{n-1}D^{n-1} + \cdots + a_1 D + a_0,$$

which is the same as the characteristic polynomial

$$p(\lambda) = a_n \lambda^n + a_{n-1}\lambda^{n-1} + \cdots + a_1\lambda + a_0$$

with D substituted for λ. Because products of polynomials in D commute (see Exercise 18), we can factor polynomials in D in the same manner we factor polynomials in λ. In particular, if $\lambda = r$ is a root of the characteristic polynomial $p(\lambda)$ so that

$$p(\lambda) = q(\lambda)(\lambda - r)$$

where $q(\lambda)$ is a polynomial in λ, then

$$p(D) = q(D)(D - r). \tag{5}$$

Equation (5) gives a very nice way of seeing why if $\lambda = r$ is a root (either real or imaginary) of the characteristic polynomial $p(\lambda)$, then e^{rx} is a solution to the differential equation $p(D)y = 0$:

$$p(D)e^{rx} = q(D)(D - r)e^{rx} = q(D)(De^{rx} - re^{rx}) = q(D)(re^{rx} - re^{rx})$$
$$= q(D)0 = 0.$$

The fact that $(D - r)e^{rx} = 0$ can be extended to obtain a justification for our method for constructing additional solutions when the characteristic polynomial has repeated roots. Notice that

$$(D - r)^2 x e^{rx} = (D - r)(D - r)x e^{rx} = (D - r)(Dx e^{rx} - rx e^{rx})$$
$$= (D - r)(rx e^{rx} + e^{rx} - rx e^{rx}) = (D - r)e^{rx} = 0.$$

Likewise,

$$(D - r)^3 x^2 e^{rx} = (D - r)^2(D - r)x^2 e^{rx} = (D - r)^2(Dx^2 e^{rx} - rx^2 e^{rx})$$
$$= (D - r)^2(rx^2 e^{rx} + 2x e^{rx} - rx^2 e^{rx}) = (D - r)^2 2x e^{rx}$$
$$= 2(D - r)^2 x e^{rx} = 0.$$

Continuing, we obtain

$$(D - r)^i x^{i-1} e^{rx} = 0 \tag{6}$$

for any positive integer i.[3]

We can use Equation (6) to prove Theorem 4.8. To refresh our memories, let us restate this theorem before doing so.

THEOREM 4.8   If $\lambda = r$ is a root of multiplicity m of the characteristic polynomial $p(\lambda)$ of the homogeneous linear differential equation

$$a_n y^{(n)} + a_{n-1} y^{(n-1)} + \cdots + a_1 y' + y = 0,$$

---

[3] A polynomial $p(D)$ in D is said to **annihilate** a function f in $C^\infty(a, b)$ if $p(D)f(x) = 0$ for all x in (a, b). Thus, in this terminology, Equation (6) says that $(D - r)^i$ annihilates $x^{i-1}e^{rx}$. The set of all polynomials $p(D)$ that annihilates a function f in $C^\infty(a, b)$ is called the **annihilator** of f. It can be shown that the annihilator of $x^{i-1}e^{rx}$ consists of all polynomials in D that have $(D - r)^i$ as a factor.

then

$$e^{rx}, xe^{rx}, x^2 e^{rx}, \ldots, x^{m-1} e^{rx}$$

are solutions to this differential equation.

Proof   Since $(D - r)^i$ is a factor of $p(D)$ for each $1 \le i \le m$, Equation (6) gives us that $p(D)x^{i-1}e^{rx} = 0$ for each $1 \le i \le m$.   ●

Our differential operator point of view also can be used to prove Theorem 4.11 that we used for finding particular solutions to nonhomogeneous linear differential equations with the method of undetermined coefficients. Recall this theorem was the following.

THEOREM 4.11   Suppose that $\lambda = r$ is a root of multiplicity m of the characteristic polynomial $p(\lambda)$ of the homogeneous linear differential equation

$$a_n y^{(n)} + \cdots + a_1 y' + a_0 y = 0.$$

Let k be a nonnegative integer and A and B be constants.

1.   If r is real, then the linear differential equation

$$a_n y^{(n)} + \cdots + a_1 y' + a_0 y = A x^k e^{rx}$$

has a particular solution of the form

$$x^m (A_k x^k + \cdots + A_1 x + A_0) e^{rx}.$$

2.   If $r = a + bi$ is imaginary, then the linear differential equation

$$a_n y^{(n)} + \cdots + a_1 y' + a_0 y = A x^k e^{ax} \cos bx + B x^k e^{ax} \sin bx$$

has a particular solution of the form

$$x^m (A_k x^k + \cdots + A_1 x + A_0) e^{ax} \cos bx + x^m (B_k x^k + \cdots + B_1 x + B_0) e^{ax} \sin bx.$$

A complete proof of Theorem 4.11 is long and arduous and will not be included here. But to give you an idea of how it goes, let us prove part (1) when $k = 1$. We must show that

$$a_n y^{(n)} + \cdots + a_1 y' + a_0 y = A x e^{rx}$$

has a solution of the form

$$x^m (A_1 x + A_0) e^{rx}.$$

Let us write $p(\lambda)$ as

$$p(\lambda) = q(\lambda)(\lambda - r)^m$$

and write $q(\lambda)$ as

$$q(\lambda) = b_{n-m} \lambda^{n-m} + \cdots + b_1 \lambda + b_0.$$

Proceeding in a manner similar to that used in obtaining Equation (6), we find

$$(D - r)^m x^m (A_1 x + A_0) e^{rx} = (D - r)^m (A_1 x^{m+1} e^{rx} + A_0 x^m e^{rx})$$
$$= (m + 1)! A_1 x e^{rx} + m! A_0 e^{rx}.$$

Doing some more calculations, we find

$$q(D)(D - r)^m x^m (A_1 x + A_0) e^{rx}$$
$$= q(D)((m + 1)! A_1 x e^{rx} + m! A_0 e^{rx})$$
$$= b_{n-m}((m + 1)! A_1 D^{n-m} x e^{rx} + m! A_0 D^{n-m} e^{rx})$$
$$\quad + \cdots + b_0((m + 1)! A_1 x e^{rx} + A_0 m! e^{rx})$$
$$= b_{n-m}((m + 1)! A_1 r^{n-m} x e^{rx} + (m + 1)! A_1 (n - m) r^{n-m-1} e^{rx}$$
$$\quad + m! A_0 r^{n-m} e^{rx}) + \cdots + b_0((m + 1)! A_1 x e^{rx} + A_0 m! e^{rx})$$
$$= (m + 1)! A_1 q(r) x e^{rx} + ((m + 1)! A_1 q'(r) + m! A_0 q(r)) e^{rx}.$$

To have a solution to the nonhomogeneous differential equation

$$a_n y^{(n)} + \cdots + a_1 y' + a_0 y = A x e^{rx},$$

we must have that the system of linear equations

$$(m + 1)! A_1 q(r) = A$$
$$(m + 1)! A_1 q'(r) + m! A_0 q(r) = 0$$

in the unknowns A_1 and A_0 has a solution. Since $q(r) \neq 0$, the first equation can be solved for A_1 and then the second equation can be solved for A_0. This completes the proof of part (1) when $k = 1$.

EXERCISES 5.2

Let $S, T : \mathbb{R}^2 \to \mathbb{R}^2$ be the linear transformations

$$S \begin{bmatrix} x \\ y \end{bmatrix} = \begin{bmatrix} 2x - y \\ x + 2y \end{bmatrix}, \quad T \begin{bmatrix} x \\ y \end{bmatrix} = \begin{bmatrix} x + 3y \\ x - y \end{bmatrix}.$$

Find:

1. $(S + T) \begin{bmatrix} x \\ y \end{bmatrix}$.

2. $(S - T) \begin{bmatrix} x \\ y \end{bmatrix}$.

3. $(2T) \begin{bmatrix} x \\ y \end{bmatrix}$.

4. $(S - 4T) \begin{bmatrix} x \\ y \end{bmatrix}$.

5. $ST \begin{bmatrix} x \\ y \end{bmatrix}$.

6. $TS \begin{bmatrix} x \\ y \end{bmatrix}$.

Let $S, T : P_1 \to P_1$ be the linear transformations

$$S(ax + b) = ax - 2a + b,$$

$$T(ax + b) = ax + 2a + b.$$

Find:

7. $(S + 3T)(ax + b)$.

8. $(TS)(ax + b)$.

9. $T^2(ax + b)$.

10. $(S + 3T)^3(ax + b)$.

Find bases for the kernels of the following linear transformations from $C^\infty(-\infty, \infty)$ to $C^\infty(-\infty, \infty)$.

11. $D^2 - 2D - 3$

12. $D^2 + 2D + 2$

13. $D^4 + D^2$

14. $D^4 + 2D^2 + 1$

15. Complete the proof of Theorem 5.6 by showing that if $T : V \to W$ is a linear transformation and c is a scalar, then cT is a linear transformation.

16. Prove the following parts of Theorem 5.9.

a) Part (1)

b) Part (2)

c) Part (3)

d) Part (4)

e) Part (5)   **f)** Part (7)

g) Part (8)   **h)** Part (9)

17. Show that if g is a function in $C^\infty(a, b)$, then $T_{g(x)}$ is a linear transformation from $C^\infty(a, b)$ to $C^\infty(a, b)$.

18. Suppose

$$p(D) = a_n D^n + a_{n-1} D^{n-1} + \cdots + a_1 D + a_0$$

and

$$q(D) = b_m D^m + b_{m-1} D^{m-1} + \cdots + b_1 D + b_0$$

where $a_n, \ldots, a_0$ and $b_m, \ldots, b_0$ are constants. Show that $p(D)q(D) = q(D)p(D)$.

19. Prove Equation (6) by using mathematical induction.

20. Suppose $T : V \rightarrow W$ is a linear transformation.

a) Show that if v is a vector in V and u is a vector in $\ker(T)$, then $T(u + v) = T(v)$.

b) Show that if u and v are vectors in V such that $T(u) = T(v)$, then $u - v$ is in $\ker(T)$.

21. Use the results of Exercise 20 to prove Theorem 4.3.

22. Show that the set of linear transformations from a vector space V to a vector space W is a vector space.

23. Let T and S be the linear transformations in Exercises 1–6.

a) Find matrices A and B so that T and S are expressed as the matrix transformations $T(X) = AX$ and $S(X) = BX$.

b) Find the matrix C so that the composite ST in Exercise 5 is expressed in the form $ST(X) = CX$. Then verify that $C = BA$.

c) Find the matrix D so that the composite TS in Exercise 6 is expressed in the form $TS(X) = DX$. Then verify that $D = AB$.

24. Parts (b) and (c) of Exercise 23 illustrate the following fact: If $T : \mathbb{R}^n \rightarrow \mathbb{R}^m$ is the linear transformation given by

$$T \begin{bmatrix} x_1 \\ x_2 \\ \vdots \\ x_n \end{bmatrix} = \begin{bmatrix} a_{11}x_1 + a_{12}x_2 + \cdots + a_{1n}x_n \\ a_{21}x_1 + a_{22}x_2 + \cdots + a_{2n}x_n \\ \vdots \\ a_{m1}x_1 + a_{m2}x_2 + \cdots + a_{mn}x_n \end{bmatrix},$$

$S : \mathbb{R}^m \rightarrow \mathbb{R}^l$ is the linear transformation given by

$$S \begin{bmatrix} x_1 \\ x_2 \\ \vdots \\ x_m \end{bmatrix} = \begin{bmatrix} b_{11}x_1 + b_{12}x_2 + \cdots + b_{1m}x_m \\ b_{21}x_1 + b_{22}x_2 + \cdots + b_{2m}x_m \\ \vdots \\ b_{l1}x_1 + b_{l2}x_2 + \cdots + b_{lm}x_m \end{bmatrix}$$

so that $T(X) = AX$ and $S(X) = BX$ where

$$A = \begin{bmatrix} a_{11} & a_{12} & \cdots & a_{1n} \\ a_{21} & a_{22} & \cdots & a_{2n} \\ \vdots & \vdots & & \vdots \\ a_{m1} & a_{m2} & \cdots & a_{mn} \end{bmatrix}$$

and

$$B = \begin{bmatrix} b_{11} & b_{12} & \cdots & b_{1m} \\ b_{21} & a_{22} & \cdots & b_{2m} \\ \vdots & \vdots & & \vdots \\ b_{l1} & b_{l2} & \cdots & b_{lm} \end{bmatrix},$$

then $ST(X) = BAX$. Prove this fact. (In Theorem 5.10 of the next section, we will generalize this fact.)

25. Suppose $T : V \rightarrow W$ is a linear transformation and suppose that T has an inverse function $T^{-1} : W \rightarrow V$. Show that T^{-1} is also a linear transformation. (You will have to use the fact that T^{-1} is defined by $T^{-1}(w) = v$ where $T(v) = w$.)

5.3   MATRICES FOR LINEAR TRANSFORMATIONS

In the first section of this chapter it was mentioned that in this section we would see that every linear transformation from $\mathbb{R}^n$ to $\mathbb{R}^m$ is a matrix transformation. Further, it was said that, more generally, we would see by using coordinate vectors that every linear transformation from one finite dimensional vector space to another is, in a sense, given by matrix multiplication. To see how we get these statements, we begin by showing you how we can associate matrices to linear transformations on nonzero finite dimension vector spaces. To avoid having to write it all the time, we assume all vector spaces are nonzero finite dimensional throughout this section and the remainder of this chapter.

Suppose $T : V \rightarrow W$ is a linear transformation. Further suppose the vectors

$$v_1, \ldots, v_n$$

form a basis α for V and the vectors

$$w_1, \ldots, w_m$$

form a basis β for W. Let us express $T(v_1), \ldots, T(v_n)$ in terms of $w_1, \ldots, w_m$:

$$T(v_1) = a_{11}w_1 + a_{21}w_2 + \cdots + a_{m1}w_m$$
$$T(v_2) = a_{12}w_1 + a_{22}w_2 + \cdots + a_{m2}w_m$$
$$\vdots$$
$$T(v_n) = a_{1n}w_1 + a_{2n}w_2 + \cdots + a_{mn}w_m.$$

The matrix comprised of the scalars a_{ij} in these equations is called the **matrix of T with respect to the bases α and β** and will be denoted $[T]_\alpha^\beta$:

$$[T]_\alpha^\beta = \begin{bmatrix} a_{11} & a_{12} & \cdots & a_{1n} \\ a_{21} & a_{22} & \cdots & a_{2n} \\ \vdots & \vdots & \cdots & \vdots \\ a_{m1} & a_{m2} & \cdots & a_{mn} \end{bmatrix}.$$

You may wish to remember how to obtain $[T]_\alpha^\beta$ by noting that its columns are the coordinate vectors of $T(v_1), \ldots, T(v_n)$ with respect to β:

$$[T]_\alpha^\beta = [[T(v_1)]_\beta \ldots [T(v_n)]_\beta].$$

A special case that comes up often is when T maps V to itself (that is, $T : V \rightarrow V$) and the same basis α is used for both the domain and the codomain. In this case, $[T]_\alpha^\alpha$ is simply called the **matrix of T with respect to α**.

So that you get a feel for these matrices, let us do an example.

EXAMPLE 1   Let $T : \mathbb{R}^3 \rightarrow \mathbb{R}^3$ be the linear transformation

$$T \begin{bmatrix} x \\ y \\ z \end{bmatrix} = \begin{bmatrix} 5x + z \\ 3x + 2y - 3z \\ 5x \end{bmatrix}.$$

For the standard basis α and the basis β given by

$$\begin{bmatrix} 1 \\ 1 \\ 2 \end{bmatrix}, \begin{bmatrix} 1 \\ -1 \\ 1 \end{bmatrix}, \begin{bmatrix} 1 \\ 1 \\ 1 \end{bmatrix}$$

of $\mathbb{R}^3$, find:

(a) The matrix of T with respect to the standard basis α of $\mathbb{R}^3$.

(b) The matrix of T with respect to the basis β.

(c) $[T]_\alpha^\beta$.

Solution

(a) We have

$$T(e_1) = T\begin{bmatrix} 1 \\ 0 \\ 0 \end{bmatrix} = \begin{bmatrix} 5 \\ 3 \\ 5 \end{bmatrix} = 5e_1 + 3e_2 + 5e_3,$$

$$T(e_2) = T\begin{bmatrix} 0 \\ 1 \\ 0 \end{bmatrix} = \begin{bmatrix} 0 \\ 2 \\ 0 \end{bmatrix} = 0e_1 + 2e_2 + 0e_3,$$

and

$$T(e_3) = T\begin{bmatrix} 0 \\ 0 \\ 1 \end{bmatrix} = \begin{bmatrix} 1 \\ -3 \\ 0 \end{bmatrix} = e_1 - 3e_2 + 0e_3.$$

Thus,

$$[T]_\alpha^\alpha = \begin{bmatrix} 5 & 0 & 1 \\ 3 & 2 & -3 \\ 5 & 0 & 0 \end{bmatrix}.$$

(b) This part is harder. We have to write

$$T\begin{bmatrix} 1 \\ 1 \\ 2 \end{bmatrix} = \begin{bmatrix} 7 \\ -1 \\ 5 \end{bmatrix}, \quad T\begin{bmatrix} 1 \\ -1 \\ 1 \end{bmatrix} = \begin{bmatrix} 6 \\ -2 \\ 5 \end{bmatrix}, \quad T\begin{bmatrix} 1 \\ 1 \\ 1 \end{bmatrix} = \begin{bmatrix} 6 \\ 2 \\ 5 \end{bmatrix}$$

in terms of the vectors in β. This means we have to solve each of the three vector equations

$$\begin{bmatrix} 7 \\ -1 \\ 5 \end{bmatrix} = c_1 \begin{bmatrix} 1 \\ 1 \\ 2 \end{bmatrix} + c_2 \begin{bmatrix} 1 \\ -1 \\ 1 \end{bmatrix} + c_3 \begin{bmatrix} 1 \\ 1 \\ 1 \end{bmatrix},$$

$$\begin{bmatrix} 6 \\ -2 \\ 5 \end{bmatrix} = c_1 \begin{bmatrix} 1 \\ 1 \\ 2 \end{bmatrix} + c_2 \begin{bmatrix} 1 \\ -1 \\ 1 \end{bmatrix} + c_3 \begin{bmatrix} 1 \\ 1 \\ 1 \end{bmatrix},$$

and

$$\begin{bmatrix} 6 \\ 2 \\ 5 \end{bmatrix} = c_1 \begin{bmatrix} 1 \\ 1 \\ 2 \end{bmatrix} + c_2 \begin{bmatrix} 1 \\ -1 \\ 1 \end{bmatrix} + c_3 \begin{bmatrix} 1 \\ 1 \\ 1 \end{bmatrix}.$$

We can simultaneously solve these three systems as follows.

$$\begin{bmatrix} 1 & 1 & 1 & | & 7 & 6 & 6 \\ 1 & -1 & 1 & | & -1 & -2 & 2 \\ 2 & 1 & 1 & | & 5 & 5 & 5 \end{bmatrix}$$

$$\rightarrow \begin{bmatrix} 1 & 1 & 1 & | & 7 & 6 & 6 \\ 0 & -2 & 0 & | & -8 & -8 & -4 \\ 0 & -1 & -1 & | & -9 & -7 & -7 \end{bmatrix}$$

$$\rightarrow \begin{bmatrix} 2 & 0 & 2 & | & 6 & 4 & 8 \\ 0 & -2 & 0 & | & -8 & -8 & -4 \\ 0 & 0 & -2 & | & -10 & -6 & -10 \end{bmatrix}$$

$$\rightarrow \begin{bmatrix} 2 & 0 & 0 & | & -4 & -2 & -2 \\ 0 & -2 & 0 & | & -8 & -8 & -4 \\ 0 & 0 & -2 & | & -10 & -6 & -10 \end{bmatrix}$$

$$\rightarrow \begin{bmatrix} 1 & 0 & 0 & | & -2 & -1 & -1 \\ 0 & 1 & 0 & | & 4 & 4 & 2 \\ 0 & 0 & 1 & | & 5 & 3 & 5 \end{bmatrix}$$

The right-hand side columns are the respective solutions to our three vector equations and hence.

$$[T]_\beta^\beta = \begin{bmatrix} -2 & -1 & -1 \\ 4 & 4 & 2 \\ 5 & 3 & 5 \end{bmatrix}.$$

(c)   Here we have to write

$$T(e_1) = \begin{bmatrix} 5 \\ 3 \\ 5 \end{bmatrix}, \quad T(e_2) = \begin{bmatrix} 0 \\ 2 \\ 0 \end{bmatrix}, \quad T(e_3) = \begin{bmatrix} 1 \\ -3 \\ 0 \end{bmatrix}$$

in terms of the vectors of β:

$$\begin{bmatrix} 5 \\ 3 \\ 5 \end{bmatrix} = c_1 \begin{bmatrix} 1 \\ 1 \\ 2 \end{bmatrix} + c_2 \begin{bmatrix} 1 \\ -1 \\ 1 \end{bmatrix} + c_3 \begin{bmatrix} 1 \\ 1 \\ 1 \end{bmatrix},$$

$$\begin{bmatrix} 0 \\ 2 \\ 0 \end{bmatrix} = c_1 \begin{bmatrix} 1 \\ 1 \\ 2 \end{bmatrix} + c_2 \begin{bmatrix} 1 \\ -1 \\ 1 \end{bmatrix} + c_3 \begin{bmatrix} 1 \\ 1 \\ 1 \end{bmatrix},$$

and

$$\begin{bmatrix} 1 \\ -3 \\ 0 \end{bmatrix} = c_1 \begin{bmatrix} 1 \\ 1 \\ 2 \end{bmatrix} + c_2 \begin{bmatrix} 1 \\ -1 \\ 1 \end{bmatrix} + c_3 \begin{bmatrix} 1 \\ 1 \\ 1 \end{bmatrix}.$$

Simultaneously solving these systems by reducing the coefficient portion of the augmented matrix

$$\begin{bmatrix} 1 & 1 & 1 & \vdots & 5 & 0 & 1 \\ 1 & -1 & 1 & \vdots & 3 & 2 & -3 \\ 2 & 1 & 1 & \vdots & 5 & 0 & 0 \end{bmatrix}$$

to reduced row-echelon form, we obtain the augmented matrix

$$\begin{bmatrix} 1 & 0 & 0 & \vdots & 0 & 0 & -1 \\ 0 & 1 & 0 & \vdots & 1 & -1 & 2 \\ 0 & 0 & 1 & \vdots & 4 & 1 & 0 \end{bmatrix}.$$

(If in doubt, perform row operations and verify this.) This gives us

$$[T]_\alpha^\beta = \begin{bmatrix} 0 & 0 & -1 \\ 1 & -1 & 2 \\ 4 & 1 & 0 \end{bmatrix}.$$ ●

You can see from Example 1 that finding a matrix of a transformation with respect to standard bases as in part (a) is easy, while parts (b) and (c) illustrate that finding matrices with respect to other bases is more involved. Shortly, we will see another way of doing problems such as parts (b) and (c) of Example 1 that many find more convenient to use. Toward this objective, we next consider the question: What happens to matrices with respect to bases when we form composites of linear transformations? You might conjecture that if composition is a type of multiplication of functions, perhaps these matrices should be multiplied. If you did so, you are exactly correct.

THEOREM 5.10   Suppose $T : V \to W$ and $S : W \to U$ are linear transformations. If α is a basis for V, β is a basis for W, and γ is a basis for U, then

$$[ST]_\alpha^\gamma = [S]_\beta^\gamma [T]_\alpha^\beta.$$

Proof   Suppose that α consists of the vectors $v_1, \ldots, v_n$, β consists of the vectors $w_1, \ldots, w_m$, and γ consists of the vectors $u_1, \ldots, u_k$. Setting

$$T(v_1) = a_{11}w_1 + \cdots + a_{m1}w_m$$

$$\vdots$$

$$T(v_n) = a_{1n}w_1 + \cdots + a_{mn}w_m$$

and

$$S(w_1) = b_{11}u_1 + \cdots + b_{k1}u_k$$

$$\vdots$$

$$S(w_m) = b_{1m}u_1 + \cdots + b_{km}u_k$$

we have

$$[T]_\alpha^\beta = \begin{bmatrix} a_{11} & \cdots & a_{1n} \\ \vdots & \cdots & \vdots \\ a_{m1} & \cdots & a_{mn} \end{bmatrix} \quad \text{and} \quad [S]_\beta^\gamma = \begin{bmatrix} b_{11} & \cdots & b_{1m} \\ \vdots & \cdots & \vdots \\ b_{k1} & \cdots & b_{km} \end{bmatrix}.$$

Notice that

$$ST(v_1) = S(T(v_1)) = S(a_{11}w_1 + \cdots + a_{m1}w_m)$$

$$= a_{11}(b_{11}u_1 + \cdots + b_{k1}u_k) + \cdots + a_{m1}(b_{1m}u_1 + \cdots + b_{km}u_k)$$

$$= (b_{11}a_{11} + \cdots + b_{1m}a_{m1})u_1 + \cdots + (b_{k1}a_{11} + \cdots + b_{km}a_{m1})u_k$$

$$\vdots$$

$$ST(v_n) = S(T(v_n)) = S(a_{1n}w_1 + \cdots + a_{mn}w_m)$$

$$= a_{1n}(b_{11}u_1 + \cdots + b_{k1}u_k) + \cdots a_{mn}(b_{1m}u_1 \cdots + b_{km}u_k)$$

$$= (b_{11}a_{1n} + \cdots + b_{1m}a_{mn})u_1 + \cdots + (b_{k1}a_{1n} + \cdots + b_{km}a_{mn})u_k$$

from which we get

$$[ST]_\alpha^\gamma = \begin{bmatrix} b_{11}a_{11} + \cdots + b_{1m}a_{m1} & \cdots & b_{11}a_{1n} + \cdots + b_{1m}a_{mn} \\ \vdots & \cdots & \vdots \\ b_{k1}a_{11} + \cdots + b_{km}a_{m1} & \cdots & b_{k1}a_{1n} + \cdots + b_{km}a_{mn} \end{bmatrix},$$

which is the same as $[S]_\beta^\gamma [T]_\alpha^\beta$.   ●

We next develop a way of changing matrices of linear transformations from one pair of bases to another. Suppose α consisting of $v_1, \ldots, v_n$ and β consisting of $w_1, \ldots, w_n$ are bases for a vector space V. Write the vectors in β in terms of those in α:

$$w_1 = p_{11}v_1 + p_{21}v_2 + \cdots + p_{n1}v_n$$
$$w_2 = p_{12}v_1 + p_{22}v_2 + \cdots + p_{n2}v_n$$
$$\vdots$$
$$w_n = p_{1n}v_1 + p_{2n}v_2 + \cdots + p_{nn}v_n.$$

The matrix

$$P = \begin{bmatrix} p_{11} & p_{12} & \cdots & p_{1n} \\ p_{21} & p_{22} & \cdots & p_{2n} \\ \vdots & \vdots & \cdots & \vdots \\ p_{n1} & p_{n2} & \cdots & p_{nn} \end{bmatrix}$$

is called the **change of basis matrix from α to β**. Note that the columns of P are the coordinate vectors of $w_1, \ldots, w_n$ with respect to α:

$$P = [[w_1]_\alpha \ [w_2]_\alpha \cdots [w_n]_\alpha].$$

But there is another way to think of P. The **identity transformation** from V to V denoted by I is defined as

$$I(v) = v.$$

It is easily verified that I is a linear transformation (Exercise 11). Notice that since

$$I(w_1) = w_1 = p_{11}v_1 + \cdots + p_{n1}v_n$$
$$\vdots$$
$$I(w_n) = w_n = p_{n1}v_1 + \cdots + p_{nn}v_n$$

we have

$$[I]_\beta^\alpha = P.$$

In the special case of $[I]_\beta^\beta$,

$$I(w_1) = w_1 = 1 \cdot w_1 + 0 \cdot w_2 + \cdots + 0 \cdot w_n$$
$$\vdots$$
$$I(w_n) = w_n = 0 \cdot w_1 + \cdots + 0 \cdot w_{n-1} + 1 \cdot w_n$$

and hence (not surprisingly!) the change of basis matrix from a basis to itself is the identity matrix:

$$[I]_\beta^\beta = \begin{bmatrix} 1 & 0 & \cdots & 0 \\ 0 & 1 & \cdots & 0 \\ \vdots & \vdots & \cdots & \vdots \\ 0 & 0 & \cdots & 1 \end{bmatrix}.$$

What would the change of basis from β to α be? We know it is $[I]_\alpha^\beta$. But notice that by

Theorem 5.10,

$$[I]_\alpha^\beta [I]_\beta^\alpha = [I]_\beta^\beta$$

and hence we have the following theorem.

THEOREM 5.11   If P is the change of basis matrix from a basis α to a basis β of a vector space, then the change of basis matrix from β to α is P^{-1}.

EXAMPLE 2   Let α be the standard basis for $\mathbb{R}^3$ and the basis β be the basis of $\mathbb{R}^3$ consisting of

$$\begin{bmatrix} 1 \\ 1 \\ 2 \end{bmatrix}, \begin{bmatrix} 1 \\ -1 \\ 1 \end{bmatrix}, \begin{bmatrix} 1 \\ 1 \\ 1 \end{bmatrix}$$

(the bases in Example 1). Find the change of basis matrices from α to β and from β to α.

Solution   Getting the change of basis matrix from α to β is easy since it is easy to write the vectors in β in terms of those in α:

$$\begin{bmatrix} 1 \\ 1 \\ 2 \end{bmatrix} = e_1 + e_2 + 2e_3, \qquad \begin{bmatrix} 1 \\ -1 \\ 1 \end{bmatrix} = e_1 - e_2 + e_3, \qquad \begin{bmatrix} 1 \\ 1 \\ 1 \end{bmatrix} = e_1 + e_2 + e_3.$$

Hence the change of basis matrix from α to β is

$$P = \begin{bmatrix} 1 & 1 & 1 \\ 1 & -1 & 1 \\ 2 & 1 & 1 \end{bmatrix}.$$

Let us use Theorem 5.11 to get the change of basis matrix from β to α, which is P^{-1}.

$$\left[\begin{array}{ccc|ccc} 1 & 1 & 1 & 1 & 0 & 0 \\ 1 & -1 & 1 & 0 & 1 & 0 \\ 2 & 1 & 1 & 0 & 0 & 1 \end{array}\right] \rightarrow \left[\begin{array}{ccc|ccc} 1 & 1 & 1 & 1 & 0 & 0 \\ 0 & -2 & 0 & -1 & 1 & 0 \\ 0 & -1 & -1 & -2 & 0 & 1 \end{array}\right]$$

$$\rightarrow \left[\begin{array}{ccc|ccc} 2 & 0 & 2 & 1 & 1 & 0 \\ 0 & -2 & 0 & -1 & 1 & 0 \\ 0 & 0 & -2 & -3 & -1 & 2 \end{array}\right] \rightarrow \left[\begin{array}{ccc|ccc} 2 & 0 & 0 & -2 & 0 & 2 \\ 0 & -2 & 0 & -1 & 1 & 0 \\ 0 & 0 & -2 & -3 & -1 & 2 \end{array}\right]$$

The change of basis matrix from β to α is then

$$P^{-1} = \begin{bmatrix} -1 & 0 & 1 \\ 1/2 & -1/2 & 0 \\ 3/2 & 1/2 & -1 \end{bmatrix}. \qquad \bullet$$

We are now ready to obtain the alternative approach mentioned after Example 1. Suppose $T : V \to W$ is a linear transformation. If α is a basis for V and β is a basis for W, we get the matrix $[T]_\alpha^\beta$. Suppose we use different bases α' for V and β' for W. Then we get the matrix $[T]_{\alpha'}^{\beta'}$. Is there a way of converting $[T]_\alpha^\beta$ to $[T]_{\alpha'}^{\beta'}$? The answer is yes, and it is shown in the following theorem.

THEOREM 5.12 Let $T : V \to W$ be a linear transformation, α and α' be bases for V, and β and β' be bases for W. If P is the change of basis matrix from α to α' and Q is the change of basis matrix from β to β', then

$$[T]_{\alpha'}^{\beta'} = Q^{-1}[T]_\alpha^\beta P.$$

Proof Since $P = [I]_{\alpha'}^\alpha$ and $Q = [I]_{\beta'}^\beta$, Theorem 5.11 and then Theorem 5.10 give us

$$Q^{-1}[T]_\alpha^\beta P = [I]_\beta^{\beta'}[T]_\alpha^\beta[I]_{\alpha'}^\alpha = [I]_\beta^{\beta'}[TI]_{\alpha'}^\beta$$
$$= [ITI]_{\alpha'}^{\beta'} = [T]_{\alpha'}^{\beta'}. \qquad \bullet$$

An important special case of Theorem 5.12 is Corollary 5.13.

COROLLARY 5.13 If $T : V \to V$ is a linear transformation, α and β are bases of V, and P is the change of basis matrix from α to β, then

$$\boxed{[T]_\beta^\beta = P^{-1}[T]_\alpha^\alpha P.}$$

EXAMPLE 3 Use the results of part (a) Example 1 and Example 2 to do part (b) of Example 1.

Solution By Corollary 5.13, we have:

$$[T]_\beta^\beta = P^{-1}[T]_\alpha^\alpha P = \begin{bmatrix} -1 & 0 & 1 \\ 1/2 & -1/2 & 0 \\ 3/2 & 1/2 & -1 \end{bmatrix} \begin{bmatrix} 5 & 0 & 1 \\ 3 & 2 & -3 \\ 5 & 0 & 0 \end{bmatrix} \begin{bmatrix} 1 & 1 & 1 \\ 1 & -1 & 1 \\ 2 & 1 & 1 \end{bmatrix}$$

$$\begin{bmatrix} -1 & 0 & 1 \\ 1/2 & -1/2 & 0 \\ 3/2 & 1/2 & -1 \end{bmatrix} \begin{bmatrix} 7 & 6 & 6 \\ -1 & -2 & 2 \\ 5 & 5 & 5 \end{bmatrix} = \begin{bmatrix} -2 & -1 & -1 \\ 4 & 4 & 2 \\ 5 & 3 & 5 \end{bmatrix}. \qquad \bullet$$

Let us look at another set of examples similar in nature to Examples 1, 2, and 3.

EXAMPLE 4 Let $T : P_2 \to P_2$ be the linear transformation

$$T(ax^2 + bx + c) = 2ax^2 + (2a + 2c)x - 2a + 2b.$$

(a) Find $[T]_\alpha^\alpha$ where α is the standard basis consisting of x^2, x, 1 for P_2.

(b) Find the change of basis matrix from α to the basis β for P_2 consisting of $x^2 + x$, $x^2 - 1$, $x - 1$.

(c) Find the change of basis matrix from β to α.

(d) Find $[T]_\beta^\beta$.

Solution

(a) As

$$T(x^2) = 2x^2 + 2x - 2, \qquad T(x) = 2 = 0x^2 + 0x + 2,$$
$$T(1) = 2x = 0x^2 + 2x + 0,$$

we see

$$[T]_\alpha^\alpha = \begin{bmatrix} 2 & 0 & 0 \\ 2 & 0 & 2 \\ -2 & 2 & 0 \end{bmatrix}.$$

(b) Writing the vectors in β in terms of the vectors of α,

$$x^2 + x = x^2 + x + 0, \qquad x^2 - 1 = x^2 + 0x - 1,$$
$$x - 1 = 0x^2 + x - 1,$$

we see

$$P = \begin{bmatrix} 1 & 1 & 0 \\ 1 & 0 & 1 \\ 0 & -1 & -1 \end{bmatrix}.$$

(c) This is P^{-1}.

$$\left[\begin{array}{rrr|rrr} 1 & 1 & 0 & 1 & 0 & 0 \\ 1 & 0 & 1 & 0 & 1 & 0 \\ 0 & -1 & -1 & 0 & 0 & 1 \end{array}\right] \rightarrow \left[\begin{array}{rrr|rrr} 1 & 1 & 0 & 1 & 0 & 0 \\ 0 & -1 & 1 & -1 & 1 & 0 \\ 0 & -1 & -1 & 0 & 0 & 1 \end{array}\right]$$

$$\rightarrow \left[\begin{array}{rrr|rrr} 1 & 0 & 1 & 0 & 1 & 0 \\ 0 & -1 & 1 & -1 & 1 & 0 \\ 0 & 0 & -2 & 1 & -1 & 1 \end{array}\right]$$

$$\rightarrow \left[\begin{array}{rrr|rrr} 2 & 0 & 0 & 1 & 1 & 1 \\ 0 & -2 & 0 & -1 & 1 & 1 \\ 0 & 0 & -2 & 1 & -1 & 1 \end{array}\right]$$

Hence our answer is

$$P^{-1} = \begin{bmatrix} 1/2 & 1/2 & 1/2 \\ 1/2 & -1/2 & -1/2 \\ -1/2 & 1/2 & -1/2 \end{bmatrix}.$$

(d) By Corollary 5.13,

$$[T]_\beta^\beta = P^{-1}[T]_\alpha^\alpha P$$

$$= \begin{bmatrix} 1/2 & 1/2 & 1/2 \\ 1/2 & -1/2 & -1/2 \\ -1/2 & 1/2 & -1/2 \end{bmatrix} \begin{bmatrix} 2 & 0 & 0 \\ 2 & 0 & 2 \\ -2 & 2 & 0 \end{bmatrix} \begin{bmatrix} 1 & 1 & 0 \\ 1 & 0 & 1 \\ 0 & -1 & -1 \end{bmatrix}$$

$$= \begin{bmatrix} 1 & 1 & 1 \\ 1 & -1 & -1 \\ 1 & -1 & 1 \end{bmatrix} \begin{bmatrix} 1 & 1 & 0 \\ 1 & 0 & 1 \\ 0 & -1 & -1 \end{bmatrix}$$

$$= \begin{bmatrix} 2 & 0 & 0 \\ 0 & 2 & 0 \\ 0 & 0 & -2 \end{bmatrix}. \qquad \bullet$$

At the end of Section 2.3, we pointed out that coordinate vectors give us a way of translating elements of a vector space into vectors with numbers, which, for example, gives us a way of working with vectors on a computer. Matrices for linear transformations are yet more of these number translations. The next theorem tells us how to compute a linear transformation applied to a vector by using matrices and coordinate vectors. The examples following the theorem illustrate how to apply this in practice.

THEOREM 5.14 Suppose $T : V \to W$ is a linear transformation. Let α be a basis for V and β be a basis for W. If v is a vector in V,

$$\boxed{[T(v)]_\beta = [T]_\alpha^\beta [v]_\alpha.}$$

Proof Suppose the basis α consists of $v_1, \ldots, v_n$ and the basis β consists of $w_1, \ldots, w_m$. We have

$$[v]_\alpha = \begin{bmatrix} c_1 \\ \vdots \\ c_n \end{bmatrix}$$

where $v = c_1 v_1 + \cdots + c_n v_n$,

$$[T]_\alpha^\beta = \begin{bmatrix} a_{11} & \cdots & a_{1n} \\ \vdots & \cdots & \vdots \\ a_{m1} & \cdots & a_{mn} \end{bmatrix}$$

where

$$T(v_1) = a_{11}w_1 + \cdots + a_{m1}w_m$$

$$\vdots$$

$$T(v_n) = a_{1n}w_1 + \cdots + a_{mn}w_m,$$

and

$$T(v) = T(c_1v_1 + \cdots + c_nv_n)$$
$$= c_1(a_{11}w_1 + \cdots + a_{m1}w_m) + \cdots + c_n(a_{1n}w_1 + \cdots + a_{mn}w_m)$$
$$= (a_{11}c_1 + \cdots + a_{1n}c_n)w_1 + \cdots + (a_{m1}c_1 + \cdots + a_{mn}c_n)w_m.$$

Hence

$$[T(v)]_\beta = \begin{bmatrix} a_{11}c_1 + \cdots + a_{1n}c_n \\ \vdots \\ a_{m1}c_1 + \cdots + a_{mn}c_n \end{bmatrix}$$

$$= \begin{bmatrix} a_{11} & \cdots & a_{1n} \\ \vdots & \cdots & \vdots \\ a_{m1} & \cdots & a_{mn} \end{bmatrix} \begin{bmatrix} c_1 \\ \vdots \\ c_n \end{bmatrix} = [T]_\alpha^\beta [v]_\alpha. \qquad \bullet$$

EXAMPLE 5   Let T and β be as in Examples 1–3.

(a) Find the coordinate vector of

$$v = \begin{bmatrix} 1 \\ 2 \\ 3 \end{bmatrix}$$

with respect to the basis β.

(b) Find $[T(v)]_\beta$.

(c) Use the result of part (b) to find $T(v)$.

Solution

(a) We have to find c_1, c_2, c_3 so that

$$c_1 \begin{bmatrix} 1 \\ 1 \\ 2 \end{bmatrix} + c_2 \begin{bmatrix} 1 \\ -1 \\ 1 \end{bmatrix} + c_3 \begin{bmatrix} 1 \\ 1 \\ 1 \end{bmatrix} = \begin{bmatrix} 1 \\ 2 \\ 3 \end{bmatrix}.$$

Solving the resulting system of equations,

$$\begin{bmatrix} 1 & 1 & 1 & \vdots & 1 \\ 1 & -1 & 1 & \vdots & 2 \\ 2 & 1 & 1 & \vdots & 3 \end{bmatrix} \rightarrow \begin{bmatrix} 1 & 1 & 1 & \vdots & 1 \\ 0 & -2 & 0 & \vdots & 1 \\ 0 & -1 & -1 & \vdots & 1 \end{bmatrix}$$

$$\rightarrow \begin{bmatrix} 2 & 0 & 2 & \vdots & 3 \\ 0 & -2 & 0 & \vdots & 1 \\ 0 & 0 & -2 & \vdots & 1 \end{bmatrix} \rightarrow \begin{bmatrix} 2 & 0 & 0 & \vdots & 4 \\ 0 & -2 & 0 & \vdots & 1 \\ 0 & 0 & -2 & \vdots & 1 \end{bmatrix}$$

$$c_1 = 2, \quad c_2 = -\frac{1}{2}, \quad c_3 = -\frac{1}{2}$$

we see

$$[v]_\beta = \begin{bmatrix} 2 \\ -1/2 \\ -1/2 \end{bmatrix}.$$

(b) By Theorem 5.14 and Example 3,

$$[T(v)]_\beta = [T]_\beta^\beta [v]_\beta = \begin{bmatrix} -2 & -1 & -1 \\ 4 & 4 & 2 \\ 5 & 3 & 5 \end{bmatrix} \begin{bmatrix} 2 \\ -1/2 \\ -1/2 \end{bmatrix} = \begin{bmatrix} -3 \\ 5 \\ 6 \end{bmatrix}.$$

(c) From the coordinates with respect to β we found in part (b),

$$T(v) = -3 \begin{bmatrix} 1 \\ 1 \\ 2 \end{bmatrix} + 5 \begin{bmatrix} 1 \\ -1 \\ 1 \end{bmatrix} + 6 \begin{bmatrix} 1 \\ 1 \\ 1 \end{bmatrix} = \begin{bmatrix} 8 \\ -2 \\ 5 \end{bmatrix}. \qquad \bullet$$

Notice the result of part (c) agrees with the result we obtain by applying the formula for T in Example 1 to v.

EXAMPLE 6 Let T and β be as in Example 4.

(a) Find the coordinate vector of $v = 3x^2 + 4x - 5$ with respect to the basis β.

(b) Find $[T]_\beta$.

(c) Use the result of part (b) to find $T(v)$.

Solution

(a) Here we need c_1, c_2, c_3 so that

$$3x^2 + 4x - 5 = c_1(x^2 + x) + c_2(x^2 - 1) + c_3(x - 1)$$
$$= (c_1 + c_2)x^2 + (c_1 + c_3)x - c_2 - c_3.$$

Solving the resulting system,

$$\begin{bmatrix} 1 & 1 & 0 & \vdots & 3 \\ 1 & 0 & 1 & \vdots & 4 \\ 0 & -1 & -1 & \vdots & -5 \end{bmatrix} \rightarrow \begin{bmatrix} 1 & 1 & 0 & \vdots & 3 \\ 0 & -1 & 1 & \vdots & 1 \\ 0 & -1 & -1 & \vdots & -5 \end{bmatrix}$$

$$\rightarrow \begin{bmatrix} 1 & 1 & 0 & \vdots & 3 \\ 0 & -1 & 1 & \vdots & 1 \\ 0 & 0 & -2 & \vdots & -6 \end{bmatrix}$$

$$c_3 = 3, \qquad c_2 = 2, \qquad c_1 = 1$$

we obtain

$$[v]_\beta = \begin{bmatrix} 1 \\ 2 \\ 3 \end{bmatrix}.$$

(b) We have

$$[T(v)]_\beta = [T]_\beta^\beta [3x^2 + 4x - 5]_\beta$$

$$= \begin{bmatrix} 2 & 0 & 0 \\ 0 & 2 & 0 \\ 0 & 0 & -2 \end{bmatrix} \begin{bmatrix} 1 \\ 2 \\ 3 \end{bmatrix} = \begin{bmatrix} 2 \\ 4 \\ -6 \end{bmatrix}.$$

(c) $T(v) = 2(x^2 + x) + 4(x^2 - 1) - 6(x - 1) = 6x^2 - 4x + 2.$ ●

Again, notice if we apply the formula for T (from Example 4 this time), we will obtain the same result as in part (c).

Theorem 5.14 is what we were referring to when we earlier made the statement that all linear transformations on finite dimensional vector spaces are, in a sense, matrix multiplication. In the special case when we have a linear transformation $T : \mathbb{R}^n \rightarrow \mathbb{R}^m$, we can see that T is in fact a matrix transformation as a corollary to Theorem 5.14.

COROLLARY 5.15   If $T : \mathbb{R}^n \rightarrow \mathbb{R}^m$ is a linear transformation and A is the matrix of T with respect to the standard bases for $\mathbb{R}^n$ and $\mathbb{R}^m$, then

$$T(X) = AX.$$

Proof   If α denotes the standard basis for $\mathbb{R}^n$ and β denotes the standard basis for $\mathbb{R}^m$, Theorem 5.14 tells us

$$[T(X)]_\beta = [T]_\alpha^\beta [X]_\alpha = A[X]_\alpha.$$

But we know that coordinate vectors relative to standard bases for these vectors are exactly the same as the vectors themselves and hence

$$T(X) = AX.$$ ●

EXERCISES 5.3

1. Let $T : \mathbb{R}^2 \to \mathbb{R}^2$ be the linear transformation

$$T\begin{bmatrix} x_1 \\ x_2 \end{bmatrix} = \begin{bmatrix} x_1 + x_2 \\ x_1 - x_2 \end{bmatrix}.$$

a) Find $[T]_\alpha^\alpha$ where α is the standard basis for $\mathbb{R}^2$.

b) Let β be the basis consisting of

$$\begin{bmatrix} 1 \\ -1 \end{bmatrix}, \begin{bmatrix} -2 \\ 1 \end{bmatrix}$$

for $\mathbb{R}^2$. Find the change of basis matrix from α to β.

c) Find the change of basis matrix from β to α.

d) Find $[T]_\beta^\beta$.

e) Find $[v]_\beta$ for

$$v = \begin{bmatrix} -2 \\ 3 \end{bmatrix}.$$

f) Find $[T(v)]_\beta$.

g) Use the result of part (f) to find $T(v)$.

In Exercises 2–8, do the parts of Exercise 1 for the given linear transformation, bases, and vector v.

2. $T : \mathbb{R}^2 \to \mathbb{R}^2$ by

$$T\begin{bmatrix} x_1 \\ x_2 \end{bmatrix} = \begin{bmatrix} 5x_1 + 3x_2 \\ -6x_1 - 4x_2 \end{bmatrix};$$

α the standard basis for $\mathbb{R}^2$; β the basis consisting of

$$\begin{bmatrix} 2 \\ 1 \end{bmatrix}, \begin{bmatrix} 1 \\ 1 \end{bmatrix};$$

$$v = \begin{bmatrix} 5 \\ -4 \end{bmatrix}.$$

3. $T : \mathbb{R}^3 \to \mathbb{R}^3$ by

$$T\begin{bmatrix} x_1 \\ x_2 \\ x_3 \end{bmatrix} = \begin{bmatrix} 17x_1 - 8x_2 - 12x_3 \\ 16x_1 - 7x_2 - 12x_3 \\ 16x_1 - 8x_2 - 11x_3 \end{bmatrix};$$

α the standard basis for $\mathbb{R}^3$; β the basis consisting of

$$\begin{bmatrix} 1 \\ 1 \\ 1 \end{bmatrix}, \begin{bmatrix} 1 \\ 2 \\ 0 \end{bmatrix}, \begin{bmatrix} 1 \\ -1 \\ 2 \end{bmatrix};$$

$$v = \begin{bmatrix} 2 \\ -1 \\ 4 \end{bmatrix}.$$

4. $T : \mathbb{R}^3 \to \mathbb{R}^3$ by

$$T\begin{bmatrix} x_1 \\ x_2 \\ x_3 \end{bmatrix} = \begin{bmatrix} x_1 - x_2 + x_3 \\ 3x_1 + 15x_2 - 16x_3 \\ 3x_1 + 13x_2 - 14x_3 \\ x_2 \end{bmatrix};$$

α the standard basis for $\mathbb{R}^3$; β the basis consisting of

$$\begin{bmatrix} 1 \\ 1 \\ 1 \end{bmatrix}, \begin{bmatrix} -2 \\ 3 \\ 2 \end{bmatrix}, \begin{bmatrix} 0 \\ 1 \\ 1 \end{bmatrix};$$

$$v = \begin{bmatrix} 2 \\ 3 \\ 1 \end{bmatrix}.$$

5. $T : P_2 \to P_2$ by

$$T(ax^2 + bx + c) = a(x + 1)^2 + b(x + 1) + c;$$

α the standard basis for P_2; β the basis consisting of $x^2 - 1, x^2 + 1, x + 1$; $v = x^2 + x + 1$.

6. $D : P_2 \to P_2$ (D the differential operator so that $D(ax^2 + bx + c) = 2ax + b$); α the standard basis for P_2; β the basis consisting of $x^2 + 1, x + 1, 2x^2 + 1$; $v = x^2 + 2x - 2$.

7. $D : V \to V$ where V is the set of solutions to the differential equation $y'' + y = 0$; α the basis of V consisting of $\sin x, \cos x$; β the basis consisting of $\sin x + \cos x, \sin x - \cos x$; $v = 3 \cos x - 2 \sin x$.

8. $S : M_{2 \times 2}(\mathbb{R}) \to M_{2 \times 2}(\mathbb{R})$ by $S(A) = A^T$; α the standard basis for $M_{2 \times 2}(\mathbb{R})$; β the basis consisting of

$$\begin{bmatrix} 1 & 0 \\ 0 & 1 \end{bmatrix}, \begin{bmatrix} 0 & 1 \\ 1 & 0 \end{bmatrix}, \begin{bmatrix} -1 & 0 \\ 0 & 1 \end{bmatrix}, \begin{bmatrix} 0 & -1 \\ 1 & 0 \end{bmatrix};$$

$$v = \begin{bmatrix} 1 & 2 \\ 3 & 4 \end{bmatrix}.$$

9. Suppose that v_1, v_2, v_3 form a basis α for a vector space V and $T : V \to V$ is a linear transformation such that

$$T(v_1) = v_1 - v_2, \qquad T(v_2) = v_2 - v_3,$$
$$T(v_3) = v_3 - v_1.$$

a) Find $[T]_\alpha^\alpha$.

b) Find $[T(v)]_\alpha$ if $v = v_1 - 2v_2 + 3v_3$.

c) Use the result of part (b) to find $T(v)$ in terms of v_1, v_2, v_3.

10. Suppose that v_1, v_2, v_3, v_4 form a basis α for a vector space V and w_1, w_2 form a basis β for a vector space W. Suppose that $T : V \to W$ is a linear transformation such that

$$T(v_1) = 2w_1 - 3w_2, \qquad T(v_2) = -w_1 + 3w_2,$$
$$T(v_3) = w_1 + 2w_2, \qquad T(v_4) = 3w_2.$$

a) Find $[T]_\alpha^\beta$.

b) Find $[T(v)]_\beta$ if $v = 4v_1 + 3v_2 + 2v_3 + v_4$.

c) Use the result of part (b) to find $T(v)$ in terms of w_1 and w_2.

11. Suppose that V is a vector space. Show that $I : V \to V$ by $I(v) = v$ is a linear transformation.

12. Suppose that $T : V \to W$ and $S : V \to W$ are linear transformations, α is a basis for V, and β is a basis for W. Show that:

a) $[T + S]_\alpha^\beta = [T]_\alpha^\beta + [S]_\alpha^\beta$.

b) If c is a scalar, $[cT]_\alpha^\beta = c[T]_\alpha^\beta$.

13. Suppose that $T : V \to V$ is a linear transformation, the vectors v_1, v_2, v_3 form a basis α for V, and

$$[T]_\alpha^\alpha = \begin{bmatrix} a_{11} & a_{12} & a_{13} \\ a_{21} & a_{22} & a_{23} \\ a_{31} & a_{32} & a_{33} \end{bmatrix}.$$

Suppose we form another basis β for V by arranging the vectors in α in the order v_3, v_1, v_2. Find the matrix $[T]_\beta^\beta$. How are $[T]_\alpha^\alpha$ and $[T]_\beta^\beta$ related?

14. Generalize the result of Exercise 13. That is, suppose that $T : V \to V$ is a linear transformation, α is a basis for V, and β is another basis for V obtained by arranging the vectors in α in another order. How are $[T]_\alpha^\alpha$ and $[T]_\beta^\beta$ related?

15. Show that every subspace of $\mathbb{R}^n$ is the set of solutions to a homogeneous system of linear equations. (*Hint:* If a subspace W consists of only the zero vector or is all of $\mathbb{R}^n$, W is the set of solutions to

$IX = 0$ or $OX = 0$, respectively. Assume W is not one of these two subspaces. Let $v_1, v_2, \ldots, v_k$ be a basis for W. Extend this to a basis $v_1, v_2, \ldots, v_k,$ $v_{k+1}, \ldots, v_n$ of $\mathbb{R}^n$. Let $T : \mathbb{R}^n \to \mathbb{R}^n$ be the linear transformation so that

$$T(v_1) = 0, \qquad T(v_2) = 0, \qquad \ldots,$$
$$T(v_k) = 0, \qquad T(v_{k+1}) = v_{k+1}, \qquad \ldots,$$
$$T(v_n) = v_n.$$

Now use T to obtain a matrix A so that W is the set of solutions to the homogeneous system $AX = 0$.)

16. In Exercise 25 of Section 5.2 we saw that if a linear transformation $T : V \to W$ has an inverse, then $T^{-1} : W \to V$ is also a linear transformation. Suppose that α is a basis for V and β is a basis for W. Show that if T has an inverse, then $[T^{-1}]_\beta^\alpha = ([T]_\alpha^\beta)^{-1}$.

In Exercises 17–20, use Maple or another appropriate software package as an aid in doing the parts of Exercise 1 for the given linear transformation, bases, and vector v.

17. $T : \mathbb{R}^4 \to \mathbb{R}^4$ by

$$T \begin{bmatrix} x_1 \\ x_2 \\ x_3 \\ x_4 \end{bmatrix} = \begin{bmatrix} x_1 - x_2 + x_3 - x_4 \\ 2x_1 - x_2 + 2x_3 + x_4 \\ 3x_1 + x_2 - x_3 + x_4 \\ -x_1 + 3x_2 - 5x_3 + x_4 \end{bmatrix};$$

α the standard basis for $\mathbb{R}^4$, β the basis consisting of

$$\begin{bmatrix} 1 \\ -1 \\ 1 \\ 0 \end{bmatrix}, \begin{bmatrix} 1 \\ 0 \\ 1 \\ 0 \end{bmatrix}, \begin{bmatrix} 1 \\ 0 \\ -1 \\ 1 \end{bmatrix}, \begin{bmatrix} 1 \\ -1 \\ 1 \\ 1 \end{bmatrix};$$

$$v = \begin{bmatrix} 2 \\ 0 \\ 4 \\ -3 \end{bmatrix}.$$

18. $T : \mathbb{R}^5 \to \mathbb{R}^5$ by

$$T \begin{bmatrix} x_1 \\ x_2 \\ x_3 \\ x_4 \\ x_5 \end{bmatrix} = \begin{bmatrix} x_1 + x_2 - x_3 - x_4 + 2x_5 \\ x_1 + 2x_2 + 3x_3 + x_4 - x_5 \\ 2x_1 + 3x_2 + x_3 + 3x_4 \\ x_1 - 3x_2 - x_4 + 2x_5 \\ x_1 - 3x_2 + 2x_5 \end{bmatrix};$$

α the standard basis for $\mathbb{R}^5$, β the basis consisting of

$$\begin{bmatrix} 1 \\ 0 \\ 1 \\ 1 \\ 1 \end{bmatrix}, \begin{bmatrix} 0 \\ 1 \\ 0 \\ 1 \\ 0 \end{bmatrix}, \begin{bmatrix} 2 \\ 1 \\ 1 \\ 1 \\ 1 \end{bmatrix}, \begin{bmatrix} -1 \\ 0 \\ 0 \\ 0 \\ 2 \end{bmatrix}, \begin{bmatrix} 2 \\ 2 \\ 2 \\ 2 \\ -1 \end{bmatrix};$$

$$v = \begin{bmatrix} 1 \\ -3 \\ -2 \\ 1 \\ 3 \end{bmatrix}.$$

19. $T : P_3 \to P_3$ by

$$T(a_3 x^3 + a_2 x^2 + a_1 x + a_0) = (a_3 + a_2)x^3$$
$$+ (a_2 - a_1)x^2 + (a_3 + a_2 + a_1)x - a_1 + a_2;$$

α the standard basis for P_3; β the basis consisting of $x^3 - x^2 + 1$, $x^2 + x + 1$, $x^3 - 1$, $x - 1$; $v = 3x^3 - 5x^2 + 2x - 1$.

20. $D : V \to V$ where V is the set of solutions to the differential equation $y^{(4)} + 2y'' + y = 0$; α the basis consisting of $\sin x$, $\cos x$, $x \sin x$, $x \cos x$ for V; β the basis consisting of $\sin x + x \sin x$, $\sin x - x \sin x$, $\cos x + x \cos x$, $\cos x - x \cos x$; $v = 3 \sin x - 2 \cos x + x \sin x - 3x \cos x$.

5.4 EIGENVALUES AND EIGENVECTORS OF MATRICES

If A is an $n \times n$ matrix, an **eigenvector** of A is a nonzero column vector v in $\mathbb{R}^n$ so that

$$\boxed{Av = \lambda v}$$

for some scalar λ; the scalar λ is called an **eigenvalue** of A. To illustrate, if

$$A = \begin{bmatrix} 1 & -3 \\ -2 & 2 \end{bmatrix},$$

the vector

$$v = \begin{bmatrix} -1 \\ 1 \end{bmatrix}$$

is an eigenvector of A with associated eigenvalue $\lambda = 4$ since

$$\begin{bmatrix} 1 & -3 \\ -2 & 2 \end{bmatrix}\begin{bmatrix} -1 \\ 1 \end{bmatrix} = \begin{bmatrix} -4 \\ 4 \end{bmatrix} = 4\begin{bmatrix} -1 \\ 1 \end{bmatrix}.$$

The terms eigenvector and eigenvalue are partial translations from the corresponding German words *Eigenvektor* and *Eigenwert*, *Wert* being the German word for value. The German adjective *eigen* has various translations, some of which are proper, inherent, special, and characteristic. Sometimes you will find eigenvectors and eigenvalues called characteristic vectors and characteristic values, respectively. As mentioned in the introduction to this chapter, eigenvalues and eigenvectors arise in a number of places in mathematics and its applications, one of which you will encounter in the next chapter. Our purpose in this and the next section is to study eigenvectors and eigenvalues of square matrices. The concepts of eigenvalues and eigenvectors extend to linear transformations from a vector space to itself, as we shall see in Section 5.6.

Let us investigate how we can find eigenvectors and eigenvalues. If λ is an eigenvalue of an $n \times n$ matrix A, the eigenvalues associated with λ will be the nonzero solutions to the equation

$$AX = \lambda X,$$

which we can rewrite as

$$\lambda X - AX = 0$$

or

$$(\lambda I - A)X = 0$$

where I is the $n \times n$ identity matrix. This last matrix equation is that of a homogeneous system of n linear equations in n unknowns with coefficient matrix $\lambda I - A$. Since we know that such a homogeneous system has nontrivial solutions if and only if its coefficient matrix is not invertible, which in turn is equivalent to its coefficient matrix having determinant zero, we have just discovered the following fact.

THEOREM 5.16   If A is an $n \times n$ matrix, a number λ is an eigenvalue of A if and only if

$$\det(\lambda I - A) = 0.$$

The equation

$$\boxed{\det(\lambda I - A) = 0}$$

is called the **characteristic equation** of the matrix A. Upon expanding the determinant $\det(\lambda I - A)$ we will have a polynomial of degree n in λ called the **characteristic polynomial** of A. Theorem 5.16 then tells us that we can find all the eigenvalues of a square matrix A by finding all of the solutions of its characteristic equation (or, equivalently, all of the roots of its characteristic polynomial).

EXAMPLE 1   Find the eigenvalues of the matrix

$$A = \begin{bmatrix} 1 & -3 \\ -2 & 2 \end{bmatrix}.$$

Solution   The characteristic equation of A is

$$\det(\lambda I - A) = \det\left(\begin{bmatrix} \lambda & 0 \\ 0 & \lambda \end{bmatrix} - \begin{bmatrix} 1 & -3 \\ -2 & 2 \end{bmatrix} \right) = \begin{bmatrix} \lambda - 1 & 3 \\ 2 & \lambda - 2 \end{bmatrix}$$

$$= (\lambda - 1)(\lambda - 2) - 6 = \lambda^2 - 3\lambda - 4 = (\lambda - 4)(\lambda + 1) = 0.$$

Its solutions, which are the eigenvalues of A, are $\lambda = 4$ and $\lambda = -1$.   ●

Now that we have a method for finding eigenvalues of a matrix A, let us consider how we may find the eigenvectors of A. Suppose λ is an eigenvalue of A. Since the eigenvectors associated with λ are the nontrivial solutions to the homogeneous system $(\lambda I - A)X = 0$, they along with the trivial solution form the nullspace of $\lambda I - A$,

$NS(\lambda I - A)$. We call this nullspace the **eigenspace** of λ and denote it by E_λ. Since we learned how to find bases of nullspaces of matrices in Section 2.4, we know how to find a basis for an eigenspace E_λ. All the nonzero linear combinations of these basis vectors give us all the eigenvectors associated with λ.

EXAMPLE 2 Find the eigenvectors of the matrix

$$A = \begin{bmatrix} 1 & -3 \\ -2 & 2 \end{bmatrix}$$

in Example 1.

Solution In Example 1, we found that A has two eigenvalues, $\lambda = 4$ and $\lambda = -1$. We individually consider each of these eigenvalues. For $\lambda = 4$, the eigenspace is the nullspace of

$$4I - A = \begin{bmatrix} 3 & 3 \\ 2 & 2 \end{bmatrix}.$$

Reducing the matrix for the associated homogeneous system,

$$\begin{bmatrix} 3 & 3 & \vdots & 0 \\ 2 & 2 & \vdots & 0 \end{bmatrix} \rightarrow \begin{bmatrix} 1 & 1 & \vdots & 0 \\ 0 & 0 & \vdots & 0 \end{bmatrix},$$

we see our solutions are

$$\begin{bmatrix} x \\ y \end{bmatrix} = \begin{bmatrix} -y \\ y \end{bmatrix} = y \begin{bmatrix} -1 \\ 1 \end{bmatrix}$$

and hence the column vector

$$\begin{bmatrix} -1 \\ 1 \end{bmatrix}$$

forms a basis for E_4. The eigenvectors associated with the eigenvalue $\lambda = 4$ are then the vectors of the form

$$c \begin{bmatrix} -1 \\ 1 \end{bmatrix}$$

where c is a nonzero scalar. For $\lambda = -1$, we need to find a basis for the nullspace of

$$-I - A = \begin{bmatrix} -2 & 3 \\ 2 & -3 \end{bmatrix}.$$

Here we find (try it) that we can use

$$\begin{bmatrix} 3/2 \\ 1 \end{bmatrix}$$

as a basis for E_{-1} (although some prefer to multiply this vector by 2 to eliminate the fraction and use

$$\begin{bmatrix} 3 \\ 2 \end{bmatrix}$$

as the basis vector). Using the former basis vector, the eigenvectors associated with $\lambda = -1$ are of the form

$$c\begin{bmatrix} 3/2 \\ 1 \end{bmatrix}$$

where c is a nonzero scalar.   ●

Since the crucial point in finding the eigenvectors of a matrix is the determination of bases for the eigenspaces, we shall henceforth content ourselves with finding these bases. Let us do some more examples.

EXAMPLE 3   Find the eigenvalues and bases for the eigenspaces of the following matrix.

$$A = \begin{bmatrix} 2 & -1 & 3 \\ 0 & -1 & 0 \\ 0 & 0 & -1 \end{bmatrix}$$

Solution   We first find the eigenvalues.

$$\det(\lambda I - A) = \begin{bmatrix} \lambda - 2 & 1 & -3 \\ 0 & \lambda + 1 & 0 \\ 0 & 0 & \lambda + 1 \end{bmatrix} = (\lambda - 2)(\lambda + 1)^2 = 0$$

The eigenvalues of A are then $\lambda = 2$ and $\lambda = -1$. Next we find bases for the eigenspaces. For $\lambda = 2$, we find a basis for the nullspace of $2I - A$.

$$\begin{bmatrix} 0 & 1 & -3 & \vdots & 0 \\ 0 & 3 & 0 & \vdots & 0 \\ 0 & 0 & 3 & \vdots & 0 \end{bmatrix} \rightarrow \begin{bmatrix} 0 & 1 & -3 & \vdots & 0 \\ 0 & 0 & 9 & \vdots & 0 \\ 0 & 0 & 3 & \vdots & 0 \end{bmatrix}$$

$$\rightarrow \begin{bmatrix} 0 & 1 & 0 & \vdots & 0 \\ 0 & 0 & 1 & \vdots & 0 \\ 0 & 0 & 0 & \vdots & 0 \end{bmatrix}$$

The homogeneous system has solutions

$$\begin{bmatrix} x \\ y \\ z \end{bmatrix} = \begin{bmatrix} x \\ 0 \\ 0 \end{bmatrix} = x \begin{bmatrix} 1 \\ 0 \\ 0 \end{bmatrix}$$

from which we see we can use

$$\begin{bmatrix} 1 \\ 0 \\ 0 \end{bmatrix}$$

as a basis vector for E_2.

Next we carry out the same procedure for $\lambda = -1$. The augmented matrix for the homogeneous system $(-I - A)X = 0$ is

$$\left[\begin{array}{ccc|c} -3 & 1 & -3 & 0 \\ 0 & 0 & 0 & 0 \\ 0 & 0 & 0 & 0 \end{array}\right]$$

from which we can see the solutions are

$$\begin{bmatrix} x \\ y \\ z \end{bmatrix} = \begin{bmatrix} y/3 - z \\ y \\ z \end{bmatrix} = y \begin{bmatrix} 1/3 \\ 1 \\ 0 \end{bmatrix} + z \begin{bmatrix} -1 \\ 0 \\ 1 \end{bmatrix}.$$

We can then use

$$\begin{bmatrix} 1/3 \\ 1 \\ 0 \end{bmatrix}, \begin{bmatrix} -1 \\ 0 \\ 1 \end{bmatrix}$$

as basis vectors for E_{-1}.   ●

EXAMPLE 4   Find the eigenvalues and bases for the eigenspaces of the following matrix.

$$A = \begin{bmatrix} 1 & -2 & -6 \\ -2 & 2 & -5 \\ 2 & 1 & 8 \end{bmatrix}$$

Solution   The characteristic equation is

$$\det(\lambda I - A) = \begin{vmatrix} \lambda - 1 & 2 & 6 \\ 2 & \lambda - 2 & 5 \\ -2 & -1 & \lambda - 8 \end{vmatrix}$$

$$= (\lambda - 1)((\lambda - 2)(\lambda - 8) + 5) - 2(2(\lambda - 8) + 10) + 6(-2 + 2(\lambda - 2))$$

$$= (\lambda - 1)(\lambda^2 - 10\lambda + 21) - 2(2\lambda - 6) + 6(2\lambda - 6)$$

$$= (\lambda - 1)(\lambda - 3)(\lambda - 7) - 4(\lambda - 3) + 12(\lambda - 3)$$

$$= (\lambda - 3)((\lambda - 1)(\lambda - 7) - 4 + 12)$$

$$= (\lambda - 3)(\lambda^2 - 8\lambda + 15) = (\lambda - 3)^2(\lambda - 5) = 0$$

from which we see the eigenvalues are $\lambda = 3$ and $\lambda = 5$. Consider $\lambda = 3$. From

$$
\left[\begin{array}{ccc|c}
2 & 2 & 6 & 0 \\
2 & 1 & 5 & 0 \\
-2 & -1 & -5 & 0
\end{array}\right]
\rightarrow
\left[\begin{array}{ccc|c}
2 & 2 & 6 & 0 \\
0 & -1 & -1 & 0 \\
0 & 1 & 1 & 0
\end{array}\right]
\rightarrow
$$

$$
\rightarrow
\left[\begin{array}{ccc|c}
2 & 0 & 4 & 0 \\
0 & -1 & -1 & 0 \\
0 & 0 & 0 & 0
\end{array}\right]
$$

we see the solutions of $(3I - A)X = 0$ are

$$
\left[\begin{array}{c}
x \\
y \\
z
\end{array}\right]
=
\left[\begin{array}{c}
-2z \\
-z \\
z
\end{array}\right]
$$

and hence the vector

$$
\left[\begin{array}{c}
-2 \\
-1 \\
1
\end{array}\right]
$$

forms a basis for E_3.

Finally, we consider $\lambda = 5$. From

$$
\left[\begin{array}{ccc|c}
4 & 2 & 6 & 0 \\
2 & 3 & 5 & 0 \\
-2 & -1 & -3 & 0
\end{array}\right]
\rightarrow
\left[\begin{array}{ccc|c}
2 & 1 & 3 & 0 \\
2 & 3 & 5 & 0 \\
-2 & -1 & -3 & 0
\end{array}\right]
$$

$$
\rightarrow
\left[\begin{array}{ccc|c}
2 & 1 & 3 & 0 \\
0 & 2 & 2 & 0 \\
0 & 0 & 0 & 0
\end{array}\right]
\rightarrow
\left[\begin{array}{ccc|c}
2 & 0 & 2 & 0 \\
0 & 1 & 1 & 0 \\
0 & 0 & 0 & 0
\end{array}\right],
$$

we see the solutions to $(5I - A)X = 0$ are

$$
\left[\begin{array}{c}
x \\
y \\
z
\end{array}\right]
=
\left[\begin{array}{c}
-z \\
-z \\
z
\end{array}\right].
$$

We can then use the vector

$$
\left[\begin{array}{c}
-1 \\
-1 \\
1
\end{array}\right]
$$

as a basis for E_5.

Our study of linear algebra (matrices, determinants, vector spaces, and linear transformations) up to this point has involved the use of the real numbers as our underlying number system. But imaginary numbers can arise as roots of the characteristic polynomial of a square matrix just as they did for the characteristic polynomial of a homogeneous linear differential equation with constant coefficients in the previous chapter. Consequently, imaginary numbers will come up in our work with eigenvalues and eigenvectors. All of the linear algebra material we have covered works equally well if complex numbers are used in place of real numbers. The final example of this section illustrates how we adapt our method for finding eigenvectors to imaginary roots of the characteristic polynomial.

EXAMPLE 5   Determine the eigenvalues and eigenvectors of

$$A = \begin{bmatrix} 1 & -1 \\ 1 & 1 \end{bmatrix}.$$

Solution   The characteristic equation is

$$\begin{bmatrix} \lambda - 1 & 1 \\ -1 & \lambda - 1 \end{bmatrix} = (\lambda - 1)^2 + 1 = \lambda^2 - 2\lambda + 2 = 0.$$

Using the quadratic formula, we find the roots of the characteristic equation are

$$\lambda = \frac{2 \pm \sqrt{4 - 8}}{2} = 1 \pm i.$$

For $\lambda = 1 + i$, we find the complex solutions to $((1 + i)I - A)X = 0$.

$$\begin{bmatrix} i & 1 & \vdots & 0 \\ -1 & i & \vdots & 0 \end{bmatrix} \begin{matrix} \\ iR_2 + R_1 \end{matrix} \rightarrow \begin{bmatrix} i & 1 & \vdots & 0 \\ 0 & 0 & \vdots & 0 \end{bmatrix} \begin{matrix} -iR_1 \\ \end{matrix} \rightarrow \begin{bmatrix} 1 & -i & \vdots & 0 \\ 0 & 0 & \vdots & 0 \end{bmatrix}$$

The complex solutions are

$$\begin{bmatrix} x \\ y \end{bmatrix} = \begin{bmatrix} iy \\ y \end{bmatrix}$$

where y is any complex number and the complex vector

$$\begin{bmatrix} i \\ 1 \end{bmatrix}$$

forms a basis for E_{1+i} over the complex numbers.

Finally, we do the corresponding work for the eigenvalue $\lambda = 1 - i$.

$$\begin{bmatrix} -i & 1 & \vdots & 0 \\ -1 & -i & \vdots & 0 \end{bmatrix} \begin{matrix} \\ iR_2 - R_1 \end{matrix} \rightarrow \begin{bmatrix} -i & 1 & \vdots & 0 \\ 0 & 0 & \vdots & 0 \end{bmatrix} \begin{matrix} iR_1 \\ \end{matrix} \rightarrow \begin{bmatrix} 1 & i & \vdots & 0 \\ 0 & 0 & \vdots & 0 \end{bmatrix}$$

The complex solutions are

$$\begin{bmatrix} x \\ y \end{bmatrix} = \begin{bmatrix} -iy \\ y \end{bmatrix}$$

where y is any complex number and the complex vector

$$\begin{bmatrix} -i \\ 1 \end{bmatrix}$$

forms a basis for E_{1-i} over the complex numbers.   ●

Notice that the entries of the basis vectors we found for the two eigenspaces E_{1+i} and E_{1-i} of the two conjugate complex roots in Example 5 are also conjugates of one another. The next theorem tells us that this is always the case for a matrix with real entries. For notational purposes, if $A = [a_{ij}]$ is a matrix, we use $\bar{A}$ to indicate the matrix with entries $[\overline{a_{ij}}]$ and call $\bar{A}$ the **conjugate matrix** of A. For example, if

$$A = \begin{bmatrix} 2+i & 3-2i \\ -1-2i & 3 \end{bmatrix},$$

then the conjugate matrix of A is

$$\bar{A} = \begin{bmatrix} 2-i & 3+2i \\ -1+2i & 3 \end{bmatrix}.$$

Properties of conjugates of complex numbers such as

$$\bar{\bar{z}} = z, \qquad \overline{z+w} = \bar{z}+\bar{w}, \qquad \text{and} \qquad \overline{zw} = \bar{z}\,\bar{w}$$

carry over to matrices:

$$\bar{\bar{A}} = A, \qquad \overline{A+B} = \bar{A}+\bar{B}, \qquad \overline{cA} = \bar{c}\bar{A}, \qquad \text{and} \qquad \overline{AB} = \bar{A}\,\bar{B}.$$

THEOREM 5.17   If A is a square matrix with real entries, $\lambda = r$ is an eigenvalue of A, and $v_1, v_2, \ldots, v_k$ form a basis for the eigenspace E_r, then $\bar{r}$ is also an eigenvalue of A and $\overline{v_1}, \overline{v_2}, \ldots, \overline{v_k}$ form a basis for the eigenspace $E_{\bar{r}}$.

Proof   Since the characteristic polynomial of A has real coefficients, we immediately have that $\bar{r}$ is also a root of the characteristic polynomial of A and hence $\bar{r}$ is an eigenvalue of A. To see why $\overline{v_1}, \overline{v_2}, \ldots, \overline{v_k}$ form a basis for $E_{\bar{r}}$, we first note that since

$$Av = rv$$

if and only if

$$\overline{Av} = \overline{rv} \quad \text{or} \quad \bar{A}\bar{v} = A\bar{v} = \bar{r}\bar{v},$$

the vectors of E_r and $E_{\bar{r}}$ are conjugates of one another. This gives us (see Exercise 19) that $\overline{v_1}, \overline{v_2}, \ldots, \overline{v_k}$ span $E_{\bar{r}}$. To see why they are linearly independent, suppose

$$c_1\overline{v_1} + c_2\overline{v_2} + \cdots + c_k\overline{v_k} = 0.$$

Then

$$\overline{c_1\overline{v_1} + c_2\overline{v_2} + \cdots + c_k\overline{v_k}} = \bar{0}$$

or

$$\overline{c_1}v_1 + \overline{c_2}v_2 + \cdots + \overline{c_k}v_k = 0.$$

Since $v_1, v_2, \ldots, v_k$ are linearly independent, $\overline{c_1} = 0$, $\overline{c_2} = 0, \ldots, \overline{c_k} = 0$ and hence $c_1 = 0$, $c_2 = 0, \ldots, c_k = 0$ giving us the required linear independence of $\overline{v_1}, \overline{v_2}, \ldots, \overline{v_k}$. ●

Because of Theorem 5.17, there is now no need to separately find a basis for the eigenspace of a conjugate root as we did in Example 5—all we have to do is conjugate the basis vectors. For instance, once we find the basis vector

$$\begin{bmatrix} i \\ 1 \end{bmatrix}$$

for E_{1+i} in Example 5, we can immediately use

$$\overline{\begin{bmatrix} i \\ 1 \end{bmatrix}} = \begin{bmatrix} -i \\ 1 \end{bmatrix}$$

as a basis for E_{1-i}.

EXERCISES 5.4

Find the eigenvalues and bases for the eigenspaces of the matrices in Exercises 1–18.

1. $\begin{bmatrix} 3 & 0 \\ 8 & -1 \end{bmatrix}$ **2.** $\begin{bmatrix} 4 & -4 \\ 1 & 0 \end{bmatrix}$ **3.** $\begin{bmatrix} 10 & -9 \\ 4 & -2 \end{bmatrix}$

4. $\begin{bmatrix} 6 & -8 \\ 4 & -6 \end{bmatrix}$ **5.** $\begin{bmatrix} 0 & 3 \\ 4 & 0 \end{bmatrix}$ **6.** $\begin{bmatrix} 1 & 2 \\ 3 & 4 \end{bmatrix}$

7. $\begin{bmatrix} 4 & 0 & 1 \\ -2 & 1 & 0 \\ -2 & 0 & 1 \end{bmatrix}$ **8.** $\begin{bmatrix} 1 & 1 & 0 \\ 0 & 1 & -1 \\ 0 & 0 & 2 \end{bmatrix}$

9. $\begin{bmatrix} 4 & 0 & 0 \\ 1 & 4 & 0 \\ 0 & 1 & 4 \end{bmatrix}$ **10.** $\begin{bmatrix} -6 & 0 & -8 \\ -4 & 2 & -4 \\ 4 & 0 & 6 \end{bmatrix}$

11. $\begin{bmatrix} 5 & 6 & 2 \\ 0 & -1 & -8 \\ 1 & 0 & -2 \end{bmatrix}$ **12.** $\begin{bmatrix} 1 & 2 & 2 \\ 1 & 2 & 3 \\ 0 & -1 & 0 \end{bmatrix}$

13. $\begin{bmatrix} 0 & 2 & 2 \\ 2 & 0 & 2 \\ 2 & 2 & 0 \end{bmatrix}$ **14.** $\begin{bmatrix} 1 & 2 & -1 \\ 0 & 0 & 2 \\ 0 & 1 & 0 \end{bmatrix}$

15. $\begin{bmatrix} 3 & -2 \\ 4 & -1 \end{bmatrix}$ **16.** $\begin{bmatrix} -4 & 5 \\ -4 & 4 \end{bmatrix}$

17. $\begin{bmatrix} 1 & 0 & 0 \\ 0 & 0 & 1 \\ 0 & -1 & 0 \end{bmatrix}$ **18.** $\begin{bmatrix} 3 & 1 & -1 \\ 0 & 0 & -2 \\ 0 & 1 & 2 \end{bmatrix}$

19. Suppose that U and W are subspaces of a vector space V over the complex numbers and suppose that the vectors of U and W are conjugates of one another. Show that if $u_1, u_2, \ldots, u_k$ span U, then $\overline{u_1}, \overline{u_2}, \ldots, \overline{u_k}$ span W.

20. If $D = \text{diag}(d_1, d_2, \ldots, d_n)$ is a diagonal matrix, what are the eigenvalues of D? Also, find bases for the eigenspaces of D involving the standard basis vectors for $\mathbb{R}^n$.

21. Show that the eigenvalues of a triangular matrix are its diagonal entries.

22. Prove that a square matrix A is not invertible if and only if zero is an eigenvalue of A.

23. Show that if A is an invertible matrix and if v is an eigenvector of A with associated eigenvalue $\lambda = r$, then v is an eigenvector of A^{-1} with associated eigenvalue $1/r$.

24. Use the result of Exercise 23 to find the eigenvalues and bases for the eigenspaces of the inverse of the matrix in Exercise 11.

25. Show that if v is an eigenvector of a square matrix A with associated eigenvalue $\lambda = r$, then v is an eigenvector of A^k with associated eigenvalue r^k for any positive integer k.

26. Show that if A is a square matrix, then A and A^T have the same eigenvalues.

In Maple, the command *eigenvalues* or *eigenvals* can be used to find the eigenvalues of a square matrix. The commands *eigenvectors* or *eigenvects* can be used to find the eigenvectors of a square matrix.[4] When using either of the latter two commands, Maple's output will include each eigenvalue, its multiplicity as a root of the characteristic polynomial, and a basis for the associated eigenspace. Use Maple or another appropriate software package to find the eigenvalues and bases for the eigenspaces of the following matrices.

27. $\begin{bmatrix} 1 & -2 & 3 \\ 3 & 2 & -4 \\ 1 & -2 & 3 \end{bmatrix}$ 28. $\begin{bmatrix} 1 & -2 & 3 \\ 3 & 2 & -4 \\ 2 & -2 & 3 \end{bmatrix}$

29. $\begin{bmatrix} 1.0 & -2 & 3 \\ 3 & 2 & -4 \\ 1 & -2 & 3 \end{bmatrix}$ 30. $\begin{bmatrix} 1.0 & -2 & 3 \\ 3 & 2 & -4 \\ 2 & -2 & 3 \end{bmatrix}$

31. $\begin{bmatrix} 3 & 10 & 16 & -26 \\ 15 & 8 & 8 & -28 \\ 0 & 10 & 17 & -24 \\ 5 & 10 & 15 & -27 \end{bmatrix}$

32. $\begin{bmatrix} 20 & 1 & -4 & -10 & -5 \\ 36 & 3 & -3 & -18 & -16 \\ 12 & 1 & -2 & -7 & -2 \\ 21 & 1 & -4 & -11 & -5 \\ 21 & 1 & -4 & -10 & -6 \end{bmatrix}$

5.5   SIMILAR MATRICES, DIAGONALIZATION, AND JORDAN CANONICAL FORM

We begin this section by assuming that we are working in the system of real numbers. Everything we are about to say, however, carries over to the system of complex numbers. We will illustrate how to adapt several of the points we are making to the complex-number setting later in this section.

Suppose that A is an $n \times n$ matrix. We say that an $n \times n$ matrix B is **similar** to A if there is an invertible $n \times n$ matrix P so that

$$\boxed{B = P^{-1}AP.}$$

One place we have already encountered similar matrices is when changing bases for matrices of linear transformations. If $T : V \to V$ is a linear transformation where V is a nonzero finite dimensional vector space, we saw in Section 5.3 that if A is the matrix of T with respect to a basis α of V (so that $A = [T]_\alpha^\alpha$) and if B is the matrix of T with respect to another basis β of V (so that $B = [T]_\beta^\beta$), then $B = P^{-1}AP$ where P is the change of basis matrix from the basis α to the basis β. In fact, all similar matrices can be obtained in this manner. To see why, let $T : \mathbb{R}^n \to \mathbb{R}^n$ be the matrix transformation

---

[4] If any entries are entered as decimal numbers, Maple will give decimal approximations in its answers; otherwise Maple will give its answers in a symbolic exact form, although these exact forms can be difficult to interpret if the roots of the characteristic polynomial are not nice.

$T(X) = AX$. If α is the standard basis for $\mathbb{R}^n$, we know that

$$[T]_\alpha^\alpha = A.$$

If P is an invertible $n \times n$ matrix, then P is row equivalent to I and hence has rank n. Consequently, the columns of P form a basis β for $\mathbb{R}^n$ and P is the change of basis matrix from α to β. Thus we have

$$P^{-1}AP = P^{-1}[T]_\alpha^\alpha P = [T]_\beta^\beta = B.$$

A square matrix is said to be **diagonalizable** if it is similar to a diagonal matrix. By the discussion of the previous paragraph, an $n \times n$ matrix A is then diagonalizable if and only if there is a basis β for $\mathbb{R}^n$ so that the matrix of $T(X) = AX$ with respect to β is a diagonal matrix. But notice that if $[T]_\beta^\beta$ is a diagonal matrix,

$$[T]_\beta^\beta = \begin{bmatrix} d_1 & 0 & \cdots & 0 \\ 0 & d_2 & \cdots & 0 \\ \vdots & \vdots & & \vdots \\ 0 & 0 & \cdots & d_n \end{bmatrix},$$

and if the basis vectors in β are $w_1, w_2, \ldots, w_n$, then

$$T(w_i) = Aw_i = d_i w_i$$

for each i. That is, each basis vector in β is an eigenvector of A. Let us record what we have just observed as

THEOREM 5.18   An $n \times n$ matrix A is diagonalizable if and only there is a basis for $\mathbb{R}^n$ consisting of eigenvectors of A.

How can we tell if a square matrix A is diagonalizable or, equivalently, if $\mathbb{R}^n$ has a basis of eigenvectors of A? As we shall see momentarily, the following lemma is one of the keys.

LEMMA 5.19   Suppose that $r_1, r_2, \ldots, r_k$ are distinct eigenvalues of a square matrix A. If the vectors $v_{11}, v_{12}, \ldots, v_{1l_1}$ form a basis for E_{r_1}, the vectors $v_{21}, v_{22}, \ldots, v_{2l_2}$ form a basis for $E_{r_2}, \ldots$, the vectors $v_{k1}, v_{k2}, \ldots, v_{kl_k}$ form a basis for E_{r_k}, then the vectors

$$v_{11}, v_{12}, \ldots, v_{1l_1}, v_{21}, v_{22}, \ldots, v_{2l_2}, \ldots, v_{k1}, v_{k2}, \ldots, v_{kl_k}$$

are linearly independent.

Proof   Suppose that

$$c_{11}v_{11} + \cdots + c_{1l_1}v_{1l_1} + c_{21}v_{21} + \cdots + c_{2l_2}v_{2l_2} + \cdots + c_{k1}v_{k1} + \cdots + c_{kl_k}v_{kl_k} = 0. \quad (1)$$

Let B_1 be the matrix

$$B_1 = (r_2 I - A)(r_3 I - A) \cdots (r_k I - A).$$

Notice that for any $j \geq 2$,

$$B_1 v_{ji} = (r_2 - r_j) \cdots (r_j - r_j) \cdots (r_k - r_j) v_{ji} = 0.$$

Thus if we multiply B_1 times Equation (1), we obtain

$$(r_2 - r_1) \cdots (r_k - r_1) c_{11} v_{11} + \cdots + (r_2 - r_1) \cdots (r_k - r_1) c_{1l_1} v_{1l_1} = 0$$

and hence

$$c_{11} v_{11} + \cdots + c_{1l_1} v_{1l_1} = 0.$$

Thus $c_{11} = 0, \ldots, c_{1l_1} = 0$ since $v_{11}, \ldots, v_{1l_1}$ are linearly independent. Repeating this procedure with the matrices

$$B_2 = (r_1 I - A)(r_3 I - A) \cdots (r_k I - A), \ldots, B_k = (r_1 I - A) \cdots (r_{k-1} I - A),$$

we obtain the remaining scalars c_{ji} are all zero, completing the proof. ●

Let us now see the impact of Lemma 5.19 on the diagonalizability of an $n \times n$ matrix A. Since the basis vectors from the distinct eigenspaces of A are linearly independent, it follows that if the total number of vectors in these eigenspace bases is n, then these vectors form a basis for $\mathbb{R}^n$ and hence A is diagonalizable. To put it another way, if, in the notation of Lemma 5.19,

$$\dim(E_{r_1}) + \dim(E_{r_2}) + \cdots + \dim(E_{r_k}) = n, \tag{2}$$

then A is diagonalizable. Conversely, if A is diagonalizable, in which case $\mathbb{R}^n$ has a basis of eigenvectors of A, then it can be shown (see Exercise 33) that the basis vectors associated with an eigenvalue r_i of A form a basis for E_{r_i}. Thus if A is diagonalizable, Equation (2) holds. Hence we can state the following theorem.

THEOREM 5.20　Suppose A is an $n \times n$ matrix with distinct eigenvalues $r_1, r_2, \ldots, r_k$. Then A is diagonalizable if and only if

$$\boxed{\dim(E_{r_1}) + \dim(E_{r_2}) + \cdots + \dim(E_{r_k}) = n.}$$

Let us illustrate how to use Theorem 5.20 along with the techniques we learned in the previous section for finding bases of eigenspaces to see if a square matrix A is diagonalizable. Further, when it is, we will illustrate how to find a diagonal matrix similar to A and a matrix P so that $P^{-1} A P$ is this diagonal matrix.

EXAMPLE 1　Determine whether the given matrix A is diagonalizable and, if it is, give a diagonal matrix similar to A as well as a matrix P so that $P^{-1} A P$ is this diagonal matrix.

$$A = \begin{bmatrix} 1 & -3 \\ -2 & 2 \end{bmatrix}$$

Solution From the solutions to Examples 1 and 2 of the previous section, we have that the eigen-values of A are $\lambda = 4$ and $\lambda = -1$ and

$$\dim(E_4) + \dim(E_{-1}) = 1 + 1 = 2.$$

Consequently, A is diagonalizable. Together, the basis vectors

$$\begin{bmatrix} -1 \\ 1 \end{bmatrix} \quad \text{and} \quad \begin{bmatrix} 3/2 \\ 1 \end{bmatrix}$$

we gave as bases for the individual eigenspaces of A form a basis

$$\begin{bmatrix} -1 \\ 1 \end{bmatrix}, \begin{bmatrix} 3/2 \\ 1 \end{bmatrix}$$

for $\mathbb{R}^2$. The matrix of the linear transformation $T(X) = AX$ with respect to this basis is

$$\begin{bmatrix} 4 & 0 \\ 0 & -1 \end{bmatrix},$$

which then is a diagonal matrix similar to A.[5] A matrix P so that $P^{-1}AP$ is this diagonal matrix is the change of basis matrix from the standard basis to our basis of eigenvectors, which is

$$P = \begin{bmatrix} -1 & 3/2 \\ 1 & 1 \end{bmatrix}. \qquad \bullet$$

EXAMPLE 2 Determine whether the given matrix A is diagonalizable and, if it is, give a diagonal matrix similar to A as well as a matrix P so that $P^{-1}AP$ is this diagonal matrix.

$$A = \begin{bmatrix} 2 & -1 & 3 \\ 0 & -1 & 0 \\ 0 & 0 & -1 \end{bmatrix}$$

Solution Let us go faster now. Look back at Example 3 in the previous section. From its solution, we can see that A is diagonalizable since

$$\dim(E_2) + \dim(E_{-1}) = 1 + 2 = 3$$

---

[5] The diagonal matrix is not unique. For instance, were we to interchange the order of the basis vectors here, the diagonal matrix would become

$$\begin{bmatrix} -1 & 0 \\ 0 & 4 \end{bmatrix}.$$

It is possible to show that the diagonal matrix is unique up to a permutation (that is, a rearrangement) of the diagonal entries.

and that A is similar to the diagonal matrix

$$\begin{bmatrix} 2 & 0 & 0 \\ 0 & -1 & 0 \\ 0 & 0 & -1 \end{bmatrix}.$$

We also see that we can take P to be

$$P = \begin{bmatrix} 1 & 1/3 & -1 \\ 0 & 1 & 0 \\ 0 & 0 & 1 \end{bmatrix}.$$ ●

EXAMPLE 3   Determine whether the given matrix A is diagonalizable and, if it is, give a diagonal matrix similar to A as well as a matrix P so that $P^{-1}AP$ is this diagonal matrix.

$$A = \begin{bmatrix} 1 & -2 & -6 \\ -2 & 2 & -5 \\ 2 & 1 & 8 \end{bmatrix}$$

Solution   From the solution to Example 4 in the previous section, we see that this matrix is not diagonalizable since

$$\dim(E_3) + \dim(E_5) = 1 + 1 \neq 3.$$ ●

As mentioned at the beginning of this section, we can adapt what we have been doing to the setting where complex numbers arise.

EXAMPLE 4   Determine if the matrix

$$A = \begin{bmatrix} 1 & -1 \\ 1 & 1 \end{bmatrix}$$

is diagonalizable. If it is, give a diagonal matrix to which A is similar as well as a matrix P so that $P^{-1}AP$ is this diagonal matrix.

Solution   Look back at the solution to Example 5 in the previous section. If we work over $\mathbb{R}$, this matrix is not diagonalizable since it has no real eigenvalues (and hence no real eigenvectors either). But if we work over the complex numbers, we have success: The matrix is diagonalizable and it is similar to the diagonal matrix

$$\begin{bmatrix} 1+i & 0 \\ 0 & 1-i \end{bmatrix}.$$

For P we may use

$$P = \begin{bmatrix} i & -i \\ 1 & 1 \end{bmatrix}.$$ ●

While not every square matrix is diagonalizable, as Example 3 illustrates, there is something close to diagonal form called the **Jordan**[6] **canonical form** of a square matrix. To describe this form, we first introduce a type of matrix called a **basic Jordan block** associated with a value λ, which is a square matrix of the form

$$\begin{bmatrix} \lambda & 1 & 0 & \cdots & 0 & 0 \\ 0 & \lambda & 1 & \cdots & 0 & 0 \\ 0 & 0 & \lambda & \cdots & 0 & 0 \\ \vdots & \vdots & \vdots & & \vdots & \vdots \\ 0 & 0 & 0 & \cdots & \lambda & 1 \\ 0 & 0 & 0 & \cdots & 0 & \lambda \end{bmatrix}.$$

The Jordan canonical form of a square matrix is comprised of such Jordan blocks.

THEOREM 5.21   Suppose that A is an $n \times n$ matrix and suppose that

$$\det(\lambda I - A) = (\lambda - r_1)^{m_1}(\lambda - r_2)^{m_2} \cdots (\lambda - r_k)^{m_k}$$

where $r_1, r_2, \ldots, r_k$ are the distinct roots of the characteristic polynomial of A. Then A is similar to a matrix of the form

$$\begin{bmatrix} B_1 & O & \cdots & O \\ O & B_2 & \cdots & O \\ \vdots & \vdots & & \vdots \\ O & O & \cdots & B_k \end{bmatrix}$$

where each B_i is an $m_i \times m_i$ matrix of the form

$$B_i = \begin{bmatrix} J_{i_1} & O & \cdots & O \\ O & J_{i_2} & \cdots & O \\ \vdots & \vdots & & \vdots \\ O & O & \cdots & J_{i_l} \end{bmatrix}$$

and each J_{i_j} is a basic Jordan block associated with r_i.

Different Jordan canonical forms of a square matrix may be obtained by permuting (rearranging) the basic Jordan blocks and these are the only possible Jordan canonical forms.[7] When writing down Jordan canonical forms, we shall write down only one such form and not concern ourselves with the others obtained by permuting the basic Jordan blocks.

---

[6] Named for the the French mathematician Camille Jordan (1838–1922).

[7] These different Jordan canonical forms correspond to permuting the vectors in a basis β for $\mathbb{R}^n$ giving us one Jordan canonical form as the matrix $[T]_\beta^\beta$ for the matrix transformation $T(X) = AX$.

EXAMPLE 5 List the possible Jordan canonical forms of a 4×4 matrix A whose characteristic polynomial is

$$\det(\lambda I - A) = (\lambda - 4)(\lambda - 2)^3.$$

Solution For the first eigenvalue, $\lambda = 4$, there is only possibility for B_1 since it must be a 1×1 matrix:

$$B_1 = [4].$$

For the second eigenvalue, $\lambda = 2$, the matrix B_2 is a 3×3 matrix and there are several possible ways we can form a 3×3 matrix consisting of basic Jordan blocks associated with $\lambda = 2$. One way is for B_2 to consist of an entire 3×3 basic Jordan block:

$$B_2 = \begin{bmatrix} 2 & 1 & 0 \\ 0 & 2 & 1 \\ 0 & 0 & 2 \end{bmatrix}.$$

Another possibility is for B_2 to consist of a 2×2 basic Jordan block and a 1×1 basic Jordan block:

$$B_2 = \begin{bmatrix} 2 & 1 & 0 \\ 0 & 2 & 0 \\ 0 & 0 & 2 \end{bmatrix}.$$

The last possibility is for B_2 to consist of three 1×1 basic Jordan blocks:

$$B_2 = \begin{bmatrix} 2 & 0 & 0 \\ 0 & 2 & 0 \\ 0 & 0 & 2 \end{bmatrix}.$$

Putting this all together, there are then three possible Jordan canonial forms for A:

$$\begin{bmatrix} 4 & 0 & 0 & 0 \\ 0 & 2 & 1 & 0 \\ 0 & 0 & 2 & 1 \\ 0 & 0 & 0 & 2 \end{bmatrix}, \begin{bmatrix} 4 & 0 & 0 & 0 \\ 0 & 2 & 1 & 0 \\ 0 & 0 & 2 & 0 \\ 0 & 0 & 0 & 2 \end{bmatrix}, \begin{bmatrix} 4 & 0 & 0 & 0 \\ 0 & 2 & 0 & 0 \\ 0 & 0 & 2 & 0 \\ 0 & 0 & 0 & 2 \end{bmatrix} \quad \bullet$$

It is interesting to observe that while there are infinitely many 4×4 matrices with characteristic polynomial $(\lambda - 4)(\lambda - 2)^3$ all of these are similar to one of the three matrices given at the end of our solution to Example 5!

The proof of Theorem 5.21 is beyond the scope of this book and is omitted. Further, there are general methods for finding the Jordan canonical form of a given square matrix, but these too are beyond the scope of this book. Let us notice, however, that it is sometimes possible to see the Jordan canonical form of an $n \times n$ matrix A by knowing the bases for the eigenspaces. We already have noted that if the sum of the dimensions of the eigenspaces is n, the matrix is diagonalizable. The resulting diagonal matrix is the Jordan canonical form of A, so we know the Jordan canonical form in this case. Thus in Examples 1, 2, and 4, we could say that we found the Jordan canonical forms of the given matrices in these examples.

Another case in which we can determine the Jordan canonical form of A is when the multiplicity of each eigenvalue is at most 2. To illustrate how, consider the matrix

$$A = \begin{bmatrix} 1 & -2 & -6 \\ -2 & 2 & -5 \\ 2 & 1 & 8 \end{bmatrix}$$

in Example 3. Since its characteristic polynomial is (see Example 4 in the previous section)

$$\det(\lambda I - A) = (\lambda - 3)^2(\lambda - 5),$$

its possible Jordan canonical forms are

$$\begin{bmatrix} 3 & 0 & 0 \\ 0 & 3 & 0 \\ 0 & 0 & 5 \end{bmatrix}, \begin{bmatrix} 3 & 1 & 0 \\ 0 & 3 & 0 \\ 0 & 0 & 5 \end{bmatrix}.$$

Because A is not diagonalizable, the first form is out and the second matrix is the Jordan canonical form of A.

While we ourselves will not study methods for finding Jordan canonical forms, software packages such as Maple employ these methods to find them. Maple also will find a change of basis matrix P so that $P^{-1}AP$ is the Jordan canonical form of an $n \times n$ matrix A. To illustrate, let A be the matrix in the previous paragraph:

$$A = \begin{bmatrix} 1 & -2 & -6 \\ -2 & 2 & -5 \\ 2 & 1 & 8 \end{bmatrix}.$$

Once the matrix has been entered on a Maple worksheet, its Jordan canonical form and the matrix P can be found by typing and entering

jordan(A,'P');

which gives us the Jordan canonical form of A as

$$\begin{bmatrix} 5 & 0 & 0 \\ 0 & 3 & 1 \\ 0 & 0 & 3 \end{bmatrix}.$$

(Notice that Maple has the basic Jordan blocks in a different order than in the previous paragraph.) To get Maple to display P, we type and enter

print(P);

and Maple gives us P to be

$$\begin{bmatrix} 1/2 & -2 & -3 \\ 1/2 & -1 & -1/2 \\ -1/2 & 1 & 3/2 \end{bmatrix}.$$

Exercises 39 and 40 ask you to use Maple or another appropriate software package to find both a Jordan canonical form of a square matrix A as well as a matrix P so that $P^{-1}AP$ is this Jordan canonical form of A.

EXERCISES 5.5

1–18. Let A be the matrix given in the corresponding exercise of Section 5.4 (see page 277). Determine if A is diagonalizable and, if it is, give a diagonal matrix similar to A as well as a matrix P so that $P^{-1}AP$ is this diagonal matrix.

List the possible Jordan canonical forms for the matrices with the given characteristic polynomials in Exercises 19–24.

19. $\lambda^2 - 3\lambda + 2$ **20.** $\lambda^2 - 3\lambda - 10$

21. $\lambda^3 - \lambda^2$ **22.** $\lambda^5 - 2\lambda^4 + \lambda^3$

23. $(\lambda - 5)(\lambda - 1)^2(\lambda + 2)^3$

24. $(\lambda - 3)^2(\lambda + 4)^4$

In Exercises 25–30, find the Jordan canonical form for the matrix in the indicated exercise of Section 5.4 (see page 277).

25. Exercise 7   **26.** Exercise 10

27. Exercise 3   **28.** Exercise 2

29. Exercise 11   **30.** Exercise 8

31. Show that similar matrices have the same characteristic polynomial.

32. Suppose that A, B, and C are square matrices of the same size.

   a) Show that A is similar to A.

   b) Show that if B is similar to A, then A is similar to B.

   c) Show that if A is similar to B and B is similar to C, then A is similar to C.

33. Show that if the square matrix B is similar to the square matrix A, then B^k is similar to A^k for any positive integer k.

34. Suppose that A is an invertible matrix and the matrix B is similar to A. Show that B is an invertible matrix and that B^{-1} is similar to A^{-1}.

35. Suppose that A is an $n \times n$ diagonalizable matrix with distinct eigenvalues $r_1, r_2, \ldots, r_k$. Further suppose that $v_1, v_2, \ldots, v_n$ are eigenvectors of A forming a basis for $\mathbb{R}^n$ arranged in such a way that $v_1, \ldots, v_{m_1}$ are eigenvectors associated with r_1, $v_{m_1+1}, \ldots, v_{m_2}$ are eigenvectors associated with $r_2, \ldots, v_{m_{k-1}+1}, \ldots, v_n$ are eigenvectors associated with r_k. Show that for each i, $v_{m_{i-1}+1}, \ldots, v_{m_i}$ span E_{r_i} and hence form a basis for E_{r_i}. (*Hint:* Let v be a vector in E_{r_i}. Write v as a linear combination of $v_1, \ldots, v_n$ and multiply by the matrix B_i in the proof of Lemma 5.19.)

36. Show that if an $n \times n$ matrix A has n distinct eigenvalues, then A is diagonalizable.

37. Suppose that A is a diagonalizable $n \times n$ matrix and $v_1, v_2, \ldots, v_n$ are eigenvectors of A with associated (not necessarily distinct) eigenvalues $r_1, r_2, \ldots, r_n$ that form a basis for $\mathbb{R}^n$. Let v be a vector in $\mathbb{R}^n$. Express v as

$$v = c_1 v_1 + c_2 v_2 + \cdots + c_n v_n.$$

Show that for any positive integer k,

$$A^k v = c_1 r_1^k v_1 + c_2 r_2^k v_2 + \cdots + c_n r_n^k v_n.$$

38. Suppose that A is square matrix that has only one eigenvalue $\lambda = r$. Show that A is diagonalizable if and only if $A = rI$.

In Exercises 39 and 40, use Maple or another appropriate software package to find a Jordan canonical form of A

and a matrix P so that $P^{-1}AP$ is this Jordan canonical form of A.

$$\textbf{39. } A = \begin{bmatrix} 21 & 1 & -4 & -11 & -4 \\ 43 & 4 & -3 & -22 & -19 \\ 10 & 1 & -1 & -7 & 0 \\ 22 & 1 & -4 & -12 & -4 \\ 22 & 1 & -4 & -11 & -5 \end{bmatrix}$$

$$\textbf{40. } A = \begin{bmatrix} 0 & 1 & -1 & 2 & 1 & -1 \\ -4 & 3 & -1 & 3 & 1 & 0 \\ 0 & 1 & 1 & 3 & -1 & -2 \\ 0 & 1 & -1 & 4 & 0 & -2 \\ 0 & 1 & -1 & 2 & 2 & -2 \\ -1 & 1 & -1 & 2 & 1 & 0 \end{bmatrix}$$

5.6   EIGENVECTORS AND EIGENVALUES OF LINEAR TRANSFORMATIONS

The concepts of eigenvectors and eigenvalues of square matrices can be extended to linear transformations from a vector space to itself. If $T : V \to V$ is a linear transformation, we say that a nonzero vector v in V is an **eigenvector** of T if

$$\boxed{T(v) = \lambda v}$$

where λ is a scalar; λ is called an **eigenvalue** of T. For example, $x^2 + 3x + 2$ is an eigenvector with associated eigenvalue $\lambda = -3$ of the linear transformation $T : P_2 \to P_2$ by

$$T(ax^2 + bx + c) = (5a - 4c)x^2 + (12a + b - 12c)x + 8a - 7c$$

since

$$T(x^2 + 3x + 2) = -3x^2 - 9x - 6 = -3(x^2 + 3x + 2).$$

If λ is an eigenvalue of T, the set of eigenvectors associated with λ along with the zero vector is the kernel of the linear transformation $\lambda I - T$ where I is the identity linear transformation $I(v) = v$ on V. We will call this subspace the **eigenspace** of λ and denote it by V_λ so that

$$V_\lambda = \ker(\lambda I - T).$$

Notice that if T is a matrix transformation $T(X) = AX$ where A is a square matrix, then the eigenvectors and eigenvalues of T are the same as the eigenvectors and eigenvalues of A.

How do we find the eigenvectors and eigenvalues if T is not a matrix transformation? The answer is to use matrices of T relative to a basis of V.

THEOREM 5.22   Suppose that $T : V \to V$ is a linear transformation. Let α be a basis consisting of v_1, $v_2, \ldots, v_n$ for V and A be the matrix of T with respect to α. Then v is an eigenvector of T with associated eigenvalue λ if and only if the coordinate vector of v with respect to α is an eigenvector of A with associated eigenvalue λ. Moreover, if

$$
\begin{bmatrix} a_{11} \\ a_{21} \\ \vdots \\ a_{n1} \end{bmatrix}, \begin{bmatrix} a_{12} \\ a_{22} \\ \vdots \\ a_{n2} \end{bmatrix}, \ldots, \begin{bmatrix} a_{1k} \\ a_{2k} \\ \vdots \\ a_{nk} \end{bmatrix}
$$

form a basis for E_λ, then

$$
u_1 = a_{11}v_1 + a_{21}v_2 + \cdots + a_{n1}v_n, \ldots, u_k = a_{1k}v_1 + a_{2k}v_2 + \cdots + a_{nk}v_n
$$

form a basis for V_λ.

Proof   Suppose v is an eigenvector of T with associated eigenvalue λ. Since $T(v) = \lambda v$ and

$$
[T(v)]_\alpha = A[v]_\alpha
$$

by Theorem 5.14, we have

$$
[\lambda v]_\alpha = A[v]_\alpha.
$$

Hence

$$
\lambda [v]_\alpha = A[v]_\alpha
$$

giving us that $[v]_\alpha$ is an eigenvector of A with associated eigenvalue λ.
Conversely, suppose the column vector

$$
\begin{bmatrix} a_1 \\ a_2 \\ \vdots \\ a_n \end{bmatrix}
$$

is an eigenvector of A with associated eigenvalue λ. Set

$$
v = a_1 v_1 + a_2 v_2 + \cdots + a_n v_n
$$

in which case

$$
[v]_\alpha = \begin{bmatrix} a_1 \\ a_2 \\ \vdots \\ a_n \end{bmatrix}.
$$

We then have

$$
[T(v)]_\alpha = A[v]_\alpha = \lambda [v]_\alpha = \begin{bmatrix} \lambda a_1 \\ \lambda a_2 \\ \vdots \\ \lambda a_n \end{bmatrix}.
$$

Hence

$$T(v) = \lambda a_1 v_1 + \lambda a_2 v_2 + \cdots + \lambda a_n v_n = \lambda v$$

giving us that v is an eigenvector of T with associated eigenvalue λ.

Finally, we must argue that $u_1, \ldots, u_k$ form a basis for V_λ. To see that they span V_λ, notice that for each v in V_λ there are scalars $c_1, \ldots, c_k$ so that

$$[v]_\alpha = c_1 \begin{bmatrix} a_{11} \\ a_{21} \\ \vdots \\ a_{n1} \end{bmatrix} + \cdots + c_k \begin{bmatrix} a_{1k} \\ a_{2k} \\ \vdots \\ a_{nk} \end{bmatrix}$$

$$= c_1[u_1]_\alpha + \cdots + c_k[u_k]_\alpha$$

from which it follows that

$$v = c_1 u_1 + \cdots + c_k u_k.$$

To see why $u_1, \ldots, u_k$ are linearly independent, notice that if

$$c_1 u_1 + \cdots + c_k u_k = 0,$$

then

$$c_1[u_1]_\alpha + \cdots + c_k[u_k]_\alpha = 0$$

or

$$c_1 \begin{bmatrix} a_{11} \\ a_{21} \\ \vdots \\ a_{n1} \end{bmatrix} + \cdots + c_k \begin{bmatrix} a_{1k} \\ a_{2k} \\ \vdots \\ a_{nk} \end{bmatrix} = 0.$$

Hence

$$c_1 = 0, \qquad c_2 = 0, \qquad \ldots, \qquad c_k = 0. \qquad \bullet$$

Following is an illustration of how we use Theorem 5.22 to find the eigenvalues and eigenvectors of a linear transformation.

EXAMPLE 1   Find the eigenvalues and bases for the eigenspaces of the linear transformation $T : P_2 \to P_2$ by

$$T(ax^2 + bx + c) = (5a - 4c)x^2 + (12a + b - 12c)x + 8a - 7c.$$

Solution   Letting α be the standard basis consisting of x^2, x, 1 for P_2, the matrix of T with respect to α is

$$[T]_\alpha^\alpha = A = \begin{bmatrix} 5 & 0 & -4 \\ 12 & 1 & -12 \\ 8 & 0 & -7 \end{bmatrix}.$$

The characteristic polynomial of A is

$$\det(\lambda I - A) = \begin{vmatrix} \lambda - 5 & 0 & 4 \\ -12 & \lambda - 1 & 12 \\ -8 & 0 & \lambda + 7 \end{vmatrix}.$$

Expanding this determinant about the second column, the characteristic equation is

$$(\lambda - 1)((\lambda - 5)(\lambda + 7) + 32) = (\lambda - 1)(\lambda^2 + 2\lambda - 3) = (\lambda - 1)^2(\lambda + 3) = 0$$

and hence the eigenvalues of both A and T are $\lambda = 1$ and $\lambda = -3$. We next find bases for the eigenspaces of A and translate them into eigenvectors for T. First consider $\lambda = 1$.

$$\begin{bmatrix} -4 & 0 & 4 & \vdots & 0 \\ -12 & 0 & 12 & \vdots & 0 \\ -8 & 0 & 8 & \vdots & 0 \end{bmatrix} \rightarrow \begin{bmatrix} 1 & 0 & -1 & \vdots & 0 \\ 0 & 0 & 0 & \vdots & 0 \\ 0 & 0 & 0 & \vdots & 0 \end{bmatrix}$$

The solutions to $(I - A)X = 0$ are

$$\begin{bmatrix} x \\ y \\ z \end{bmatrix} = \begin{bmatrix} z \\ y \\ z \end{bmatrix}$$

from which we see we may use the vectors

$$\begin{bmatrix} 0 \\ 1 \\ 0 \end{bmatrix}, \begin{bmatrix} 1 \\ 0 \\ 1 \end{bmatrix}$$

as a basis for E_1. These two vectors in $\mathbb{R}^3$ are the coordinate vectors of the polynomials $x, x^2 + 1$, which form a basis for V_1 where $V = P_2$.

Finally, consider $\lambda = -3$.

$$\begin{bmatrix} -8 & 0 & 4 & \vdots & 0 \\ -12 & -4 & 12 & \vdots & 0 \\ -8 & 0 & 4 & \vdots & 0 \end{bmatrix} \rightarrow \begin{bmatrix} 2 & 0 & -1 & \vdots & 0 \\ 0 & -4 & 6 & \vdots & 0 \\ 0 & 0 & 0 & \vdots & 0 \end{bmatrix}$$

The solutions to $(-3I - A)X = 0$ are

$$\begin{bmatrix} x \\ y \\ z \end{bmatrix} = \begin{bmatrix} z/2 \\ 3z/2 \\ z \end{bmatrix}$$

from which we see we may use the vector

$$\begin{bmatrix} 1/2 \\ 3/2 \\ 1 \end{bmatrix}$$

as a basis for E_{-3}. This vector in $\mathbb{R}^3$ is the coordinate vector of the polynomial

$$\frac{x^2}{2} + \frac{3x}{2} + 1,$$

which forms a basis for V_{-3} where again $V = P_2$.   ●

    The theory of diagonalizability for square matrices extends to linear transformations $T : V \to V$. We say that T is **diagonalizable** if there is a basis β of V where each vector in β is an eigenvector of T or, equivalently, if V has a basis β so that $[T]_\beta^\beta$ is a diagonal matrix. We can see from our solution to Example 1 that the linear transformation of this example is diagonalizable. The notion of Jordan canonical form also extends to such linear transformations $T : V \to V$ by using the Jordan canonical form of a matrix of T with respect to any basis of V.

EXERCISES 5.6

Do the following for each linear transformation in Exercises 1–6.

  (a) Find the eigenvalues and bases for the eigenspaces of the linear transformation.

  (b) Determine whether the linear transformation is diagonalizable.

  (c) Find the Jordan canonical form of the linear transformation.

1. $T : \mathbb{R}^3 \to \mathbb{R}^3$ by

$$T\begin{bmatrix} x \\ y \\ z \end{bmatrix} = \begin{bmatrix} 7x - 5z \\ 15x + 2y - 15z \\ 10x - 8z \end{bmatrix}.$$

2. $T : \mathbb{R}^3 \to \mathbb{R}^3$ by $T\begin{bmatrix} x \\ y \\ z \end{bmatrix} = \begin{bmatrix} 3x + y - 4z \\ 3x - 3z \\ 4x + y - 5z \end{bmatrix}.$

3. $T : P_1 \to P_1$ by $T(ax + b) = 2bx + a + b.$

4. $T : P_2 \to P_2$ by $T(ax^2 + bx + c) = ax^2 + (2b + 2c - 3a)x + 3c.$

5. $D : V \to V$ where V is the vector space of solutions to the differential equation $y'' - y = 0.$

6. $D^2 : V \to V$ where V is the vector space of solutions to the differential equation $y'' + y = 0.$

7. The **determinant of a linear transformation** $T : V \to V$, denoted $\det(T)$, is defined to be

$$\det(T) = \det([T]_\alpha^\alpha)$$

where α is a basis for V. Show that this definition of $\det(T)$ is independent of the choice of basis of V; that is, if β is another basis for V, then $\det([T]_\alpha^\alpha) = \det([T]_\beta^\beta)$. (This is described by saying that the definition of the determinant of a linear transformation from a nonzero finite dimensional vector space to itself is *well-defined*).

8. Find the determinant of the linear transformation in Exercise 2.

9. Find the determinant of the linear transformation in Exercise 3.

10. Suppose that $T : V \to V$ is a diagonalizable linear transformation. Show that if T has only one eigenvalue $\lambda = r$, then $T(v) = rv$ for all v in V.

11. You may have observed in every example and exercise involving eigenvalues and eigenvectors in this chapter that if $\lambda = r$ is an eigenvalue of a $n \times n$ matrix A, then $\dim(E_r)$ is less than or equal to the multiplicity of the root r in the characteristic polynomial of A. This is always true. Prove it as follows:

a) Let $v_1, v_2, \ldots, v_k$ be a basis for E_r. Extend this to a basis β consisting of $v_1, v_2, \ldots, v_k$, $v_{k+1}, \ldots, v_n$ for $\mathbb{R}^n$. Show that the matrix of the linear transformation $T(X) = AX$ with respect to β has the form

$$[T]_\beta^\beta = \begin{bmatrix} rI_k & B_{12} \\ O_{(n-k) \times k} & B_{22} \end{bmatrix}$$

where B_{12} is a $k \times (n-k)$ matrix and B_{22} is an $(n-k) \times (n-k)$ matrix.

b) Show that $(\lambda - r)^k$ is a factor of the characteristic polynomial of $[T]_\beta^\beta$. This then gives us the desired result. Why is this?

12. Let $T : P_3 \to P_3$ be the linear transformation given by

$$T(ax^3 + bx^2 + cx + d) = (3a + 6b - 7c + 7d)x^3$$
$$+ (c - a - 3b)x^2 + (26a + 27b - 31c + 29d)x$$
$$+ 26a + 26b - 27c + 24d.$$

Use Maple or another appropriate software package to aid in finding the eigenvalues and bases for the eigenspaces of T and determining whether T is diagonalizable.

Systems of Differential Equations

We begin this chapter with an example that, upon first glance, may appear to be a problem you already have encountered in your calculus classes.

An object is moving in the xy-plane with velocity vector

$$v = \begin{bmatrix} 2x - 5y \\ x - 2y \end{bmatrix}.$$

Find the position of the object at time t if the initial position of the object is the point (1,1).

Upon closer examination, however, you will notice that this problem is new to you. Realizing the velocity vector v is

$$v = \begin{bmatrix} \dfrac{dx}{dt} \\ \dfrac{dy}{dt} \end{bmatrix},$$

the vector equation given in the statement of the problem is the same as the system of equations:

$$\frac{dx}{dt} = 2x - 5y$$

$$\frac{dy}{dt} = x - 2y.$$

This is an example of a system of differential equations. Because we know the initial postion of the object is the point $(1, 1)$, we have initial conditions

$$x(0) = 1, \qquad y(0) = 1$$

giving us an initial value problem. A solution to this initial value problem would, of course, consist of functions of the form $x = x(t)$, $y = y(t)$ satisfying the system of differential equations and the initial conditions. Software packages such as Maple can be used to graph phase portraits to an initial value problem such as the one we have here. The phase portrait in Figure 6.1 for this initial value problem, which was obtained by typing and entering

```
phaseportrait([D(x)(t)= 2*x(t)-5*y(t),
     D(y)(t)=x(t)-2*y(t)], [x(t),y(t)],
   t=0..7,[[x(0)=1,y(0)=1]], arrows = none);
```

illustrates the path of the object.

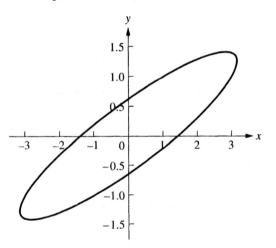

Figure 6.1

The main purpose of this chapter is to develop techniques for solving systems of differential equations and to consider some applications of them. The system of differential equations

$$\frac{dx}{dt} = 2x - 5y$$

$$\frac{dy}{dt} = x - 2y$$

we just considered is an example of a first order linear system of differential equations. While there will be a few places in this chapter where we will consider some higher order systems of linear differential equations and some nonlinear systems of differential equations, we will focus most of our attention on first order systems of linear differential equations. As you are about to see, there is a great deal of similarity between first order systems of linear differential equations and the linear differential equations we studied in Chapter 4. Indeed, we begin this chapter with a section on the theory of these systems that parallels the first section of Chapter 4.

6.1   THE THEORY OF SYSTEMS OF LINEAR DIFFERENTIAL EQUATIONS

By a **system of first order linear differential equations** we mean a system of first order differential equations that can be written in the form:

$$
\begin{aligned}
y_1' &= a_{11}(x)y_1 + a_{12}(x)y_2 + \cdots + a_{1n}(x)y_n + g_1(x) \\
y_2' &= a_{21}(x)y_1 + a_{22}(x)y_2 + \cdots + a_{2n}(x)y_n + g_2(x) \\
&\vdots \\
y_n' &= a_{n1}(x)y_1 + a_{n2}(x)y_2 + \cdots + a_{nn}(x)y_n + g_n(x).
\end{aligned}
\tag{1}
$$

(Note that we are using x as the independent variable in our functions rather than t as we had in the example in the introduction of this chapter.) If $g_1(x) = 0, \ldots, g_n(x) = 0$, the system is called **homogeneous**; otherwise it is called **nonhomogeneous.** Throughout this chapter we assume each a_{ij} and each g_j is continous on an interval (a, b). If we add the initial conditions

$$
y_1(x_0) = b_1, \qquad y_2(x_0) = b_2, \qquad \ldots, \qquad y_n(x_0) = b_n
$$

for fixed real numbers $b_1, b_2, \ldots, b_n$, we call the system of linear differential equations an **initial value problem.**

A **solution** to the system of linear differential equations in Equation (1) is an $n \times 1$ column vector

$$
Y = \begin{bmatrix} y_1 \\ y_2 \\ \vdots \\ y_n \end{bmatrix}
$$

where each y_i is a function of x so that $y_1, y_2, \ldots, y_n$ satisfy each equation of the system. Further, Y is a solution to the initial value problem if it satisfies the initial conditions. That is,

$$
Y(x_0) = \begin{bmatrix} y_1(x_0) \\ y_2(x_0) \\ \vdots \\ y_n(x_0) \end{bmatrix} = \begin{bmatrix} b_1 \\ b_2 \\ \vdots \\ b_n \end{bmatrix}.
$$

We can use matrices to write our system of linear differential equations as

$$
\begin{bmatrix} y_1' \\ y_2' \\ \vdots \\ y_n' \end{bmatrix} = \begin{bmatrix} a_{11}(x) & a_{12}(x) & \cdots & a_{1n}(x) \\ a_{21}(x) & a_{22}(x) & \cdots & a_{2n}(x) \\ \vdots & \vdots & \vdots & \vdots \\ a_{n1}(x) & a_{n2}(x) & \cdots & a_{nn}(x) \end{bmatrix} \begin{bmatrix} y_1 \\ y_2 \\ \vdots \\ y_n \end{bmatrix} + \begin{bmatrix} g_1(x) \\ g_2(x) \\ \vdots \\ g_n(x) \end{bmatrix}.
$$

Letting $A(x)$ be the $n \times n$ matrix $[a_{ij}(x)]$, Y be the $n \times 1$ vector $[y_i]$, $G(x)$ be the $n \times 1$ vector $[g_i(x)]$, and Y' be the $n \times 1$ vector $[y_i'],$[1] our system then becomes

$$
\boxed{Y' = A(x)Y + G(x).}
$$

The homogeneous system of linear differential equations takes on the form

$$
\boxed{Y' = A(x)Y.}
$$

As mentioned in the introduction to this chapter, the theory of first order systems of linear differential equations parallels the theory of linear differential equations we learned in Chapter 4. Corresponding to Theorem 4.1, we have the following uniqueness and existence theorem.

THEOREM 6.1   If $a_{i,j}(x)$ and $g_i(x)$ are continuous on the interval (a, b) containing x_0 for $1 \leq i \leq n$ and $1 \leq j \leq n$, then the initial value problem

$$
Y' = A(x)Y + G(x); \qquad Y(x_0) = \begin{bmatrix} b_1 \\ b_2 \\ \vdots \\ b_n \end{bmatrix}
$$

has a unique solution on (a, b).

Recall from Theorem 4.2 and its proof that the solutions to an nth order homogeneous linear differential equation form an n-dimensional subspace of the vector space $C^n(a, b)$. The solutions to a homogeneous system of n linear differential equations $Y' = AY$ form a subset of the set of all $n \times 1$ column vector functions of the form

$$
\begin{bmatrix} f_1(x) \\ f_2(x) \\ \vdots \\ f_n(x) \end{bmatrix}
$$

---

[1] The derivative of a matrix $M(x) = [f_{ij}(x)]$ is the matrix

$$
M'(x) = \lim_{\Delta x \to 0} \frac{(M(x + \Delta x) - M(x))}{\Delta x} = \left[\lim_{\Delta x \to 0} \frac{(f_{ij}(x + \Delta x) - f_{ij}(x))}{\Delta x} \right] = [f_{ij}'(x)].
$$

Many properties of derivatives of functions extend to matrices. (See Exercise 23.)

where each f_i is in $C^1(a, b)$. We have an addition and scalar multiplication on these column functions (add and multiply them by scalars entrywise just as we do with column vectors) making the set of such column vector functions into a vector space. Viewing the solutions to $Y' = AY$ as a subset of the vector space of these column vector functions, we can proceed along the lines of the proof of Theorem 4.2 (see Exercise 18) to obtain Theorem 6.2.

THEOREM 6.2   The solutions to a homogeneous system of n first order linear differential equations $Y' = A(x)Y$ form a vector space of dimension n.

As we did in Chapter 4, we will call a set of n linearly independent solutions $Y_1, Y_2, \ldots, Y_n$ to a homogeneous system of n first order linear differential equations a **fundamental set of solutions.** If $Y_1, Y_2, \ldots, Y_n$ form a fundamental set of solutions, they form a basis for the vector space of solutions of the homogeneous system, and the **general solution** to the homogeneous system of n first order linear differential equations is given by

$$Y_H = c_1 Y_1 + c_2 Y_2 + \cdots + c_n Y_n$$

where $c_1, c_2, \ldots, c_n$ are constants.

If we let M be the matrix

$$M = \begin{bmatrix} Y_1 & Y_2 & \cdots & Y_n \end{bmatrix}$$

and C be the vector

$$C = \begin{bmatrix} c_1 \\ c_2 \\ \vdots \\ c_n \end{bmatrix},$$

then the general solution to the homogeneous system of first order linear differential equations can be written as

$$Y_H = MC.$$

We call M a **matrix of fundamental solutions.**

In the nonhomogeneous case we have the following theorem corresponding to Theorem 4.3.

THEOREM 6.3   Suppose that $Y_1, Y_2, \ldots, Y_n$ form a fundamental set of solutions to the homogeneous system of first order linear differential equations

$$Y' = A(x)Y$$

and that Y_P is a solution to the nonhomogeneous system of first order linear differential equations

$$Y' = A(x)Y + G(x).$$

Then every solution to this nonhomogeneous system of first order linear differential equations has the form

$$\boxed{Y = Y_H + Y_P = c_1 Y_1 + \cdots + c_n Y_n + Y_P = MC + Y_P.}$$

The proof of Theorem 6.3 is similar to the proof of Theorem 4.3 and is left as Exercise 19. As in Chapter 4, we call Y_P a **particular solution** to the nonhomogeneous system.

For our last part of the theory of systems of first order linear differential equations, we discuss how the Wronskian is modified to the system setting. Given column vector functions

$$Y_1(x) = \begin{bmatrix} y_{11}(x) \\ y_{21}(x) \\ \vdots \\ y_{n1}(x) \end{bmatrix}, \quad Y_2(x) = \begin{bmatrix} y_{12}(x) \\ y_{22}(x) \\ \vdots \\ y_{n2}(x) \end{bmatrix}, \quad \ldots, \quad Y_n(x) = \begin{bmatrix} y_{1n}(x) \\ y_{2n}(x) \\ \vdots \\ y_{nn}(x) \end{bmatrix},$$

we define the **Wronskian** of $Y_1(x), Y_2(x), \ldots, Y_n(x)$, denoted

$$w(Y_1(x), Y_2(x), \ldots, Y_n(x)),$$

to be the determinant

$$w(Y_1(x), Y_2(x), \ldots, Y_n(x)) = \begin{vmatrix} y_{11}(x) & y_{12}(x) & \cdots & y_{1n}(x) \\ y_{21}(x) & y_{22}(x) & \cdots & y_{2n}(x) \\ \vdots & \vdots & & \vdots \\ y_{n1}(x) & y_{n2}(x) & \cdots & y_{nn}(x) \end{vmatrix}.$$

In Theorem 2.15 of Chapter 2 we saw that if a Wronskian of a set of n functions was nonzero on an interval, then the functions were linearly independent on that interval. The same holds in this setting. Just so you do not get upset with us leaving all the proofs to you, we include this one.

THEOREM 6.4   If $w(Y_1(x), Y_2(x), \ldots, Y_n(x)) \neq 0$ for some x in (a, b), then $Y_1(x), Y_2(x), \ldots, Y_n(x)$ are linearly independent on (a, b).

Proof   Suppose that

$$c_1 Y_1(x) + c_2 Y_2(x) + \cdots + c_n Y_n(x) = 0.$$

Let x_0 be a value in (a, b) for which $w(Y_1(x_0), Y_2(x_0), \ldots, Y_n(x_0)) \neq 0$. The system

$$y_{11}(x_0)c_1 + y_{12}(x_0)c_2 + \cdots + y_{1n}(x_0)c_n = 0$$
$$y_{21}(x_0)c_1 + y_{22}(x_0)c_2 + \cdots + y_{2n}(x_0)c_n = 0$$
$$\vdots$$
$$y_{n1}(x_0)c_1 + y_{n2}(x_0)c_2 + \cdots + y_{nn}(x_0)c_n = 0$$

then has only the trivial solution $c_1 = 0, c_2 = 0, \ldots, c_n = 0$, completing the proof. ●

Recall that we had a converse to Theorem 2.15, provided the functions were solutions to a homogeneous linear differential equation in Theorem 4.4. Here too we get a converse, provided the functions are solutions of a homogeneous first order linear system.

THEOREM 6.5   If $Y_1, Y_2, \ldots, Y_n$ are solutions to a homogeneous first order linear system $Y' = A(x)Y$ on an interval (a, b) such that $w(Y_1(x_0), Y_2(x_0), \ldots, Y_n(x_0)) = 0$ for some x_0 in (a, b), then $Y_1, Y_2, \ldots, Y_n$ are linearly dependent.

See Exercises 20 and 21 for modifying the proof of Theorem 4.4 to obtain Theorem 6.5. An immediate consequence of Theorem 4.4 was that the Wronskian of a fundamental set of solutions to a homogeneous linear differential equation is never zero. Here we obtain the same result.

COROLLARY 6.6   If $Y_1, Y_2, \ldots, Y_n$ form a fundamental set of solutions to a homogeneous linear system $Y' = A(x)Y$ on an interval (a, b), then $w(Y_1(x), Y_2(x), \ldots, Y_n(x)) \neq 0$ for all x in (a, b).

We now turn our attention to solving systems of linear differential equations. In Chapter 4, we first looked at homogeneous linear differential equations with constant coefficients. We take the same approach here; that is, we consider systems of the form

$$Y' = AY$$

where each entry of A is a constant. One case where it is especially easy to solve such a system is when A is a diagonal matrix. To illustrate, consider the system

$$\begin{bmatrix} y_1' \\ y_2' \end{bmatrix} = \begin{bmatrix} 3 & 0 \\ 0 & -1 \end{bmatrix} \begin{bmatrix} y_1 \\ y_2 \end{bmatrix}.$$

Considering each equation

$$y_1' = 3y_1$$

and

$$y_2' = -y_2$$

of the system, using the methods of either Chapter 3 or 4 we determine that the general solutions to these two equations are

$$y_1 = c_1 e^{3x}$$

and

$$y_2 = c_2 e^{-x}.$$

It then follows that

$$Y = \begin{bmatrix} y_1 \\ y_2 \end{bmatrix} = \begin{bmatrix} c_1 e^{3x} \\ c_2 e^{-x} \end{bmatrix} = c_1 \begin{bmatrix} e^{3x} \\ 0 \end{bmatrix} + c_2 \begin{bmatrix} 0 \\ e^{-x} \end{bmatrix} = \begin{bmatrix} e^{3x} & 0 \\ 0 & e^{-x} \end{bmatrix} \begin{bmatrix} c_1 \\ c_2 \end{bmatrix}$$

is a solution to this system of linear differential equations. If we let

$$Y_1 = \begin{bmatrix} e^{3x} \\ 0 \end{bmatrix}$$

and

$$Y_2 = \begin{bmatrix} 0 \\ e^{-x} \end{bmatrix},$$

Y_1 and Y_2 form a fundamental set of solutions to this system of linear differential equations. The corresponding matrix of fundamental solutions is

$$M = \begin{bmatrix} Y_1 & Y_2 \end{bmatrix} = \begin{bmatrix} e^{3x} & 0 \\ 0 & e^{-x} \end{bmatrix}.$$

In general, if A is the diagonal matrix

$$A = \begin{bmatrix} d_1 & 0 & 0 & \cdots & 0 \\ 0 & d_2 & 0 & \cdots & 0 \\ \vdots & \vdots & \vdots & \vdots & \vdots \\ \vdots & \vdots & \vdots & \vdots & \vdots \\ 0 & 0 & \cdots & 0 & d_n \end{bmatrix},$$

the matrix

$$M = \begin{bmatrix} e^{d_1 x} & 0 & 0 & \cdots & 0 \\ 0 & e^{d_2 x} & 0 & \cdots & 0 \\ \vdots & \vdots & \vdots & \vdots & \vdots \\ \vdots & \vdots & \vdots & \vdots & \vdots \\ 0 & 0 & \cdots & 0 & e^{d_n x} \end{bmatrix}$$

is a matrix of fundamental solutions to $Y' = AY$. We let you verify this fact in Exercise 17.

In the next section we will consider the case in which A is diagonalizable. We will make use of our ability to solve diagonal systems along with the technique for diagonalizing matrices we learned in Chapter 5 to solve $Y' = AY$ in this case.

EXERCISES 6.1

1. Show that

$$Y = \begin{bmatrix} c_1 e^{2x} + c_2 e^{3x} \\ 2c_1 e^{2x} + c_2 e^{3x} \end{bmatrix}$$

is a solution of

$$Y' = \begin{bmatrix} 4 & -1 \\ 2 & 1 \end{bmatrix} Y.$$

2. Show that

$$Y = \begin{bmatrix} c_1 e^{2x} + c_2 e^{3x} - \frac{x}{6} - \frac{11}{36} \\ 2c_1 e^{2x} + c_2 e^{3x} - \frac{2x}{3} - \frac{1}{18} \end{bmatrix}$$

is a solution of

$$Y' = \begin{bmatrix} 4 & -1 \\ 2 & 1 \end{bmatrix} Y + \begin{bmatrix} 1 \\ x \end{bmatrix}.$$

In Exercises 3 and 4, show that the given columns of functions are linearly independent on the interval $(-\infty, \infty)$.

3. $\begin{bmatrix} e^{2x} \cos 3x \\ -e^{2x} \sin 3x \end{bmatrix}, \begin{bmatrix} e^{2x} \sin 3x \\ e^{2x} \cos 3x \end{bmatrix}$

4. $\begin{bmatrix} e^{-2x} \\ 0 \\ 0 \end{bmatrix}, \begin{bmatrix} 0 \\ 3\cos 5x \\ -3\sin 5x \end{bmatrix}, \begin{bmatrix} 0 \\ \sin 5x \\ \cos 5x \end{bmatrix}$

In Exercises 5–8, determine the general solution to $Y' = AY$ for the given matrix A. Also give a matrix of fundamental solutions.

5. $A = \begin{bmatrix} 1 & 0 \\ 0 & -2 \end{bmatrix}$ **6.** $A = \begin{bmatrix} 0 & 0 \\ 0 & 3 \end{bmatrix}$

7. $A = \begin{bmatrix} -1 & 0 & 0 \\ 0 & 0 & 0 \\ 0 & 0 & 4 \end{bmatrix}$

8. $A = \begin{bmatrix} -3 & 0 & 0 & 0 \\ 0 & -2 & 0 & 0 \\ 0 & 0 & 2 & 0 \\ 0 & 0 & 0 & 5 \end{bmatrix}$

In Exercises 9–12, determine the solution to the initial value problem $Y' = AY$, $Y(0) = Y_0$ for the system of the indicated exercise of this section and the given initial condition $Y(0)$.

9. Exercise 5; $Y(0) = \begin{bmatrix} 2 \\ 1 \end{bmatrix}$

10. Exercise 6; $Y(0) = \begin{bmatrix} 1 \\ -1 \end{bmatrix}$

11. Exercise 7; $Y(0) = \begin{bmatrix} 2 \\ 1 \\ 0 \end{bmatrix}$

12. Exercise 8; $Y(0) = \begin{bmatrix} 2 \\ 1 \\ -1 \\ 0 \end{bmatrix}$

In Exercises 13–16, determine the general solution to the nonhomogeneous equation $Y' = AY + G(x)$ by individually solving each equation of the system. Compare your solution in each of these exercises with your solution to each of Exercises 5–8, respectively, and observe how these solutions illustrate Theorem 6.3.

13. $Y' = \begin{bmatrix} 1 & 0 \\ 0 & -2 \end{bmatrix} Y + \begin{bmatrix} 2 \\ x \end{bmatrix}$

14. $Y' = \begin{bmatrix} 0 & 0 \\ 0 & 3 \end{bmatrix} Y + \begin{bmatrix} \sin x \\ e^{2x} \end{bmatrix}$

15. $Y' = \begin{bmatrix} -1 & 0 & 0 \\ 0 & 0 & 0 \\ 0 & 0 & 4 \end{bmatrix} Y + \begin{bmatrix} 1 - 2x \\ xe^{-x} \\ \cos 2x \end{bmatrix}$

16. $Y' = \begin{bmatrix} -3 & 0 & 0 & 0 \\ 0 & -2 & 0 & 0 \\ 0 & 0 & 2 & 0 \\ 0 & 0 & 0 & 5 \end{bmatrix} Y + \begin{bmatrix} e^{3x} \\ 0 \\ 1 - 2x^2 \\ \sin 3x \end{bmatrix}$

17. Show that a matrix of fundamental solutions to

$$Y' = \begin{bmatrix} d_1 & 0 & 0 & \cdots & 0 \\ 0 & d_2 & 0 & \cdots & 0 \\ \vdots & \vdots & \vdots & \vdots & \vdots \\ \vdots & \vdots & \vdots & \vdots & \vdots \\ 0 & 0 & \cdots & 0 & d_n \end{bmatrix} Y$$

is

$$M = \begin{bmatrix} e^{d_1 x} & 0 & 0 & \cdots & 0 \\ 0 & e^{d_2 x} & 0 & \cdots & 0 \\ \vdots & \vdots & \vdots & \vdots & \vdots \\ \vdots & \vdots & \vdots & \vdots & \vdots \\ 0 & 0 & \cdots & 0 & e^{d_n x} \end{bmatrix}.$$

18. Prove Theorem 6.2.   **19.** Prove Theorem 6.3.

20. Modify the proof of Theorem 4.4 in the second order case given in Section 4.1 to obtain a proof of Theorem 6.5 for a system of two linear differential equations.

21. Prove Theorem 6.5 for a system of n linear differential equations.

22. Let V be the set of all column functions

$$Y = \begin{bmatrix} y_1 \\ \vdots \\ y_n \end{bmatrix}$$

where $y_1, \ldots, y_n$ are in $F(a, b)$.

a) Prove that V is a vector space.

b) Let W be the set of all column functions

$$Y = \begin{bmatrix} y_1 \\ \vdots \\ y_n \end{bmatrix}$$

where $y_1, \ldots, y_n$ are in $C^\infty(a, b)$. Prove that W is a subspace of V.

c) Define the function $L : W \to W$ by

$$L(Y) = Y' - A(x)Y$$

where each entry of $A(x)$ is in $C^\infty(a, b)$. Show that L is a linear transformation and that $\ker(L)$ is the set of solutions to the homogeneous system of linear differential equations

$$Y' = A(x)Y.$$

23. Assuming that the indicated operations are defined, prove the following where $A(x)$, $B(x)$, and C are matrices, k a constant, and all the entries of C are constants.

a) $(A(x) \pm B(x))' = A'(x) \pm B'(x)$

b) $(kA(x))' = kA'(x)$

c) $(CA(x))' = CA'(x)$

d) $(A(x)B(x))' = A(x)B'(x) + A'(x)B(x)$

For

$$A(x) = \begin{bmatrix} 4x & -e^x \\ 2e^{-x} & x \end{bmatrix}, \qquad B(x) = \begin{bmatrix} e^{2x} & -x \\ 2x & e^{3x} \end{bmatrix},$$

$$C = \begin{bmatrix} 4 & -1 \\ 2 & 1 \end{bmatrix},$$

find the following.

24. $(A + B)'$ **25.** $(3A)'$

26. $(CA)'$ **27.** $(AB)'$

28. Use Maple or another appropriate software package to obtain a phase portrait of the solution to the initial value problem in Exercise 9.

29. Use Maple or another appropriate software package to obtain a phase portrait of the solution to the initial value problem consisting of the system of differential equations in Exercise 14 with initial conditions $y_1(0) = 1$, $y_2(0) = 2$.

6.2 HOMOGENEOUS SYSTEMS WITH CONSTANT COEFFICIENTS: THE DIAGONALIZABLE CASE

In Section 6.1 we saw how to solve the homogeneous system of linear differential equations

$$Y' = AY$$

for a diagonal matrix A. In Chapter 5 we saw that some matrices are similar to a diagonal matrix. Theorem 6.7 shows how solutions to linear systems with constant coefficient similar matrices are related. We will use this theorem later in this section to solve $Y' = AY$ when A is similar to a diagonal matrix. In the next section we will see how to use it to solve such systems when A is not similar to a diagonal matrix.

THEOREM 6.7   Suppose that A and B are similar $n \times n$ matrices with $B = P^{-1}AP$ where P is an invertible $n \times n$ matrix. If Z is a solution of $Y' = BY$, then PZ is a solution of $Y' = AY$. Moreover, if $Z_1, Z_2, \ldots, Z_n$ forms a fundamental set of solutions to $Y' = BY$, then $PZ_1, PZ_2, \ldots, PZ_n$ forms a fundamental set of solutions to $Y' = AY$.

Proof   Since Z is a solution to $Y' = BY$, we have

$$Z' = BZ.$$

Multiplying this equation on the left by P, we get

$$PZ' = PBZ.$$

Since $PZ' = (PZ)'$ and $PB = AP$,

$$(PZ)' = APZ$$

and hence PZ is a solution of $Y' = AY$.

Now suppose $Z_1, Z_2, \ldots, Z_n$ forms a fundamental set of solutions to $Y' = BY$. By the first part of this theorem, we have that $PZ_1, PZ_2, \ldots, PZ_n$ are solutions to $Y' = AY$. To complete this proof, we must show $PZ_1, PZ_2, \ldots, PZ_n$ are linearly independent. Considering their Wronskian, we have

$$
\begin{aligned}
w(PZ_1, PZ_2, \ldots, PZ_n) &= \det([\; PZ_1 \quad PZ_2 \quad \cdots \quad PZ_n \;]) \\
&= \det(P [\; Z_1 \quad Z_2 \quad \cdots \quad Z_n \;]) \\
&= \det(P)w(Z_1, Z_2, \ldots, Z_n) \neq 0
\end{aligned}
$$

since $\det(P) \neq 0$ and $w(Z_1, Z_2, \ldots, Z_n) \neq 0$. Thus we have that $PZ_1, PZ_2, \ldots, PZ_n$ are linearly independent.   ●

Another way to state the last part of Theorem 6.7 is to say if

$$M_B = [\; Z_1 \quad Z_2 \quad \cdots \quad Z_n \;]$$

is a matrix of fundamental solutions for $Y' = BY$, then

$$M_A = PM_B = [\; PZ_1 \quad PZ_2 \quad \cdots \quad PZ_n \;]$$

is a matrix of fundamental solutions to $Y' = AY$. Notice further that if the general solution to $Y' = BY$ is expressed as

$$c_1 Z_1 + \cdots + c_n Z_n,$$

then the general solution to $Y' = AY$ can be expressed as

$$c_1 PZ_1 + \cdots + c_n PZ_n = P(c_1 Z_1 + \cdots + c_n Z_n).$$

Let us now focus our attention on a constant coefficient homogeneous system of linear differential equations

$$Y' = AY,$$

when the $n \times n$ matrix A is diagonalizable. Suppose P is an $n \times n$ invertible matrix so that

$$D = P^{-1}AP$$

is the diagonal matrix

$$D = \begin{bmatrix} d_1 & 0 & 0 & \cdots & 0 \\ 0 & d_2 & 0 & \cdots & 0 \\ \vdots & \vdots & \vdots & \vdots & \vdots \\ \vdots & \vdots & \vdots & \vdots & \vdots \\ 0 & 0 & \cdots & 0 & d_n \end{bmatrix}$$

where each of $d_1, d_2, \ldots, d_n$ is a real number. (We will consider the case where all or some of $d_1, d_2, \ldots, d_n$ are imaginary numbers later in this section.) We know from the last section that

$$\begin{bmatrix} c_1 e^{d_1 x} \\ c_2 e^{d_2 x} \\ \vdots \\ c_n e^{d_n x} \end{bmatrix}$$

is the general solution to the system

$$Y' = DY.$$

It follows from Theorem 6.7 that the general solution to $Y' = AY$ is then

$$P \begin{bmatrix} c_1 e^{d_1 x} \\ c_2 e^{d_2 x} \\ \vdots \\ c_n e^{d_n x} \end{bmatrix}.$$

The next example illustrates how we apply this in practice.

EXAMPLE 1   Determine the general solution to $Y' = AY$ for

$$A = \begin{bmatrix} 1 & -3 \\ -2 & 2 \end{bmatrix}.$$

Solution   The matrix A is the matrix of Examples 1 and 2 of Section 5.4 and Example 1 of Section 5.5. In these examples we found that A has eigenvalues $\lambda = 4$ and $\lambda = -1$ (Example 1 of Section 5.4), E_4 has

$$\begin{bmatrix} -1 \\ 1 \end{bmatrix}$$

as a basis vector, E_{-1} has

$$\begin{bmatrix} 3 \\ 2 \end{bmatrix}$$

as a basis vector (Example 2 of Section 5.4), and A is similar to the diagonal matrix

$$D = P^{-1}AP = \begin{bmatrix} 4 & 0 \\ 0 & -1 \end{bmatrix}$$

where

$$P = \begin{bmatrix} -1 & 3 \\ 1 & 2 \end{bmatrix}$$

(Example 1 of Section 5.5). We know that

$$\begin{bmatrix} c_1 e^{4x} \\ c_2 e^{-x} \end{bmatrix}$$

is the general solution to $Y' = DY$. Therefore, the general solution to $Y' = AY$ is

$$P \begin{bmatrix} c_1 e^{4x} \\ c_2 e^{-x} \end{bmatrix} = \begin{bmatrix} -1 & 3 \\ 1 & 2 \end{bmatrix} \begin{bmatrix} c_1 e^{4x} \\ c_2 e^{-x} \end{bmatrix} = \begin{bmatrix} -c_1 e^{4x} + 3c_2 e^{-x} \\ c_1 e^{4x} + 2c_2 e^{-x} \end{bmatrix}. \qquad \bullet$$

In the next example we solve an initial value problem.

EXAMPLE 2   Solve the initial value problem

$$Y' = \begin{bmatrix} 2 & -3 & -3 \\ 2 & -2 & -2 \\ -2 & 1 & 1 \end{bmatrix} Y, \qquad Y(0) = \begin{bmatrix} 1 \\ 0 \\ 0 \end{bmatrix}.$$

Solution   We leave it to you to show that the eigenvalues of

$$A = \begin{bmatrix} 2 & -3 & -3 \\ 2 & -2 & -2 \\ -2 & 1 & 1 \end{bmatrix}$$

are $-1, 0$, and 2 and the vectors

$$\begin{bmatrix} 1 \\ 0 \\ 1 \end{bmatrix}, \begin{bmatrix} 0 \\ 1 \\ -1 \end{bmatrix}, \quad \text{and} \quad \begin{bmatrix} -1 \\ -1 \\ 1 \end{bmatrix}$$

are bases for the eigenspaces E_{-1}, E_0, and E_2, respectively. Letting

$$P = \begin{bmatrix} 1 & 0 & -1 \\ 0 & 1 & -1 \\ 1 & -1 & 1 \end{bmatrix}$$

gives us

$$D = P^{-1}AP = \begin{bmatrix} -1 & 0 & 0 \\ 0 & 0 & 0 \\ 0 & 0 & 2 \end{bmatrix}.$$

The general solution to $Y' = AY$ is then

$$Y = P \begin{bmatrix} c_1 e^{-x} \\ c_2 \\ c_3 e^{2x} \end{bmatrix} = \begin{bmatrix} 1 & 0 & -1 \\ 0 & 1 & -1 \\ 1 & -1 & 1 \end{bmatrix} \begin{bmatrix} c_1 e^{-x} \\ c_2 \\ c_3 e^{2x} \end{bmatrix} = \begin{bmatrix} c_1 e^{-x} - c_3 e^{2x} \\ c_2 - c_3 e^{2x} \\ c_1 e^{-x} - c_2 + c_3 e^{2x} \end{bmatrix}.$$

Using the initial conditions,

$$Y(0) = \begin{bmatrix} c_1 - c_3 \\ c_2 - c_3 \\ c_1 - c_2 + c_3 \end{bmatrix} = \begin{bmatrix} 1 \\ 0 \\ 0 \end{bmatrix}.$$

Solving, we find that $c_1 = 0, c_2 = -1, c_3 = -1$. The solution to the initial value problem is then

$$Y = \begin{bmatrix} e^{2x} \\ -1 + e^{2x} \\ 1 - e^{2x} \end{bmatrix}. \qquad \bullet$$

Up to this point, all of our matrices A have had real eigenvalues. If A has imaginary eigenvalues, we proceed in a manner similar to the method we used in Section 4.2 when the characteristic equation of a homogeneous linear differential equation with constant coefficients had imaginary roots. Corresponding to Theorem 4.9, which told us that if $w(x) = u(x) + iv(x)$ is a complex-valued solution to a homogeneous linear differential equation, then $u(x)$ and $v(x)$ are real-valued solutions, we have the following theorem.

THEOREM 6.8   If $U(x) + iV(x)$ is a solution to $Y' = A(x)Y$, then $U(x)$ and $V(x)$ are solutions to $Y' = A(x)Y$.

The proof of Theorem 6.8 is straightforward and is left as an exercise (Exercise 29).

Recall how we used the identity

$$e^{(a+bi)x} = e^{ax}\cos bx + ie^{ax}\sin bx$$

along with Theorem 4.9 to obtain the solutions $e^{ax}\cos bx$ and $e^{ax}\sin bx$ to a constant coefficient homogeneous linear differential equation whose characteristic equation has $a + bi$ as a root. The next example illustrates how we again use this identity along with Theorem 6.8 to solve a system $Y' = AY$ when A has imaginary eigenvalues.

EXAMPLE 3   Find the general solution to $Y' = AY$ for

$$A = \begin{bmatrix} 1 & -1 \\ 1 & 1 \end{bmatrix}.$$

Solution   Here A is the matrix in Example 5 of Section 5.4 and Example 4 of Section 5.5. The eigenvalues of A are $\lambda = 1 + i$ and $\lambda = 1 - i$. The vector

$$\begin{bmatrix} i \\ 1 \end{bmatrix}$$

forms a basis for E_{1+i}, and the vector

$$\begin{bmatrix} -i \\ 1 \end{bmatrix}$$

forms a basis for E_{1-i} (Example 5 of Section 5.4). The matrix A is similar to the diagonal matrix

$$D = P^{-1}AP = \begin{bmatrix} 1+i & 0 \\ 0 & 1-i \end{bmatrix}$$

where

$$P = \begin{bmatrix} i & -i \\ 1 & 1 \end{bmatrix}$$

(Example 4 of Section 5.5). One complex-valued solution to $Y' = DY$ is

$$Z = \begin{bmatrix} e^{(1+i)x} \\ 0 \end{bmatrix} = \begin{bmatrix} e^x\cos x + ie^x\sin x \\ 0 \end{bmatrix}.$$

(We get another solution

$$\begin{bmatrix} 0 \\ e^{(1-i)x} \end{bmatrix}$$

to $Y' = DY$ from the conjugate eigenvalue $\lambda = 1 - i$. However, as occurred in Chapter 4, we will not have to use this conjugate to obtain a fundamental set of real-valued solutions to $Y' = AY$.) By the first part of Theorem 6.7,

$$PZ = \begin{bmatrix} i & -i \\ 1 & 1 \end{bmatrix} \begin{bmatrix} e^x \cos x + ie^x \sin x \\ 0 \end{bmatrix} = \begin{bmatrix} -e^x \sin x + ie^x \cos x \\ e^x \cos x + ie^x \sin x \end{bmatrix}$$

$$= \begin{bmatrix} -e^x \sin x \\ e^x \cos x \end{bmatrix} + i \begin{bmatrix} e^x \cos x \\ e^x \sin x \end{bmatrix}$$

is a (complex-valued) solution to $Y' = AY$. By Theorem 6.8,

$$\begin{bmatrix} -e^x \sin x \\ e^x \cos x \end{bmatrix}, \begin{bmatrix} e^x \cos x \\ e^x \sin x \end{bmatrix}$$

are real-valued solutions to $Y' = AY$. Calculating their Wronskian (try it), we find these solutions are linearly independent. Hence they form a fundamental set of solutions to $Y' = AY$. The general solution to $Y' = AY$ is then

$$c_1 \begin{bmatrix} -e^x \sin x \\ e^x \cos x \end{bmatrix} + c_2 \begin{bmatrix} e^x \cos x \\ e^x \sin x \end{bmatrix} = \begin{bmatrix} -c_1 e^x \sin x + c_2 e^x \cos x \\ c_1 e^x \cos x + c_2 e^x \sin x \end{bmatrix}. \qquad \bullet$$

Notice that we did not multiply the matrix P (whose columns consist of the eigenvectors) times the general solution to the diagonal system in Example 3 as we did in the case of only real eigenvalues in Examples 1 and 2. Instead, we multiplied P times one of the complex-valued solutions to the diagonal system from which we obtained two real-valued solutions. This process may be continued if we have more eigenvalues some of which are imaginary. To do so, multiply P times each solution to the diagonal system. As we do this for imaginary eigenvalues, however, we can omit the conjugate eigenvalues since each imaginary eigenvalue produces two real-valued solutions. The final example of this section illustrates this procedure.

EXAMPLE 4   Find the general solution of $Y' = AY$ for

$$A = \begin{bmatrix} 1 & 1 & 1 \\ 0 & 1 & -1 \\ 0 & 1 & 1 \end{bmatrix}.$$

Solution   The characteristic equation of A is

$$\det(\lambda I - A) = \begin{vmatrix} \lambda - 1 & -1 & -1 \\ 0 & \lambda - 1 & 1 \\ 0 & -1 & \lambda - 1 \end{vmatrix}$$

$$= (\lambda - 1)((\lambda - 1)(\lambda - 1) + 1) = (\lambda - 1)(\lambda^2 - 2\lambda + 2) = 0$$

from which we find the eigenvalues are

$$\lambda = 1, \qquad \lambda = 1 \pm i.$$

When $\lambda = 1$, the augmented matrix for the homogeneous system $(\lambda I - A)X = 0$ is

$$\left[\begin{array}{ccc|c} 0 & -1 & -1 & 0 \\ 0 & 0 & 1 & 0 \\ 0 & -1 & 0 & 0 \end{array}\right].$$

The solutions to this system have the form

$$\begin{bmatrix} x \\ 0 \\ 0 \end{bmatrix}$$

from which we see

$$\begin{bmatrix} 1 \\ 0 \\ 0 \end{bmatrix}$$

forms a basis for E_1. When $\lambda = 1 + i$, the augmented matrix for the homogeneous system $(\lambda I - A)X = 0$ is

$$\left[\begin{array}{ccc|c} i & -1 & -1 & 0 \\ 0 & i & 1 & 0 \\ 0 & -1 & i & 0 \end{array}\right].$$

Solving this system (try it) leads us to

$$\begin{bmatrix} 1 - i \\ i \\ 1 \end{bmatrix}$$

as a basis vector for E_{1+i}. It now follows that

$$\begin{bmatrix} 1 + i \\ -i \\ 1 \end{bmatrix}$$

is a basis vector for E_{1-i}. Thus if we set

$$P = \begin{bmatrix} 1 & 1 - i & 1 + i \\ 0 & i & -i \\ 0 & 1 & 1 \end{bmatrix},$$

$$D = P^{-1}AP = \begin{bmatrix} 1 & 0 & 0 \\ 0 & 1 + i & 0 \\ 0 & 0 & 1 - i \end{bmatrix}.$$

From the diagonal entry 1, we obtain one solution to $Y' = DY$:

$$\begin{bmatrix} e^x \\ 0 \\ 0 \end{bmatrix}.$$

This gives us the solution

$$P \begin{bmatrix} e^x \\ 0 \\ 0 \end{bmatrix} = \begin{bmatrix} 1 & 1-i & 1+i \\ 0 & i & -i \\ 0 & 1 & 1 \end{bmatrix} \begin{bmatrix} e^x \\ 0 \\ 0 \end{bmatrix} = \begin{bmatrix} e^x \\ 0 \\ 0 \end{bmatrix} \quad (1)$$

to $Y' = AY$. From the diagonal entry $1 + i$, we obtain the complex-valued solution

$$\begin{bmatrix} 0 \\ e^{(1+i)x} \\ 0 \end{bmatrix} = \begin{bmatrix} 0 \\ e^x \cos x + i e^x \sin x \\ 0 \end{bmatrix}$$

to $Y' = DY$. This gives us the complex-valued solution

$$P \begin{bmatrix} 0 \\ e^x \cos x + i e^x \sin x \\ 0 \end{bmatrix} = \begin{bmatrix} 1 & 1-i & 1+i \\ 0 & i & -i \\ 0 & 1 & 1 \end{bmatrix} \begin{bmatrix} 0 \\ e^x \cos x + i e^x \sin x \\ 0 \end{bmatrix}$$

$$= \begin{bmatrix} e^x \cos x + e^x \sin x + i(e^x \sin x - e^x \cos x) \\ -e^x \sin x + i e^x \cos x \\ e^x \cos x + i e^x \sin x \end{bmatrix}$$

from which we obtain the two real-valued solutions

$$\begin{bmatrix} e^x \cos x + e^x \sin x \\ -e^x \sin x \\ e^x \cos x \end{bmatrix}, \quad \begin{bmatrix} e^x \sin x - e^x \cos x \\ e^x \cos x \\ e^x \sin x \end{bmatrix} \quad (2)$$

to $Y' = AY$. The solutions in (1) and (2) are linearly independent (verify this) and hence the general solution to $Y' = AY$ is

$$c_1 \begin{bmatrix} e^x \\ 0 \\ 0 \end{bmatrix} + c_2 \begin{bmatrix} e^x \cos x + e^x \sin x \\ -e^x \sin x \\ e^x \cos x \end{bmatrix} + c_3 \begin{bmatrix} e^x \sin x - e^x \cos x \\ e^x \cos x \\ e^x \sin x \end{bmatrix}$$

$$= \begin{bmatrix} c_1 e^x + c_2(e^x \cos x + e^x \sin x) + c_3(e^x \sin x - e^x \cos x) \\ -c_2 e^x \sin x + c_3 e^x \cos x \\ c_2 e^x \cos x + c_3 e^x \sin x \end{bmatrix}. \quad \bullet$$

EXERCISES 6.2

In Exercises 1–12, determine the general solution to $Y' = AY$ where A is the matrix in the indicated exercise of Section 5.4 (see page 277).

1. Exercise 1	**2.** Exercise 4
3. Exercise 5	**4.** Exercise 6
5. Exercise 7	**6.** Exercise 10
7. Exercise 13	**8.** Exercise 14
9. Exercise 15	**10.** Exercise 16
11. Exercise 17	**12.** Exercise 18

In Exercises 13–18, determine the solution to the initial value problem $Y' = AY$, $Y(0) = Y_0$ for the given column vector Y_0 and the system in the indicated exercise of this section.

13. Exercise 1; $Y(0) = \begin{bmatrix} 2 \\ 1 \end{bmatrix}$

14. Exercise 2; $Y(0) = \begin{bmatrix} 0 \\ -1 \end{bmatrix}$

15. Exercise 5; $Y(0) = \begin{bmatrix} 0 \\ 1 \\ -1 \end{bmatrix}$

16. Exercise 8; $Y(0) = \begin{bmatrix} 0 \\ 1 \\ -1 \end{bmatrix}$

17. Exercise 9; $Y(0) = \begin{bmatrix} 2 \\ 1 \end{bmatrix}$

18. Exercise 12; $Y(0) = \begin{bmatrix} 1 \\ 0 \\ -1 \end{bmatrix}$

In Exercises 19–26, find the general solution to the given system of linear differential equations.

19. $Y' = \begin{bmatrix} 8 & -10 \\ 5 & -7 \end{bmatrix} Y$ **20.** $Y' = \begin{bmatrix} 4 & 2 \\ -1 & 7 \end{bmatrix} Y$

21. $\begin{aligned} y_1' &= 3y_1 + y_2 \\ y_2' &= 9y_1 + y_2 \end{aligned}$

22. $\begin{aligned} \frac{dy_1}{dt} &= 4y_1 + 3y_2 \\ \frac{dy_2}{dt} &= 3y_1 - 4y_2 \end{aligned}$

23. $\begin{aligned} \frac{dy_1}{dt} &= 5y_1 - 3y_2 + 6y_3 \\ \frac{dy_2}{dt} &= 6y_1 - 4y_2 + 24y_3 \\ \frac{dy_3}{dt} &= 5y_3 \end{aligned}$

24. $Y' = \begin{bmatrix} 1 & 3 & -3 \\ 0 & 1 & 0 \\ 6 & 3 & -8 \end{bmatrix} Y$

25. $Y' = \begin{bmatrix} 1 & -3 \\ 3 & 1 \end{bmatrix} Y$ **26.** $\begin{aligned} y_1' &= 2y_1 - 5y_2 \\ y_2' &= 2y_1 - 4y_2 \\ y_3' &= 3y_3 \end{aligned}$

27. Solve the initial value problem in the example given in the introduction to this chapter.

28. An object is moving in 3-space with velocity vector

$$v = \begin{bmatrix} 2x + y \\ 3x + 4y \\ 5x - 6y + 3z \end{bmatrix}.$$

Find the position of the object at time t if the initial position of the object is the point $(5, 3, 4)$.

29. Prove Theorem 6.8.

30. Let A be a diagonalizable matrix. Determine the conditions on the eigenvalues of A so that if

$$Y(x) = \begin{bmatrix} y_1(x) \\ y_2(x) \\ \vdots \\ y_n(x) \end{bmatrix}$$

is a solution to $Y' = AY$, then $\lim_{x \to \infty} y_i(x) = 0$ for each $i = 1, \ldots, n$.

6.3   HOMOGENEOUS SYSTEMS WITH CONSTANT COEFFICIENTS: THE NONDIAGONALIZABLE CASE

We know from Chapter 5 that not all matrices are diagonalizable. Consequently, we need to develop additional methods in order to be able to solve all homogeneous constant coefficient linear systems. In preparation for this, we first consider systems of the form $Y' = AY$ where A is an upper triangular matrix. These are easily solved by backtracking through the system individually solving each equation in the system. The following example illustrates this.

EXAMPLE 1   Determine the general solution to the following system of differential equations

$$Y' = \begin{bmatrix} 3 & 1 & -1 \\ 0 & 2 & 1 \\ 0 & 0 & 2 \end{bmatrix} Y = AY.$$

Solution   Writing our system with individual equations, our system has the form:

$$y_1' = 3y_1 + y_2 - y_3$$
$$y_2' = 2y_2 + y_3$$
$$y_3' = 2y_3.$$

Using the techniques of either Section 3.5 or Chapter 4, the last equation has general solution

$$y_3 = c_3 e^{2x}.$$

Substituting this into the second equation, we have

$$y_2' = 2y_2 + c_3 e^{2x} \quad \text{or} \quad y_2' - 2y_2 = c_3 e^{2x}.$$

Again using the techniques of either Section 3.5 or Chapter 4, we find the second equation has general solution

$$y_2 = c_2 e^{2x} + c_3 x e^{2x}.$$

Now substituting our results for y_2 and y_3 into the first equation gives us

$$y_1' = 3y_1 + c_2 e^{2x} + c_3 x e^{2x} - c_3 e^{2x}$$

or

$$y_1' - 3y_1 = c_2 e^{2x} + c_3 x e^{2x} - c_3 e^{2x}.$$

Finally using the techniques of either Section 3.5 or Chapter 4 one more time, we find

$$y_1 = c_1 e^{3x} - c_2 e^{2x} - c_3 x e^{2x}.$$

Therefore, the general solution is given by

$$Y_H = \begin{bmatrix} c_1 e^{3x} - c_2 e^{2x} - c_3 x e^{2x} \\ c_2 e^{2x} + c_3 x e^{2x} \\ c_3 e^{2x} \end{bmatrix} = \begin{bmatrix} e^{3x} & -e^{2x} & -x e^{2x} \\ 0 & e^{2x} & x e^{2x} \\ 0 & 0 & e^{2x} \end{bmatrix} \begin{bmatrix} c_1 \\ c_2 \\ c_3 \end{bmatrix}.$$

Note that a matrix of fundamental solutions is

$$M = \begin{bmatrix} e^{3x} & -e^{2x} & -x e^{2x} \\ 0 & e^{2x} & x e^{2x} \\ 0 & 0 & e^{2x} \end{bmatrix}. \qquad \bullet$$

Now that we know how to solve linear systems $Y' = AY$ where A is an upper triangular matrix, let us consider solving $Y' = AY$ when A is not upper triangular. If we could find an upper triangular matrix B similar to A and an invertible matrix P so that $B = P^{-1}AP$, we could find the general solution to $Y' = BY$ and then use Theorem 6.7 to obtain the general solution of $Y' = AY$. How could we find such a matrix B? One way is to use the Jordan canonical form of A (which is an upper triangular matrix) for B. This is the approach we take in the following example.

EXAMPLE 2   Find the general solution of $Y' = AY$ for

$$A = \begin{bmatrix} 1 & -2 & -6 \\ -2 & 2 & -5 \\ 2 & 1 & 8 \end{bmatrix}.$$

Solution   The matrix A in this example is the matrix of Example 3 of Section 5.5 where we saw that A is not diagonalizable. At the end of Section 5.5 we used Maple to find a Jordan canonical form of A to be

$$B = \begin{bmatrix} 5 & 0 & 0 \\ 0 & 3 & 1 \\ 0 & 0 & 3 \end{bmatrix}$$

and found a change of basis matrix P so that $P^{-1}AP = B$ to be

$$P = \begin{bmatrix} 1/2 & -2 & -3 \\ 1/2 & -1 & -1/2 \\ -1/2 & 1 & 3/2 \end{bmatrix}.$$

Solving the triangular system $Y' = BY$, which is

$$y_1' = 5y_1$$
$$y_2' = 3y_2 + y_3$$
$$y_3' = 3y_3$$

we find its general solution to be

$$Z = \begin{bmatrix} c_1 e^{5x} \\ c_2 e^{3x} + c_3 x e^{3x} \\ c_3 e^{3x} \end{bmatrix}.$$

It follows from Theorem 6.7 that the general solution to $Y' = AY$ is then

$$Y = PZ = \begin{bmatrix} 1/2 & -2 & -3 \\ 1/2 & -1 & -1/2 \\ -1/2 & 1 & 3/2 \end{bmatrix} \begin{bmatrix} c_1 e^{5x} \\ c_2 e^{3x} + c_3 x e^{3x} \\ c_3 e^{3x} \end{bmatrix}$$

$$= \begin{bmatrix} \frac{1}{2} c_1 e^{5x} - 2 c_2 e^{3x} - 2 c_3 x e^{3x} - 3 c_3 e^{3x} \\ \frac{1}{2} c_1 e^{5x} - c_2 e^{3x} - c_3 x e^{3x} - \frac{1}{2} c_3 e^{3x} \\ -\frac{1}{2} c_1 e^{5x} + c_2 e^{3x} + c_3 x e^{3x} + \frac{3}{2} c_3 e^{3x} \end{bmatrix}. \qquad \bullet$$

Using the Jordan canonical form is just one way to solve these systems of linear differential equations. We will see another way in Section 9.3.

EXERCISES 6.3

In Exercises 1–8, determine the general solution to $Y' = AY$ where A is the matrix in the indicated exercise of Section 5.4 (see pages 277–278) by using Maple or an appropriate software package to find a Jordan canonical form for A and a matrix P so that $P^{-1}AP$ is this Jordan canonical form.

1. Exercise 3

2. Exercise 2

3. Exercise 9

4. Exercise 8

5. Exercise 11

6. Exercise 12

7. Exercise 31

8. Exercise 32

In Exercises 9–12, determine the solution to the initial value problem $Y' = AY$, $Y(0) = Y_0$ for the system $Y' = AY$ in the indicated exercise of this section and the given column vector $Y(0)$.

9. Exercise 1; $Y(0) = \begin{bmatrix} 2 \\ 1 \end{bmatrix}$

10. Exercise 2; $Y(0) = \begin{bmatrix} 0 \\ -1 \end{bmatrix}$

11. Exercise 5; $Y(0) = \begin{bmatrix} 0 \\ 2 \\ -1 \end{bmatrix}$

12. Exercise 6; $Y(0) = \begin{bmatrix} 0 \\ 1 \\ 4 \end{bmatrix}$

In Exercises 13–16, use the *dsolve* command in Maple or the corresponding command in another software package to find the general solution to the system in the indicated exercise of this section and compare this result with your solution to the exercise.

13. Exercise 5

14. Exercise 6

15. Exercise 7

16. Exercise 8

17. An object is moving with velocity vector

$$v = \begin{bmatrix} -5x - y \\ x - 7y \end{bmatrix}.$$

Find the position of the object at time t if the initial position of the object is $(1, 0)$.

18. Let A be a matrix. Determine the conditions on the eigenvalues of A so that if

$$Y(x) = \begin{bmatrix} y_1(x) \\ y_2(x) \\ \vdots \\ y_n(x) \end{bmatrix}$$

is a solution to $Y' = AY$, then $\lim_{x \to \infty} y_i(x) = 0$ for $i = 1, \ldots, n$.

6.4   NONHOMOGENEOUS LINEAR SYSTEMS

In this section we determine a particular solution to the nonhomogeneous system of linear differential equations

$$Y' = A(x)Y + G(x).$$

The technique is similar to the method of variation of parameters presented in Section 4.4. Suppose M is a matrix of fundamental solutions for the homogeneous system of linear differential equations

$$Y' = A(x)Y.$$

We are going to determine a vector of functions V so that

$$Y_P = MV$$

is a particular solution to

$$Y' = A(x)Y + G(x).$$

Differentiating, we have

$$Y_P' = (MV)' = M'V + MV'.$$

Substituting this into the nonhomogeneous system of linear differential equations gives us

$$M'V + MV' = A(x)MV + G(x).$$

Rearranging leads to the equation

$$(M' - A(x)M)V + MV' = G(x). \tag{1}$$

Notice that if

$$M = \begin{bmatrix} Y_1 & Y_2 & \cdots & Y_n \end{bmatrix},$$

then

$$M' = \begin{bmatrix} Y_1' & Y_2' & \cdots & Y_n' \end{bmatrix} = \begin{bmatrix} A(x)Y_1 & A(x)Y_2 & \cdots & A(x)Y_n \end{bmatrix}$$
$$= A(x)\begin{bmatrix} Y_1 & Y_2 & \cdots & Y_n \end{bmatrix} = A(x)M$$

so that

$$M' - A(x)M = 0.$$

Thus Equation (1) reduces to

$$MV' = G(x).$$

Since $\det(M)$ is the Wronskian of a fundamental set of solutions $Y_1, Y_2, \ldots, Y_n$ to $Y' = A(x)Y$ on an interval (a, b), $\det(M) \neq 0$ for each x in (a, b) by Corollary 6.6 and hence M is invertible for each x in (a, b). We can then solve for V' by multiplying by M^{-1} on the left, obtaining

$$V' = M^{-1}G(x).$$

We can now find V by integrating each of the entries in the column vector

$$M^{-1}G(x),$$

which we will indicate by writing

$$V = \int M^{-1}G(x)\,dx.$$

Hence, a particular solution is given by

$$Y_P = MV = M \int M^{-1}G(x)\,dx.$$

Let us summarize what we have just done with the following theorem.

THEOREM 6.9   Suppose that the entries of $A(x)$ and $G(x)$ are continuous on an interval (a, b). Further suppose that $Y_1, Y_2, \ldots, Y_n$ form a fundamental set of solutions of the homogeneous system of linear differential equations

$$Y' = A(x)Y$$

on (a, b). If M is the matrix of fundamental solutions

$$\boxed{M = \begin{bmatrix} Y_1 & Y_2 & \cdots & Y_n \end{bmatrix},}$$

then a particular solution to

$$Y' = A(x)Y + G(x)$$

on (a, b) is given by

$$\boxed{Y_P = M \int M^{-1}G(x)\,dx.}$$

EXAMPLE 1   Determine the general solution to the following system of linear differential equations

$$Y' = \begin{bmatrix} 1 & 2 \\ -1 & 4 \end{bmatrix} Y + \begin{bmatrix} 2 \\ x \end{bmatrix}.$$

Solution   We leave it to you to show that the general solution to the corresponding homogeneous equation is

$$Y_H = \begin{bmatrix} 2c_1e^{2x} + c_2e^{3x} \\ c_1e^{2x} + c_2e^{3x} \end{bmatrix}.$$

A matrix of fundamental solutions is

$$M = \begin{bmatrix} 2e^{2x} & e^{3x} \\ e^{2x} & e^{3x} \end{bmatrix}.$$

We have

$$\begin{bmatrix} 2e^{2x} & e^{3x} & \vdots & 1 & 0 \\ e^{2x} & e^{3x} & \vdots & 0 & 1 \end{bmatrix} \rightarrow \begin{bmatrix} 2e^{2x} & e^{3x} & \vdots & 1 & 0 \\ 0 & e^{3x} & \vdots & -1 & 2 \end{bmatrix}$$

$$\rightarrow \begin{bmatrix} 2e^{2x} & 0 & \vdots & 2 & -2 \\ 0 & e^{3x} & \vdots & -1 & 2 \end{bmatrix},$$

which leads to

$$M^{-1} = \begin{bmatrix} e^{-2x} & -e^{-2x} \\ -e^{-3x} & 2e^{-3x} \end{bmatrix}.$$

The product inside the integral is then

$$M^{-1}G(x) = \begin{bmatrix} e^{-2x} & -e^{-2x} \\ -e^{-3x} & 2e^{-3x} \end{bmatrix}\begin{bmatrix} 2 \\ x \end{bmatrix} = \begin{bmatrix} (2-x)e^{-2x} \\ (-2+2x)e^{-3x} \end{bmatrix}.$$

Integrating each component of this column vector gives us

$$\int M^{-1}G(x)\,dx = \begin{bmatrix} \int(2-x)e^{-2x}\,dx \\ \int(-2+2x)e^{-3x}dx \end{bmatrix} = \begin{bmatrix} (-\frac{3}{4}+\frac{x}{2})e^{-2x} \\ (\frac{4}{9}-\frac{2}{3}x)e^{-3x} \end{bmatrix}.$$

We thus have

$$Y_P = M\int M^{-1}G(x)\,dx = \begin{bmatrix} 2e^{2x} & e^{3x} \\ e^{2x} & e^{3x} \end{bmatrix}\begin{bmatrix} (-\frac{3}{4}+\frac{x}{2})e^{-2x} \\ (\frac{4}{9}-\frac{2}{3}x)e^{-3x} \end{bmatrix} = \begin{bmatrix} -\frac{19}{18}+\frac{x}{3} \\ -\frac{11}{36}-\frac{x}{6} \end{bmatrix}.$$

The general solution is then

$$Y = Y_H + Y_P = \begin{bmatrix} 2c_1e^{2x} + c_2e^{3x} \\ c_1e^{2x} + c_2e^{3x} \end{bmatrix} + \begin{bmatrix} -\frac{19}{18} + \frac{x}{3} \\ -\frac{11}{36} - \frac{x}{6} \end{bmatrix}$$

$$= \begin{bmatrix} 2c_1e^{2x} + c_2e^{3x} - \frac{19}{18} + \frac{x}{3} \\ c_1e^{2x} + c_2e^{3x} - \frac{11}{36} - \frac{x}{6} \end{bmatrix}.$$

●

EXERCISES 6.4

In Exercises 1–10, determine the general solution to $Y' = AY + G(x)$ for the given $G(x)$ where $Y' = AY$ is the homogeneous system in the indicated exercise of Section 6.2 (see page 311) or 6.3 (see page 314).

1. Exercise 1 of Section 6.2; $G(x) = \begin{bmatrix} 2 \\ x \end{bmatrix}$

2. Exercise 4 of Section 6.2; $G(x) = \begin{bmatrix} x \\ e^{2x} \end{bmatrix}$

3. Exercise 9 of Section 6.2; $G(x) = \begin{bmatrix} 5 \\ 1-x \end{bmatrix}$

4. Exercise 10 of Section 6.2; $G(x) = \begin{bmatrix} 3x \\ 0 \end{bmatrix}$

5. Exercise 5 of Section 6.2; $G(x) = \begin{bmatrix} 1-2x \\ xe^{-x} \\ 0 \end{bmatrix}$

6. Exercise 12 of Section 6.2; $G(x) = \begin{bmatrix} 0 \\ e^{-2x} \\ 0 \end{bmatrix}$

7. Exercise 1 of Section 6.3; $G(x) = \begin{bmatrix} 2 \\ x \end{bmatrix}$

8. Exercise 2 of Section 6.3; $G(x) = \begin{bmatrix} x \\ e^{2x} \end{bmatrix}$

9. Exercise 5 of Section 6.3; $G(x) = \begin{bmatrix} 0 \\ e^{-2x} \\ 0 \end{bmatrix}$

10. Exercise 6 of Section 6.3; $G(x) = \begin{bmatrix} 1-2x \\ xe^{-x} \\ 0 \end{bmatrix}$

In Exercises 11–14, determine the solution to the initial value problem $Y' = AY+G(x)$, $Y(0) = Y_0$ for the given column vector $Y(0)$ and the system $Y' = A(x)Y+G(x)$ in the indicated exercise of this section.

11. Exercise 1; $Y_0 = \begin{bmatrix} 2 \\ 1 \end{bmatrix}$

12. Exercise 2; $Y_0 = \begin{bmatrix} 0 \\ -1 \end{bmatrix}$

13. Exercise 5; $Y_0 = \begin{bmatrix} 0 \\ 1 \\ -1 \end{bmatrix}$

14. Exercise 6; $Y_0 = \begin{bmatrix} 0 \\ 1 \\ -1 \end{bmatrix}$

15. An object is moving with velocity vector

$$v = \begin{bmatrix} 2x + y - t \\ 2x + 3y + t \end{bmatrix}.$$

Find the position of the object at time t if the initial position of the object is the point $(1,1)$.

16. An object is moving with velocity vector

$$v = \begin{bmatrix} 2x + y + 2 \\ 3x + 4y - t \\ 5x - 6y + 3z \end{bmatrix}.$$

Find the position of the object at time t if the initial position of the object is the point $(5, 3, 4)$.

6.5 CONVERTING DIFFERENTIAL EQUATIONS TO FIRST ORDER SYSTEMS

Every linear differential equation can be converted to a system of first order linear differential equations. The following examples illustrate how this can be done in the homogeneous case.

EXAMPLE 1 Convert the following linear differential equation to a system of linear equations.

$$y'' + 3y' + 2y = 0$$

Solution We let $v_1 = y$ and $v_2 = y'$. Using the differential equation, this gives us

$$v_1' = y' = v_2$$
$$v_2' = y'' = -3y' - 2y = -2v_1 - 3v_2,$$

which can be rewritten as the system

$$\begin{bmatrix} v_1' \\ v_2' \end{bmatrix} = \begin{bmatrix} 0 & 1 \\ -2 & -3 \end{bmatrix} \begin{bmatrix} v_1 \\ v_2 \end{bmatrix}.$$

The characteristic equation of $y'' + 3y' + 2y = 0$ is $\lambda^2 + 3\lambda + 2 = 0$, which is also the characteristic equation of the matrix

$$\begin{bmatrix} 0 & 1 \\ -2 & -3 \end{bmatrix}$$

of the system of differential equations obtained from the substitutions $v_1 = y$ and $v_2 = y'$.

EXAMPLE 2 Convert the following linear differential equation to a system of linear equations.

$$y''' + 4y'' - y' - 4y = 0$$

Solution We let $v_1 = y$, $v_2 = y'$, and $v_3 = y''$. Proceeding as in Example 1, we see that

$$v_1' = y' = v_2$$
$$v_2' = y'' = v_3$$
$$v_3' = y''' = -4y'' + y' + 4y = 4v_1 + v_2 - 4v_3,$$

which can be rewritten as the system

$$\begin{bmatrix} v_1' \\ v_2' \\ v_3' \end{bmatrix} = \begin{bmatrix} 0 & 1 & 0 \\ 0 & 0 & 1 \\ 4 & 1 & -4 \end{bmatrix} \begin{bmatrix} v_1 \\ v_2 \\ v_3 \end{bmatrix}.$$

The characteristic equation of $y''' + 4y'' - y' - 4y = 0$ is $\lambda^3 + 4\lambda^2 - \lambda - 4 = 0$, which is also the characteristic equation of the matrix

$$\begin{bmatrix} 0 & 1 & 0 \\ 0 & 0 & 1 \\ 4 & 1 & -4 \end{bmatrix}$$

of the system of differential equations formed from the substitutions $v_1 = y$, $v_2 = y'$, and $v_3 = y''$.

In Exercises 1 and 2 we ask you to solve the systems obtained in Examples 1 and 2, respectively. Notice that the first entries of the general solutions to these systems will be v_1 and, since $v_1 = y$, they will give us the general solutions to the original differential equations. ●

Examples 1 and 2 lead to the following theorem. We leave the proof of this theorem for Exercise 10.

THEOREM 6.10 The nth order homogeneous linear differential equation

$$q_n(x)y^{(n)} + q_{n-1}(x)y^{(n-1)} + \cdots + q_0(x)y = 0$$

is equivalent to the system of n homogeneous linear differential equations

$$V' = Q(x)V$$

where

$$Q(x) = \begin{bmatrix} 0 & 1 & 0 & \cdots & 0 \\ 0 & 0 & 1 & \cdots & 0 \\ & & \vdots & & \\ 0 & & \cdots & 0 & 1 \\ -\frac{q_0(x)}{q_n(x)} & -\frac{q_1(x)}{q_n(x)} & \cdots & & -\frac{q_{n-1}(x)}{q_n(x)} \end{bmatrix} \quad \text{and } V = \begin{bmatrix} y \\ y' \\ \vdots \\ y^{(n-1)} \end{bmatrix}.$$

Furthermore, if $q_0, q_1, \ldots, q_n$ are constant functions, then the characteristic equation of the nth order constant coefficient homogeneous linear differential equation and the characteristic equation of the matrix $Q(x)$ are the same. The general solution to the nth order homogeneous linear differential equation is the first entry of the general solution of the system $V' = Q(x)V$.

Not only can we solve homogeneous linear differential equations by converting them to systems of first order linear homogeneous differential equations, but we can use a similar conversion approach to solve nonhomogeneous equations in conjunction with our method of solving nonhomogeneous systems from the last section. The following example illustrates how we do the conversion.

EXAMPLE 3 Convert the following nonhomogeneous linear differential equation to a nonhomogeneous system of linear equations.

$$y'' - y' - 2y = \sin x$$

Solution   Letting $v_1 = y$ and $v_2 = y'$ leads to

$$v_1' = y' = v_2$$
$$v_2' = y'' = y' + 2y + \sin x = 2v_1 + v_2 + \sin x,$$

which can be rewritten as the system

$$\begin{bmatrix} v_1' \\ v_2' \end{bmatrix} = \begin{bmatrix} 0 & 1 \\ 2 & 1 \end{bmatrix} \begin{bmatrix} v_1 \\ v_2 \end{bmatrix} + \begin{bmatrix} 0 \\ \sin x \end{bmatrix}.$$

We will let you solve this system in Exercise 3. As was the case in the homogeneous setting, the first entry of the general solution to this system will be the general solution to the original differential equation.   ●

Our technique for converting single linear differential equations into first order linear systems can also be used to convert higher order systems of linear differential equations into first order ones as the next example illustrates. Indeed, the fact that we may do these conversions gives us the sufficiency of studying first order linear systems when considering systems of linear differential equations in general.

EXAMPLE 4   Convert the following system into a first order system of linear differential equations:

$$y_1'' = y_1 + y_2$$
$$y_2'' = y_1 + y_1' + y_2 + y_2'.$$

Solution   Letting

$$v_1 = y_1$$
$$v_2 = y_1' = v_1'$$
$$v_3 = y_2$$
$$v_4 = y_2' = v_3'$$

we obtain the first order system:

$$v_1' = v_2$$
$$v_2' = v_1 + v_3$$
$$v_3' = v_4$$
$$v_4' = v_1 + v_2 + v_3 + v_4.$$

In matrix form, our system is:

$$\begin{bmatrix} v_1' \\ v_2' \\ v_3' \\ v_4' \end{bmatrix} = \begin{bmatrix} 0 & 1 & 0 & 0 \\ 1 & 0 & 1 & 0 \\ 0 & 0 & 0 & 1 \\ 1 & 1 & 1 & 1 \end{bmatrix} \begin{bmatrix} v_1 \\ v_2 \\ v_3 \\ v_4 \end{bmatrix}.$$

We will let you find the solution to the system in this example in Exercise 12.   ●

EXERCISES 6.5

In Exercises 1–3, determine the general solution to the system obtained in the indicated example of this section. Use your general solution to the system to find the general solution to the original differential equation in the example.

  1. Example 1   **2.** Example 2   **3.** Example 3

In Exercises 4–9, do the indicated exercise of the section from Chapter 4 (see pages 201 and 211) by converting to a first order system of linear differential equations.

  4. Section 4.2, Exercise 1

  5. Section 4.2, Exercise 5

  6. Section 4.2, Exercise 9

  7. Section 4.2, Exercise 13

  8. Section 4.3, Exercise 1

  9. Section 4.3, Exercise 11

  10. Prove Theorem 6.10

  11. Reformulate Theorem 6.10 for the nonhomogeneous case and prove this result.

  12. Find the general solution of the system in Example 4.

  13. Find the general solution of the system:

$$y_1'' = 2y_1' + y_2$$
$$y_2'' = 6y_1' + 3y_2.$$

6.6   APPLICATIONS INVOLVING SYSTEMS OF LINEAR DIFFERENTIAL EQUATIONS

There are many applications where systems of linear differential equations arise besides the one we saw in the introduction of this chapter involving a given velocity vector. In this section we consider a few of these. Our first example is a mixing problem. Recall that we considered mixing problems involving a single tank in Section 3.6. Mixing problems with more than one tank lead to systems of first order linear differential equations.

EXAMPLE 1   Consider two tanks each with volume 100 gallons connected together by two pipes. The first tank initially contains a well-mixed solution of 5 lb salt in 50 gal water. The second tank initially contains 100 gal salt-free water. A pipe from tank 1 to tank 2 allows the solution in tank 1 to enter tank 2 at a rate of 5 gal/min. A second pipe from tank 2 to tank 1 allows the solution from tank 2 to enter tank 1 at a rate of 5 gal/min. (See Figure 6.2.) Assume that the salt mixture in each tank is well-stirred. How much salt is in each tank after 5 min? Graph the amount of salt in each tank for the first 25 min.

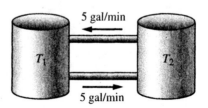

5 gal/min

T_1 T_2

5 gal/min

Figure 6.2

Solution   First note that the volume of solution in each tank remains constant over time. Let q_1 and q_2 be the amount of salt in the first tank and the second tank, respectively. We have that

$$\frac{dq_1}{dt} = \text{the rate of salt going into tank 1} - \text{the rate of salt going out of tank 1,}$$

$$\frac{dq_2}{dt} = \text{the rate of salt going into tank 2} - \text{the rate of salt going out of tank 2.}$$

Using the flow rates given, it follows that

$$\frac{dq_1}{dt} = \frac{q_2 \text{ lb}}{100 \text{ gal}} \cdot 5 \frac{\text{gal}}{\text{min}} - \frac{q_1 \text{ lb}}{50 \text{ gal}} \cdot 5 \frac{\text{gal}}{\text{min}}$$

and

$$\frac{dq_2}{dt} = \frac{q_1 \text{ lb}}{50 \text{ gal}} \cdot 5 \frac{\text{gal}}{\text{min}} - \frac{q_2 \text{ lb}}{100 \text{ gal}} \cdot 5 \frac{\text{gal}}{\text{min}}.$$

Letting

$$Q = \begin{bmatrix} q_1 \\ q_2 \end{bmatrix}$$

we have

$$Q' = \begin{bmatrix} -1/10 & 1/20 \\ 1/10 & -1/20 \end{bmatrix} Q = AQ.$$

We leave it to you to show that the eigenvalues of A are 0 and $-3/20$ and the vectors

$$\begin{bmatrix} 1 \\ 2 \end{bmatrix} \quad \text{and} \quad \begin{bmatrix} -1 \\ 1 \end{bmatrix}$$

are bases for the eigenspaces E_0 and $E_{-3/20}$, respectively. Using

$$P = \begin{bmatrix} 1 & -1 \\ 2 & 1 \end{bmatrix},$$

we see the general solution is

$$Q = P \begin{bmatrix} c_1 \\ c_2 e^{-\frac{3}{20}t} \end{bmatrix} = \begin{bmatrix} 1 & -1 \\ 2 & 1 \end{bmatrix} \begin{bmatrix} c_1 \\ c_2 e^{-\frac{3}{20}t} \end{bmatrix} = \begin{bmatrix} c_1 - c_2 e^{-\frac{3}{20}t} \\ 2c_1 + c_2 e^{-\frac{3}{20}t} \end{bmatrix}.$$

The initial conditions $q_1(0) = 5$ and $q_2(0) = 0$ give us

$$Q(0) = \begin{bmatrix} c_1 - c_2 \\ 2c_1 + c_2 \end{bmatrix} = \begin{bmatrix} 5 \\ 0 \end{bmatrix}.$$

Solving for c_1 and c_2, we have $c_1 = 5/3$ and $c_2 = -10/3$. Consequently,

$$q_1(t) = \frac{5}{3} + \frac{10}{3} e^{-\frac{3}{20}t}$$

$$q_2(t) = \frac{10}{3} - \frac{10}{3} e^{-\frac{3}{20}t}.$$

When $t = 5$ min,

$$q_1(5) = \frac{5}{3} + \frac{10}{3}e^{-\frac{3}{4}} \approx 3.24 \text{ lb}$$

$$q_2(5) = \frac{10}{3} - \frac{10}{3}e^{-\frac{3}{4}} \approx 1.76 \text{ lb.}$$

Graphs of the two salt amounts obtained by letting Maple graph the functions q_1 and q_2 appear in Figure 6.3. Notice that q_1 and q_2 approach 5/3 and 10/3, respectively, as $t \to \infty$. ●

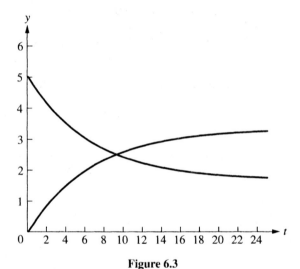

Figure 6.3

For our next application involving a system of linear differential equations, we consider a problem involving a circuit. In Section 4.5 we presented a single loop closed circuit and found a differential equation modeling the circuit using Kirchoff's law. Suppose we have a double loop closed circuit as in Figure 6.4. In this setting Kirchoff's law leads to the following system of differential equations for the currents i_1 and i_2 where the derivatives are with respect to t:

$$Li_1' + R(i_1 - i_2) = E(t)$$

$$R(i_2' - i_1') + \frac{1}{C}i_2 = 0.$$

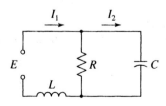

Figure 6.4

In matrix form, this system becomes

$$\begin{bmatrix} L & 0 \\ -R & R \end{bmatrix} \begin{bmatrix} i_1' \\ i_2' \end{bmatrix} = \begin{bmatrix} -R & R \\ 0 & -1/C \end{bmatrix} \begin{bmatrix} i_1 \\ i_2 \end{bmatrix} + \begin{bmatrix} E(t) \\ 0 \end{bmatrix}.$$

Solving this system for i_1' and i_2' by multiplying each side of this equation on the left by

$$\begin{bmatrix} L & 0 \\ -R & R \end{bmatrix}^{-1} = \begin{bmatrix} 1/L & 0 \\ 1/L & 1/R \end{bmatrix}$$

gives us the nonhomogeneous system

$$\begin{bmatrix} i_1' \\ i_2' \end{bmatrix} = \begin{bmatrix} -R/L & R/L \\ -R/L & R/L - 1/RC \end{bmatrix} \begin{bmatrix} i_1 \\ i_2 \end{bmatrix} + \begin{bmatrix} E(t)/L \\ E(t)/L \end{bmatrix}$$

modeling the circuit. Here is an example where we use this model.

EXAMPLE 2   Determine i_1 and i_2 for a circuit in the configuration of Figure 6.4 if $R = 2$, $L = 2$, $C = 1/9$, and $E(t) = 2 \sin t$ and there is no initial current.

Solution   Substituting for R, L, C, and $E(t)$ into the system gives us

$$\begin{bmatrix} i_1' \\ i_2' \end{bmatrix} = \begin{bmatrix} -1 & 1 \\ -1 & -7/2 \end{bmatrix} \begin{bmatrix} i_1 \\ i_2 \end{bmatrix} + \begin{bmatrix} \sin t \\ \sin t \end{bmatrix}.$$

Letting

$$Y = \begin{bmatrix} i_1 \\ i_2 \end{bmatrix} \quad \text{and} \quad G = \begin{bmatrix} \sin t \\ \sin t \end{bmatrix}$$

we have

$$Y' = \begin{bmatrix} -1 & 1 \\ -1 & -7/2 \end{bmatrix} Y + \begin{bmatrix} \sin t \\ \sin t \end{bmatrix} = AY + G.$$

We again leave it to you to find the eigenvalues and bases for the eigenspaces of A. Doing so you will find the eigenvalues of A are -3 and $-3/2$ and the vectors

$$\begin{bmatrix} 1 \\ -2 \end{bmatrix} \quad \text{and} \quad \begin{bmatrix} -2 \\ 1 \end{bmatrix}$$

are bases for the eigenspaces E_{-3} and $E_{-3/2}$, respectively. Solving the homogeneous system, we take P to be

$$P = \begin{bmatrix} 1 & -2 \\ -2 & 1 \end{bmatrix}$$

from which it follows that

$$Y_H = P \begin{bmatrix} c_1 e^{-3t} \\ c_2 e^{-\frac{3}{2}t} \end{bmatrix} = \begin{bmatrix} 1 & -2 \\ -2 & 1 \end{bmatrix} \begin{bmatrix} c_1 e^{-3t} \\ c_2 e^{-\frac{3}{2}t} \end{bmatrix} = \begin{bmatrix} c_1 e^{-3t} - 2c_2 e^{-\frac{3}{2}t} \\ -2c_1 e^{-3t} + c_2 e^{-\frac{3}{2}t} \end{bmatrix}$$

$$= \begin{bmatrix} e^{-3t} & -2e^{-\frac{3}{2}t} \\ -2e^{-3t} & e^{-\frac{3}{2}t} \end{bmatrix} \begin{bmatrix} c_1 \\ c_2 \end{bmatrix} = MC$$

where

$$M = \begin{bmatrix} e^{-3t} & -2e^{-\frac{3}{2}t} \\ -2e^{-3t} & e^{-\frac{3}{2}t} \end{bmatrix}$$

is a matrix of fundamental solutions. We leave it to you to find that

$$M^{-1} = \begin{bmatrix} -\frac{1}{3}e^{3t} & -\frac{2}{3}e^{3t} \\ -\frac{2}{3}e^{\frac{3}{2}t} & -\frac{1}{3}e^{\frac{3}{2}t} \end{bmatrix}.$$

A particular solution (leaving out the calculations) is

$$Y_P = M \int M^{-1} G(t)\, dt = \begin{bmatrix} -\frac{67}{130}\cos t + \frac{81}{130}\sin t \\ \frac{7}{65}\cos t + \frac{9}{65}\sin t \end{bmatrix}$$

from which we obtain the general solution

$$Y = Y_H + Y_P = \begin{bmatrix} e^{-3t} & -2e^{-\frac{3}{2}t} \\ -2e^{-3t} & e^{-\frac{3}{2}t} \end{bmatrix} \begin{bmatrix} c_1 \\ c_2 \end{bmatrix} + \begin{bmatrix} -\frac{67}{130}\cos t + \frac{81}{130}\sin t \\ \frac{7}{65}\cos t + \frac{9}{65}\sin t \end{bmatrix}.$$

The initial conditions $i_1(0) = 0$ and $i_2(0) = 0$ give us

$$Y(0) = \begin{bmatrix} c_1 - 2c_2 - \frac{67}{130} \\ -2c_1 + c_2 + \frac{7}{65} \end{bmatrix} = \begin{bmatrix} 0 \\ 0 \end{bmatrix}.$$

Solving for c_1 and c_2 we have $c_1 = -1/10$ and $c_2 = -4/13$ and hence

$$i_1(t) = -\frac{1}{10}e^{-3t} + \frac{8}{13}e^{-\frac{3}{2}t} - \frac{67}{130}\cos t + \frac{81}{130}\sin t$$

$$i_2(t) = \frac{1}{5}e^{-3t} - \frac{4}{13}e^{-\frac{3}{2}t} + \frac{7}{65}\cos t + \frac{9}{65}\sin t.$$

Graphs of the currents appear in Figure 6.5. So that we can see which curve is which, we individually graphed i_1 and i_2 in Figures 6.5(a) and (b), respectively, before graphing them together in Figure 6.5(c). ●

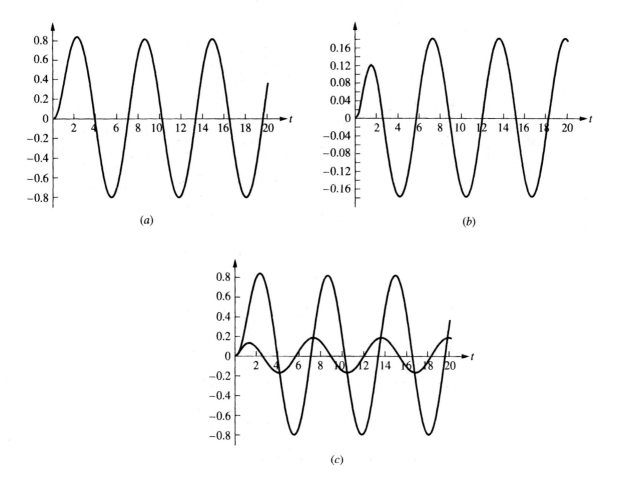

Figure 6.5

Both Examples 1 and 2 involved first order linear systems. Following is an example with a second order linear system.

EXAMPLE 3   A force given by the vector

$$F = \begin{bmatrix} 36x \\ 12x + 16y \end{bmatrix}$$

is acting upon a mass of 4 kg in the xy-plane. Find the position of the mass at time t if the initial position of the mass is the point $(10, 8)$ and its initial velocity vector is the zero vector.

Solution Since

$$F = ma \quad \text{or} \quad a = \frac{F}{m}$$

where a is the acceleration vector and m is the mass of the object, and since

$$a = \begin{bmatrix} \dfrac{d^2x}{dt^2} \\ \dfrac{d^2y}{dt^2} \end{bmatrix},$$

we have the second order system:

$$\frac{d^2x}{dt^2} = 9x$$

$$\frac{d^2y}{dt^2} = 3x + 4y$$

This gives us the first order system

$$v_1' = v_2$$
$$v_2' = 9v_1$$
$$v_3' = v_4$$
$$v_4' = 3v_1 + 4v_3$$

where $v_1 = x$, $v_2 = dx/dt$, $v_3 = y$, and $v_4 = dy/dt$. The eigenvalues of the matrix

$$A = \begin{bmatrix} 0 & 1 & 0 & 0 \\ 9 & 0 & 0 & 0 \\ 0 & 0 & 0 & 1 \\ 3 & 0 & 4 & 0 \end{bmatrix}$$

for the system of differential equations are

$$\lambda = 2, -2, 3, -3$$

and bases for the eigenspaces E_2, E_{-2}, E_3, E_{-3}, each of which are one-dimensional spaces, are

$$\begin{bmatrix} 0 \\ 0 \\ 1/2 \\ 1 \end{bmatrix}, \begin{bmatrix} 0 \\ 0 \\ -1/2 \\ 1 \end{bmatrix}, \begin{bmatrix} 5/9 \\ 5/3 \\ 1/3 \\ 1 \end{bmatrix}, \begin{bmatrix} -5/9 \\ 5/3 \\ -1/3 \\ 1 \end{bmatrix},$$

respectively. (Check and see if we did this correctly.) The general solution to our first order system is

$$
\begin{bmatrix} v_1 \\ v_2 \\ v_3 \\ v_4 \end{bmatrix} = \begin{bmatrix} 0 & 0 & 5/9 & -5/9 \\ 0 & 0 & 5/3 & 5/3 \\ 1/2 & -1/2 & 1/3 & -1/3 \\ 1 & 1 & 1 & 1 \end{bmatrix} \begin{bmatrix} c_1 e^{2t} \\ c_2 e^{-2t} \\ c_3 e^{3t} \\ c_4 e^{-3t} \end{bmatrix}
$$

$$
= \begin{bmatrix} \frac{5}{9} c_3 e^{3t} - \frac{5}{9} c_4 e^{-3t} \\ \frac{5}{3} c_3 e^{3t} + \frac{5}{3} c_4 e^{-3t} \\ \frac{1}{2} c_1 e^{2t} - \frac{1}{2} c_2 e^{-2t} + \frac{1}{3} c_3 e^{3t} - \frac{1}{3} c_4 e^{-3t} \\ c_1 e^{2t} + c_2 e^{-2t} + c_3 e^{3t} + c_4 e^{-3t} \end{bmatrix}.
$$

The initial conditions in this problem tell us $v_1(0) = 10$, $v_2(0) = 0$, $v_3(0) = 8$, and $v_4(0) = 0$. Using these initial conditions, we have the system

$$\frac{5}{9} c_3 - \frac{5}{9} c_4 = 10$$

$$\frac{5}{9} c_3 + \frac{5}{9} c_4 = 0$$

$$\frac{1}{2} c_1 - \frac{1}{2} c_2 + \frac{1}{3} c_3 - \frac{1}{3} c_4 = 8$$

$$c_1 + c_2 + c_3 + c_4 = 0$$

whose solution is $c_1 = 2$, $c_2 = -2$, $c_3 = 9$, $c_4 = -9$. The position of the mass at time t is then:

$$x = v_1 = 5e^{3t} + 5e^{-3t},$$
$$y = v_3 = e^{2t} + e^{-2t} + 3e^{3t} + 3e^{-3t}.$$

We will not graph each of the functions for x and y as we have done in the previous examples of this section. Instead we have graphed the curve given by the parametric equations for x and y in Figure 6.6. This curve is more meaningful in this example since it represents the path of the particle. We authors had to do some experimenting, as we used Maple to graph this curve. Plotting for values of t from $t = 0$ to as small as $t = 0.01$ resulted in a graph that appeared to be a line, as is illustrated in Figure 6.6(a). However, this set of parametric equations for x and y does not give us a line. (Why is this?) By plotting from $t = 0$ to $t = 0.0005$, Maple gives us the graph in Figure 6.6(b) exhibiting a nonlinear shape. Even here Maple is giving us only a rough approximation of this graph since it will not consist of pieced-together line segments. ●

The final application we consider involves masses and springs. In Section 4.5 we studied a mass-spring system involving a single mass m and a single spring. Using Newton's and Hooke's laws, we obtained that the motion of the mass was determined by the differential equation

$$mu'' + fu' + ku = h(t)$$

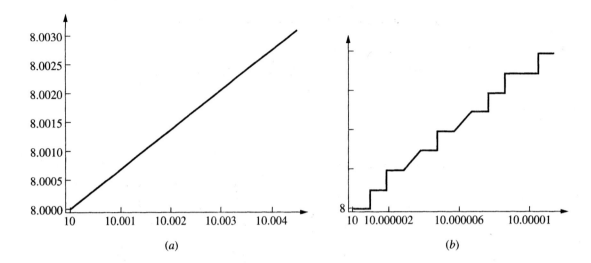

(a)   (b)

Figure 6.6

where k is the spring constant, f is the frictional constant, and $h(t)$ is the external force. The reasoning we used can be extended to mass-spring systems such as the one illustrated in Figure 6.7 involving two masses, m_1 and m_2, and two springs. Here it can be argued that the motion of the two masses is governed by the system of second order linear differential equations

$$m_1 x_1'' = -k_1 x_1 - f_1 x_1' + k_2(x_2 - x_1) + h_1(t)$$
$$m_2 x_2'' = -k_2(x_2 - x_1) - f_2 x_2' + h_2(t)$$

where k_1 and k_2 are the associated spring constants, f_1 and f_2 are the associated frictional constants, and $h_1(t)$ and $h_2(t)$ are the associated external forces. Converting this system to a system of first order linear differential equations by letting $v_1 = x_1$, $v_2 = x_1'$,

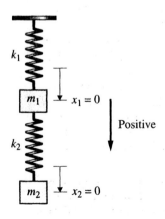

Figure 6.7

$v_3 = x_2$, and $v_4 = x_2'$, we have the first order system:

$$v_1' = v_2$$

$$v_2' = -\frac{k_1 + k_2}{m_1} v_1 - \frac{f_1}{m_1} v_2 + \frac{k_2}{m_1} v_3 + \frac{1}{m_1} h_1(t)$$

$$v_3' = v_4$$

$$v_4' = \frac{k_2}{m_2} v_1 - \frac{k_2}{m_2} v_3 - \frac{f_2}{m_2} v_4 + \frac{1}{m_2} h_2(t).$$

In matrix form, this system is

$$V' = \begin{bmatrix} 0 & 1 & 0 & 0 \\ -\frac{k_1 + k_2}{m_1} & -\frac{f_1}{m_1} & \frac{k_2}{m_1} & 0 \\ 0 & 0 & 0 & 1 \\ \frac{k_2}{m_2} & 0 & -\frac{k_2}{m_2} & -\frac{f_2}{m_2} \end{bmatrix} V + \begin{bmatrix} 0 \\ \frac{1}{m_1} h_1(t) \\ 0 \\ \frac{1}{m_2} h_2(t) \end{bmatrix}.$$

In Exercises 11–18, we will give you some mass-spring systems. You will quickly discover that these problems become long and difficult to solve by hand. Our advice is to use a software package such as Maple to solve the resulting first order systems in these problems.

EXERCISES 6.6

1. Tank 1 and tank 2 are arranged as in Figure 6.2. Each has volume 500 gal and is filled with a liquid fertilizer. The solution in tank 1 contains 10 lb ammonia and the solution in tank 2 contains 1 lb ammonia. Assume the flow rate in the pipes is 25 gal/min. Determine the amount of ammonia in each tank at any time.

2. Suppose in Example 1 that tank 1 has 100 gal and tank 2 has 50 gal. Determine the amount of salt in each tank under these conditions.

3. Suppose in Example 1 that both tanks contain 100 gal and the initial amount of salt in the example. Salt water containing 1 lb/gal is added at a rate of 2 gal/min to tank 1, tank 2 is allowed to drain at 2 gal/min, solution flows in the pipe from tank 1 to tank 2 at a rate of 6 gal/min, and solution flows in the pipe from tank 2 to tank 1 at a rate of 4 gal/min. Determine the amount of salt in each tank under these conditions.

4. Two perfectly insulated rooms are separated by a hallway. The two rooms are connected by air vents. Each room also has an outside air vent. The air vent into room 1 allows air to enter at $r_{1,1}$ gal/min. The air vent from room 1 to room 2 allows air to enter room 2 at $r_{1,2}$ gal/min. The air vent from room 2 to room 1 allows air to enter room 1 at $r_{2,1}$ gal/min. The air vent out of room 2 allows air to leave room 2 at $r_{2,2}$ gal/min. The air contains carbon monoxide.

   a) Set up a system of differential equations that gives the amount of carbon monoxide in each room at any time.

   b) Determine the conditions for $r_{1,1}, r_{1,2}, r_{2,1}$, and $r_{2,2}$ so that the air volume in each room stays constant.

   c) Solve the system for $r_{1,1} = 3$, $r_{1,2} = 5$, $r_{2,1} = 2$, and $r_{2,2} = 3$ with room 1 having volume 1500 cubic meters and room 2 having volume 1000 cubic meters. Assume initially that room 1 has no carbon monoxide and the air in room 2 is 10% carbon monoxide.

5. Using Kirchoff's law, determine the system of differential equations describing the currents for the circuit in Figure 6.8. Solve this system if $L = 1$, $R = 2$, $C = 1/16$, $i_1(0) = 1$, $i_2(0) = 0$.

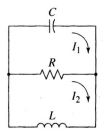

Figure 6.8

6. Using Kirchoff's law, determine the system of differential equations describing the currents for the circuit in Figure 6.9. Solve this system if $L = 1.5$, $R = 1$, $C = 1/3$, $i_1(0) = 0$, $i_2(0) = 1$.

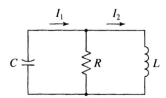

Figure 6.9

7. Using Kirchoff's law, determine the system of differential equations describing the currents for the circuit in Figure 6.10. Solve this system if $R = 2$, $L = 2$, $C = 1/4$, $i_1(0) = 0$, $i_2(0) = 0$, $E(t) = 10 \sin t$.

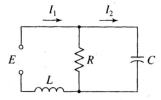

Figure 6.10

8. Using Kirchoff's law, determine the system of differential equations describing the currents for the circuit in Figure 6.11. Solve this system if $R_1 = 2$, $R_2 = 8$, $L = 8$, $C = 1/2$, $i_1(0) = 0$, $i_2(0) = 0$, $E(t) = 150$.

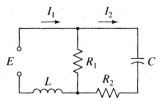

Figure 6.11

9. A force given by the vector

$$F = \begin{bmatrix} 500x - 250y \\ 125y \end{bmatrix}$$

is acting upon a mass of 5 kg in the xy-plane. Find the position of the mass at time t if the initial position of the mass is the point (2, 6) and its initial velocity vector is the zero vector.

10. A force given by the vector

$$F = \begin{bmatrix} 10x \\ 10y \\ 10z \end{bmatrix}$$

is acting upon a mass of 10 kg in the 3-space. Find the position of the mass at time t if the initial position of the mass is the point (2, 4, 6) and its initial velocity vector is the zero vector.

In Exercises 11–18, refer to Figure 6.7. Use $g = 10$ m/s^2.

11. Determine the motion of the two masses if $m_1 = 1$ kg, $m_2 = 1$ kg, and the springs have spring constants $k_1 = 3$, $k_2 = 2$. Assume there is no friction or external forces and that the two masses are given initial velocities at equilibrium of 1 m/sec.

12. Determine the motion of the two masses if $k_1 = 2k_2$, $m_1 = 2m_2$, and the two masses are pulled down 1 m and released from rest. Experiment with different values for m_2 and k_2.

13. A 1-kg mass stretches the first spring 1.25 m and a 1-kg mass stretches the second spring 10/3 m. Assuming there is no friction or external forces and each mass is pulled down 1/10 m from equilibrium and released, determine the motion of the masses.

14. A 2-kg mass stretches the first spring 20/9 m and a 2-kg mass stretches the second spring 10/3 m.

Assuming there is no friction or external forces and each mass is given an initial velocity up of 3/10 m from equilibrium, determine the motion of the masses.

15. Determine the motion of the masses in Exercise 11 if each of the masses is subject to frictional forces with frictional constants $f_1 = f_2 = 1$.

16. Determine the motion of the masses in Exercise 14 if each of the masses is subject to frictional forces with frictional constants $f_1 = f_2 = 2$.

17. Determine the motion of the masses in Exercise 11 if the first mass is subject to an external force of $\cos 2t$ kg m/sec^2.

18. Determine the motion of the masses in Exercise 13 if the second mass is connected to an external force of $3 \sin t$ kg m/sec^2.

19. Show that the characteristic polynomial of the coefficient matrix of the first order system of linear differential equations associated with a force vector

$$F = \left[\begin{array}{c} a_{11}x + a_{12}y \\ a_{21}x + a_{22}y \end{array} \right]$$

is

$$\lambda^2(\lambda^2 - a_{22}/m) - (a_{11}/m)(\lambda^2 - a_{22}/m) - a_{12}a_{21}/m^2.$$

20. Show that the characteristic polynomial of the coefficient matrix of the first order system of linear differential equations associated with a mass-spring system as in Figure 6.7 with no frictional or external forces has four imaginary roots.

21. The line integral of a differential form $f(x, y)\, dx + g(x, y)\, dy$ over a curve C in the xy-plane, denoted $\int_C (f(x, y)\, dx + g(x, y)\, dy)$, is

$$\int_C (f(x, y)\, dx + g(x, y)\, dy)$$

$$= \int_a^b (f(x(t), y(t))x'(t) + g(x(t), y(t))y'(t))\, dt$$

where $x = x(t), y = y(t), a \le t \le b$ is a parame-

terization of C. If

$$F = \left[\begin{array}{c} f(x, y) \\ g(x, y) \end{array} \right]$$

is a force vector, the line integral $\int_C (f(x, y)\, dx + g(x, y)\, dy)$ is the work done by the force along the curve C. Find the work done by the force in Exercise 9 from $t = 0$ to $t = 1$.

22. For a differential form $f(x, y, z)\, dx + g(x, y, z)\, dy + h(x, y, z)\, dz$ and a curve C in 3-space given parametrically as $x = x(t), y = y(t), z = z(t), a \le t \le b$ the line integral of this differential form along C is

$$\int_C (f(x, y, z)\, dx + g(x, y, z)\, dy + h(x, y, z)\, dz)$$

$$= \int_a^b (f(x(t), y(t), z(t))x'(t)$$

$$+ g(x(t), y(t), z(t))y'(t)$$

$$+ h(x(t), y(t), z(t))z'(t))\, dt.$$

As in Exercise 21, the line integral corresponding to a force vector

$$F = \left[\begin{array}{c} f(x, y, z) \\ g(x, y, z) \\ h(x, y, z) \end{array} \right]$$

in 3-space is the work done by the force along the curve C. Find the work done by the force in Exercise 10 from $t = 1$ to $t = 2$.

23. Recall from calculus that the largest increase in a function of two or three variables occurs in the direction of the gradient of the function.[2] Thus if $z = f(x, y)$, an object starting at a point (x_0, y_0) will follow the curve

$$x = x(t), \qquad y = y(t), \qquad z = f(x(t), y(t))$$

in the direction f increases most rapidly when $x =$

---

[2] For a function $z = f(x, y)$, the gradient is $\nabla f(x, y) = f_x(x, y)\mathbf{i} + f_y(x, y)\mathbf{j}$; for a function of three variables $w = f(x, y, z)$, it is $\nabla f(x, y, z) = f_x(x, y, z)\mathbf{i} + f_y(x, y, z)\mathbf{j} + f_z(x, y, z)\mathbf{k}$.

$x(t)$, $y = y(t)$ solve the initial value problem

$$x' = f_x(x, y)$$
$$y' = f_y(x, y)$$
$$x(0) = x_0, \qquad y(0) = y_0.$$

Suppose the elevation at a point is

$$f(x, y) = x^2 + 4xy + y^2$$

and a hiker is at the point with $x = 10$ and $y = 15$. Find the curve the hiker follows in ascending most rapidly.

24. The largest decrease in a function of two or three variables occurs in the direction of the negative of the gradient of the function. Suppose the temperature T at a point (x, y, z) in space is

$$T = f(x, y, z) = x^2 + 4y^2 + 3yz.$$

If a bug is initially at the point $(0, 0, 1)$, find the curve along which the bug must fly for the temperature to decrease most rapidly.

6.7   2 × 2 SYSTEMS OF NONLINEAR DIFFERENTIAL EQUATIONS

In this section we consider nonlinear systems of differential equations

$$\frac{dx}{dt} = F(x, y)$$
$$\frac{dy}{dt} = G(x, y)$$

or

$$\begin{aligned} x' &= F(x, y) \\ y' &= G(x, y) \end{aligned} \tag{1}$$

where x and y are functions of t such as:

$$x' = -4x^2 + xy,$$

and

$$y' = 2y - xy.$$

We will refer to these as simply **2 × 2 systems** throughout this section. Two places where these systems arise are in modeling the interaction of two competing species and in modeling the motion of a pendulum, as will be seen later in this section.

The theory of Taylor polynomials and series for functions of one variable that you studied in calculus can be extended to functions of several variables. A full-scale development of the topic is left for an advanced calculus course. In this section we are going to make use of the fact that (under appropriate conditions) a function f in two variables x and y has a first degree Taylor polynomial representation of the form

$$f(x, y) = f(x_0, y_0) + f_x(x_0, y_0)(x - x_0) + f_y(x_0, y_0)(y - y_0) + R_f(x - x_0, y - y_0).$$

The function R_f is called the remainder and has the property that

$$\lim_{(x,y)\to(x_0,y_0)} \frac{R_f(x - x_0, y - y_0)}{\sqrt{(x - x_0)^2 + (y - y_0)^2}} = 0.$$

As you are about to see, the idea will be to approximate the functions F and G in the 2×2 system shown in Equation (1) by their first degree Taylor polynomials. This will result in a linear system that we know we can solve. Solving it, we then obtain a solution that gives us qualitative information about the original 2×2 system.

An **equilibrium solution** to the 2×2 system is a constant solution $x = x_0$, $y = y_0$. We will denote this solution as (x_0, y_0). At an equilibrium solution (x_0, y_0), we have $F(x_0, y_0) = 0$ and $G(x_0, y_0) = 0$ since the derivatives of the constant functions

$$x(t) = x_0, \qquad y(t) = y_0$$

are 0. The first degree Taylor polynomial representation of the 2×2 system at (x_0, y_0) is

$$x' = F(x, y) = F_x(x_0, y_0)(x - x_0) + F_y(x_0, y_0)(y - y_0) + R_F(x - x_0, y - y_0)$$
$$y' = G(x, y) = G_x(x_0, y_0)(x - x_0) + G_y(x_0, y_0)(y - y_0) + R_G(x - x_0, y - y_0).$$

To simplify notation, we let $u = x - x_0$ and $v = y - y_0$. We then have $u' = x'$ and $v' = y'$. Our system becomes

$$u' = F_x(x_0, y_0)u + F_y(x_0, y_0)v + R_F(u, v)$$
$$v' = G_x(x_0, y_0)u + G_y(x_0, y_0)v + R_G(u, v).$$

In this notation

$$\lim_{(x,y) \to (x_0, y_0)} \frac{R_F(x - x_0, y - y_0)}{\sqrt{(x - x_0)^2 + (y - y_0)^2}} = \lim_{(u,v) \to (0,0)} \frac{R_F(u, v)}{\sqrt{u^2 + v^2}} = 0$$

and

$$\lim_{(x,y) \to (x_0, y_0)} \frac{R_G(x - x_0, y - y_0)}{\sqrt{(x - x_0)^2 + (y - y_0)^2}} = \lim_{(u,v) \to (0,0)} \frac{R_G(u, v)}{\sqrt{u^2 + v^2}} = 0.$$

These equations imply that $R_F(u, v)$ and $R_G(u, v)$ are small for (u, v) close to $(0, 0)$ (why?) so we ignore them to obtain the linear system:

$$\boxed{\begin{aligned} u' &= F_x(x_0, y_0)u + F_y(x_0, y_0)v \\ v' &= G_x(x_0, y_0)u + G_y(x_0, y_0)v. \end{aligned}}$$

In matrix form we have

$$\boxed{\begin{bmatrix} u' \\ v' \end{bmatrix} = \begin{bmatrix} F_x(x_0, y_0) & F_y(x_0, y_0) \\ G_x(x_0, y_0) & G_y(x_0, y_0) \end{bmatrix} \begin{bmatrix} u \\ v \end{bmatrix}.}$$

The matrix

$$\boxed{A = \begin{bmatrix} F_x(x_0, y_0) & F_y(x_0, y_0) \\ G_x(x_0, y_0) & G_y(x_0, y_0) \end{bmatrix}}$$

is called the **linear part** of the 2×2 system at the equilibrium solution (x_0, y_0). We can solve this linear system using the techniques of this chapter.

To demonstrate consider:

$$x' = -4x^2 + xy,$$
$$y' = 2y - xy.$$

We have

$$F(x, y) = -4x^2 + xy = x(y - 4x)$$

and

$$G(x, y) = 2y - xy = y(2 - x).$$

We see that

$$F(x, y) = 0 \quad \text{if} \quad x = 0 \quad \text{or} \quad y = 4x$$

and that

$$G(x, y) = 0 \quad \text{if} \quad y = 0 \quad \text{or} \quad x = 2.$$

We can now see that $(0, 0)$ and $(2, 8)$ are the equilibrium solutions. Figure 6.12 illustrates the graphs of $F(x, y) = 0$ and $G(x, y) = 0$. The equilibrium solutions are the intersection points of these two curves and the origin.

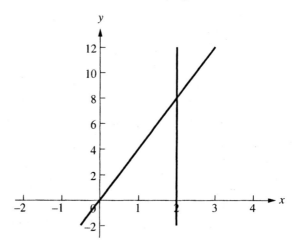

Figure 6.12

To obtain our Taylor polynomials, we note that

$$F_x(x, y) = -8x + y,$$
$$F_y(x, y) = x,$$
$$G_x(x, y) = -y,$$

and

$$G_y(x, y) = 2 - x.$$

We consider the equilibrium solution $(2, 8)$ first and then the equilibrium solution $(0, 0)$. At $(2, 8)$ our first degree Taylor polynomial representation is

$$F(x, y) = F(2, 8) + F_x(2, 8)(x - 2) + F_y(2, 8)(y - 8) + R_F(x - 2, y - 8)$$
$$= -8(x - 2) + 2(y - 8) + R_F(x - 2, y - 8),$$

and

$$G(x, y) = G(2, 8) + G_x(2, 8)(x - 2) + G_y(2, 8)(y - 8) + R_G(x - 2, y - 8)$$
$$= -8(x - 2) + R_G(x - 2, y - 8).$$

We leave it as Exercise 7 to show that $R_F(x - 2, y - 8) = -4(x - 2)^2 + (x - 2)(y - 8)$ and $R_G(x - 2, y - 8) = -(x - 2)(y - 8)$ and that $R_F(x - 2, y - 8)$ and $R_G(x - 2, y - 8)$ satisfy the limit conditions.

Using the substitutions $u = x - 2$ and $y = x - 8$, we transform

$$x' = -8(x - 2) + 2(y - 8) + R_F(x - 2, y - 8)$$
$$y' = -8(x - 2) + R_G(x - 2, y - 8)$$

into the system:

$$u' = -8u + 2v + R_F(u, v)$$
$$v' = -8u + R_G(u, v).$$

If we ignore $R_F(u, v)$ and $R_G(u, v)$, we obtain the linear system:

$$u' = -8u + 2v$$
$$v' = -8u.$$

The solution to this system is:

$$u = ((c_2 - 4c_1) - 4c_2 t)e^{-4t},$$
$$v = -8(c_1 + c_2 t)e^{-4t}.$$

Notice that

$$\lim_{t \to \infty} u(t) = \lim_{t \to \infty} x(t) - 2 = 0$$

and

$$\lim_{t \to \infty} v(t) = \lim_{t \to \infty} y(t) - 8 = 0.$$

As a consequence, if we have an initial value problem with initial conditions $x(0) = b_1$, $y(0) = b_2$ at $t = 0$ that are sufficiently close to $(2, 8)$, the solution $(x(t), y(t))$ to the initial value problem will also approach $(2, 8)$ as $t \to \infty$.

We now consider the equilibrium solution $(0, 0)$. At $(0, 0)$ we obtain

$$F(x, y) = F_x(0, 0)x + F_y(0, 0)y + R_F(x, y)$$
$$= 0x + 0y - 4x^2 + xy$$

and

$$G(x, y) = G_x(0, 0)x + G_y(0, 0)y + R_G(x, y)$$
$$= 0x + 2y - xy$$

where $R_F(x, y) = -4x^2 + xy$ and $R_G(x, y) = -xy$. If we ignore $R_F(x, y)$ and $R_G(x, y)$, we have

$$x' = 0$$
$$y' = 2y.$$

The solution to this system is $x = c_1$, $y = c_2 e^{2t}$. We have

$$\lim_{t \to \infty} x(t) = c_1$$

and

$$\lim_{t \to \infty} y(t) = \infty.$$

When either x or y approaches infinity as $t \to \infty$ for the solution of a system of differential equations, we say the solution is **unstable**; otherwise, the solution is said to be **stable.** An in-depth study of stability must be left for future courses. We are merely giving you a brief introduction to this here. At the equilibium solution $(0, 0)$ in the example we just considered, the linear system has an unstable solution; at the equilibrium solution $(2, 8)$, we have a stable solution for the linear system. Unstable solutions to linear systems are characterized by the fact that the real part of at least one of the associated eigenvalues of the linear system is positive. If all of the eigenvalues of the linear system do not have a positive real part, the solution is stable. The equilibrium solution of the nonlinear system and the corresponding linear system are the same, and the eigenvalues of the linear system help determine the stability of solutions to the nonlinear system. It can be shown that except when the eigenvalues are purely imaginary, the solutions to the nonlinear system have the same stability conditions as the linear system if the solution to the 2×2 nonlinear system is sufficiently close to the equilibrium solution. Figure 6.13 contains the direction field for the original nonlinear system

$$x' = -4x^2 + xy$$
$$y' = 2y - xy$$

along with the graphs of the curves $F(x, y) = 0$ and $G(x, y) = 0$ that appeared in Figure 6.12. Notice the behavior of this direction field near the point $(2, 8)$ where the linear system has a stable solution and and its behavior near the point $(0, 0)$ where the linear system has an unstable solution.

We now consider two applications that involve 2×2 systems.

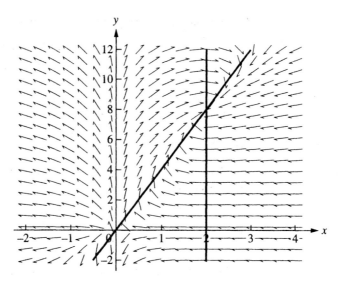

Figure 6.13

6.7.1 Predator-Prey

Wildlife biologists have studied the interaction of moose and wolves on the island Isle Royale in Lake Superior to determine the population cycle of these two species. The moose and wolves are confined to the island. The wolves prey on the moose. This problem is known as a predator-prey problem. Various systems of differential equations are used to model these types of interactions. The Austrian biologist A. J. Lotka (1880–1949) and the Italian mathematician Vito Volterra (1860–1940) are credited with the first system of differential equations models describing the predator-prey population relationship.

To illustrate one such model, let $w(t)$ represent the population of the predator and $m(t)$ represent the population of the prey at time t. Assume the moose have a birthrate of $\alpha m(t)$ and die at a rate β depending on the number of interactions between the moose and wolves. Suppose the number of interactions between the moose and wolves is proportional to $m(t)w(t)$, the product of the populations of each species. We assume the wolves die off at a rate proportional to the square of the population of the wolves (wolves die off due to interactions with themselves) and that their population increases at a rate depending on the number of interactions of the moose and wolves since the wolves get to eat when a moose is killed during an interaction. The equations describing these populations are then:

$$m'(t) = \alpha m(t) - \beta m(t)w(t) = m(t)(\alpha - \beta w(t))$$

and

$$w'(t) = -\gamma w(t)^2 + \delta m(t)w(t) = w(t)(-\gamma w(t) + \delta m(t)).$$

The equilibrium solutions of this predator-prey model are

$$(0, 0)$$

and

$$\left(\frac{\gamma\alpha}{\delta\beta}, \frac{\alpha}{\beta}\right).$$

The equilibrium solution $(0, 0)$ implies that the predators and prey both die off. The equilibrium solution $(\gamma\alpha/\delta\beta, \alpha/\beta)$ implies that the predators and prey coexist at constant populations with the population of the moose being $\gamma\alpha/\delta\beta$ and the population of the wolves being α/β.

The linear part of the predator-prey problem at the equilibrium $(0, 0)$ is

$$A = \begin{bmatrix} \alpha & 0 \\ 0 & 0 \end{bmatrix}.$$

The eigenvalues of A are α and 0. Since $\alpha > 0$, the equilibrium solution $(0, 0)$ is unstable. Hence this model indicates that if the moose and wolf populations are initially small they will increase. The linear part of the predator-prey problem at the equilibrium $(\gamma\alpha/\delta\beta, \alpha/\beta)$ is

$$A = \begin{bmatrix} 0 & -\dfrac{\gamma\alpha}{\delta} \\ \dfrac{\alpha\delta}{\beta} & -\dfrac{\alpha\delta}{\beta} \end{bmatrix}.$$

The eigenvalues of this system are more complicated to analyze for stability. In Figure 6.14 we present a direction field for $\alpha = 0.1$, $\beta = 0.0025$, $\delta = 0.025$, and $\gamma = 0.05$. In Exercises 8–11, we will ask you to determine the eigenvalues for different sets of α, β, γ, and δ.

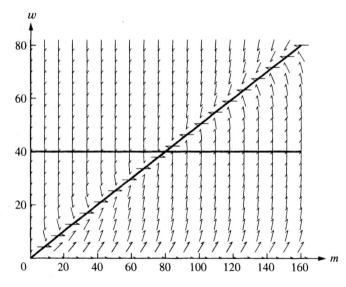

Figure 6.14

6.7.2 The Undamped Pendulum

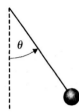

Figure 6.15

The equation describing the angle θ a pendulum bob makes with the vertical under undamped (frictionless) motion without an external applied force as illustrated in Figure 6.15 is given by

$$ml\theta'' = -mg \sin \theta,$$

where m is the mass of the bob, l is the length of the pendulum, and g is the acceleration of gravity. In Exercise 12 you are asked to derive these equations.

If we let $x = \theta$ and $y = \theta'$ we obtain the system:

$$x' = y$$

$$y' = -\frac{g}{l} \sin x.$$

The equilibrium solutions of this system are

$$(\pm k\pi, 0) \text{ for } k = 0, 1, 2, 3, \ldots$$

(why?). The equilibrium position of the pendulum bob for

$$k = 0, 2, 4, \ldots$$

is vertically down and the equilibrium position of the pendulum bob for

$$k = 1, 3, 5, \ldots$$

is vertically up (why?). Therefore, we expect that the equilibrium solutions are stable if k is even and unstable if k is odd (why?). We consider only the cases $k = 0$ and $k = 1$ (why?).

For $k = 0$ we have the equilibrium solution $(0, 0)$. It is easy to see that the linear part is

$$A = \begin{bmatrix} 0 & 1 \\ -g/l & 0 \end{bmatrix}.$$

The eigenvalues are

$$\sqrt{\frac{g}{l}} i \quad \text{and} \quad -\sqrt{\frac{g}{l}} i.$$

Since the real parts of the eigenvalues are 0, we have a stable solution. The solution to the linear system, which is

$$x = c_1 \cos \sqrt{\frac{g}{l}} t + c_2 \sin \sqrt{\frac{g}{l}} t,$$

indicates that if $x = \theta$ and $y = \theta'$ are initially small, the pendulum bob will oscillate at a fixed frequency.

For $k = 1$ we have the equilibrium solution $(\pi, 0)$ and the linear part is

$$A = \begin{bmatrix} 0 & 1 \\ g/l & 0 \end{bmatrix}.$$

The eigenvalues are

$$\sqrt{\frac{g}{l}} \quad \text{and} \quad -\sqrt{\frac{g}{l}}.$$

Since one of the eigenvalues is a positive real number, we have an unstable solution. In Exercises 13 and 14 we ask you to determine the motion of the pendulum under specific initial conditions.

EXERCISES 6.7

In Exercises 1–6, determine (a) the equilibrium solutions of the given 2×2 system, (b) the linear system at each equilibrium solution, (c) the general solutions of these linear systems, and (d) the stability of these linear systems. Use Maple or another appropriate software package to (e) graph the curves $F(x, y) = 0$ and $G(x, y) = 0$ and the direction field for the nonlinear system. Observe the stable or unstable behavior of the solutions of the nonlinear systems near the equilibrium solutions.

1. $x' = x + xy, \; y' = 2y - xy$

2. $x' = x - x^2 - xy, \; y' = 2y - y^2 - 3xy$

3. $x' = x - 2xy + xy^2, \; y' = y + xy$

4. $x' = 2y - xy, \; y' = x^2 - y^2$

5. $x' = 2 + x - 2e^{-2y}, \; y' = x - \sin y$

6. $x' = \sin y, \; y' = x^2 + y$

7. For $F(x, y) = -4x^2 + xy$ and $G(x, y) = 2y - xy$, show that

$$R_F(x - 2, y - 8) = -4(x - 2)^2 + (x - 2)(y - 8)$$

and

$$R_G(x - 2, y - 8) = -(x - 2)(y - 8).$$

Then show that $R_F(x-2, y-8)$ and $R_G(x-2, y-8)$ satisfy the limit conditions

$$\lim_{(x,y) \to (2,8)} \frac{R(x - 2, y - 8)}{\sqrt{(x - 2)^2 + (y - 8)^2}}$$

$$= \lim_{(u,v) \to (0,0)} \frac{R(u, v)}{\sqrt{u^2 + v^2}} = 0.$$

In Exercises 8–11, do parts (a–e) of Exercises 1–6 for the predator-prey problem

$$m'(t) = \alpha m(t) - \beta m(t)w(t) = m(t)(\alpha - \beta w(t))$$
$$w'(t) = -\gamma w(t)^2 + \delta m(t)w(t)$$
$$= w(t)(-\gamma w(t) + \delta m(t))$$

with the given values.

8. $\alpha = 1/2, \beta = 1/4, \gamma = 1/4, \delta = 1/4$

9. $\alpha = 1/2, \beta = 1/4, \gamma = 1/4, \delta = 3/4$

10. $\alpha = 1/2, \beta = 1/4, \gamma = 1/2, \delta = 3/4$

11. $\alpha = 3/4, \beta = 1/4, \gamma = 1/4, \delta = 1/4$

12. Derive the differential equation $ml\theta'' = -mg \sin \theta$ for the motion of the pendulum in Figure 6.15.

13. For the system

$$x' = y$$
$$y' = -\frac{g}{l} \sin x$$

of the undamped pendulum problem, determine the relation between x and y by solving the differential equation

$$\frac{dy}{dx} = \frac{dy/dt}{dx/dt} = \frac{y'}{x'}.$$

Use Maple or another appropriate software package to graph the solution to the initial value problem with initial conditions $x(0) = \pi/4$ and $y(0) = 0$ if the length of the pendulum is 6 in. Since the eigenvalues of the linearized undamped pendulum system are purely imaginary for the equilibrium $(0, 0)$, this is a way to show the solution is stable. Also, use Maple or another appropriate software package to graph the direction field of the system and observe the stable or unstable behavior near the equilibrium solutions.

14. For the undamped pendulum, determine the solution of the linear system at the equilibrium solution $(0, \pi)$ for $\theta(0) = 0$ and $\theta'(0) = 1$ if the length of the pendulum is 6 in.

15. The motion of a pendulum damped by friction is given by the differential equation $ml\theta'' + c\theta' + mg \sin\theta = 0$ where $c > 0$ is the friction constant. Convert this into a 2×2 system and determine the equilibrium solutions. Find the linear systems at the equilibrium solutions. Also, use Maple or another appropriate software package to graph the direction field of the system and observe the stable or unstable behavior near the equilibrium solutions for $m = 0.0125$ kg, $c = 0.02$, and $l = 10$ cm.

16. The original Lotka-Volterra equations for a predator-prey problem are

$$x' = -\alpha x + \beta xy$$
$$y' = \gamma y - \delta xy$$

where x is the predator and y is the prey. Determine the equilibrium solutions of this system. Find the linear systems at the equilibrium solutions, solve these systems, and determine the stability of these solutions. Finally, determine the relation between x

and y by solving the differential equation

$$\frac{dy}{dx} = \frac{dy/dt}{dx/dt} = \frac{y'}{x'}.$$

17. Do Exercise 16 with $\alpha = 3/4$, $\beta = 1/4$, $\gamma = 1$, and $\delta = 1/4$. Also, use Maple or another appropriate software package to graph the direction field of the system and observe the stable or unstable behavior near the equilibrium solutions.

18. One model for describing the competition between two species in which neither species is predator or prey is:

$$x' = (\alpha - \beta x - \gamma y)x$$
$$y' = (\delta - \mu x - \nu y)y.$$

Determine the equilibrium solutions of this system. Find the linear systems at the equilibrium solutions, solve these systems, and determine the stability of these solutions.

19. Do Exercise 18 with $\alpha = 1$, $\beta = 1/4$, $\gamma = 1/2$, $\delta = 1$, $\mu = 1/2$, and $\nu = 1/4$. Also, use Maple or another appropriate software package to graph the direction field of the system and observe the stable or unstable behavior near the equilibrium solutions.

The Laplace Transform

In Chapter 4, you learned techniques for solving an initial value problem of the form

$$a_n y^{(n)} + a_{n-1} y^{(n-1)} + \cdots + a_1 y' + a_0 y = g(t);$$

$$y^{(n-1)}(t_0) = k_{n-1}, \ldots, y'(t_0) = k_1, \qquad y(t_0) = k_0$$

containing a linear differential equation. In this chapter you are going to see another method that is often more convenient to use for solving these initial value problems than the Chapter 4 techniques. It is especially effective in the case where the function g has discontinuities, a case that we judiciously avoided in Chapter 4. Many engineering problems involve initial value problems where g has discontinuties, so having a method for such a case is quite valuable. The method involves a function first introduced by the French mathematician Pierre Simon Marquis de Laplace (1749–1827) called the Laplace transform. Later, the English electrical engineer Oliver Heaviside (1850–1925) used the Laplace transform to solve initial value problems. In the first section of this chapter, we acquaint you with the basic aspects of this transform. Then, in the second section and continuing throughout the next several sections of this chapter, we shall apply the Laplace transform to solve various initial value problems. The chapter concludes with a section on the use of the Laplace transform to solve initial value problems for systems of linear differential equations.

7.1 DEFINITION AND PROPERTIES OF THE LAPLACE TRANSFORM

The **Laplace transform,** denoted $\mathcal{L}$, is defined in terms of an improper integral as

$$\mathcal{L}(f) = \int_0^\infty e^{-st} f(t) \, dt = \lim_{b \to \infty} \int_0^b e^{-st} f(t) \, dt.$$

When we apply the Laplace transform $\mathcal{L}$ to a function f, $\mathcal{L}(f)$ is a function of s, as the following examples illustrate.

EXAMPLE 1   Determine the Laplace transform of the following function.

$$f(t) = 1$$

Solution   We have

$$\mathcal{L}(f) = \mathcal{L}(1) = \int_0^\infty e^{-st}(1)\,dt = \lim_{b\to\infty}\int_0^b e^{-st}(1)\,dt = \lim_{b\to\infty} -\frac{1}{s}e^{-st}\Big|_0^b$$

$$= \lim_{b\to\infty}\left(-\frac{1}{s}e^{-sb} + \frac{1}{s}\right) = \frac{1}{s}$$

if $s > 0$. If $s \le 0$, the integral diverges.   ●

EXAMPLE 2   Determine the Laplace transform of the following function.

$$f(t) = t$$

Solution   Using integration by parts,

$$\mathcal{L}(f) = \mathcal{L}(t) = \int_0^\infty e^{-st}t\,dt = \lim_{b\to\infty}\left(-\frac{t}{s}e^{-st}\Big|_0^b + \frac{1}{s}\int_0^b e^{-st}\,dt\right)$$

$$= \lim_{b\to\infty} -\frac{1}{s^2}e^{-st}\Big|_0^b = \frac{1}{s^2}$$

if $s > 0$. Again, if $s \le 0$, the integral diverges.   ●

EXAMPLE 3   Determine the Laplace transform of the following function.

$$f(t) = e^{at}$$

Solution   Here we obtain

$$\mathcal{L}(f) = \mathcal{L}(e^{at}) = \int_0^\infty e^{-st}e^{at}\,dt = \int_0^\infty e^{(a-s)t}\,dt = \lim_{b\to\infty}\frac{1}{a-s}e^{(a-s)t}\Big|_0^b = \frac{1}{s-a}$$

if $s > a$. If $s \le a$, the integral diverges.   ●

EXAMPLE 4   Determine the Laplace transform of the following function.

$$f(t) = \begin{cases} t, & 0 \le t \le 1, \\ 1, & t > 1 \end{cases}$$

Solution The graph of f appears in Figure 7.1. We have

$$\mathcal{L}(f) = \int_0^1 e^{-st} t \, dt + \int_1^\infty e^{-st} \cdot 1 \, dt.$$

Using integration by parts on the first integral,

$$\mathcal{L}(f) = \left(-\frac{t}{s} e^{-st} - \frac{1}{s^2} e^{-st} \right) \Big|_0^1 + \lim_{b \to \infty} \left(-\frac{1}{s} e^{-st} \Big|_1^b \right)$$

$$= -\frac{1}{s} e^{-s} - \frac{1}{s^2} e^{-s} + \frac{1}{s^2} + \frac{1}{s} e^{-s} = \frac{1}{s^2} - \frac{1}{s^2} e^{-s}$$

if $s > 0$; otherwise, it diverges. ●

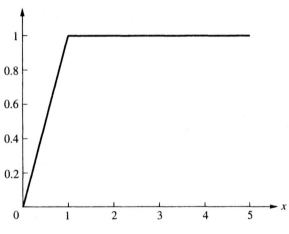

Figure 7.1

We shall consider the domain of $\mathcal{L}$ to consist of all functions f defined on the interval $[0, \infty)$ for which the improper integral

$$\int_0^\infty e^{-st} f(t) \, dt$$

converges for all values of s in some open interval of the form (a, ∞). Notice that the value of a depends on f as Examples 1–4 (and especially Example 3) illustrate. If we use V to denote the domain of $\mathcal{L}$ and recall that $F[0, \infty)$ is our notation for the vector space of all functions defined on the interval $[0, \infty)$, we will let you show that V is a subspace of $F[0, \infty)$ (see Exercise 43). We also shall let you show that $\mathcal{L}$ is a linear transformation on V (see Exercise 44).

We can use the fact that $\mathcal{L}$ is a linear transformation to find $\mathcal{L}$ of a linear combination of functions when we know the action of $\mathcal{L}$ on each of the functions. For instance, suppose we want to find

$$\mathcal{L}(5 - 3t + 6e^{2t}).$$

Since we know

$$\mathcal{L}(1) = \frac{1}{s}, \qquad \mathcal{L}(t) = \frac{1}{s^2}, \qquad \text{and } \mathcal{L}(e^{2t}) = \frac{1}{s-2}$$

from Examples 1, 2, and 3, we have

$$\mathcal{L}(5 - 3t + 6e^{2t}) = 5\mathcal{L}(1) - 3\mathcal{L}(t) + 6\mathcal{L}(e^{2t}) = \frac{5}{s} - \frac{3}{s^2} + \frac{6}{s-2} \quad \text{for } s > 2.$$

Besides the properties of a linear transformation possessed by the Laplace transform, here are some other important properties of $\mathcal{L}$:

THEOREM 7.1   If $\mathcal{L}(f) = F(s)$, then:

> **(1)** $\mathcal{L}(e^{at} f(t)) = F(s - a)$.
> **(2)** $\mathcal{L}(\int_0^t f(\tau) \, d\tau) = \frac{1}{s} F(s)$.
> **(3)** $F'(s) = -\mathcal{L}(t f(t))$.

Proof   We will show property 1 and leave the remaining two as exercises (see Exercises 45 and 46). Note that since

$$F(s) = \int_0^\infty e^{-st} f(t) \, dt,$$

$$F(s - a) = \int_0^\infty e^{-(s-a)t} f(t) \, dt = \int_0^\infty e^{-st} e^{at} f(t) \, dt = \mathcal{L}(e^{at} f(t)). \qquad \bullet$$

One use of the properties in Theorem 7.1 is that they sometimes can be used to more easily compute the Laplace transform of a function. Consider the following example.

EXAMPLE 5   Determine the Laplace transforms of $f(t) = t e^{at}$.

Solution   We could calculate

$$\mathcal{L}(t e^{at}) = \int_0^\infty e^{-st} \cdot t e^{at} \, dt = \int_0^\infty t e^{(a-s)t} \, dt$$

directly. Doing so, we will find we have to use integration by parts. (Try it.) Let us instead use property 3 of Theorem 7.1 and the result of Example 3:

$$\mathcal{L}(t e^{at}) = -\frac{d}{ds} \mathcal{L}(e^{at}) = -\frac{d}{ds} \left[\frac{1}{s-a} \right] = \frac{1}{(s-a)^2}. \qquad \bullet$$

Another way to do Example 5 is to use property 1 of Theorem 7.1 with $f(t) = t$ in conjunction with Example 2.

We do not consider the Laplace transform of a function f to exist if the integral $\int_0^\infty e^{-st} f(t) \, dt$ does not converge for all s in some open interval of the form (a, ∞). Exercise 48 has an example of a function f for which $\mathcal{L}(f)$ does not exist. The next theorem gives us a wide class of functions that do have a Laplace transform.

THEOREM 7.2   If f is continuous on $[0, \infty)$ with $|f(t)| \le Me^{at}$ for all t in $[0, \infty)$ where a is a constant and M is a positive constant, then

$$F(s) = \mathcal{L}(f) = \int_0^\infty e^{-st} f(t) \, dt$$

exists for all s in the interval (a, ∞).

We leave the proof of Theorem 7.2 as Exercise 47. A function f satisfying the condition $|f(t)| \le Me^{at}$ on $[0, \infty)$ for some $M > 0$ of Theorem 7.2 is called an **exponentially bounded** function on $[0, \infty)$. It can be shown that functions of the form $t^n e^{at}$, $t^n e^{at} \cos bt$, $t^n e^{at} \sin bt$ are exponentially bounded. We will use the fact that these functions have Laplace transforms in the next section when we use Laplace transforms to solve initial value problems.

We will also use the fact that the Laplace transform has an inverse, provided we impose some restrictions on the functions in the domain of $\mathcal{L}$. We develop this with the help of the following theorem whose proof is beyond the scope of this text.

THEOREM 7.3   If f and g are continuous on $[0, \infty)$ with

$$\mathcal{L}(f(t)) = \mathcal{L}(g(t)),$$

then

$$f(t) = g(t)$$

for $0 \le t \le \infty$.

We obtain an inverse Laplace transform much like we obtain the inverse trigonometric functions. Recall that we restrict the domain of the trigonometric functions in order to make them one-to-one to obtain inverse trigonometric functions. We will do something similar here by letting W be the set of continuous functions in the domain V of $\mathcal{L}$. Because $\mathcal{L}$ restricted to W is a one-to-one function by Theorem 7.3, the Laplace transform has an inverse on the range of $\mathcal{L}$ on W back to W defined as follows: If f is continuous on the interval $[0, \infty)$ with convergent integral $\int_0^\infty e^{-st} f(t) \, dt$ for all s in an open interval (a, ∞) and if

$$F(s) = \mathcal{L}(f(t)),$$

then

$$\mathcal{L}^{-1}(F(s)) = f(t).$$

To illustrate, from Examples 1–4, respectively, we have

$$\mathcal{L}^{-1}\left(\frac{1}{s}\right) = 1,$$

$$\mathcal{L}^{-1}\left(\frac{1}{s^2}\right) = t,$$

$$\mathcal{L}^{-1}\left(\frac{1}{s-a}\right) = e^{at},$$

and

$$\mathcal{L}^{-1}\left(\frac{1}{s^2} - \frac{1}{s^2}e^{-s}\right) = \begin{cases} t, & 0 \le t \le 1, \\ 1, & t > 1. \end{cases}$$

Since $\mathcal{L}$ satisfies the properties of a linear transform, so does $\mathcal{L}^{-1}$ by Exercise 25 of Section 5.2. Consequently, if $F_1(s)$, $F_2(s)$, ..., $F_n(s)$ are functions in the domain of $\mathcal{L}^{-1}$ and $c_1, c_2, \ldots, c_n$ are real numbers, then

$$\mathcal{L}^{-1}(c_1 F_1(s) + \cdots + c_n F_n(s)) = c_1 \mathcal{L}^{-1}(F_1(s)) + \cdots + c_n \mathcal{L}^{-1}(F_n(s)).$$

This linear combination property is frequently used along with tables of Laplace transforms of commonly occurring functions to determine the inverse Laplace transforms. Table 7.1 contains the Laplace transforms (and their corresponding inverses) of some functions frequently arising when solving initial value problems.

	$f(t) = \mathcal{L}^{-1}(F(s))$	$F(s) = \mathcal{L}(f(t))$
1	1	$\dfrac{1}{s}$
2	t^n	$\dfrac{n!}{s^{n+1}}, n = 0, 1, 2, 3, \ldots$
3	e^{at}	$\dfrac{1}{s - a}$, a any real number
4	$\cos bt$	$\dfrac{s}{s^2 + b^2}$
5	$\sin bt$	$\dfrac{b}{s^2 + b^2}$

Table 7.1

We use Table 7.1 along with our linear combination property for $\mathcal{L}^{-1}$ in the following examples to determine some inverse Laplace transforms.

EXAMPLE 6   Determine the inverse Laplace transform of the following function.

$$F(s) = \frac{2}{s^4}$$

Solution $\mathcal{L}^{-1}\left(\dfrac{2}{s^4}\right) = \mathcal{L}^{-1}\left(\dfrac{1}{3} \cdot \dfrac{3!}{s^4}\right) = \dfrac{1}{3}\mathcal{L}^{-1}\left(\dfrac{3!}{s^4}\right) = \dfrac{1}{3}t^3$ ●

EXAMPLE 7   Determine the inverse Laplace transform of the following function.

$$F(s) = \frac{1}{s^2 + 2s - 3}$$

Solution   In this example we use the partial fraction method you saw in calculus to write $F(s)$ in the form

$$F(s) = \frac{1}{s^2 + 2s - 3} = \frac{A}{s+3} + \frac{B}{s-1}.$$

Solving for A and B, we find $A = -1/4$, $B = 1/4$. Therefore, we have

$$F(s) = \frac{1}{s^2 + 2s - 3} = -\frac{1/4}{s+3} + \frac{1/4}{s-1}.$$

and hence

$$\mathcal{L}^{-1}(F(s)) = \mathcal{L}^{-1}\left(-\frac{1/4}{s+3} + \frac{1/4}{s-1}\right)$$

$$= -\frac{1}{4}\mathcal{L}^{-1}\left(\frac{1}{s+3}\right) + \frac{1}{4}\mathcal{L}^{-1}\left(\frac{1}{s-1}\right) = -\frac{1}{4}e^{-3t} + \frac{1}{4}e^{t}. \qquad \bullet$$

There are more extensive tables of Laplace transforms than the one in Table 7.1. However, as is now the case with many tables, they are falling into disuse since mathematical software packages such as Maple can be used to obtain Laplace transforms and inverse Laplace transforms of many functions. See Exercises 49–60 for some problems asking you to use a software package to do this. Table 7.1 has been designed to be large enough for our needs as we apply the Laplace transform and its inverse to solve the initial value problems of this chapter.

EXERCISES 7.1

Use the definition of the Laplace transform to determine the Laplace transform of the functions in Exercises 1–6.

1. 4   **2.** $3t$ **3.** $2t^2$

4. $t^2 - t$ **5.** e^{-4t} **6.** $2e^{3t}$

Use Table 7.1 and the properties of Theorem 7.1 if necessary to find the Laplace transform of the functions in Exercises 7–16.

7. $\cos 3t$ **8.** $\sin 4t$ **9.** $t \sin t$

10. $t \cos 3t$ **11.** $t^2 e^{3t}$ **12.** $3t e^{-2t}$

13. $e^{2t} \sin 3t$ **14.** $te^{-t} \cos t$ **15.** $e^{at} \cos bt$

16. $e^{at} \sin bt$

The hyperbolic cosine, cosh, is defined by

$$\cosh t = \frac{e^t + e^{-t}}{2}$$

and the hyperbolic sine, sinh, is defined by

$$\sinh t = \frac{e^t - e^{-t}}{2}.$$

In Exercises 17–22 use Table 7.1 and properties of the Laplace transform to find the Laplace transform of the given functions.

17. $\cosh 3t$ **18.** $\sinh 2t$ **19.** $2t \sinh t$

20. $3t \cosh 2t$ **21.** $e^{at} \cosh bt$ **22.** $e^{at} \sinh bt$

23. Show that if f is continuous on $[0, \infty)$ and if $\lim_{t\to\infty} f(t)/e^{at}$ is a finite number L for some constant a, then f is exponentially bounded on $[0, \infty)$.

In Exercises 24–34 use the result of Exercise 23 and l'Hôpital's rule to show that the given functions are exponentially bounded on $[0, \infty)$.

24. t^2 **25.** $4t^3$ **26.** $2t \cos t$

27. $t^2 \sin 3t$ **28.** $3t \cosh 2t$ **29.** $t^2 \sinh 3t$

30. $-2te^t \sin 4t$ **31.** $2t^3 e^{-7t}$ **32.** $t^n e^{at}$

33. $t^n e^{at} \cos bt$ **34.** $t^n e^{at} \sin bt$

Find the inverse Laplace transform of the functions in Exercises 35–42 with the aid of Table 7.1.

35. $\dfrac{4}{s}$ **36.** $\dfrac{8}{s^2}$ **37.** $\dfrac{4}{s-6}$

38. $\dfrac{2s}{s+1}$

39. $\dfrac{4}{s^2+2s-8}$

40. $\dfrac{3s-2}{s^2-4s-5}$

41. $\dfrac{2s}{s^2+9}$

42. $\dfrac{4}{s^3+4s}$

43. Show that the domain V we use for $\mathcal{L}$ is a subspace of $F[0, \infty)$.

44. Show the Laplace transform is a linear transformation on V.

45. Prove part (2) of Theorem 7.1.

46. Prove part (3) of Theorem 7.1.

47. a) Show that if f and g are continuous functions on an interval $[a, \infty)$ such that $|f(x)| \le g(x)$ for all x in this interval and if $\int_a^\infty g(x)\,dx$ converges, then $\int_a^\infty f(x)\,dx$ also converges.

b) Use the result of part (a) to prove Theorem 7.2.

48. Show the Laplace transform of e^{t^2} does not exist for any value of s.

To use Maple to find Laplace and inverse Laplace transforms, type and enter

```
with(inttrans);
```

first. The commands

```
laplace and invlaplace
```

are then used for finding Laplace and inverse Laplace tranforms of functions, respectively. Use Maple to find the Laplace transforms of the given functions in Exercises 49–54.

49. $\cos 2t$ **50.** $\sin 3t$ **51.** $3t \sin t$

52. $t^2 \cos 2t$ **53.** $-t^2 e^{3t} \cos 4t$ **54.** $4te^{-2t} \sin 3t$

In Exercises 55–60, use Maple to find the inverse Laplace transform of the given function.

55. $\dfrac{2}{s^2}$ **56.** $\dfrac{3}{2s-1}$ **57.** $\dfrac{s-2}{s^2-3s-4}$

58. $\dfrac{2s}{s^2+16}$ **59.** $\dfrac{2}{s+2} + \dfrac{3s}{s^2-3s-18}$

60. $\dfrac{2s}{3s-2} - \dfrac{4s^2}{s^2+2s+10}$

7.2   SOLVING CONSTANT COEFFICIENT LINEAR INITIAL VALUE PROBLEMS WITH LAPLACE TRANSFORMS

We are now going to see how the Laplace transform can be used to solve initial value problems of the form

$$a_n y^{(n)} + a_{n-1} y^{(n-1)} + \cdots + a_1 y' + a_0 y = g(t);$$
$$y^{(n-1)}(0) = k_{n-1}, \quad \cdots, \quad y'(0) = k_1, \quad y(0) = k_0.$$

The approach involves applying the Laplace tranform to each side of the differential equation. As we do this, we will use formulas for the Laplace transform applied to derivatives, which we develop now.

Applying integration by parts to the Laplace transform of y' with $u = e^{-st}$ and $dv = y'(t)\,dt$, we have

$$\mathcal{L}(y') = \int_0^\infty e^{-st} y'(t)\,dt = \lim_{b \to \infty} e^{-st} y(t)\Big|_0^b + s \int_0^\infty e^{-st} y(t)\,dt \qquad (1)$$
$$= -y(0) + s\mathcal{L}(y).$$

Next, replacing y by y' in Equation (1), we have

$$\mathcal{L}(y'') = -y'(0) + s\mathcal{L}(y')$$

which, upon using Equation (1) for $\mathcal{L}(y')$, gives us

$$\boxed{\mathcal{L}(y'') = -y'(0) + s(-y(0) + s\mathcal{L}(y)) = -y'(0) - sy(0) + s^2\mathcal{L}(y).} \qquad (2)$$

As exercises (see Exercises 24 and 25), we will let you continue this process and show that

$$\boxed{\mathcal{L}(y''') = -y''(0) - sy'(0) - s^2y(0) + s^3\mathcal{L}(y)} \qquad (3)$$

and more generally that

$$\boxed{\mathcal{L}(y^{(n)}) = -y^{(n-1)}(0) - sy^{(n-2)}(0) - \cdots - s^{n-1}y(0) + s^n\mathcal{L}(y).} \qquad (4)$$

Let us now apply the Laplace transform to each side of the differential equation

$$a_n y^{(n)} + a_{n-1} y^{(n-1)} + \cdots + a_1 y' + a_0 y = g(t).$$

Doing so we get

$$\mathcal{L}(a_n y^{(n)} + a_{n-1} y^{(n-1)} + \cdots + a_1 y' + a_0 y) = \mathcal{L}(g(t))$$

or

$$\boxed{a_n \mathcal{L}\left(y^{(n)}\right) + a_{n-1} \mathcal{L}\left(y^{(n-1)}\right) + \cdots + a_1 \mathcal{L}(y') + a_0 \mathcal{L}(y) = \mathcal{L}(g(t)).} \qquad (5)$$

The idea at this point is to insert the results of Equations (1–4) into Equation (5). This will give us an equation that we can solve for $\mathcal{L}(y)$ and then, by using the inverse Laplace tranform, find y. The following examples illustrate how this looks in practice.

EXAMPLE 1   Determine the solution to the following initial value problem.

$$y' - 2y = 0, \qquad y(0) = 2$$

Solution   Applying the Laplace transform to each side of the differential equation, we have

$$\mathcal{L}(y' - 2y) = \mathcal{L}(0)$$

or

$$\mathcal{L}(y') - 2\mathcal{L}(y) = 0.$$

Using Equation (1), we obtain

$$-y(0) + s\mathcal{L}(y) - 2\mathcal{L}(y) = 0$$

or

$$(s - 2)\mathcal{L}(y) = 2.$$

We thus have

$$\mathcal{L}(y) = \frac{2}{s - 2}.$$

Applying the inverse Laplace transform gives us

$$y = 2e^{2t}.$$

●

EXAMPLE 2 Determine the solution to the following initial value problem.

$$y'' + y = 0; \qquad y(0) = 1, \qquad y'(0) = 0$$

Solution Applying the Laplace transform to each side of the differential equation gives us

$$\mathcal{L}(y'') + \mathcal{L}(y) = \mathcal{L}(0).$$

Using Equations (1) and (2), this equation becomes

$$-y'(0) - sy(0) + s^2\mathcal{L}(y) + \mathcal{L}(y) = -s + (s^2 + 1)\mathcal{L}(y) = 0.$$

Solving for $\mathcal{L}(y)$ gives us

$$\mathcal{L}(y) = \frac{s}{s^2 + 1}.$$

Finally, using the inverse Laplace transform, we find our solution is

$$y = \cos t.$$

●

EXAMPLE 3 Determine the solution to the following initial value problem.

$$2y'' + 3y' - 2y = 1; \qquad y(0) = 0, \qquad y'(0) = \frac{1}{2}$$

Solution After applying the Laplace transform and simplifying (try it), we are led to

$$-1 + (2s^2 + 3s - 2)\mathcal{L}(y) = \frac{1}{s}$$

and hence

$$\mathcal{L}(y) = \frac{1}{s(2s^2 + 3s - 2)} + \frac{1}{2s^2 + 3s - 2}.$$

We now use the method of partial fractions to determine the inverse Laplace transform. Let us work on the second fraction first. Its partial fraction decomposition has the form

$$\frac{1}{2s^2 + 3s - 2} = \frac{1}{(2s - 1)(s + 2)} = \frac{A}{2s - 1} + \frac{B}{s + 2},$$

where we find $A = 2/5$ and $B = -1/5$. Thus

$$\frac{1}{2s^2 + 3s - 2} = \frac{2/5}{2s - 1} + \frac{-1/5}{s + 2} = \frac{1/5}{s - 1/2} + \frac{-1/5}{s + 2}.$$

The inverse Laplace transform of this expression is

$$\frac{1}{5}e^{t/2} - \frac{1}{5}e^{-2t}.$$

Using property 2 of Theorem 7.1, we find that the inverse Laplace transform of

$$\frac{1}{s(2s^2 + 3s - 2)} = \frac{1}{s} \cdot \frac{1}{2s^2 + 3s - 2}$$

is

$$\int_0^t \left(\frac{1}{5}e^{\tau/2} - \frac{1}{5}e^{-2\tau} \right) d\tau = \frac{2}{5}e^{t/2} + \frac{1}{10}e^{-2t} - \frac{1}{2}.$$

Therefore, adding the two parts gives us

$$y = \frac{1}{5}e^{t/2} - \frac{1}{5}e^{-2t} + \frac{2}{5}e^{t/2} + \frac{1}{10}e^{-2t} - \frac{1}{2}$$

$$= \frac{3}{5}e^{t/2} - \frac{1}{10}e^{-2t} - \frac{1}{2}. \qquad \bullet$$

Each of the initial value problems in Examples 1–3 can be solved by the methods of Chapter 4, which involve first finding the general solution to the differential equation and then using the initial conditions to find the constants in the general solution. Notice one advantage of the Laplace transform approach: We do not have to solve for the constants! In the introduction to this chapter it was mentioned that the Laplace transform approach is also useful for solving initial value problems with constant coefficient linear differential equations when the function $g(t)$ on the right-hand side is discontinuous. We will begin to study this in the next section.

EXERCISES 7.2

In Exercises 1–16, solve the initial value problems using Laplace transforms.

1. $y' - 2y = 0,\ y(0) = 1$

2. $y' + 3y = 0,\ y(0) = 2$

3. $y'' + 8y' = 0;\ y(0) = -1,\ y'(0) = 0$

4. $y'' - y' - 6y = 0;\ y(0) = 2,\ y'(0) = 1$

5. $y'' + 6y' + 9y = 0;\ y(0) = 1,\ y'(0) = 2$

6. $y'' - 4y' + 4y = 0;\ y(0) = 0,\ y'(0) = 1$

7. $4y'' + 16y = 0;\ y(0) = 0,\ y'(0) = 1$

8. $2y'' - 8y' + 14y = 0;\ y(0) = -1,\ y'(0) = 0$

9. $y''' + 4y'' - y' - 4y = 0;\ y(0) = -1,\ y'(0) = 0,$
  $y''(0) = 1$

10. $y''' - 2y' + 4y = 0;\ y(0) = 0,\ y'(0) = 2,\ y''(0) = 0$

11. $y' + 3y = 2,\ y(0) = 0$

12. $y' - 4y = 5,\ y(0) = 1$

13. $y'' + 4y' + 3y = 2 - e^{3t};\ y(0) = 0,\ y'(0) = 0$

14. $y'' - y = 5e^t;\ y(0) = 0,\ y'(0) = -1$

15. $y''' - 4y'' + 5y' = 7 - 2\cos t;\ y(0) = 1,\ y'(0) = 0,$
  $y''(0) = 2$

16. $y''' - 5y'' + 3y' + 9y = 2 - 3e^{-t};\ y(0) = 0,$
  $y'(0) = -1,\ y''(0) = 1$

Use Laplace transforms to solve the initial value problems that arise in Exercises 17–23.

17. An amoeba population in a jar of water starts at 1000 amoeba and doubles in 30 minutes. After this time, amoeba are removed continuously from the jar at a rate of 1000 amoeba per hour. Assuming continuous growth, what will the amoeba population in the jar be after 4 hr?

18. A 500-gal tank contains 200 gal fresh water. Salt water containing 1/2 lb salt per gallon enters the tank at 4 gal/min. The well-stirred solution leaves the tank at the same rate. How much salt is in the tank after 10 min?

19. A cup of liquid is put in a microwave oven and heated to 180° F. The cup is then taken out of the microwave and put into a room with a constant temperature of 70° F. After 2 min the liquid has cooled to 160° F. What will the temperature of the liquid be at 10 min?

20. A 2-lb object stretches a spring 6 in. The mass-spring system is placed on a horizontal track that has a cushion of air eliminating friction. The object is given an initial velocity of 2 ft/sec. An external force of 3 lb is applied to this system. Determine the position of the object at time t.

21. A 5-kg mass stretches a spring 25 cm. The mass-spring system is hung vertically in a tall tank filled with oil offering a resistance to the motion of 20 kg/sec. The object is pulled 25 cm from rest and given an initial velocity of 50 cm/sec. If an external force of $3e^{-t/10}$ newtons is applied to this system, determine the position of the object at time t.

22. Determine the current at time t in an RLC circuit that has $L = 0.5$, $R = 2$, $C = 1$, and $E(t) = 0$ with initial charge 1 and initial current 1.

23. Determine the current at time t in an RLC circuit that has $L = 0.5$, $R = 2$, $C = 2/3$, and $E(t) = 3\sin(2t)$ with no initial charge or current.

24. Verify Equation (3).   25. Verify Equation (4).

26. Verify that when the Laplace transform is applied to the differential equation in the initial value problem

$$a_n y^{(n)} + a_{n-1} y^{(n-1)} + \cdots + a_1 y' + a_0 y = g(t);$$
$$y^{(n-1)}(0) = k_{n-1}, \quad \ldots, \quad y'(0) = k_1, \, y(0) = k_0$$

we obtain the equation

$$-a_n y^{(n-1)}(0) - (a_n s + a_{n-1}) y^{(n-2)}(0) - \cdots$$
$$- (a_n s^{n-1} + a_{n-1} s^{n-2} + \cdots + a_1) y(0)$$
$$+ p(s)\mathcal{L}(y) = \mathcal{L}(g(t))$$

where $p(\lambda)$ is the characteristic polynomial of the corresponding homogeneous differential equation.

7.3 STEP FUNCTIONS, IMPULSE FUNCTIONS, AND THE DELTA FUNCTION

One of the strengths of the Laplace transform lies in the fact that we can use it to solve constant coefficient linear differential equations

$$a_n y^{(n)} + a_{n-1} y^{(n-1)} + \cdots + a_1 y' + a_0 y = g(t)$$

with initial conditions

$$y^{(n-1)}(0) = k_{n-1}, \quad \ldots, \quad y'(0) = k_1, \quad y(0) = k_0$$

for a large class of g. As promised at the end of the previous section, in this section you will see that we can use the Laplace transform to solve such initial value problems for some discontinuous functions g. In many applications g is discontinuous. For example, in a circuit problem g might model a switch that can be turned on or off in a discontinuous manner.

One case of practical significance where discontinuous functions g arise involves a type of function called a **unit step function** or **Heaviside function**. These are functions of the form

$$u_a(t) = \begin{cases} 0, & t < a, \\ 1, & t \geq a \end{cases}$$

where a is a positive constant. We can think of $u_a(t)$ as a switch that is off (with $u_a(t) = 0$) before time a and is on (with $u_a(t) = 1$) at and after time a. Notice that these unit step functions are discontinuous at $t = a$. The graph of u_4 appearing in Figure 7.2 was obtained using Maple. Observe how Maple puts in a vertical line segment at the discontinuity $t = 4$. This vertical line segment is not part of the graph of u_4. Graphing devices frequently do this since they graph by "connecting the dots"; that is, they plot points and sketch the graph connnecting these plotted points whether the points should be connected or not. As we use Maple to graph functions in this text, it is to be understood that any vertical line segments at discontinuities should be erased to obtain a correct sketch of the graph.

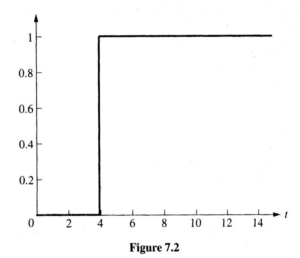

Figure 7.2

Calculating the Laplace transform of u_a, we have

$$\mathcal{L}(u_a(t)) = \int_0^a e^{-st} \cdot 0\, dt + \int_a^\infty e^{-st}\, dt = \int_a^\infty e^{-st}\, dt = \lim_{b \to \infty} -\frac{1}{s} e^{-st}\bigg|_a^b = \frac{1}{s} e^{-as}$$

for $s > 0$. For example,

$$\mathcal{L}(u_2(t)) = \frac{1}{s} e^{-2s}.$$

Cases where a switch is repeatedly turned on and off can be expressed in terms of unit step functions. For example, suppose a switch initially off is turned on at $t = 1$ and then turned off again at $t = 2$. We can express this with the function w given as

$$w(t) = u_1(t) - u_2(t).$$

(Why?) The graph of w appears in Figure 7.3.

For its Laplace transform, we have

$$\mathcal{L}(w(t)) = \mathcal{L}(u_1(t)) - \mathcal{L}(u_2(t)) = \frac{1}{s} e^{-s} - \frac{1}{s} e^{-2s}.$$

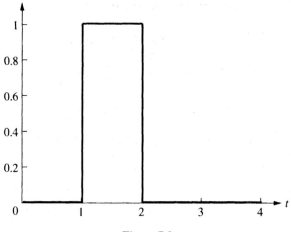

Figure 7.3

As another illustration of the use of these unit step functions, suppose the switch follows a pattern given by a function f at time t. If the switch is turned on at time $t = a$, then its behavior is described by the function g where

$$g(t) = \begin{cases} 0, & t < a, \\ f(t-a), & t \geq a. \end{cases}$$

An example of a graph of such a g appears in Figure 7.4 where the portion of the graph of g for $t \geq 4$ is the graph of $f(t)$ with $t \geq 0$ shifted to the right 4 units. Notice that our unit step function allows us to write g as

$$g(t) = u_a(t) f(t-a).$$

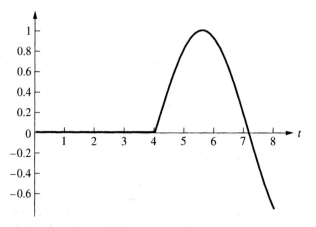

Figure 7.4

The Laplace transform of g is

$$\mathcal{L}(g(t)) = \mathcal{L}(u_a(t)f(t-a)) = \int_0^\infty e^{-st} u_a(t)f(t-a)\,dt$$

$$= \int_a^\infty e^{-st} u_a(t)f(t-a)\,dt = \int_a^\infty e^{-st} f(t-a)\,dt.$$

Substituting $\tau = t - a$ gives us

$$\int_a^\infty e^{-st} f(t-a)\,dt = \int_0^\infty e^{-s(\tau+a)} f(\tau)\,d\tau = e^{-as}\int_0^\infty e^{-s\tau} f(\tau)\,d\tau = e^{-as}\mathcal{L}(f(t)).$$

Hence

$$\boxed{\mathcal{L}(u_a(t)f(t-a)) = e^{-as}\mathcal{L}(f(t)).} \qquad (1)$$

For example,

$$\mathcal{L}(u_\pi(t)\cos t) = \mathcal{L}(-u_\pi(t)\cos(t-\pi)) = -e^{-\pi s} \cdot \frac{s}{s^2+1}.$$

As another example, suppose we have to find the inverse Laplace transform[1] of

$$F(s) = \frac{4e^{-2s}}{s-2}.$$

We write $F(s)$ as

$$F(s) = e^{-2s} \cdot \frac{4}{s-2}$$

and note that

$$\mathcal{L}(4e^{2t}) = \frac{4}{s-2}.$$

If we let $g(t) = 4e^{2t}$, we have that

$$f(t) = \mathcal{L}^{-1}(F(s)) = u_2(t)g(t-2) = 4u_2(t)e^{2(t-2)}.$$

We now solve some initial value problems that involve the unit step functions and Laplace transforms.

---

[1] We are abusing notation and terminology here. Since we are dealing with discontinuous functions, Theorem 7.3 no longer applies and the Laplace transform is no longer a one-to-one function. Consequently, we no longer have an inverse function in the usual sense. Nevertheless, we still will speak of the inverse Laplace transform and, if f is a function so that $F(s) = \mathcal{L}(f(t))$, will write $\mathcal{L}^{-1}(F(s)) = f(t)$. Keep in mind, however, that f need not be unique; that is, there may be other functions g so that $\mathcal{L}^{-1}(F(s)) = g(t)$. Some like to describe this by saying that $\mathcal{L}^{-1}$ is a *multivalued function*.

EXAMPLE 1   Determine a solution to the intial value problem

$$y'' + 2y' - 3y = \begin{cases} t, & 1 \le t < 2, \\ 0, & 0 \le t < 1 \text{ or } t \ge 2 \end{cases}$$

$$y(0) = 2, \quad y'(0) = 0.$$

Solution   We express the right-hand side in terms of unit step functions as

$$g(t) = tu_1(t) - tu_2(t).$$

In order to apply the Laplace formula for the unit step functions in Equation (1), we rewrite this as

$$g(t) = (t-1)u_1(t) - (t-2)u_2(t) + u_1(t) - 2u_2(t).$$

Applying the Laplace transform, we have

$$\mathcal{L}(g(t)) = \mathcal{L}((t-1)u_1(t) - (t-2)u_2(t) + u_1(t) - 2u_2(t))$$

$$= e^{-s}\mathcal{L}(t) - e^{-2s}\mathcal{L}(t) + \frac{e^{-s}}{s} - 2 \cdot \frac{e^{-2s}}{s} = \frac{e^{-s}}{s^2} - \frac{e^{-2s}}{s^2} + \frac{e^{-s}}{s} - \frac{2e^{-2s}}{s}.$$

When we apply the Laplace transform to the differential equation, we obtain

$$(s^2 + 2s - 3)\mathcal{L}(y) = 2s + 4 + \frac{e^{-s}}{s^2} - \frac{e^{-2s}}{s^2} + \frac{e^{-s}}{s} - \frac{2e^{-2s}}{s},$$

which gives us

$$\mathcal{L}(y) = \frac{2s+4}{s^2+2s-3} + \frac{e^{-s}}{s^2(s^2+2s-3)} - \frac{e^{-2s}}{s^2(s^2+2s-3)}$$

$$+ \frac{e^{-s}}{s(s^2+2s-3)} - 2\frac{e^{-2s}}{s(s^2+2s-3)}.$$

Next (leaving out the details) we find

$$\mathcal{L}^{-1}\left(\frac{2s+4}{s^2+2s-3}\right) = \frac{3}{2}e^t + \frac{1}{2}e^{-3t},$$

$$\mathcal{L}^{-1}\left(\frac{1}{s(s^2+2s-3)}\right) = -\frac{1}{3} + \frac{1}{4}e^t + \frac{1}{12}e^{-3t},$$

and

$$\mathcal{L}^{-1}\left(\frac{1}{s^2(s^2+2s-3)}\right) = -\frac{1}{3}t - \frac{2}{9} + \frac{1}{4}e^t - \frac{1}{36}e^{-3t}.$$

Hence using Equation (1), we have that

$$y(t) = \frac{3}{2}e^t + \frac{1}{2}e^{-3t} - \frac{5}{9}u_1(t) + \frac{1}{18}u_1(t)e^{-3(t-1)} + \frac{1}{2}u_1(t)e^{t-1} - \frac{t-1}{3}u_1(t)$$

$$+ \frac{8}{9}u_2(t) - \frac{5}{36}u_2(t)e^{-3(t-2)} - \frac{3}{4}u_2(t)e^{t-2} + \frac{t-2}{3}u_2(t).$$ ●

In Section 4.5 we saw that if a mass-spring system with no friction was put in motion by a sinusoidal external force with the natural frequency, then resonance occurred. We now consider a model where we turn the force off.

EXAMPLE 2   Suppose a mass-spring system is on a horizontal track and that the mass is kept off the track by a cushion of air from a compressor. The mass is an object that weighs 4 lb and stretches the spring 6 in. Determine the motion of the object if the object is released 2 in. from equilibrium and an external force $\sin 8t$ is applied for the first π sec.

Solution   If we let y be the position of the mass, the differential equation with initial conditions describing the motion of the object attached to the spring is

$$\frac{1}{8}y'' + 8y = \begin{cases} \sin 8t, & 0 \le t \le \pi, \\ 0, & t > \pi \end{cases}$$

$$y(0) = \frac{1}{6}, \quad y'(0) = 0.$$

Let us rewrite the differential equation as

$$y'' + 64y = \begin{cases} 8 \sin 8t, & 0 \le t \le \pi, \\ 0, & t > \pi. \end{cases} \tag{2}$$

We can express the right-hand side of Equation (2) as

$$g(t) = 8 \sin 8t - 8u_\pi(t) \sin 8(t - \pi).$$

Applying the Laplace transform to Equation (2) gives us

$$(s^2 + 64)\mathcal{L}(y) = \frac{s}{6} + \frac{64}{s^2 + 64} - \frac{64e^{-\pi s}}{s^2 + 64},$$

so that

$$\mathcal{L}(y) = \frac{s}{6(s^2 + 64)} + \frac{64}{(s^2 + 64)^2} - \frac{64e^{-\pi s}}{(s^2 + 64)^2}.$$

We then find

$$y(t) = \frac{1}{6} \cos 8t + \frac{1}{16} \sin 8t - \frac{1}{2}t \cos 8t$$

$$- \frac{1}{16}u_\pi(t) \sin 8(t - \pi) + \frac{1}{2}u_\pi(t)(t - \pi) \cos 8(t - \pi).$$

We graph the solution in Figure 7.5. Note that the amplitudes of the oscillations do not continue to grow over time, but instead stabilize. (Compare this result with that of Example 4 in Section 4.5.)   ●

We are next going to consider a function that arises in modeling impulses (for instance, electrical impulses). These functions also will be used to obtain a function whose Laplace transform is 1 (that is, a function for $\mathcal{L}^{-1}(1)$), which is something we do not have at this point.

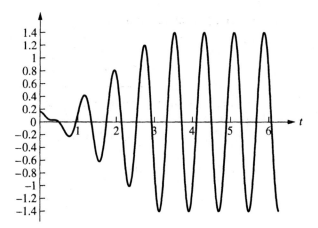

Figure 7.5

If k is a natural number, then the kth unit impulse function is defined by

$$\delta_{a,k}(t) = \begin{cases} k, & a < t < a + 1/k, \\ 0, & 0 \le t \le a \text{ or } t \ge a + 1/k \end{cases}$$
$$= ku_a(t) - ku_{a+1/k}(t).$$

The graph of the unit impulse function $\delta_{2,10}$ appears in Figure 7.6.

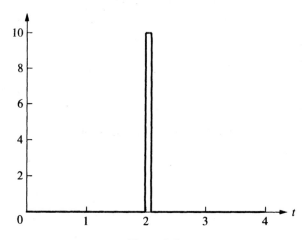

Figure 7.6

It is easy to see that

$$\int_0^\infty \delta_{a,k}(t)\, dt = 1$$

and that its Laplace transform is

$$\mathcal{L}(\delta_{a,k}(t)) = \int_a^{a+1/k} ke^{-st}\, dt = -k \cdot \frac{e^{-s(a+1/k)} - e^{-sa}}{s}.$$

Using l'Hôpital's rule (try it), we have

$$\lim_{k\to\infty} \mathcal{L}(\delta_{a,k}(t)) = \lim_{k\to\infty} e^{-as} \cdot \frac{1 - e^{-s/k}}{s/k} = e^{-as}.$$

In particular, we note that

$$\lim_{k\to\infty} \mathcal{L}(\delta_{0,k}(t)) = 1.$$

At this point we are going to think of the limit $\lim_{k\to\infty} \mathcal{L}(\delta_{0,k}(t))$ as becoming the Laplace transform of a function denoted by $\delta(t)$ called the **Dirac delta function**[2], or simply the **delta function**; that is,

$$\lim_{k\to\infty} \mathcal{L}(\delta_{0,k}(t)) = \mathcal{L}(\delta(t)).$$

The details justifying our ability to do this are left for future courses; we will simply assume it is possible to do so. This then gives us a function $\delta(t)$ so that

$$\mathcal{L}(\delta(t)) = 1 \quad \text{or} \quad \mathcal{L}^{-1}(1) = \delta(t).$$

EXAMPLE 3   Determine the solution to the initial value problem

$$y'' + y = \delta(t-2); \quad y(0) = 0, \quad y'(0) = 0.$$

Solution   Applying the Laplace transform gives us

$$\mathcal{L}(y) = \frac{e^{-2s}}{s^2 + 1}.$$

From the inverse Laplace transform we find

$$y = u_2(t)\sin(t-2) = \begin{cases} 0, & 0 \le t \le 2, \\ \sin(t-2), & t > 2. \end{cases}$$

Notice that the derivative of the solution in Example 3 is

$$y' = \begin{cases} 0, & 0 \le t < 2, \\ \cos(t-2), & t > 2. \end{cases}$$

---

[2] This is named after the Nobel physicist Paul Dirac (1902–1984).

The graphs of y and y' appear in Figures 7.7 and 7.8, respectively. Observe that the solution to Example 3 is continuous, but that its derivative is discontinuous at $t = 2$. The initial value problem in Example 3 describes the motion of a frictionless mass-spring system at rest that is hit with an impulse force at 2 sec. The solution in Example 3 indicates that a discontinuous forcing function causes a jump in the velocity of the mass. Exercise 23 also asks you to consider the jump in the derivative caused by $\delta(t)$ for a specific initial value problem.

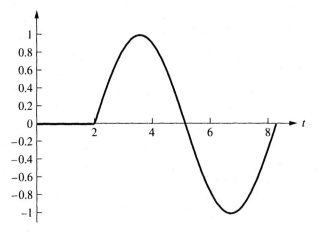

Figure 7.7

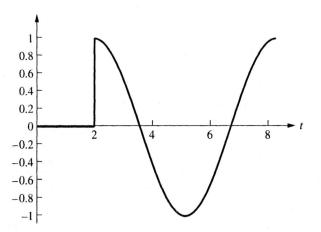

Figure 7.8

EXERCISES 7.3

In Exercises 1–6, (a) express the given function in terms of unit step functions, (b) graph the function, and (c) determine the Laplace transform of the function.

1. $f(t) = \begin{cases} t, & 0 \le t < 2, \\ 0, & t > 2 \end{cases}$

2. $g(t) = \begin{cases} e^t, & 0 \le t < 1, \\ 0, & t > 1 \end{cases}$

3. $\alpha(t) = \begin{cases} e^{2t}, & 2 \le t < 4, \\ 0, & 0 \le t < 2 \text{ or } t \ge 4 \end{cases}$

4. $v(t) = \begin{cases} \sin t, & 0 \le t < \pi, \\ 0, & t \ge \pi \end{cases}$

5. $h(t) = \begin{cases} \cos 2t, & \pi \le t < 2\pi, \\ 0, & 0 \le t < \pi \text{ or } t \ge 2\pi \end{cases}$

6. $\beta(t) = \begin{cases} 0, & 0 \le t < 2\pi, \\ 3 \sin 3t, & t \ge 2\pi \end{cases}$

In Exercises 7–10, (a) graph the given function and (b) determine the Laplace transform of the function.

7. $t u_2(t) - 2t u_4(t)$ **8.** $u_1(t) e^3 e^{-3t}$

9. $u_\pi(t) \cos t$ **10.** $t u_\pi(t) \sin t$

In Exercises 11–14, find a function f so that $f(t) = \mathcal{L}^{-1}(F(s))$ for the given function $F(s)$.

11. $\dfrac{e^{-2s}}{s}$ **12.** $\dfrac{e^{-3s}}{s^2}$

13. $\dfrac{e^{-s}}{s^2 - 4s}$ **14.** $\dfrac{e^{-2s}}{s^2 + 4}$

In Exercises 15–22, solve the initial value problems using Laplace transforms.

15. $y' + 3y = \begin{cases} t, & 0 \le t \le 2, \\ 0, & t > 2 \end{cases}$ $y(0) = 0$

16. $y' - 4y = 5t u_1(t), \; y(0) = 1$

17. $y'' + 4y' + 3y = e^{3t-3} u_1(t); \; y(0) = 0, \; y'(0) = 0$

18. $y'' - y = \begin{cases} 5e^t, & 0 \le t < 3, \\ 0, & t \ge 3 \end{cases}$
$y(0) = 0, \; y'(0) = -1$

19. $y'' - 4y' + 5y = \begin{cases} 7 - 3\cos 2t, & 0 \le t < 2\pi, \\ 0, & t \ge 2\pi \end{cases}$
$y(0) = 1, \; y'(0) = 0$

20. $y'' - 5y' - 36y = \begin{cases} 0, & 0 \le t < 2, \\ 2 - 3e^{-t}, & t \ge 2 \end{cases}$
$y(0) = 0, \; y'(0) = -1$

21. $y'' + 2y' - 8y = \delta(t); \; y(0) = 0, \; y'(0) = 0$

22. $y'' - 4y' - 3y = \delta(t - 2); \; y(0) = 0, \; y'(0) = 0$

23. Compare the solution to the initial value problem $y'' + y = 0; \; y(0) = 0, \; y'(0) = 1$ to the solution to the initial value problem $y'' + y = \delta(t); \; y(0) = 0, \; y'(0) = 0$. Discuss your findings with respect to the jump in the derivative caused by the delta function.

Use Laplace transforms to solve the problems in Exercises 24–31.

24. A 500-gal tank contains 200 gal fresh water. A saltwater water solution enters the tank at a rate of 4 gal/min and the well-stirred solution leaves the tank at the same rate. Suppose for the first 10 min the salt concentration of the solution entering the tank is 1/2 lb/gal and then the salt concentration of the entering solution is reduced to 1/4 lb/gal. How much salt is in the tank after 30 min?

25. An amoeba population in a jar of water starts at 1000 amoeba and doubles in 30 minutes. After this time, amoeba are removed continuously from the jar at a rate of 1000 amoeba per hour for 2 hr. After the end of these 2 hr, no more amoeba are removed from the jar. Assuming continuous growth, how many amoeba will be in the jar after 4 hr?

26. A 2-lb object stretches a spring 6 in. The mass-spring system is placed on a horizontal track that has a cushion of air eliminating friction. The object is given an initial velocity of 2 ft/sec. An external force of 3 lb is applied to this system for 5 sec and then shut off. Determine the position of the object.

27. A 5-kg mass stretches a spring 25 cm. The mass-spring system is hung vertically in a tall tank filled with oil offering a resistance to the motion of 20 kg/sec. The object is pulled 25 cm from rest and given an initial velocity of 50 cm/sec. If an external force of $3e^{-t/10}$ newtons is applied to this system for 3 sec and then stopped, determine the motion of the mass.

28. An *RLC* circuit that has $L = 0.5$, $R = 2$, $C = 2/3$ with no initial charge or current is connected to a generator that supplies the voltage $E(t) = 3\sin(2t)$. If the generator is turned on for π sec and then turned off, what is the voltage in the system?

29. An *RLC* circuit that has $L = 0.5$, $R = 2$, $C = 2/3$ with no initial charge or current is connected to a 6-volt battery through a switch. If the switch is turned on for 2 sec and then turned off, what is the voltage in the system?

30. Using the delta function, determine the position of the object in Exercise 26 if the external force of 3 lb is an impulse at $t = 0$ sec.

31. An *RLC* circuit that has $L = 0.5$, $R = 2$, $C = 2/3$ with no initial charge or current is connected to a 6-volt battery with a wire containing a switch. Use the delta function to find the charge in the system if the switch is turned on and off at π sec.

In Maple the unit step function $u_a(t)$ is typed as *Heaviside(t − a)* and the Dirac δ function typed as *Dirac(t)*. In Exercises 32–35, use Maple or another appropriate software package to determine the Laplace transform of the function.

32. $u_1(t)$

33. $tu_2(t)$

34. $e^t u_2(t) + \delta(t)$

35. $u_\pi(t)\cos t + 2\delta(t-2)$

In Exercises 36–39, use Maple or another appropriate software package to find a function f so that $f(t) = \mathcal{L}^{-1}(F(s))$ for the given function $F(s)$.

36. $\dfrac{e^{-2s}}{s-2}$

37. $\dfrac{e^{-s}}{s^2-9}$

38. $\dfrac{e^{-\pi s}}{s^2+4} - 4$

39. $e^{-3s}\left(\dfrac{2s}{s^2+1} + 2\right)$

7.4   CONVOLUTION INTEGRALS

Suppose we need to determine the inverse Laplace transform of

$$H(s) = \frac{1}{s^2 - 2s - 3}.$$

As we have seen, one way to do this is to determine the partial fraction decomposition of

$$H(s) = \frac{1}{s^2 - 2s - 3} = \frac{1}{(s-3)(s+1)},$$

which has the form

$$H(s) = \frac{A}{s-3} + \frac{B}{s+1}.$$

Doing so, we find $A = 1/4$ and $B = -1/4$ from which we have

$$\mathcal{L}^{-1}(H(s)) = \frac{1}{4}\mathcal{L}^{-1}(F(s)) - \frac{1}{4}\mathcal{L}^{-1}(G(s))$$

where

$$F(s) = \frac{1}{s-3} \quad \text{and} \quad G(s) = \frac{1}{s+1}.$$

We now are going to develop another way of determining $\mathcal{L}^{-1}(H(s))$ in terms of $\mathcal{L}^{-1}(F(s))$ and $\mathcal{L}^{-1}(G(s))$.

We do this by letting $f(t) = \mathcal{L}^{-1}(F(s))$ and $g(t) = \mathcal{L}^{-1}(G(s))$ and note that

$$e^{-s\tau}G(s) = \mathcal{L}(g(t-\tau)u_\tau(t))$$

$$= \int_0^\infty e^{-st}g(t-\tau)u_\tau(t)\,dt \qquad (1)$$

$$= \int_\tau^\infty e^{-st}g(t-\tau)\,dt.$$

We have that

$$F(s)G(s) = G(s)\mathcal{L}(f(\tau))$$

$$= G(s) \int_0^\infty e^{-s\tau} f(\tau)\, d\tau$$

$$= \int_0^\infty f(\tau) e^{-s\tau} G(s)\, d\tau.$$

Using Equation (1) to replace $e^{-s\tau} G(s)$ gives us

$$F(s)G(s) = \int_0^\infty f(\tau) \int_\tau^\infty e^{-st} g(t-\tau)\, dt\, d\tau = \int_0^\infty \int_\tau^\infty f(\tau) e^{-st} g(t-\tau)\, dt\, d\tau.$$

This is a double integral over the region shown in Figure 7.9. We can change the order of integration to obtain

$$\int_0^\infty \int_\tau^\infty f(\tau) e^{-st} g(t-\tau)\, dt\, d\tau = \int_0^\infty e^{-st} \int_0^t f(\tau) g(t-\tau)\, d\tau\, dt$$

$$= \mathcal{L}\left(\int_0^t f(\tau) g(t-\tau)\, d\tau \right).$$

We have, therefore, shown that

$$\mathcal{L}\left(\int_0^t f(\tau) g(t-\tau)\, d\tau \right) = \mathcal{L}(f(t))\mathcal{L}(g(t)) = F(s)G(s).$$

The corresponding inverse Laplace transform rule is

$$\boxed{\mathcal{L}^{-1}(F(s)G(s)) = \int_0^t f(\tau) g(t-\tau)\, d\tau.}$$

This integral is called the **convolution integral** of f and g and is a function of t. This function, denoted $f * g$, is called the **convolution product** of f and g; that is, the

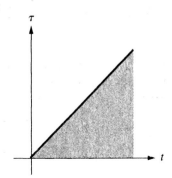

Figure 7.9

convolution product of f and g is the function $f * g$ given by

$$(f * g)(t) = \int_0^t f(\tau)g(t - \tau)\, d\tau.$$

We will also write $(f * g)(t)$ as $f(t) * g(t)$.

We now apply the convolution integral to some examples. We consider the problem introduced at the beginning of this section.

EXAMPLE 1   Determine the inverse Laplace transform of

$$\frac{1}{s^2 - 2s - 3}.$$

Solution   We rewrite this as

$$\frac{1}{s + 1} \cdot \frac{1}{s - 3}.$$

Since

$$\mathcal{L}(e^{-t}) = \frac{1}{s + 1}$$

and

$$\mathcal{L}(e^{3t}) = \frac{1}{s - 3},$$

we have that

$$\mathcal{L}^{-1}\left(\frac{1}{s^2 - 2s - 3}\right) =$$

$$e^{-t} * e^{3t} = \int_0^t e^{-\tau} e^{3(t-\tau)}\, d\tau = e^{3t} \int_0^t e^{-4\tau}\, d\tau = -e^{3t}\frac{e^{-4t} - 1}{4} = \frac{e^{3t}}{4} - \frac{e^{-t}}{4}.$$

(Compare this to the problem at the beginning of the section.)   ●

EXAMPLE 2   Solve the following initial value problem.

$$y' + y = \sin t, \qquad y(0) = 0.$$

Solution   Applying the Laplace transform gives us

$$(s + 1)\mathcal{L}(y) = \frac{1}{s^2 + 1}.$$

We have

$$\mathcal{L}(y) = \frac{1}{s + 1} \cdot \frac{1}{s^2 + 1}.$$

Therefore, our convolution rule tells us

$$y = e^{-t} * \sin t = \int_0^t e^{-\tau} \sin(t - \tau)\, d\tau = \frac{1}{2}(e^{-t} + \sin t - \cos t).$$ ●

Convolution integrals are also useful for solving some integral equations, which are equations similar to differential equations except that they involve integrals instead of derivatives. See Exercises 13–17 for details about this.

EXERCISES 7.4

In Exercises 1–4, use convolution integrals to determine the inverse Laplace transform of the given function.

1. $\dfrac{1}{s^2 - s}$

2. $\dfrac{3s}{s^3 + 4s}$

3. $\dfrac{1}{(s + 1)(s^2 + 1)}$

4. $\dfrac{3s - 1}{s^2 - 2s - 3}$

In Exercises 5 and 6, find the Laplace transform of the given function.

5. $\displaystyle\int_0^t (t - \tau)e^\tau\, d\tau$

6. $\displaystyle\int_0^t e^\tau \cos(t - \tau)\, d\tau$

In Exercises 7–10, use convolution integrals to solve the initial value problems.

7. $y'' + y = t;\ y(0) = 0,\ y'(0) = 1$
8. $y'' - 5y' + 4y = 0;\ y(0) = 1,\ y'(0) = 0$
9. $y'' + 4y = \sin 2t;\ y(0) = 1,\ y'(0) = 0$
10. $y'' - y' = 2;\ y(0) = 0,\ y'(0) = 0$
11. Find the convolution products $1 * (t * t)$ and $(1 * t) * t$.
12. Use the convolution integral to show that

$$\mathcal{L}\left(\int_0^t f(\tau)\, d\tau\right) = \frac{F(s)}{s}$$

where $F(s) = \mathcal{L}(f(t))$.

Convolution integrals can be used to solve a type of integral equation introduced by the mathematical biologist V. Volterra (1860–1940) in the early 1900s of the form

$$y(t) = h(t) + \int_0^t w(t - \tau)y(\tau)\, d\tau.$$

These can be rewritten as

$$y(t) = h(t) + w(t) * y(t).$$

Applying the Laplace transform gives us

$$\mathcal{L}(y(t)) = \mathcal{L}(h(t)) + \mathcal{L}(w(t))\mathcal{L}(y(t))$$

or

$$Y(s) = H(s) + W(s)Y(s)$$

where

$$Y(s) = \mathcal{L}(y(t)),\quad H(s) = \mathcal{L}(h(t)),$$

$$\text{and}\quad W(s) = \mathcal{L}(w(t)).$$

Solving for $Y(s)$, we obtain

$$Y(s) = \frac{H(s)}{1 - W(s)}.$$

Applying the inverse Laplace transform then gives us $y(t)$. In Exercises 13–16, solve the integral equations in this manner.

13. $y(t) = 3 + \int_0^t e^{t-\tau} y(\tau)\, d\tau$

14. $y(t) = t - \int_0^t (t - \tau)y(\tau)\, d\tau$

15. $y(t) = e^t + 2\int_0^t e^{2(t-\tau)} y(\tau)\, d\tau$

16. $y(t) = 4 - \int_0^t \cos(t - \tau)y(\tau)\, d\tau$

17. An initial value problem involving both integrals and derivatives such as

$$y'(t) = 4 + \int_0^t (t - \tau)y(\tau)\, d\tau,\qquad y(0) = 1$$

is called an integro-differential equation. Use the Laplace transform to solve this equation.

7.5 SYSTEMS OF LINEAR DIFFERENTIAL EQUATIONS

In this section we will see how to use the Laplace transform to solve initial value problems for systems of linear differential equations with constant coefficients. We start with a first order system.

EXAMPLE 1   Solve

$$y_1' = 2y_1 - y_2$$

$$y_2' = 5y_1 - 4y_2 + \begin{cases} 0, & 0 \le t < 1, \\ 1, & t \ge 1 \end{cases}$$

$$y_1(0) = 1, \qquad y_2(0) = 0.$$

Solution   We use the unit step function to first write the system as:

$$y_1' = 2y_1 - y_2$$
$$y_2' = 5y_1 - 4y_2 + u_1(t).$$

Now applying the Laplace transform to each equation in the system gives us

$$\mathcal{L}(y_1') = \mathcal{L}(2y_1 - y_2)$$
$$\mathcal{L}(y_2') = \mathcal{L}(5y_1 - 4y_2 + u_1(t))$$

or

$$\mathcal{L}(y_1') = 2\mathcal{L}(y_1) - \mathcal{L}(y_2)$$
$$\mathcal{L}(y_2') = 5\mathcal{L}(y_1) - 4\mathcal{L}(y_2) + \frac{1}{s}e^{-s}$$

Using the property

$$\mathcal{L}(y') = -y(0) + s\mathcal{L}(y)$$

in Equation (1) of Section 7.2, we have

$$-1 + s\mathcal{L}(y_1) = 2\mathcal{L}(y_1) - \mathcal{L}(y_2)$$
$$s\mathcal{L}(y_2) = 5\mathcal{L}(y_1) - 4\mathcal{L}(y_2) + \frac{1}{s}e^{-s}$$

or

$$(s - 2)\mathcal{L}(y_1) + \mathcal{L}(y_2) = 1$$
$$-5\mathcal{L}(y_1) + (s + 4)\mathcal{L}(y_2) = \frac{1}{s}e^{-s}.$$

Solving this system for $\mathcal{L}(y_1)$ and $\mathcal{L}(y_2)$, we find

$$\mathcal{L}(y_1) = \frac{s + 4}{(s + 3)(s - 1)} - \frac{1}{s(s + 3)(s - 1)}e^{-s}$$

$$\mathcal{L}(y_2) = \frac{5}{(s + 3)(s - 1)} + \frac{s - 2}{s(s + 3)(s - 1)}e^{-s}.$$

Applying the inverse Laplace transform leads to

$$y_1 = -\frac{1}{4}e^{-3t} + \frac{5}{4}e^t + \frac{1}{3}u_1(t) - \frac{1}{12}u_1(t)e^{-3t+3} - \frac{1}{4}u_1(t)e^{t-1}$$

$$y_2 = \frac{5}{4}e^t - \frac{5}{4}e^{-3t} + \frac{2}{3}u_1(t) - \frac{1}{4}u_1(t)e^{t-1} - \frac{5}{12}u_1(t)e^{-3t+3}.$$

The graph of the solutions is in Figure 7.10. (Which one is the graph of y_1 and which one is the graph of y_2? *Hint*: Use the initial conditions.) ●

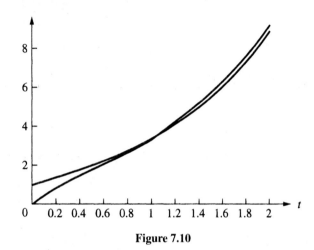

Figure 7.10

We now solve a second order system using the Laplace transform.

EXAMPLE 2 Solve

$$x_1'' = -10x_1 + 4x_2$$
$$x_2'' = 4x_1 - 4x_2 + \delta(t)$$

$$x_1(0) = 0, \quad x_1'(0) = 0; \quad x_2(0) = 0, \quad x_2'(0) = 0.$$

Solution Applying the Laplace transform and using the property

$$\mathcal{L}(y'') = -y'(0) - sy(0) + s^2\mathcal{L}(y)$$

in Equation (2) of Section 7.2 gives us

$$s^2\mathcal{L}(x_1) = -10\mathcal{L}(x_1) + 4\mathcal{L}(x_2)$$
$$s^2\mathcal{L}(x_2) = 4\mathcal{L}(x_1) - 4\mathcal{L}(x_2) + 1.$$

We rewrite this system as

$$(s^2 + 10)\mathcal{L}(x_1) - 4\mathcal{L}(x_2) = 0$$
$$4\mathcal{L}(x_1) - (s^2 + 4)\mathcal{L}(x_2) = -1.$$

Solving for $\mathcal{L}(x_1)$ and $\mathcal{L}(x_2)$ we find

$$\mathcal{L}(x_1) = \frac{4}{(s^2 + 2)(s^2 + 12)}$$

$$\mathcal{L}(x_2) = \frac{s^2 + 10}{(s^2 + 2)(s^2 + 12)}.$$

Applying the inverse Laplace transform gives us

$$x_1 = \frac{\sqrt{2}}{5} \sin \sqrt{2}t - \frac{\sqrt{3}}{15} \sin 3\sqrt{2}t$$

$$x_2 = \frac{2\sqrt{2}}{5} \sin \sqrt{2}t + \frac{\sqrt{3}}{30} \sin 3\sqrt{2}t.$$

The graph of the solutions appears in Figure 7.11. The lighter curve is the graph of x_2. ●

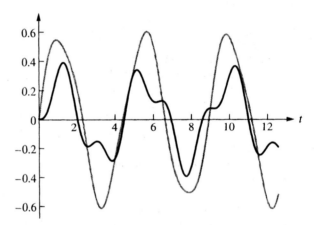

Figure 7.11

In the exercises we will ask you to solve systems that arose from applications considered in Chapter 6. Some of these will now involve the unit step function or the delta function.

EXERCISES 7.5

In Exercises 1–8, solve the initial value problems using the Laplace transform.

1. $y_1' = y_1 - y_2$

$$y_2' = 2y_1 + 4y_2 + \begin{cases} 0, & 0 \le t < 2, \\ 4, & t \ge 2 \end{cases}$$

$$y_1(0) = 0, \ y_2(0) = 1$$

2. $y_1' = 3y_1 + 5y_2$

$y_2' = y_1 - y_2 + t$

$y_1(0) = 1, \ y_2(0) = 1$

3. $y_1' = y_1 + y_2$

$y_2' = 4y_1 - 2y_2 + e^t u_2(t)$

$y_1(0) = 1, \ y_2(0) = 0$

4. $y_1' = 2y_1 - y_2$
$y_2' = 5y_1 - 4y_2 + \delta(t - 1)$
$y_1(0) = 0, y_2(0) = 0$

5. $x_1'' = 10x_1 - 5x_2$
$x_2'' = -12x_1 + 6x_2 + tu_1(t)$
$x_1(0) = 0, x_1'(0) = 0; x_2(0) = 0, x_2'(0) = 0$

6. $x_1'' = 4x_1 + \delta(t)$
$x_2'' = -3x_1 + x_2$
$x_1(0) = 1, x_1'(0) = 0; x_2(0) = 0, x_2'(0) = 0$

7. $x_1'' = -3x_1 + 2x_2 + \delta(t)$
$x_2'' = x_1 - 4x_2 + \delta(t)$
$x_1(0) = 0, x_1'(0) = 1; x_2(0) = 0, x_2'(0) = 1$

8. $x_1'' = -10x_1 + 13x_2 + tu_2(t)$
$x_2'' = 52x_1 - 10x_2 + \delta(t)$
$x_1(0) = 0, x_1'(0) = 0; x_2(0) = 0, x_2'(0) = 0$

Use the Laplace transform to solve Exercises 9–14.

9. Two tanks of volume 100 gal are connected together by two pipes. The first tank initially contains a well-mixed solution of 4 lb salt in 50 gal water. The second tank initially contains 100 gal salt-free water. A pipe from tank 1 to tank 2 allows the solution in tank 1 to enter tank 2 at a rate of 3 gal/min. A second pipe from tank 2 to tank 1 allows the solution from tank 2 to enter tank 1 at a rate of 3 gal/min. Assume that the salt mixture in each tank is always well-stirred. How much salt is in each tank after 5 min?

10. Two tanks of volume 500 gal are connected together by two pipes. Each tank is filled with a liquid fertilizer. The fertilizer in tank 1 contains 10 lb crushed corn and the fertilizer in tank 2 contains 1 lb crushed corn. Assume the flow rate in the pipes is 25 gal/min. Determine the amount of corn in each tank at any time.

11. Using Kirchoff's law, determine the system of differential equations describing the charges for the circuit in Figure 6.10 if $R = 2, L = 2, C = 1/4$, and

$$E(t) = \begin{cases} 100, & 0 \le t \le 1, \\ 0, & t > 1. \end{cases}$$

Solve this system with initial conditions $Q_1(0) = 0$, $Q_2(0) = 0$; $i_1(0) = 0, i_2(0) = 0$.

12. Using Kirchoff's law determine the system of differential equations describing the voltages for the circuit in Figure 6.11 if $R_1 = 2, R_2 = 8, L = 8, C = 1/2$, and

$$E(t) = \begin{cases} 10 \sin t, & 0 \le t < \pi, \\ 0, & t \ge \pi. \end{cases}$$

Solve this system with initial conditions $Q_1(0) = 0$, $Q_2(0) = 0$; $i_1(0) = 0, i_2(0) = 0$.

13. Assuming no friction, determine the motion of the two masses in Figure 6.7 if $m_1 = 1$ kg, $m_2 = 1$ kg, $k_1 = 3, k_2 = 2$, and the two masses are at rest in their equilibrium positions and the delta function is applied to the lower mass at $t = 2$ sec.

14. Determine the motion of the two masses in Figure 6.7 if $k_1 = 2k_2, m_1 = 2m_2$, and the two masses are at rest in their equilibrium positions and the bottom mass is hooked up to a machine generating the external force $\sin t$ for 2 sec beginning at $t = 0$.

Power Series Solutions to Linear Differential Equations

The techniques we have seen for solving higher order linear differential equations in Chapters 4 and 7 center around the constant coefficient case. One commonly used approach for solving nonconstant coefficient linear differential equations involves finding power series solutions to the differential equations. Initially you might react to this by thinking that power series are unnatural and would not be useful solutions to know to a differential equation. But this is not a proper perspective. While you may feel some apprehension about dealing with functions represented as power series, in many instances power series are the best way to represent functions. One instance of this you have seen in your calculus classes occurs when approximating values of functions. A standard method for approximating $\sin 1° = \sin(\pi/180)$, for example, is to use a Maclaurin polynomial of $\sin x$ evaluated at $\pi/180$, which is the same as a truncation of the Maclaurin series of $\sin x$ evaluated at $\pi/180$.

From the approximation point of view, if you have studied Section 3.9 you might ask: "Why not leave out power series techniques and just apply numerical methods?" One reason is that there are points at which numerical methods do not work well for approximating solutions to differential equations. Also power series techniques are important in the study of numerical solutions since many numerical methods are based on power series techniques. Further, numerical considerations are not the only reason for studying power series solutions. For instance, many interesting phenomena in mathematical physics have been discovered through the power series methods.

There are entire texts that deal with the complexities and peculiarities of power series methods for differential equations. We give only an overview here and leave many of the technical issues for future courses. We begin the first section of this chapter with a review of power series before we consider power series solutions to differential equations.

8.1   INTRODUCTION TO POWER SERIES SOLUTIONS

Recall that the **Taylor polynomial** of a function $f(x)$ of degree n about a fixed value x_0 is

$$
\begin{aligned}
T_n(x) &= f(x_0) + f'(x_0)(x - x_0) + \frac{f''(x_0)}{2}(x - x_0)^2 + \frac{f'''(x_0)}{3!}(x - x_0)^3 \\
&\quad + \cdots + \frac{f^{(n)}(x_0)}{n!}(x - x_0)^n \\
&= \sum_{k=0}^{n} \frac{f^{(k)}(x_0)}{k!}(x - x_0)^k.
\end{aligned}
$$

(Of course, for this nth degree Taylor polynomial to exist, f must have an nth derivative at x_0. Throughout this chapter we will assume any indicated derivatives exist.) In the particular case when $x_0 = 0$, the Taylor polynomial is called the **Maclaurin polynomial** of degree n. Taylor polynomials are useful for approximating values of functions and

$$
R_n(x, x_0) = f(x) - T_n(x),
$$

which is called the **remainder** of the Taylor polynomial of degree n at x_0, measures the error when we approximate $f(x)$ by $T_n(x)$. Various formulas are known for $R_n(x, x_0)$ that are used for estimating the error in such an approximation. The most commonly used one is that there is a value c between x_0 and x so that

$$
R_n(x_0) = \frac{f^{(n+1)}(c)}{(n+1)!}(x - x_0)^n.
$$

This is known as the **Lagrange form** of the remainder.

If we let n approach infinity, the nth degree Taylor polynomial becomes the **Taylor series** of f about x_0:

$$
\begin{aligned}
f(x_0) &+ f'(x_0)(x - x_0) + \frac{f''(x_0)}{2}(x - x_0)^2 + \frac{f'''(x_0)}{3!}(x - x_0)^3 \\
&+ \cdots + \frac{f^{(n)}(x_0)}{n!}(x - x_0)^n + \cdots \\
&= \sum_{n=0}^{\infty} \frac{f^{(n)}(x_0)}{n!}(x - x_0)^n.
\end{aligned}
$$

In the special case when $x_0 = 0$, we obtain the **Maclaurin series** of $f(x)$:

$$f(0) + f'(0)x + \frac{f''(0)}{2}x^2 + \frac{f'''(0)}{3!}x^3 + \cdots + \frac{f^{(n)}(0)}{n!}x^n + \cdots$$

$$= \sum_{n=0}^{\infty} \frac{f^{(n)}(0)}{n!}x^n.$$

If

$$\lim_{n \to \infty} R_n(x, x_0) = 0$$

for all x in an interval, then $f(x)$ is equal to the Taylor series on this interval. This often (but not always) occurs on the interval where the Taylor series converges. Some typical examples of this equality of a function and its Taylor series that you have seen in your calculus courses are

$$e^x = 1 + x + \frac{x^2}{2} + \frac{x^3}{3!} + \frac{x^4}{4!} + \cdots, \qquad -\infty < x < \infty,$$

$$\sin x = x - \frac{x^3}{3!} + \frac{x^5}{5!} - \frac{x^7}{7!} + \cdots, \qquad -\infty < x < \infty,$$

$$\cos x = 1 - \frac{x^2}{2} + \frac{x^4}{4!} - \frac{x^6}{6!} + \cdots, \qquad -\infty < x < \infty,$$

and

$$\ln x = (x-1) - \frac{(x-1)^2}{2} + \frac{(x-1)^3}{3} - \frac{(x-1)^4}{4} + \cdots, \qquad 0 < x \leq 2.$$

Taylor series are examples of power series. A **power series** about a fixed value x_0 is a series of the form

$$\sum_{n=0}^{\infty} a_n(x - x_0)^n.$$

Let us recall some facts about these series from your calculus courses. First, concerning their convergence, we have Theorem 8.1.

THEOREM 8.1   For a power series

$$y = \sum_{n=0}^{\infty} a_n(x - x_0)^n$$

there is a value $0 \leq R \leq \infty$ so that this power series converges absolutely[1] for all

---

[1] Recall that a series $\sum_{n=0}^{\infty} a_n$ converges absolutely if $\sum_{n=0}^{\infty} |a_n|$ converges. Absolute convergence is stronger than convergence; that is, if $\sum_{n=0}^{\infty} |a_n|$ converges, then $\sum_{n=0}^{\infty} a_n$ converges. There are series that converge but do not converge absolutely. Series such as the alternating harmonic series that converge but do not converge absolutely are called **conditionally convergent** series.

real values of x satisfying $|x - x_0| < R$ and diverges for all real values of x satisfying $|x - x_0| > R$.

The value of R in Theorem 8.1 is called the **radius of convergence** of the power series. The set of all values for which the power series converges is called the **interval of convergence** of the power series. Since the real values of x satisfying $|x - x_0| < R$ are the same as those in the open interval $(x_0 - R, x_0 + R)$, the interval of convergence always consists of this open interval with or without its endpoints depending on whether or not we have convergence at an endpoint. In the case when $R = 0$, we do have convergence at the (only) endpoint x_0 and the interval of convergence is the closed interval $[x_0, x_0]$ consisting of the single value x_0. In the case when $0 < R < \infty$, the interval of convergence has one of the forms $(x_0 - R, x_0 + R)$, $[x_0 - R, x_0 + R)$, $(x_0 - R, x_0 + R]$, or $[x_0 - R, x_0 + R]$ depending on whether or not the power series converges at the endpoints $x_0 - R$ and $x_0 + R$. In the case when $R = \infty$, the interval of convergence is $(-\infty, \infty)$. The value of R is often found by applying the ratio test or the root test to the absolute values of the terms of the power series.

One property that power series possess that will enter into our work here is that, between the endpoints of the interval of convergence, these infinite sums can be differentiated term by term just like finite sums.

THEOREM 8.2   Suppose that the power series

$$y = \sum_{n=0}^{\infty} a_n (x - x_0)^n$$

has radius of convergence $0 < R \leq \infty$. Then for all x in the open interval $|x - x_0| < R$,

$$y' = \frac{d}{dx}\left(\sum_{n=0}^{\infty} a_n (x - x_0)^n\right) = \sum_{n=0}^{\infty} \frac{d}{dx}(a_n (x - x_0)^n)$$

$$= \sum_{n=1}^{\infty} n a_n (x - x_0)^{n-1} = \sum_{n=0}^{\infty} (n + 1) a_{n+1} (x - x_0)^n.$$

Further, the power series $\sum_{n=1}^{\infty} n a_n (x - x_0)^{n-1}$ also has radius of convergence R.

We are going to use the fact that we can differentiate power series term by term to find solutions to initial value problems for linear differential equations where the initial conditions are specified at a value x_0 as a power series about x_0; that is, we will seek a solution of the form

$$y = \sum_{n=0}^{\infty} a_n (x - x_0)^n.$$

Another way to view this is to realize that we are going to specify the solution in terms of its Taylor series about x_0 since, as the following corollary to Theorem 8.2 tells us, power series and Taylor series are one and the same.

COROLLARY 8.3   If

$$f(x) = \sum_{n=0}^{\infty} a_n(x - x_0)^n$$

has radius of convergence $0 < R \leq \infty$, then

$$\boxed{a_n = \frac{f^{(n)}(x_0)}{n!}}$$

and hence the Taylor series of $f(x)$ expanded about x_0 is $\sum_{n=0}^{\infty} a_n(x - x_0)^n$.

A function f is said to be **analytic** at a point x_0 if $f(x)$ is equal to a power series on some open interval $|x - x_0| < R$ about x_0 (that is, if $f(x)$ can be expressed as

$$f(x) = \sum_{n=0}^{\infty} a_n(x - x_0)^n$$

for all x satisfying $|x - x_0| < R$) for some $R > 0$. By Corollary 8.3, saying that a function f is analytic at x_0 is equivalent to saying that $f(x)$ equals its Taylor series about x_0 on some open interval $|x - x_0| < R$ about x_0.

Finding the desired solution

$$y = \sum_{n=0}^{\infty} a_n(x - x_0)^n$$

to an initial value problem at x_0 involves finding the coefficients a_n of the terms to the power series. These coefficients are found by using the initial conditions and the differential equation as the following examples illustrate.

EXAMPLE 1   Determine a power series solution to the following linear initial value problem.

$$y' = 2y, \qquad y(0) = 1$$

Solution   In this problem, $x_0 = 0$ so that our desired solution will have the form $y = \sum_{n=0}^{\infty} a_n x^n$. Writing our differential equation in the form $y' - 2y = 0$ and substituting $y = \sum_{n=0}^{\infty} a_n x^n$ into it using our formula for y' from Theorem 8.2, we obtain

$$y' - 2y = \sum_{n=0}^{\infty} (n + 1)a_{n+1}x^n - 2\sum_{n=0}^{\infty} a_n x^n = 0.$$

This leads to

$$\sum_{n=0}^{\infty} ((n + 1)a_{n+1} - 2a_n)x^n = 0.$$

This equation holds if and only if

$$(n + 1)a_{n+1} - 2a_n = 0 \quad \text{for} \quad n = 0, 1, 2, \dots.$$

Solving for a_{n+1} gives us

$$a_{n+1} = \frac{2a_n}{n+1}.$$

From this equation we can generate all the coefficients in the power series for y. This equation is called a **recurrence relation.** Using $n = 0$ in Corollary 8.3, we have that

$$a_0 = \frac{y^{(0)}(0)}{0!} = y(0) = 1.$$

Now, from the recurrence relation we have:

$$a_1 = \frac{2a_0}{1} = 2$$

$$a_2 = \frac{2a_1}{2} = 2$$

$$a_3 = \frac{2a_2}{3} = \frac{4}{3}$$

$$\vdots$$

Clearly, we can generate as many coefficients of the power series as we want using the recurrence relation. We can even use the recurrence relation to obtain a formula for a_n. Notice that

$$a_1 = \frac{2}{1!}a_0$$

$$a_2 = \frac{2a_1}{2} = \frac{2^2}{2!}a_0$$

$$a_3 = \frac{2a_2}{3} = \frac{2^3}{3!}a_0$$

$$a_4 = \frac{2a_3}{4} = \frac{2^4}{4!}a_0$$

from which we see

$$a_n = \frac{2^n}{n!}a_0.$$

Since $a_0 = 1$, this gives us

$$a_n = \frac{2^n}{n!}.$$

We now have

$$y = 1 + 2x + \frac{2^2}{2!}x^2 + \frac{2^3}{3!}x^3 + \cdots = \sum_{n=0}^{\infty} \frac{2^n}{n!}x^n = \sum_{n=0}^{\infty} \frac{1}{n!}(2x)^n,$$

which is the Maclaurin series for e^{2x}. This is the same solution that would have been obtained using the techniques of Chapter 3 or 4. ●

EXAMPLE 2   Determine a power series solution to the following linear initial value problem.

$$y' = (x - 1)^2 y, \qquad y(1) = -1$$

Solution   In this problem, $x_0 = 1$ and we substitute in $y = \sum_{n=0}^{\infty} a_n (x-1)^n$ into $y' - (x-1)^2 y = 0$ to obtain

$$y' - (x-1)^2 y = \sum_{n=0}^{\infty} (n+1)a_{n+1}(x-1)^n - \sum_{n=0}^{\infty} a_n (x-1)^{n+2} = 0.$$

The powers of $x-1$ do not agree in the two power series in this equation since the exponent in the first power series starts at 0 while in the second power series the exponent starts at 2. Let us rewrite the first power series to get the exponents to start at 2 in the summation part, which gives us

$$a_1 + 2a_2(x-1) + \sum_{n=2}^{\infty} (n+1)a_{n+1}(x-1)^n - \sum_{n=0}^{\infty} a_n(x-1)^{n+2} = 0.$$

We now reindex the first summation so that the index n starts at 0 rather than 2,

$$a_1 + 2a_2(x-1) + \sum_{n=0}^{\infty} (n+3)a_{n+3}(x-1)^{n+2} - \sum_{n=0}^{\infty} a_n(x-1)^{n+2} = 0,$$

giving us

$$a_1 + 2a_2(x-1) + \sum_{n=0}^{\infty} ((n+3)a_{n+3} - a_n)(x-1)^{n+2} = 0.$$

Therefore,

$$a_1 = 0, \qquad a_2 = 0, \qquad \text{and} \qquad (n+3)a_{n+3} - a_n = 0$$

and our recurrence relation is

$$a_{n+3} = \frac{a_n}{n+3}.$$

Consequently we have:

$$a_1 = 0$$
$$a_2 = 0$$
$$a_3 = \frac{a_0}{3}$$
$$a_4 = \frac{a_1}{4} = 0$$
$$a_5 = \frac{a_2}{5} = 0$$
$$a_6 = \frac{a_3}{6} = \frac{a_0}{3 \cdot 6}$$

$$a_7 = \frac{a_4}{7} = 0$$

$$a_8 = \frac{a_5}{8} = 0$$

$$a_9 = \frac{a_6}{9} = \frac{a_0}{3 \cdot 6 \cdot 9}$$

$$\vdots$$

It follows that only coefficients of the form a_{3n} for $n = 1, 2, 3, \ldots$ are nonzero and that

$$a_{3n} = \frac{1}{3^n n!} a_0.$$

Since $a_0 = y(1) = -1$, we have that

$$y = -\sum_{n=0}^{\infty} \frac{1}{3^n n!} (x - 1)^{3n} = -\sum_{n=0}^{\infty} \frac{[(x - 1)^3/3]^n}{n!}.$$

This is the Maclaurin series for $-e^{(x-1)^3/3}$, which is the same solution as obtained by using the methods of Chapter 3.   ●

EXAMPLE 3   Determine a power series solution to the following linear initial value problem.

$$y'' + y = 0; \qquad y(0) = 1, \qquad y'(0) = 0$$

Solution   Here $x_0 = 0$ and we substitute $y = \sum_{n=0}^{\infty} a_n x^n$ into $y'' + y = 0$. In order to do so, we need y'' which, applying Theorem 8.2 twice, is

$$y'' = \sum_{n=2}^{\infty} n(n - 1)a_n x^{n-2} = \sum_{n=0}^{\infty} (n + 2)(n + 1)a_{n+2} x^n.$$

Now substituting into the differential equation we obtain

$$y'' + y = \sum_{n=0}^{\infty} (n + 2)(n + 1)a_{n+2} x^n + \sum_{n=0}^{\infty} a_n x^n = 0.$$

Combining the power series we have

$$\sum_{n=0}^{\infty} ((n + 2)(n + 1)a_{n+2} + a_n) x^n = 0.$$

From this equation we see that

$$(n + 2)(n + 1)a_{n+2} + a_n = 0$$

giving us the recurrence relation

$$a_{n+2} = -\frac{a_n}{(n + 2)(n + 1)}.$$

Consequently:

$$a_2 = -\frac{a_0}{2} = -\frac{a_0}{2!}$$

$$a_3 = -\frac{a_1}{2 \cdot 3} = -\frac{a_1}{3!}$$

$$a_4 = -\frac{a_2}{3 \cdot 4} = \frac{a_0}{4!}$$

$$a_5 = -\frac{a_3}{4 \cdot 5} = \frac{a_1}{5!}$$

$$\vdots$$

The pattern arising here is

$$a_{2n} = (-1)^n \frac{a_0}{(2n)!} \quad \text{and} \quad a_{2n+1} = (-1)^n \frac{a_1}{(2n+1)!}$$

where n is a positive integer. Viewing y in the form

$$y = a_0 + a_2 x^2 + a_4 x^4 + \cdots + a_{2n} x^{2n} + \cdots + a_1 x + a_3 x^3 + \cdots + a_{2n+1} x^{2n+1} + \cdots$$

$$= \sum_{n=0}^{\infty} a_{2n} x^{2n} + \sum_{n=0}^{\infty} a_{2n+1} x^{2n+1},$$

we have

$$y = \sum_{n=0}^{\infty} (-1)^n \frac{a_0}{(2n)!} x^{2n} + \sum_{n=0}^{\infty} (-1)^n \frac{a_1}{(2n+1)!} x^{2n+1}.$$

Using Corollary 8.3 and the initial conditions,

$$a_0 = y(0) = 1 \quad \text{and} \quad a_1 = \frac{y'(0)}{1!} = y'(0) = 0.$$

Thus

$$y = \sum_{n=0}^{\infty} (-1)^n \frac{1}{(2n)!} x^{2n}.$$

This is the Maclaurin series for $\cos x$ about $x_0 = 0$, which is the same solution obtained using the methods of Chapter 4.   ●

EXERCISES 8.1

In Exercises 1–8, find the power series solution to the initial value problem.

1. $y' + y = 0$, $y(0) = 2$ **2.** $y' = y$, $y(0) = -1$

3. $y' = (2x - 2)y$, $y(1) = 1$

4. $y' + (x + 1)^2 y = 0$, $y(-1) = 0$

5. $y'' - y = 0$; $y(0) = 1$, $y'(0) = 0$

6. $y'' + 9y = 0$; $y(0) = 0$, $y'(0) = 2$

7. $y'' + y' = 0$; $y(0) = 1$, $y'(0) = -1$

8. $y'' - xy' = 0$; $y(0) = 2$, $y'(0) = 1$

In Exercises 9–12, find the power series solution to the initial value problem and determine the interval of convergence of the power series solution.

9. $y' = xy$, $y(0) = 1$

10. $y' + (x - 2); y = 0$, $y(2) = 1$

11. $y'' + 4y = 0$; $y(0) = 1$, $y'(0) = 0$

12. $y'' + y' = 0$; $y(0) = 0$, $y'(0) = 1$

Viewing the Taylor series solution

$$y = \sum_{n=0}^{\infty} a_n (x - x_0)^n$$

of an initial value problem in the form

$$\sum_{n=0}^{\infty} \frac{y^{(n)}(x_0)}{n!} (x - x_0)^n$$

gives rise to an alternative method for finding the Taylor series solution: Use the initial conditions to get the first terms of the series and the differential equation and its derivatives to get the remaining terms. To illustrate, we will generate the first three terms of the Maclaurin series of the solution to the initial value problem

$$y' = 2y, \qquad y(0) = 1$$

in Example 1. We have $a_0 = y(0) = 1$ from the initial condition. From the differential equation,

$$a_1 = y'(0) = 2y(0) = 2.$$

Differentiating the differential equation, we have

$$y'' = 2y'$$

and hence

$$a_2 = \frac{y''(0)}{2} = \frac{2y'(0)}{2} = 2.$$

Differentiating again,

$$y''' = 2y''$$

so that

$$a_3 = \frac{y'''(0)}{3!} = \frac{2y''(0)}{3!} = \frac{4}{3}.$$

The Maclaurin series is then

$$1 + 2x + 2x^2 + \frac{4}{3}x^3 + \cdots.$$

Continuing this procedure, we obtain more terms of the series. Use this alternative method to obtain the first four nonzero terms of the Taylor series solutions in Exercises 13–16.

13. $y' = 2xy$, $y(0) = 1$

14. $y' + x^2 y = 0$, $y(0) = 1$

15. $y'' = 4(x - 2)y$; $y(0) = 1$, $y'(0) = 0$

16. $y'' + (x + 1)y = 0$; $y(-1) = 0$, $y'(-1) = 1$

In Exercises 17–20, use the *dsolve* command in Maple with the *series option* or the corresponding command in another appropriate software package to find the first six terms of the Taylor series solutions of the initial value problems.

17. $y'' = 2xy$, $y(0) = 1$

18. $y'' + x^2 y = 0$, $y(0) = 1$

19. $y'' = xy' - y$; $y(0) = 1$, $y'(0) = 0$

20. $y'' - y' + (x + 1)y = 0$; $y(-1) = 0$, $y'(-1) = 1$

8.2   SERIES SOLUTIONS FOR SECOND ORDER LINEAR DIFFERENTIAL EQUATIONS

You might have noticed that every example and problem with a linear differential equation in the last section contained a differential equation that could be solved by the methods of Chapter 3 or 4. We did this so you could get a feel for power series solutions. Now we are going to move on and solve some initial value problems where the techniques of Chapters 3 and 4 do not apply. In this section we are going to consider second order initial value problems

$$y'' + q(x)y' + r(x)y = g(x); \qquad y(x_0) = k_0, \qquad y'(x_0) = k_1$$

when q, r, and g are analytic at x_0. (Recall that this means that $q(x)$, $r(x)$, and $g(x)$ are equal to their Taylor series about x_0 in some open interval about x_0.) By Theorem

4.1 we know there is a unique solution to this initial value problem. We first consider the case where q, r, and g are polynomials. (Polynomials are analytic functions whose Taylor series have only a finite number of terms—see Exercise 25.)

EXAMPLE 1   Determine a power series solution to the following initial value problem.

$$y'' + xy = 0; \qquad y(0) = 1, \qquad y'(0) = 0$$

Solution   We substitute in $y = \sum_{n=0}^{\infty} a_n x^n$ and obtain

$$y'' + xy = \sum_{n=0}^{\infty} (n+2)(n+1)a_{n+2}x^n + \sum_{n=0}^{\infty} a_n x^{n+1}$$

$$= 2a_2 + \sum_{n=1}^{\infty} (n+2)(n+1)a_{n+2}x^n + \sum_{n=0}^{\infty} a_n x^{n+1} = 0.$$

Combining the power series we have

$$2a_2 + \sum_{n=0}^{\infty} ((n+3)(n+2)a_{n+3} + a_n)x^{n+1} = 0,$$

giving us

$$2a_2 = 0 \quad \text{and} \quad (n+3)(n+2)a_{n+3} + a_n = 0$$

or

$$a_2 = 0 \quad \text{and} \quad a_{n+3} = -\frac{a_n}{(n+3)(n+2)}.$$

Consequently:

$$a_3 = -\frac{a_0}{3!}$$

$$a_4 = -\frac{a_1}{12} = -\frac{2a_1}{4!}$$

$$a_5 = -\frac{a_2}{20} = 0$$

$$a_6 = -\frac{a_3}{6 \cdot 5} = \frac{4a_0}{6!}$$

$$a_7 = -\frac{a_4}{7 \cdot 6} = \frac{2 \cdot 5a_1}{7!}$$

$$a_8 = 0$$

$$a_9 = -\frac{a_6}{9 \cdot 8} = -\frac{4 \cdot 7a_0}{9!}$$

$$a_{10} = -\frac{a_7}{10 \cdot 9} = -\frac{2 \cdot 5 \cdot 8a_1}{10!}$$

$$a_{11} = -\frac{a_8}{11 \cdot 10} = 0$$

$$\vdots$$

Here are the patterns developing:

$$a_{3n-1} = 0 \quad \text{for} \quad n = 1, 2, 3, \ldots$$

$$a_3 = -\frac{a_0}{3!} \quad \text{and} \quad a_{3n} = (-1)^n \frac{4 \cdot 7 \cdots (3n-2)}{(3n)!} a_0 \quad \text{for} \quad n = 2, 3, \ldots$$

$$a_4 = \frac{-2a_1}{4!} \quad \text{and} \quad a_{3n+1} = (-1)^n \frac{2 \cdot 5 \cdots (3n-1)}{(3n+1)!} a_1 \quad \text{for} \quad n = 2, 3, \ldots$$

We have

$$y = a_0 - \frac{a_0}{3!}x^3 + \sum_{n=2}^{\infty} (-1)^n \frac{4 \cdot 7 \cdots (3n-2)}{(3n)!} a_0 x^{3n}$$

$$+ a_1 x - \frac{2a_1}{4!}x^4 + \sum_{n=2}^{\infty} (-1)^n \frac{2 \cdot 5 \cdots (3n-1)}{(3n+1)!} a_1 x^{3n+1}.$$

Using $a_0 = y(0) = 1$ and $a_1 = y'(0) = 0$, gives us

$$y = 1 - \frac{1}{3!}x^3 + \sum_{n=2}^{\infty} (-1)^n \frac{4 \cdot 7 \cdots (3n-2)}{(3n)!} x^{3n}.$$

 ●

EXAMPLE 2   Determine a power series solution to the following initial value problem.

$$y'' - 2xy' + y = 0; \qquad y(0) = 1, \qquad y'(0) = 1$$

Solution   Substituting $y = \sum_{n=0}^{\infty} a_n x^n$ leads to

$$y'' - 2xy' + y = \sum_{n=0}^{\infty} (n+2)(n+1)a_{n+2}x^n - 2x \sum_{n=0}^{\infty} (n+1)a_{n+1}x^n + \sum_{n=0}^{\infty} a_n x^n$$

$$= \sum_{n=0}^{\infty} (n+2)(n+1)a_{n+2}x^n - \sum_{n=0}^{\infty} 2(n+1)a_{n+1}x^{n+1} + \sum_{n=0}^{\infty} a_n x^n = 0.$$

Rewriting the power series, we have

$$a_0 + 2a_2 + \sum_{n=0}^{\infty} ((n+3)(n+2)a_{n+3} - 2(n+1)a_{n+1} + a_{n+1})x^{n+1} = 0.$$

This gives us

$$a_2 = -\frac{a_0}{2} \quad \text{and} \quad a_{n+3} = \frac{(2n+1)a_{n+1}}{(n+3)(n+2)}.$$

Thus:

$$a_3 = \frac{a_1}{3!}$$

$$a_4 = \frac{3}{4 \cdot 3}a_2 = -\frac{3}{4!}a_0$$

$$a_5 = \frac{5}{5 \cdot 4}a_3 = \frac{5}{5!}a_1$$

$$a_6 = \frac{7}{6 \cdot 5}a_4 = -\frac{3 \cdot 7}{6!}a_0$$

$$a_7 = \frac{9}{7 \cdot 6}a_5 = \frac{5 \cdot 9}{7!}a_1$$

$$\vdots$$

Hence we see that:

$$a_2 = -\frac{a_0}{2} \quad \text{and} \quad a_{2n} = -\frac{3 \cdot 7 \cdots (4n-5)}{(2n)!}a_0 \quad \text{for} \quad n = 2, 3, \ldots$$

$$a_{2n+1} = \frac{1 \cdot 5 \cdots (4n-3)}{(2n+1)!}a_1 \quad \text{for} \quad n = 1, 2, 3, \ldots$$

The initial conditions tell us that $a_0 = y(0) = 1$ and $a_1 = y'(0) = 1$. Therefore, we have that

$$y = a_0 - \frac{a_0 x^2}{2} - \sum_{n=2}^{\infty} \frac{3 \cdot 7 \cdots (4n-5)}{(2n)!}a_0 x^{2n} + a_1 x + \sum_{n=1}^{\infty} \frac{1 \cdot 5 \cdots (4n-3)}{(2n+1)!}a_1 x^{2n+1}$$

$$= 1 - \frac{x^2}{2} - \sum_{n=2}^{\infty} \frac{3 \cdot 7 \cdots (4n-5)}{(2n)!}x^{2n} + x + \sum_{n=1}^{\infty} \frac{1 \cdot 5 \cdots (4n-3)}{(2n+1)!}x^{2n+1}.$$ ●

Notice that the power series solution can be expressed as two sums with one depending on a_0 and the other depending on a_1 in Examples 1 and 2. (Of course, the second sum dropped out in Example 1 since $a_1 = 0$.) This is always true for a homogeneous equation

$$y'' + q(x)y' + r(x)y = 0$$

when q and r are analytic at x_0. When q and r are analytic at x_0, we say x_0 is an **ordinary point** of this differential equation. Indeed the comments just made lead to the following theorem.

THEOREM 8.4   If x_0 is an ordinary point of

$$y'' + q(x)y' + r(x)y = 0,$$

then the solution to the initial value problem

$$y'' + q(x)y' + r(x)y = 0; \qquad y(x_0) = k_0, \qquad y'(x_0) = k_1$$

is given by

$$y = \sum_{n=0}^{\infty} a_n(x - x_0)^n = a_0 y_1(x) + a_1 y_2(x)$$

where $y_1(x)$ and $y_2(x)$ are power series in $x - x_0$ and $a_0 = y(x_0)$ and $a_1 = y'(x_0)$.

But we can say more. Concerning the radii of convergence of $y_1(x)$ and $y_2(x)$ we have Theorem 8.5.

THEOREM 8.5   Under the hypotheses and notation of Theorem 8.4, the radius of convergence of each power series $y_1(x)$ and $y_2(x)$ is at least R where R is the smaller of the radii of convergence of the Taylor series of q and r about x_0.

Finally, we can even use these power series to obtain general solutions. Omitting specific values for k_0 and k_1 in Theorem 8.4 means that a_0 and a_1 become arbitrary constants. This leads us to Theorem 8.6.

THEOREM 8.6   If x_0 is an ordinary point of

$$y'' + q(x)y' + r(x)y = 0$$

and if $y_1(x)$ and $y_2(x)$ are as in Theorem 8.4 and R is as in Theorem 8.5, then $y_1(x)$ and $y_2(x)$ are linearly independent on the interval $|x - x_0| < R$ and consequently the general solution of

$$y'' + q(x)y' + r(x)y = 0$$

is

$$a_0 y_1(x) + a_1 y_2(x).$$

As an illustration of Theorem 8.6, we have that the general solution to the differential equation

$$y'' - 2xy' + y = 0$$

in Example 2 is given by

$$y = a_0 \left(1 - \frac{x^2}{2} - \sum_{n=2}^{\infty} \frac{3 \cdot 7 \cdots (4n - 5)}{(2n)!} x^{2n} \right) + a_1 \left(x + \sum_{n=1}^{\infty} \frac{1 \cdot 5 \cdots (4n - 3)}{(2n + 1)!} x^{2n+1} \right).$$

In Exercise 26 we indicate how one goes about proving these theorems.

Up to this point in this section we have only looked at homogeneous equations. But the approach of Examples 1 and 2 also works for nonhomogeneous equations, as the following example illustrates.

EXAMPLE 3   Determine the power series solution to

$$y'' + 2y' - xy = 1 + 4x; \qquad y(0) = 0, \qquad y'(0) = 0.$$

Solution   Substituting the power series $y = \sum_{n=0}^{\infty} a_n x^n$, we obtain

$$y'' + 2y' - xy = \sum_{n=0}^{\infty} (n+2)(n+1)a_{n+2}x^n + 2\sum_{n=0}^{\infty} (n+1)a_{n+1}x^n - x\sum_{n=0}^{\infty} a_n x^n$$

$$= 2a_2 + 2a_1 + \sum_{n=0}^{\infty} ((n+3)(n+2)a_{n+3} + 2(n+2)a_{n+2} - a_n)x^{n+1}$$

$$= 2a_2 + 2a_1 + (6a_3 + 4a_2 - a_0)x$$

$$+ \sum_{n=1}^{\infty} ((n+3)(n+2)a_{n+3} + 2(n+2)a_{n+2} - a_n)x^{n+1}$$

$$= 1 + 4x.$$

Equating coefficients, we have

$$2a_2 + 2a_1 = 1, \qquad (6a_3 + 4a_2 - a_0) = 4, \quad \text{and}$$

$$((n+3)(n+2)a_{n+3} + 2(n+2)a_{n+2} - a_n) = 0 \quad \text{for} \quad n = 1, 2, 3, \ldots$$

From these and the initial conditions we obtain:

$$a_0 = 0$$

$$a_1 = 0$$

$$a_2 = \frac{1 - 2a_1}{2} = \frac{1}{2}$$

$$a_3 = \frac{4 + a_0 - 4a_2}{6} = \frac{1}{3}$$

$$a_{n+3} = \frac{-2(n+2)a_{n+2} + a_n}{(n+3)(n+2)} \quad \text{for } n = 2, 3, 4, \ldots$$

Unlike Examples 1 and 2, no pattern is apparent for the the general term here. Instead, we generate a few of them:

$$a_4 = -\frac{a_3}{2} = -\frac{1}{6}$$

$$a_5 = -\frac{2}{5}a_4 + \frac{1}{20}a_2$$

$$a_6 = -\frac{1}{3}a_5 + \frac{1}{30}a_3$$

Listing the first seven terms of the power series, we obtain:

$$y = \frac{1}{2}x^2 + \frac{1}{3}x^3 - \frac{1}{6}x^4 + \frac{11}{120}x^5 - \frac{7}{360}x^6 + \cdots$$

●

The last example is not atypical. In many cases it is not possible to determine the general term of the power series solution. This is especially true if either q or r are not polynomials. In this case we have to express $q(x)$ or $r(x)$ as power series to find power series solutions as the final example of this section illustrates.

EXAMPLE 4   Determine the first four terms of the power series solution about $x_0 = 0$ to

$$y'' + (\cos x)y = 0$$

in terms of a_0 and a_1.

Solution   We substitute the Maclaurin series for $\cos x$,

$$\cos x = 1 - \frac{x^2}{2} + \frac{x^4}{4!} - \frac{x^6}{6!} + \cdots,$$

and $\sum_{n=0}^{\infty} a_n x^n$ for y to get

$$\sum_{n=0}^{\infty}(n+2)(n+1)a_{n+2}x^n + \left(1 - \frac{x^2}{2} + \frac{x^4}{4!} - \frac{x^6}{6!} + \cdots\right)\sum_{n=0}^{\infty} a_n x^n = 0. \qquad \textbf{(1)}$$

In order to get the first four terms of the power series solution, we only need to look at terms involving 1, x, x^2, and x^3 in Equation (1). Consequently, terms involving powers of four or more of x in the Maclaurin series of $\cos x$ will not come into play. This enables us to view Equation (1) as

$$\sum_{n=0}^{\infty}(n+2)(n+1)a_{n+2}x^n + \sum_{n=0}^{\infty} a_n x^n - \frac{x^2}{2}\sum_{n=0}^{\infty} a_n x^n + \cdots = 0.$$

Next we rewrite the preceding equation as

$$2a_2 + 6a_3x + 12a_4x^2 + 20a_5x^3 + \sum_{n=4}^{\infty}(n+2)(n+1)a_{n+2}x^n + a_0 + a_1x + a_2x^2$$

$$+ a_3x^3 + \sum_{n=4}^{\infty} a_n x^n - \frac{1}{2}a_0x^2 - \frac{1}{2}a_1x^3 - \frac{x^2}{2}\sum_{n=2}^{\infty} a_n x^n + \cdots = 0.$$

From the constant term we see

$$a_0 + 2a_2 = 0 \quad \text{or} \quad a_2 = -\frac{a_0}{2}.$$

From the first degree term, we see

$$6a_3 + a_1 = 0 \quad \text{or} \quad a_3 = -\frac{a_1}{6}.$$

Now that we have found a_2 and a_3 we can stop and give our answer as

$$y = a_0 + a_1 x - \frac{a_0}{2} x^2 - \frac{a_1}{6} x^3 + \cdots$$

●

Software packages such as Maple can be used to generate terms of power series solutions about a value x_0 when q, r, and g are analytic at x_0. Exercises 27–30 ask you to use Maple or another appropriate software package to generate terms of power series solutions to some second order linear differential equations of this type.

We conclude this section with the study of a type of differential equation that arises in many problems in mathematical physics called a **Legendre equation**.[2] These are differential equations of the form

$$\boxed{(1 - x^2)y'' - 2xy' + \nu(\nu + 1)y = 0}$$

where ν is a real constant. (The constant ν is determined by the application.) We are going to find power series solutions about $x_0 = 0$, which is an ordinary point of the Legendre equation. Substituting

$$y = \sum_{n=0}^{\infty} a_n x^n$$

into the Legendre equation, we have

$$(1 - x^2)y'' - 2xy' + \nu(\nu + 1)y = (1 - x^2) \sum_{n=0}^{\infty} (n + 2)(n + 1)a_{n+2} x^n$$

$$-2x \sum_{n=0}^{\infty} (n + 1)a_{n+1} x^n + \nu(\nu + 1) \sum_{n=0}^{\infty} a_n x^n$$

$$= 2a_2 + 6a_3 x + \sum_{n=2}^{\infty} (n + 2)(n + 1)a_{n+2} x^n$$

$$- \sum_{n=0}^{\infty} (n + 2)(n + 1)a_{n+2} x^{n+2}$$

$$- 2a_1 x - \sum_{n=1}^{\infty} 2(n + 1)a_{n+1} x^{n+1} + \nu(\nu + 1)a_0$$

$$+ \nu(\nu + 1)a_1 x + \sum_{n=2}^{\infty} \nu(\nu + 1)a_n x^n$$

---

[2] Adrien Marie Legendre (1752–1833) was a French mathematician who made significant contributions in the areas of special functions.

$$= 2a_2 + \nu(\nu + 1)a_0 + ([\nu(\nu + 1) - 2]a_1 + 6a_3)x$$

$$+ \sum_{n=0}^{\infty} [(n + 4)(n + 3)a_{n+4}$$

$$-((n + 2)(n + 3) - \nu(\nu + 1))a_{n+2}]x^{n+2} = 0.$$

From this we see that

$$a_2 = -\frac{\nu(\nu + 1)}{2}a_0,$$

$$a_3 = \frac{2 - \nu(\nu + 1)}{6}a_1,$$

and

$$a_{n+4} = \frac{(n + 2)(n + 3) - \nu(\nu + 1)}{(n + 4)(n + 3)}a_{n+2}.$$

The recurrence relation gives us two linearly independent power series solutions to the Legendre equation as described in Theorem 8.4. An interesting result occurs in the case when

$$\nu = k + 2$$

for some nonnegative integer k. Notice that, in this case, $a_{k+4} = 0$, which then forces coefficients depending on a_{k+4}, which are the coefficients

$$a_{k+6}, a_{k+8}, a_{k+10}, \ldots,$$

to be 0. Consequently one of the linearly independent power series solutions in Theorem 8.4 is a polynomial $p_{k+2}(x)$ of degree $k + 2$. (Why?) When the condition $p_{k+2}(1) = 1$ is imposed, $p_{k+2}(x)$ is called the **Legendre polynomial** of degree $k + 2$. In this setting, it is sometimes easier to obtain the two solutions by first finding the Legendre polynomial solution and then using the method of reduction of order introduced in Chapter 4 to find the other solution. Exercises 17–20 ask you to apply this approach.

EXERCISES 8.2

In Exercises 1–8, determine the power series solution to the initial value problems.

1. $y'' - xy = 0$; $y(0) = 1$, $y'(0) = 0$

2. $y'' + x^2y = 0$; $y(0) = 0$, $y'(0) = -1$

3. $y'' - xy' + 2y = 0$; $y(0) = 0$, $y'(0) = 2$

4. $y'' - 2y' - xy = 0$; $y(0) = 2$, $y'(0) = 0$

5. $y'' - 2y' + 2xy = 4 - 2x$; $y(0) = 0$, $y'(0) = 0$

6. $y'' + xy' - y = x$; $y(0) = 0$, $y'(0) = 0$

7. $y'' - xy = 1 + x$; $y(0) = 1$, $y'(0) = 0$

8. $y'' + x^2y = -2x$; $y(0) = 0$, $y'(0) = -1$

In Exercises 9–12, determine the power series solution to the differential equation.

9. $y'' - xy = 0$ **10.** $y'' + x^2y = 0$

11. $y'' - xy' + 2y = 0$ **12.** $y'' - 2y' + x^2y = 0$

In Exercises 13–16, determine the first three terms of the power series solution to the differential equation.

13. $y'' - y \cos x = 0$ **14.** $y'' + e^x y = 0$

15. $e^x y'' + xy = 0$ **16.** $y'' - 2y' + y \sin x = 0$

In Exercises 17–20, determine the power series solution to the Legendre equation. If one of the solutions is a polynomial, use reduction of order to find the second linearly independent solution.

17. $(1 - x^2)y'' - 2xy' + 3/4y = 0$

18. $(1 - x^2)y'' - 2xy' - 1/4y = 0$

19. $(1 - x^2)y'' - 2xy' + 2y = 0$

20. $(1 - x^2)y'' - 2xy' + 6y = 0$

In Exercises 21–22, determine the polynomial solution to the Legendre equation.

21. $(1 - x^2)y'' - 2xy' + 20y = 0$

22. $(1 - x^2)y'' - 2xy' + 30y = 0$

23. The equation

$$y'' - 2xy' + \lambda y = 0$$

where λ is a constant is known as the Hermite[3] equation. This equation is important in physics and numerical analysis.

a) Show that the power series solution is a polynomial (called a Hermite polynomial) if λ is an even natural number.

b) Determine the Hermite polynomials for $\lambda = 0, 2, 4, 6$, and 8.

24. The equation

$$(1 - x^2)y'' - xy' + \alpha^2 y = 0$$

where α is a constant is known as the Chebyshev[4] equation.

a) Determine the power series solution to this equation for $|x| < 1$.

b) Show that if α is a natural number, then there is a polynomial solution.

c) Determine the Chebyshev polynomials for $\alpha = 0, 1$, and 2.

25. For a polynomial $p(x) = a_n x^n + a_{n-1} x^{n-1} + \cdots + a_1 x + a_0$ and a number x_0, determine $b_0, b_1, \ldots, b_n$ so that

$$p(x) = b_0 + b_1(x - x_0) + b_2(x - x_0)^2 + \cdots$$
$$+ b_n(x - x_0)^n + 0(x - x_0)^{n+1}$$
$$+ 0(x - x_0)^{n+2} + \cdots$$
$$= b_0 + b_1(x - x_0) + b_2(x - x_0)^2 + \cdots$$
$$+ b_n(x - x_0)^n$$

for all x and hence conclude that every polynomial is equal to its Taylor series about x_0 for all x and also that its Taylor series has a finite number of terms.

26. a) Show that in Theorem 8.4 $a_0 = k_0$, $a_1 = k_1$, $a_2 = y''(0)/2 = -(q(0)k_1 + r(0)k_0)/2$.

b) Continuing the process in part (a), show that we can generate a power series solution for Theorem 8.4 that satisfies $y(0) = k_0$ and $y'(0) = k_1$.

To complete the proof of Theorem 8.4 we need to show this power series converges.

In Exercises 27–30, use the *dsolve* command in Maple with the `type=series` option or the corresponding command in another appropriate software package to obtain the first six terms of the power series solution. Compare this answer to your results for Exercises 13–16.

27. $y'' - y \cos x = 0$ **28.** $y'' + e^x y = 0$

29. $e^x y'' + xy = 0$ **30.** $y'' - 2y' + y \sin x = 0$

8.3   EULER TYPE EQUATIONS

In the last section we considered power series solutions about x_0 when x_0 is an ordinary point. Our next objective is to study power series solutions when x_0 is not an ordinary point. In this study a type of linear differential equation called an **Euler type equation** (sometimes called an **equidimensional equation**) will arise. In the second order homo-

---

[3] Charles Hermite (1822–1901) was a French mathematician who studied algebra and analysis.

[4] Pafnuty Chebyshev (1821–1894) was one of the most influential Russian mathematicians of all time and is well known for his work in polynomial approximations, number theory, and probability.

geneous case, these Euler type equations are differential equations that can be written in the form

$$(x - x_0)^2 y'' + \alpha(x - x_0)y' + \beta y = 0$$

where α and β are constants. When we write one of these Euler type equations in the form

$$y'' + q(x)y' + r(x)y = 0,$$

we have the differential equation

$$y'' + \frac{\alpha}{x - x_0} y' + \frac{\beta}{(x - x_0)^2} y = 0.$$

The value x_0 is not an ordinary point of this differential equation since neither $\alpha/(x - x_0)$ nor $\beta/(x - x_0)^2$ is defined at x_0 much less analytic at x_0. For the moment we are going to put aside power series and see how to find closed form solutions to the Euler type equations. Once we see how to solve Euler type equations, we will employ their solutions in the next section to find power series solutions about x_0 of other differential equations $y'' + q(x)y' + r(x)y = 0$ where q and r are not analytic at x_0.

Actually, one method of solving Euler type equations when $x_0 = 0$ is discussed in Exercise 43 of Section 4.2. We are going to use a different method here. As in Exercise 43 of Section 4.2, we will only develop the method when $x_0 = 0$. It is easy, however, to extend either of these two methods to cases when $x_0 \neq 0$.

Consider the Euler type equation

$$x^2 y'' + \alpha x y' + \beta y = 0.$$

We will determine solutions for $x > 0$. It is also possible to determine solutions for $x < 0$. See Exercise 21 for details about this. Since differentiating reduces the power on the exponent by one, it makes sense to try a solution of the form

$$y = x^r.$$

Substituting this into the differential equation gives us

$$x^2 r(r - 1)x^{r-2} + \alpha x r x^{r-1} + \beta x^r = 0$$

or

$$(r(r - 1) + \alpha r + \beta)x^r = 0.$$

Consequently,

$$r(r - 1) + \alpha r + \beta = 0.$$

This is a polynomial equation in r similar to the characteristic equations we had for constant coefficient linear differential equations in Chapter 4. We will call this equation the **Euler indicial equation**. Solving this equation for r gives us

$$r = \frac{1 - \alpha \pm \sqrt{(1 - \alpha)^2 - 4\beta}}{2}.$$

As with constant coefficient linear differential equations, we have to consider distinct real, repeated real, and imaginary root cases. Here these respective cases arise as follows:

$$\text{Case 1. } (1 - \alpha)^2 - 4\beta > 0$$
$$\text{Case 2. } (1 - \alpha)^2 - 4\beta = 0$$
$$\text{Case 3. } (1 - \alpha)^2 - 4\beta < 0.$$

In Case 1 we have two distinct real roots, r_1 and r_2. We leave it for you in Exercise 15 to show that

$$x^{r_1} \quad \text{and} \quad x^{r_2}$$

are linearly independent for $x > 0$. Therefore, in this case, if $x > 0$ the general solution to the Euler equation is

$$y = c_1 x^{r_1} + c_2 x^{r_2}.$$

In Case 2 we have one real root, r, and hence only one solution,

$$x^r,$$

when $x > 0$. In Exercise 16 we will ask you to obtain that

$$x^r \ln x$$

is a second solution linearly independent of x^r. Consequently, the general solution for $x > 0$ is

$$c_1 x^r + c_2 x^r \ln x$$

in Case 2.

In Case 3 we have two complex roots, $r = a \pm ib$. As in Chapters 4 and 6, we will only need the root $r = a + ib$ to produce two linearly independent solutions and will be able to ignore the conjugate root, $r = a - ib$. If $x > 0$ we have

$$x^r = x^{a+ib} = x^a x^{ib} = x^a e^{\ln x^{ib}} = x^a e^{ib \ln x} = x^a (\cos(b \ln x) + i \sin(b \ln x)).$$

(We are using laws of exponents here that extend from real exponents to complex ones. Proofs that these exponent laws do in fact hold for complex numbers are left for a course in complex variables.) We now proceed in much the same manner as we did in Chapter 4 when we had complex roots. It can be shown (see Exercise 17) that the general solution in this case for $x > 0$ is

$$\boxed{y = x^a(c_1 \cos(b \ln x) + c_2 \sin(b \ln x)).}$$

Let us do some examples involving Euler type equations.

EXAMPLE 1   Find the general solution of $x^2 y'' + 4xy' + 2y = 0$.

Solution   Here the equation $r(r - 1) + \alpha r + \beta = 0$ is

$$r(r - 1) + 4r + 2 = 0.$$

Solving for r, we see the solutions are $r = -2, -1$. The general solution is therefore

$$y = c_1 x^{-1} + c_2 x^{-2}.$$

●

EXAMPLE 2   Solve the initial value problem $x^2 y'' + 3xy' + y = 0$; $y(1) = 0$, $y'(1) = 1$.

Solution   In this example the equation $r(r - 1) + \alpha r + \beta = 0$ is

$$r(r - 1) + 3r + 1 = 0,$$

which has $r = -1$ as a repeated root. The general solution is

$$y = \frac{c_1}{x} + \frac{c_2 \ln x}{x}.$$

Using the initial conditions, we obtain

$$c_1 = 0, \qquad c_2 = 1.$$

The solution to this initial value problem is then

$$y = \frac{\ln x}{x}.$$

●

EXAMPLE 3   Determine the general solution of $x^2 y'' + 2xy' + y = 0$.

Solution   The equation with r for this example is

$$r(r - 1) + 2r + 1 = 0.$$

Its solutions are

$$r = -\frac{1}{2} \pm \frac{\sqrt{3}}{2} i.$$

The general solution is

$$y = x^{-1/2}\left(c_1 \cos\left(\frac{\sqrt{3}}{2} \ln x\right) + c_2 \sin\left(\frac{\sqrt{3}}{2} \ln x\right)\right).$$

●

EXERCISES 8.3

In Exercises 1–8, determine the general solution of the differential equation.

1. $x^2 y'' - 3xy' + 3y = 0$ **2.** $x^2 y'' + xy' - y = 0$

3. $x^2 y'' + xy' + y = 0$ **4.** $x^2 y'' + xy' + 4y = 0$

5. $y'' + (5/x)y' + (4/x^2)y = 0$

6. $y'' + (7/x)y' + (9/x^2)y = 0$

7. $x^2 y'' + 3xy' + 2y = 0$ **8.** $x^2 y'' + 5xy' + 5y = 0$

In Exercises 9–10, solve the initial value problem.

9. $y'' - y'/x + (2/x^2)y = 0$; $y(1) = 1$, $y'(1) = 0$

10. $x^2 y'' + 3xy' + 5y = 0$; $y(1) = 0$, $y'(1) = 1$

In Exercises 11–12, solve the nonhomogeneous equation.

11. $x^2 y'' - 3xy' + 3y = x$

12. $x^2 y'' + 3xy' + 5y = 8x$

In Exercises 13–14, adapt our method for solving Euler type equations with $x_0 = 0$ to the given Euler type equation to obtain its general solution.

13. $(x - 1)^2 y'' + (5x - 5)y' + 3y = 0$

14. $(x + 2)^2 y'' - 2(x + 2)y'2y = 0$

Exercises 15–17 deal with the three cases for the roots of the Euler indicial equation.

15. Show that in Case 1 the solutions are linearly independent for $x > 0$.

16. Use the reduction of order method in Section 4.2 to find a second linearly independent solution for $x > 0$ in Case 2.

17. Show that the two solutions we gave in Case 3 are solutions of the Euler type equation and are linearly independent.

18. Determine the behavior of the solutions to the Euler type equations as $x \to 0$ if the real parts of r_1 and r_2 are positive.

19. Determine the behavior of the solutions to the Euler type equations as $x \to 0$ if the real parts of r_1 and r_2 are negative.

20. Determine the behavior of the solutions to the Euler type equations as $x \to 0$ if r_1 and r_2 are purely imaginary.

21. a) To determine the solutions to the Euler type equation $x^2 y'' + \alpha xy' + \beta y = 0$ for $x < 0$, we make the substition $z = -x$. Using the Chain Rule, show that this Euler type equation takes the form

$$z^2 \frac{d^2 y}{dz^2} + \alpha z \frac{dy}{dz} + \beta y = 0$$

under this substitution. Describe the general solution to this Euler type equation in terms of z and then in terms of x in Cases 1–3.

b) Obtain the solutions to the Euler type equation $x^2 y'' + \alpha xy' + \beta y = 0$ for $x \neq 0$ in terms of $|x|$.

22. a) Let L be the linear transformation $L = x^2 D^2 + \alpha x D + \beta$ so that kernel of L consists of the solutions to the Euler type equation $x^2 y'' + \alpha xy' + \beta y = 0$. Further, let $F(r) = r(r-1) + \alpha r + \beta$, which is the polynomial in the Euler indicial equation of this Euler type equation. Show that

$$L(x^r \ln x) = F(r)x^r \ln x + F'(r)x^r.$$

b) Suppose that the indicial equation has a repeated root r_1 so that $F(r) = (r - r_1)^2$. Use the result of part (a) to show that $L(x^{r_1} \ln x) = 0$ and hence conclude that $x^{r_1} \ln x$ is another solution to the Euler type equation.

8.4 SERIES SOLUTIONS NEAR A REGULAR SINGULAR POINT

As mentioned in the introduction of Section 8.3, one reason we studied the Euler type equations

$$(x - x_0)^2 y'' + \alpha(x - x_0)y' + \beta y = 0$$

or

$$y'' + \frac{\alpha}{x - x_0} y' + \frac{\beta}{(x - x_0)^2} y = 0$$

where α and β are constants was to prepare us for the study of power series solutions to differential equations of the type

$$y'' + p(x)y' + q(x)y = 0$$

when x_0 is not an ordinary point (that is, either p or q is not analytic at x_0). Such points x_0 are called **singular points** of the differential equation $y'' + p(x)y' + q(x)y = 0$.

In this section we are going to study differential equations

$$\boxed{y'' + p(x)y' + q(x)y = 0}$$

where

$$\boxed{(x - x_0)p(x) \quad \text{and} \quad (x - x_0)^2 q(x)}$$

are analytic at x_0. When this occurs, x_0 is called a **regular singular point** of the differential equation $y'' + p(x)y' + q(x)y = 0$. Notice that the Euler type equation

$$y'' + \frac{\alpha}{x - x_0} y' + \frac{\beta}{(x - x_0)^2} y = 0$$

has a regular singular point at x_0 since $(x - x_0)p(x) = \alpha$ and $(x - x_0)^2 q(x) = \beta$ are analytic at x_0. Power series solutions about x_0 to $y'' + p(x)y' + q(x)y = 0$ when this differential equation has a regular singular point at x_0 are related to the solutions of Euler type equations. As we did for Euler type equations in the previous section, we will restrict our attention to the case when $x_0 = 0$ and our solutions will be for $x > 0$. The method we develop here can be easily modified to handle nonzero values for x_0 and other values of x.

Suppose that $x_0 = 0$ is a regular singular point of $y'' + p(x)y' + q(x)y = 0$. Then $xp(x)$ and $x^2 q(x)$ are equal to their Maclaurin series in some open interval $|x| < R$. To put it another way, we can express $xp(x)$ and $x^2 q(x)$ as

$$xp(x) = p_0 + p_1 x + p_2 x^2 + p_3 x^3 + \cdots = \sum_{n=0}^{\infty} p_n x^n \tag{1}$$

and

$$x^2 q(x) = q_0 + q_1 x + q_2 x^2 + q_3 x^3 + \cdots = \sum_{n=0}^{\infty} q_n x^n \tag{2}$$

for $|x| < R$. In the case of the Euler type equation $y'' + (\alpha/x)y' + (\beta/x^2)y = 0$, $xp(x) = \alpha$, $x^2 q(x) = \beta$, and the solutions x^r are x^r times the Maclaurin series of the constant function $f(x) = 1$. Could it be that solutions to $y'' + p(x)y' + q(x)y = 0$ are of the form x^r times the Maclaurin series of some function? That is, are there solutions of the form

$$y = x^r \sum_{n=0}^{\infty} a_n x^n = \sum_{n=0}^{\infty} a_n x^{n+r} ?$$

Let us try an example and see what happens.

EXAMPLE 1   If possible, determine solutions to the differential equation

$$y'' + \frac{5}{2x}y' - \frac{2+x}{2x^2}y = 0$$

of the form

$$y = x^r \sum_{n=0}^{\infty} a_n x^n = \sum_{n=0}^{\infty} a_n x^{n+r}.$$

Solution   Note that $x = 0$ is a regular singular point for this differential equation. Indeed, Equation (1) is

$$xp(x) = \frac{5}{2} = \frac{5}{2} + 0 \cdot x + 0 \cdot x^2 + \cdots,$$

and Equation (2) is

$$x^2 q(x) = -\frac{2+x}{2} = -1 - \frac{1}{2}x + 0 \cdot x^2 + 0 \cdot x^3 + \cdots.$$

It will turn out to be convenient if we first multiply our differential equation by x^2 before we attempt to determine series solutions. Doing so, our differential equation becomes

$$x^2 y'' + \frac{5}{2}xy' - \frac{2+x}{2}y = 0. \tag{3}$$

For our desired solution y we have

$$y' = \sum_{n=0}^{\infty} a_n(n+r)x^{n+r-1}$$

and

$$y'' = \sum_{n=0}^{\infty} a_n(n+r)(n+r-1)x^{n+r-2}.$$

Substituting into Equation (3) gives us

$$\sum_{n=0}^{\infty} a_n(n+r)(n+r-1)x^{n+r} + \sum_{n=0}^{\infty} \frac{5}{2}a_n(n+r)x^{n+r} - \sum_{n=0}^{\infty} a_n x^{n+r} - \sum_{n=0}^{\infty} \frac{1}{2}a_n x^{n+r+1} = 0.$$

We now have

$$\sum_{n=0}^{\infty} \left(a_n(n+r)(n+r-1) + \frac{5}{2}a_n(n+r) - a_n \right) x^{n+r} - \sum_{n=0}^{\infty} \frac{1}{2}a_n x^{n+r+1} = 0.$$

Separating off the term with $n = 0$ in the first sum and reindexing, we can rewrite this as

$$\left(r(r-1) + \frac{5}{2}r - 1 \right) a_0 x^r + \sum_{n=0}^{\infty} \left(\left((n+1+r)(n+r) + \frac{5}{2}(n+1+r) - 1 \right) a_{n+1} \right.$$

$$\left. - \left(\frac{1}{2} \right) a_n \right) x^{n+r+1} = 0.$$

From the first term we see that we want

$$r(r-1) + \frac{5}{2}r - 1 = 0. \tag{4}$$

Do you see a similarity between Equation (4) and the equation

$$r(r-1) + \alpha r + \beta = 0 \tag{5}$$

we had for Euler type equations? Observe that they are the same with

$$\alpha = \frac{5}{2},$$

which is the term p_0 of the Maclaurin series of $xp(x)$, and

$$\beta = -1,$$

which is the term q_0 of the Maclaurin series of $x^2 q(x)$. Shortly we will see that this is always the case for a differential equation $y'' + p(x)y' + q(x)y = 0$ with a regular singular point at $x_0 = 0$. Equation (4) is called the **indicial equation** of $y'' + p(x)y' + q(x)y = 0$.

Our indicial equation (4) has $r = 1/2$ and $r = -2$ as solutions, each of which will give us a power series solution. When $r = 1/2$, we have

$$\left(\left(n + \frac{3}{2} \right) \left(n + \frac{1}{2} \right) + \frac{5}{2} \left(n + \frac{3}{2} \right) - 1 \right) a_{n+1} - \frac{1}{2} a_n = 0.$$

Solving this for a_{n+1}, we find

$$a_{n+1} = \frac{a_n}{2n^2 + 9n + 7}.$$

Consequently:

$$a_1 = \frac{a_0}{7}$$

$$a_2 = \frac{a_1}{18} = \frac{a_0}{126}$$

$$a_3 = \frac{a_2}{33} = \frac{a_0}{4158}$$

$$a_4 = \frac{a_3}{52} = \frac{a_0}{216216}$$

Therefore, for $r = 1/2$ we obtain the solution

$$y = x^{1/2}(a_0 + a_1 x + a_2 x^2 + a_3 x^3 + \cdots)$$

$$= a_0 x^{1/2} \left(1 + \frac{1}{7}x + \frac{1}{126}x^2 + \frac{1}{4158}x^3 + \frac{1}{216216}x^4 + \cdots \right).$$

We now consider $r = -2$. We have

$$\left((n-1)(n-2) + \frac{5}{2}(n-1) - 1 \right) a_{n+1} - \frac{1}{2} a_n = 0.$$

Thus

$$a_{n+1} = \frac{a_n}{2n^2 - n - 3}$$

and hence:

$$a_1 = -\frac{a_0}{3}$$

$$a_2 = -\frac{a_1}{2} = \frac{a_0}{6}$$

$$a_3 = \frac{a_2}{3} = \frac{a_0}{18}$$

$$a_4 = \frac{a_3}{12} = \frac{a_0}{216}$$

Therefore, for $r = -2$ we obtain the solution

$$y = x^{-2}(a_0 + a_1 x + a_2 x^2 + a_3 x^3 + \cdots)$$

$$= a_0 x^{-2}\left(1 - \frac{1}{3}x + \frac{1}{6}x^2 + \frac{1}{18}x^3 + \frac{1}{216}x^4 + \cdots\right). \qquad \bullet$$

It can be verified that the two solutions we obtained with $r = 1/2$ and $r = -2$ in Example 1 when $a_0 = 1$ are linearly independent solutions for $x > 0$. Thus, replacing a_0 by c_1 in the solution with $r = -1/2$ and a_0 by c_2 in the second solution with $r = -2$, the general solution to the second order homogeneous linear differential equation in Example 1 is

$$y = c_1 x^{1/2}\left(1 + \frac{1}{7}x + \frac{1}{126}x^2 + \frac{1}{4158}x^3 + \frac{1}{216216}x^4 + \cdots\right)$$

$$+ c_2 x^{-2}\left(1 - \frac{1}{3}x + \frac{1}{6}x^2 + \frac{1}{18}x^3 + \frac{1}{216}x^4 + \cdots\right)$$

for $x > 0$.

Let us now apply the procedure of Example 1 to a general differential equation

$$\boxed{y'' + p(x)y' + q(x)y = 0}$$

with a regular singular point at $x_0 = 0$. First, multiply this differential equation by x^2 obtaining

$$\boxed{x^2 y'' + x^2 p(x)y' + x^2 q(x)y = 0.} \qquad (6)$$

Viewing $x^2 p(x)$ as $x \cdot xp(x)$ and substituting the series in Equation (1) for $xp(x)$ into Equation (6), substituting the series in Equation (2) for $x^2 q(x)$ into Equation (6), and substituting

$$\boxed{y = x^r \sum_{n=0}^{\infty} a_n x^n = \sum_{n=0}^{\infty} a_n x^{n+r}}$$

into Equation (6) leads to

$$\sum_{n=0}^{\infty} a_n[(n+r)(n+r-1) + (p_0 + p_1x + p_2x^2 + p_3x^3 + \cdots)(n+r)$$
$$+ (q_0 + q_1x + q_2x^2 + q_3x^3 + \cdots)]x^{n+r} = 0$$

or

$$a_0(r(r-1) + p_0r + q_0)x^r + (a_1((1+r)r + p_0(1+r) + q_0) + a_0(p_1r + q_1))x^{r+1}$$
$$+ (a_2((2+r)(1+r) + p_0(2+r) + q_0) + a_1(p_1(1+r) + q_1) + a_0(p_2r + q_2))x^{r+2}$$
$$+ (a_3((3+r)(2+r) + p_0(3+r) + q_0) + a_2(p_1(2+r) + q_1) + a_1(p_2(1+r) + q_2)$$
$$+ a_0(p_3r + q_3))x^{r+3} + \cdots = 0.$$

$$(7)$$

From the first term with a_0x^r, we obtain the indicial equation

$$\boxed{r(r-1) + p_0r + q_0 = 0}$$

corresponding to the equation in r for the Euler type equation with $\alpha = p_0$ and $\beta = q_0$. The expression on the left-hand side of the indicial equation appears repeatedly so we shall denote it as

$$\boxed{F(r) = r(r-1) + p_0r + q_0}$$

to save writing. Indeed, with this notation, observe that Equation (7) now becomes:

$$a_0F(r)x^r + (a_1F(1+r) + a_0(p_1r + q_1))x^{r+1}$$
$$+ (a_2F(2+r) + a_1(p_1(1+r) + q_1) + a_0(p_2r + q_2))x^{r+2}$$
$$+ (a_3F(3+r) + a_2(p_1(2+r) + q_1) + a_1(p_2(1+r) + q_2)$$
$$+ a_0(p_3r + q_3))x^{r+3} + \cdots = 0.$$

We then have:

$$F(r) = 0$$
$$a_1F(1+r) + a_0(p_1r + q_1) = 0$$
$$a_2F(2+r) + a_1(p_1(1+r) + q_1) + a_0(p_2r + q_2) = 0$$
$$a_3F(3+r) + a_2(p_1(2+r) + q_1) + a_1(p_2(1+r) + q_2) + a_0(p_3r + q_3) = 0$$
$$\vdots$$

From these equations we obtain:

$$F(r) = 0$$

$$a_1 = -\frac{a_0(p_1 r + q_1)}{F(1 + r)}$$

$$a_2 = -\frac{a_0(p_2 r + q_2) + a_1(p_1(1 + r) + q_1)}{F(2 + r)}$$

$$a_3 = -\frac{a_2(p_1(2 + r) + q_1) + a_1(p_2(1 + r) + q_2) + a_0(p_3 r + q_3)}{F(3 + r)}$$

$$\vdots$$

This process gives us a power series solution as long as $F(n + r) \neq 0$ for $n = 1, 2,$ $3, \ldots .$ To put it another way, if whenever r_1 is a root of the indicial equation $F(r) = 0$, the indicial equation does not have a second root r_2 of the form $r_2 = n + r_1$ where n is a positive integer, then we are assured our process leads to a power series solution with $r = r_1$. As a consequence, if the roots of the indicial equation $F(r)$ are r_1 and r_2 with $r_1 > r_2$, we then have that each root r_1 and r_2 will produce power series solutions whenever $r_1 - r_2$ is not a positive integer. Let us state this as a theorem. Since the coefficients a_n of the power series solutions depend on the roots r_1 and r_2 of the indicial equation, we are going to indicate these coefficients as $a_n(r_1)$ and $a_n(r_2)$ as we state this theorem.

THEOREM 8.7   Suppose that $x_0 = 0$ is a regular singular point of the differential equation

$$y'' + p(x)y' + q(x)y = 0$$

and r_1 and r_2 are the roots of its indicial equation with $r_1 > r_2$. If $r_1 - r_2$ is not a positive integer, then this differential equation has a solution of the form

$$y_1 = x^{r_1} \sum_{n=0}^{\infty} a_n(r_1)x^n$$

and a second solution of the form

$$y_2 = x^{r_2} \sum_{n=0}^{\infty} a_n(r_2)x^n.$$

Looking back at Example 1, notice that it illustrates Theorem 8.7 with $r_1 = 1/2$ and $r_2 = -2$.

It can be argued that y_1 and y_2 in Theorem 8.7 converge to a solution for $0 < x < R$ when the series for $xp(x)$ and $x^2 q(x)$ in Equations (1) and (2) converge to $xp(x)$ and $x^2 q(x)$, respectively, on this interval. Further, if r_1 are r_2 are real numbers and we choose nonzero values of $a_0(r_1)$ and $a_0(r_2)$, y_1 and y_2 are linearly independent on this interval

$0 < x < R$ and consequently the general solution to $y'' + p(x)y' + q(x)y = 0$ is $c_1 y_1 + c_2 y_2$ on this interval. We will only use this theorem in the case when the indicial equation has real roots. The complex case as well as a complete proof of Theorem 8.7 including the facts concerning the interval of convergence and independence of the solutions are left for another course.

We have begun a study carried out by the German mathematician Georg Frobenius.[5] His study does not end here. What do we do if the hypothesis on r_1 and r_2 in Theorem 8.7 is not satisfied? We consider two cases. First, suppose $r_1 = r_2$; that is, suppose the indicial equation has a repeated root. In this case our procedure leads to one power series solution. Can we get a second one? The next theorem of Frobenius tells us how to do so.

THEOREM 8.8   Suppose that $x_0 = 0$ is a regular singular point of the differential equation

$$y'' + p(x)y' + q(x)y = 0$$

and its indicial equation has a root r_1 of multiplicity 2. Then this differential equation has a solution of the form

$$y_1 = x^{r_1} \sum_{n=0}^{\infty} a_n x^n$$

and a second solution of the form

$$y_2 = y_1 \ln x + x^{r_1} \sum_{n=0}^{\infty} b_n x^n.$$

If $r_1 - r_2$ is a positive integer, our process does produce a power series solution for the larger root, but breaks down for the smaller root, r_2. Here again, can we produce a second solution? Frobenius found a positive answer in this case too, as described in Theorem 8.9.

THEOREM 8.9   Suppose that $x_0 = 0$ is a regular singular point of the differential equation

$$y'' + p(x)y' + q(x)y = 0$$

and its indicial equation has roots $r_1 > r_2$ where $r_1 - r_2$ is a positive integer. Then this differential equation has a solution of the form

$$y_1 = x^{r_1} \sum_{n=0}^{\infty} a_n(r_1)x^n$$

[5] Ferdinand Georg Frobenius (1849–1917) was a student of the noted German mathematican Karl Weierstrass (1815–1897). Frobenius made significant contributions to many areas of mathematics, but is best known for his seminal work in group theory.

and there is a constant a so that it has a second solution of the form

$$y_2 = ay_1 \ln|x| + x^{r_2} \sum_{n=1}^{\infty} a_n(r_2)x^n.$$

The coefficients a_n in Theorem 8.8 and a and a_n in Theorem 8.9 for the second solutions, y_2, are found by substituting the form of the solution for y_2 into the differential equation in the same manner as we do for y_1. Proofs of Theorems 8.8 and 8.9 will be omitted. As is the case in Theorem 8.7, y_1 and y_2 in these two theorems converge to solutions for $0 < x < R$ when the series for $xp(x)$ and $x^2q(x)$ in Equations (1) and (2) converge to $xp(x)$ and $x^2q(x)$, respectively, on this interval. Also, y_1 and y_2 are linearly independent on this interval when we choose nonzero values of a_0 and a in the case of Theorem 8.9 so that the general solution is $c_1y_1 + c_2y_2$.

Let us next look at an example where we have to determine whether we are in the case of Theorem 8.7, 8.8, or 8.9.

EXAMPLE 2   Determine which of Theorem 8.7, 8.8, or 8.9 applies in the following differential equation,

$$x^2y'' + 2xy' + (x - x^3)y = 0,$$

and give the forms of the power series solutions for this case.

Solution   When written in the form $y'' + p(x)y' + q(x)y = 0$, this differential equation is

$$y'' + \frac{2}{x} + \frac{1 - x^2}{x}y = 0.$$

We have

$$xp(x) = 2 \quad \text{and} \quad x^2q(x) = x - x^3$$

so that $p_0 = 2$ and $q_0 = 0$. The indicial equation is then

$$r(r - 1) + 2r = 0$$

and its solutions are $r_1 = 0$ and $r_2 = -1$. Since $r_1 - r_2 = 1$, we are in the case of Theorem 8.9 and the solutions have the forms

$$y_1 = x^0\left(\sum_{n=0}^{\infty} a_nx^n\right) = \sum_{n=0}^{\infty} a_n(0)x^n$$

and

$$y_2 = ay_1 \ln x + x^{-1}\left(\sum_{n=0}^{\infty} a_n(-1)x^n\right). \qquad\bullet$$

Finding the power series solutions in Example 2 by hand takes quite some time. Fortunately, software packages such as Maple can be used to do this work.

We conclude this section with a type of differential equation that has a regular singular point at $x_0 = 0$ that arises in many physical applications called a **Bessel equation.**[6] These differential equations have the form

$$x^2 y'' + xy' + (x^2 - v^2)y = 0$$

where $v \geq 0$ is a real constant. The value v is called the **order of the Bessel equation.** Notice that the Bessel equation has $x = 0$ as a regular singular point. Also, $xp(x) = 1$ and $x^2 q(x) = x^2 - v^2$. Since $p_0 = 1$ and $q_0 = -v^2$, the indicial equation of the Bessel equation is

$$r(r - 1) + r - v^2 = r^2 - v^2 = 0.$$

The roots of the indicial equation are

$$r_1 = v, \qquad r_2 = -v.$$

Observe that we are then in the case of Theorem 8.7 if $2v$ is not a nonnegative integer, in the case of Theorem 8.8 if $v = 0$, and in the case of Theorem 8.9 if $2v$ is a positive integer.

Our final set of examples involve power series solutions to Bessel equations in some special cases.

EXAMPLE 3   Determine the series solutions to the following Bessel equation: the Bessel equation of order 0,

$$x^2 y'' + xy' + x^2 y = 0.$$

Solution   Here $v = 0$ and we are in the case of Theorem 8.8 with $r = 0$. Let us find the solution of the form

$$y_1 = x^r \sum_{n=0}^{\infty} a_n x^n = \sum_{n=0}^{\infty} a_n x^n.$$

Substituting into the differential equation,

$$\sum_{n=0}^{\infty} (n-1)n a_n x^n + \sum_{n=0}^{\infty} n a_n x^n + \sum_{n=0}^{\infty} a_n x^{n+2} = \sum_{n=0}^{\infty} [(n-1)n + n] a_n x^n + \sum_{n=0}^{\infty} a_n x^{n+2} = 0.$$

Simplifying and reindexing leads to

$$a_1 x + \sum_{n=0}^{\infty} [(n+2)^2 a_{n+2} + a_n] x^{n+2} = 0.$$

---

[6] Friedrich Wilhelm Bessel (1784–1846) was a German astronomer and mathematician who served as the director of the Königsberg observatory.

The recurrence relation is

$$a_{n+2} = -\frac{a_n}{(n+2)^2}$$

so that

$$a_2 = -\frac{a_0}{2^2}$$

$$a_4 = -\frac{a_2}{4^2} = \frac{a_0}{2^4(2!)^2}$$

$$a_6 = -\frac{a_4}{6^2} = -\frac{a_0}{2^6(3!)^2}$$

$$a_8 = -\frac{a_6}{8^2} = \frac{a_0}{2^8(4!)^2}$$

$$\vdots$$

It appears that

$$a_{2n} = \frac{(-1)^n a_0}{2^{2n}(n!)^2}.$$

We will let you verify that this is true in general in Exercise 9(a).
     The first solution in Theorem 8.8 is then

$$y_1 = \sum_{n=0}^{\infty} \frac{(-1)^n a_0}{2^{2n}(n!)^2} x^{2n}.$$

When $a_0 = 1$, we obtain the function

$$J_0(x) = \sum_{n=0}^{\infty} \frac{(-1)^n}{2^{2n}(n!)^2} x^{2n}$$

called the **Bessel function of the first kind of order 0.** We leave it for you to find the second solution, y_2, in Theorem 8.8 in Exercise 9(b).   ●

EXAMPLE 4   Determine the series solutions to the following Bessel equations: the Bessel equation of order $1/2$,

$$x^2 y'' + xy' + \left(x^2 - \frac{1}{4}\right) y = 0.$$

Solution   Here $\nu = 1/2$, the roots of the indicial equation are $r_1 = 1/2$ and $r_2 = -1/2$. We are then in the case of Theorem 8.9. We again find the first solution

$$y_1 = x^{1/2} \sum_{n=0}^{\infty} a_n x^n = \sum_{n=0}^{\infty} a_n x^{n+1/2}.$$

Substituting, we have

$$\sum_{n=0}^{\infty}\left(n-\frac{1}{2}\right)\left(n+\frac{1}{2}\right)a_n x^{n+1/2} + \sum_{n=0}^{\infty}\left(n+\frac{1}{2}\right)a_n x^{n+1/2} + \left(x^2-\frac{1}{4}\right)\sum_{n=0}^{\infty}a_n x^{n+1/2}$$

$$= 2a_1 x^{3/2} + \sum_{n=0}^{\infty}\left(\left[\left(n+\frac{5}{2}\right)^2 - \left(\frac{1}{4}\right)\right]a_{n+2} + a_n\right)x^{n+5/2} = 0.$$

We can now see that

$$a_1 = 0$$

and

$$a_{n+2} = -\frac{a_n}{\left(n+\frac{5}{2}\right)^2 - \frac{1}{4}}.$$

Thus,

$$a_1 = 0 = a_3 = a_5 = a_7 = \cdots$$

and

$$a_2 = -\frac{a_0}{6} = -\frac{a_0}{3!}$$

$$a_4 = -\frac{a_2}{20} = \frac{a_0}{5!}$$

$$a_6 = -\frac{a_4}{42} = -\frac{a_0}{7!}$$

$$\vdots$$

The second set of equations suggest that

$$a_{2n} = \frac{(-1)^n a_0}{(2n+1)!}.$$

We will let you prove that this in fact is the case in Exercise 9(c). We then have

$$y_1 = x^{1/2}\sum_{n=0}^{\infty}\frac{(-1)^n a_0}{(2n+1)!}x^{2n} = a_0 x^{-1/2}\sum_{n=0}^{\infty}\frac{(-1)^n}{(2n+1)!}x^{2n+1} = a_0 x^{-1/2}\sin x.$$

When $a_0 = \sqrt{2/\pi}$, the solution

$$J_{1/2}(x) = \left(\frac{2}{\pi x}\right)^{1/2}\sin x$$

is called the **Bessel function of the first kind of order 1/2.** We will let you determine the second solution, y_2, in Exercise 9(d).   ●

EXAMPLE 5   Determine the series solutions to the following Bessel equation: the Bessel equation of order 1,

$$x^2 y'' + xy' + (x^2 - 1)y = 0.$$

Solution   For this Bessel equation, $\nu = 1$, $r_1 = 1$, and $r_2 = -1$. We are then in the case of Theorem 8.9. The first solution has the form

$$y_1 = x \sum_{n=0}^{\infty} a_n x^n = \sum_{n=0}^{\infty} a_n x^{n+1}.$$

Substituting it in,

$$\sum_{n=0}^{\infty} n(n+1)a_n x^{n+1} + \sum_{n=0}^{\infty} (n+1)a_n x^{n+1} + (x^2 - 1)\sum_{n=0}^{\infty} a_n x^{n+1}$$

$$= 3a_1 x^2 + \sum_{n=0}^{\infty} ([(n+3)^2 - 1]a_{n+2} + a_n)x^{n+3} = 0$$

from which we obtain

$$a_1 = 0$$

and

$$a_{n+2} = -\frac{a_n}{(n+3)^2 - 1}.$$

We have

$$a_1 = 0 = a_3 = a_5 = a_7 = \cdots$$

and

$$a_2 = -\frac{a_0}{8} = -\frac{a_0}{2^2(2!)(1!)}$$

$$a_4 = -\frac{a_2}{24} = \frac{a_0}{2^4(3!)(2!)}$$

$$a_6 = -\frac{a_4}{48} = -\frac{a_0}{2^6(4!)(3!)}$$

$$\vdots$$

We will let you verify in Exercise 9(e) that the second set of equations generalizes to

$$a_{2n} = \frac{(-1)^n a_0}{2^{2n}(n+1)!n!}.$$

Thus

$$y_1 = x \sum_{n=0}^{\infty} \frac{(-1)^n a_0}{2^{2n}(n+1)!n!} x^{2n} = a_0 \sum_{n=0}^{\infty} \frac{(-1)^n}{2^{2n}(n+1)!n!} x^{2n+1}.$$

When $a_0 = 1/2$, we obtain the solution

$$J_1(x) = \frac{1}{2} \sum_{n=0}^{\infty} \frac{(-1)^n}{2^{2n}(n+1)!n!} x^{2n+1}$$

called the **Bessel function of the first kind of order 1.** Finding the second solution, y_2, is best done by using a software package such as Maple. We will ask you to do this in Exercise 14.

EXERCISES 8.4

In Exercises 1–2, determine the regular singular points for the differential equation.

1. $x^2 y'' + \sin x y' + y = 0$

2. $(1 - x^2)y'' - 2xy' + 20y = 0$

In Exercises 3–4, determine the indicial equation for the differential equation at $x_0 = 0$.

3. $x(x-1)y'' + (3x-1)y' + y = 0$

4. $(x^2 - x)y'' - xy' + 4y = 0$

In Exercises 5–8, determine the general solution of the differential equation.

5. $x^2 y'' + 3xy' + 4x^4 y = 0$

6. $x^2 y'' + x^3 y' + x^2 y = 0$

7. $6x^2 y'' + 4xy' + 2x^2 y = 0$

8. $xy'' + 2xy' + xy = 0$

9. These problems are related to the Bessel equations in Examples 3–5.

a) Show that

$$a_{2n} = \frac{(-1)^n a_0}{2^{2n}(n!)^2}$$

in Example 3.

b) Determine the second linearly independent solution for the Bessel equation of order 0 in Example 3.

c) Show that

$$a_{2n} = \frac{(-1)^n a_0}{(2n+1)!}$$

in Example 4.

d) Determine the second linearly independent solution for the Bessel equation of order 1/2 in Example 4.

e) Show that

$$a_{2n} = \frac{(-1)^n a_0}{2^{2n}(n+1)!n!}$$

in Example 5.

In Exercises 10–11, determine the general solution for the Bessel equation.

10. $x^2 y'' + xy' + (x^2 - 16/9)y = 0$

11. $x^2 y'' + xy' + (x^2 - 9/4)y = 0$

12. Consider the Legendre equation

$$(1 - x^2)y'' - 2xy' + v(v+1)y = 0.$$

a) Show that $x = \pm 1$ are regular singular points of this differential equation.

b) Find a solution in powers of $x - 1$ for $x > 1$. (*Hint:* Let $t = x - 1$ and apply the methods of this section.)

13. Consider the Bessel equation

$$x^2 y'' + xy' + (x^2 - v^2)y = 0.$$

a) Show that the substitution $y(x) = x^r u(x)$ where $r = \pm v$ converts this Bessel equation into the form

$$x^2 u'' + (1 + 2r)xu' + x^2 u = 0.$$

b) Use part (a) to determine the two linearly independent solutions to the Bessel equation of order 1/2. (*Hint:* Use $r = -1/2$.)

c) Using the result of part (b), solve the nonhomogeneous equation
$$x^2 y'' + xy' + (x^2 - 1/4)y = x^{3/2}.$$

14. Use Maple or another appropriate software package to obtain a second solution to the Bessel equation in Example 5.

In Exercises 15–18 determine one solution to the differential equation at $x_0 = 0$ and use Maple to determine a second linearly independent solution at $x_0 = 0$.

15. $x^2 y'' - x(2 + x)y' + (2 + x^2)y = 0$

16. $x(x + 1)y'' + y' + xy = 0$

17. $x^2 y'' + xy' + (x^2 - 4)y = 0$

18. $x^2 y'' + xy' + (x^2 - 9)y = 0$

19. The differential equation
$$x(1 - x)y'' + (\alpha - (1 + \beta + \gamma)x)y' - \beta\gamma y = 0$$

occurs often in mathematical physics.

a) Show $x_0 = 0$ and $x_0 = 1$ are regular singular points of this differential equation.

b) Determine the indicial equation for the regular singular point $x_0 = 0$.

c) Determine the recurrence relations at the regular singular point $x_0 = 0$.

d) Use the result of part (c) and Maple or another appropriate software package to determine two linearly independent solutions at the regular singular point $x_0 = 0$ for $\alpha = 3/2$.

20. Consider the differential equation
$$x^2 y'' + \cos xy = 0.$$

a) Show $x_0 = 0$ is a regular singular point of this differential equation.

b) Determine the indicial equation of this differential equation.

c) Determine the first four terms of one of the solutions of this differential equation guaranteed by Theorems 8.7–8.9.

Inner Product Spaces

The dot product of two vectors u and v in $\mathbb{R}^2$ or in $\mathbb{R}^3$ (usually denoted $u \cdot v$) that you have seen in your calculus courses is an operation on pairs of vectors that yields a real number. An inner product space is a vector space on which we have an operation (called an inner product) along the lines of a dot product. (We will state exactly what is meant by this shortly.) The vector spaces $\mathbb{R}^2$ and $\mathbb{R}^3$ with their dot products are just two examples of inner product spaces. The first two sections of this chapter are devoted to the study of inner product spaces. In the third and final section of this chapter we will look at some connections between inner products and matrices. Among the results we will obtain in this third section is that every symmetric matrix is diagonalizable. Since dot products on $\mathbb{R}^2$ and $\mathbb{R}^3$ are our motivational examples for inner product spaces, we begin by refreshing your memory about them.

9.1 INNER PRODUCT SPACES

Following our convention of writing vectors in $\mathbb{R}^n$ as column vectors, the **dot product** (or **scalar product**) of two vectors

$$u = \begin{bmatrix} a_1 \\ a_2 \end{bmatrix} \quad \text{and} \quad v = \begin{bmatrix} b_1 \\ b_2 \end{bmatrix}$$

in $\mathbb{R}^2$, denoted $u \cdot v$, is

$$u \cdot v = a_1 b_1 + a_2 b_2.$$

If u and v are vectors in $\mathbb{R}^3$,

$$u = \begin{bmatrix} a_1 \\ a_2 \\ a_3 \end{bmatrix} \quad \text{and} \quad v = \begin{bmatrix} b_1 \\ b_2 \\ a_3 \end{bmatrix},$$

their dot product is

$$u \cdot v = a_1 b_1 + a_2 b_2 + a_3 b_3.$$

If we identify a 1×1 matrix with its entry by leaving out the brackets about the entry, notice that our dot products in either $\mathbb{R}^2$ and $\mathbb{R}^3$ can be expressed as the matrix product of the tranpose of u and v:

$$u \cdot v = u^T v.$$

The length (also called magnitude or norm) of a vector

$$v = \begin{bmatrix} b_1 \\ b_2 \end{bmatrix}$$

in $\mathbb{R}^2$, which we shall denote by $\|v\|$, is

$$\|v\| = \sqrt{b_1^2 + b_2^2}.$$

Of course, if we follow the usual convention of indicating such a vector v graphically by drawing a directed line segment from the origin to the point (b_1, b_2), as in Figure 9.1, then $\|v\|$ is the length of this directed line segment. If v is a vector in $\mathbb{R}^3$,

$$v = \begin{bmatrix} b_1 \\ b_2 \\ b_3 \end{bmatrix},$$

then its length (magnitude, or norm) is

$$\|v\| = \sqrt{b_1^2 + b_2^2 + b_3^2},$$

which is the length of the directed line segment in 3-space from the origin to the point (b_1, b_2, b_3).

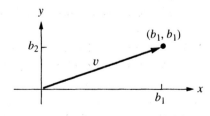

Figure 9.1

Notice that we can equally well describe the length of a vector v in $\mathbb{R}^2$ or $\mathbb{R}^3$ as

$$\|v\| = \sqrt{v \cdot v} = \sqrt{v^T v}.$$

Besides using dot products to describe lengths of vectors, another use of them is for finding angles between two vectors in $\mathbb{R}^2$ or $\mathbb{R}^3$. By the angle between two vectors u and v, we mean the smallest nonnegative angle θ between them, as is illustrated in Figure 9.2.[1] Using the law of cosines (see a calculus book for the details), it can be shown that

$$u \cdot v = \|u\| \, \|v\| \cos \theta.$$

If u and v are nonzero vectors, we can solve this equation for θ obtaining

$$\theta = \cos^{-1} \frac{u \cdot v}{\|u\| \, \|v\|}.$$

Also notice that the vectors u and v are orthogonal (or perpendicular) if and only if $u \cdot v = 0$.

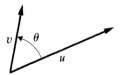

Figure 9.2

The dot product on $\mathbb{R}^2$ or $\mathbb{R}^3$ is a function from the set of pairs of vectors in $\mathbb{R}^2$ or $\mathbb{R}^3$ to the set of real numbers $\mathbb{R}$. Some easily verified properties of it are

$$u \cdot v = v \cdot u$$
$$(u + v) \cdot w = u \cdot w + v \cdot w$$
$$(cu) \cdot v = c(u \cdot v)$$
$$v \cdot v = 0 \quad \text{if and only if} \quad v = 0$$

where c is a scalar. Any vector space possessing such a function satisfying these properties is called an inner product space.

DEFINITION An **inner product space** is a vector space V with a function from the set of all pairs of vectors of V to $\mathbb{R}$ that assigns to each pair of vectors u and v of V a corresponding real number denoted $\langle u, v \rangle$ satisfying the following properties:
1. $\langle u, v \rangle = \langle v, u \rangle$ for all u and v in V.
2. $\langle u + v, w \rangle = \langle u, w \rangle + \langle v, w \rangle$ for all u, v, and w in V.

---

[1] The vectors u and v in Figure 9.2 are nonzero. Should either u or v be the zero vector, we take the angle between u and v to be zero.

3. $\langle cu, v \rangle = c \langle u, v \rangle$ for all u and v in V and all scalars c.

4. $\langle v, v \rangle = 0$ if and only if $v = 0$.

The function in this definition assigning to the pair of vectors u and v the value $\langle u, v \rangle$ is called an **inner product** on V. We read $\langle u, v \rangle$ as "the inner product of u and v." The dot products on $\mathbb{R}^2$ and $\mathbb{R}^3$ give us two examples of inner product spaces (where $\langle u, v \rangle = u \cdot v$). This naturally extends to higher dimensions, which we make as our next example.

EXAMPLE 1   For vectors

$$
u = \begin{bmatrix} a_1 \\ a_2 \\ \vdots \\ a_n \end{bmatrix} \quad \text{and} \quad v = \begin{bmatrix} b_1 \\ b_2 \\ \vdots \\ b_n \end{bmatrix}
$$

in $\mathbb{R}^n$, define $\langle u, v \rangle$ to be

$$
\langle u, v \rangle = a_1 b_1 + a_2 b_2 + \cdots + a_n b_n.
$$

Of course, as we did for $\mathbb{R}^2$ and $\mathbb{R}^3$, we could equally well give this formula for $\langle u, v \rangle$ as

$$
\langle u, v \rangle = u^T v
$$

provided we identify a 1×1 matrix with its entry. Since

$$
\langle u, v \rangle = a_1 b_1 + a_2 b_2 + \cdots + a_n b_n = b_1 a_1 + b_2 a_2 + \cdots + b_n a_n = \langle v, u \rangle,
$$
$$
\langle u + v, w \rangle = (u + v)^T w = (u^T + v^T) w = u^T w + v^T w = \langle u, w \rangle + \langle v, w \rangle,
$$
$$
\langle cu, v \rangle = (cu)^T v = c(u^T v) = c \langle u, v \rangle,
$$

and

$$
\langle v, v \rangle = b_1^2 + b_2^2 + \cdots + b_n^2 = 0 \quad \text{if and only if} \quad v = \begin{bmatrix} 0 \\ 0 \\ \vdots \\ 0 \end{bmatrix},
$$

we have an inner product space.   ●

You will see the inner product in Example 1 called the **dot product, standard inner product,** or **Euclidean inner product** on $\mathbb{R}^n$, and $\langle u, v \rangle$ is usually denoted as $u \cdot v$ as is done in $\mathbb{R}^2$ and $\mathbb{R}^3$.

Dot products are not the only type of inner products. Here is another one that arises frequently in mathematics and its applications.

EXAMPLE 2   Recall that the set of continuous functions on a closed interval $[a, b]$, denoted $C[a, b]$, is a vector space. For two functions f and g in $C[a, b]$, define

$$\langle f, g \rangle = \int_a^b f(x)g(x)\,dx.$$

We leave it to you to verify that the four required properties for this to be an inner product are satisfied (Exercise 7). Hence $C[a, b]$ with this inner product involving integration is an inner product space.   ●

Other examples of some inner products you may encounter in future courses can be found in Exercises 8, 9, 11, 12, and 13.

The next two theorems contain some basic properties of inner products.

THEOREM 9.1   If u, v, and w are vectors in an inner product space V and c is a scalar, then:

1. $\langle 0, v \rangle = \langle v, 0 \rangle = 0$.
2. $\langle v, u + w \rangle = \langle v, u \rangle + \langle v, w \rangle$.
3. $\langle v, cu \rangle = c\langle v, u \rangle$.
4. $\langle v - u, w \rangle = \langle v, w \rangle - \langle u, w \rangle$.
5. $\langle v, u - w \rangle = \langle v, u \rangle - \langle v, w \rangle$.

Proof   We prove the first two parts here and leave the proofs of the remaining parts as exercises (Exercise 14). For part (1), observe that

$$\langle 0, v \rangle = \langle 0 + 0, v \rangle = \langle 0, v \rangle + \langle 0, v \rangle.$$

Subtracting $\langle 0, v \rangle$ from the far left- and far right-hand sides of this equation gives us the desired result that $\langle 0, v \rangle = 0$. The fact that $\langle v, 0 \rangle = 0$ now follows from the commutativity property $\langle v, u \rangle = \langle u, v \rangle$ of inner products.

The commutativity property along with the right-hand distributive property $\langle v + u, w \rangle = \langle v, w \rangle + \langle u, w \rangle$ give us the left-hand distributive property in part (2) of this theorem:

$$\langle v, u + w \rangle = \langle u + w, v \rangle = \langle u, v \rangle + \langle w, v \rangle = \langle v, u \rangle + \langle v, w \rangle.   ●$$

THEOREM 9.2   If $v_1, v_2, \ldots, v_n, u$ are vectors in an inner product space V and if $c_1, c_2, \ldots, c_n$ are scalars, then:

1. $\langle c_1 v_1 + c_2 v_2 + \cdots + c_n v_n, u \rangle = c_1 \langle v_1, u \rangle + c_2 \langle v_2, u \rangle + \cdots + c_n \langle v_n, u \rangle$.
2. $\langle u, c_1 v_1 + c_2 v_2 + \cdots + c_n v_n \rangle = c_1 \langle u, v_1 \rangle + c_2 \langle u, v_2 \rangle + \cdots + c_n \langle u, v_n \rangle$.

The proof of Theorem 9.2 is Exercise 15.

Recall that we earlier noted that the length of a vector v in $\mathbb{R}^2$ or $\mathbb{R}^3$ is $\|v\| = \sqrt{v \cdot v}$. We use this same approach to define lengths of vectors in inner product spaces: If V is an inner product space, we define the **length (magnitude, or norm)** of a vector v in V, denoted $\|v\|$, to be

$$\boxed{\|v\| = \sqrt{\langle v, v \rangle}.}$$

In $\mathbb{R}^2$ or $\mathbb{R}^3$ where $v \cdot u = \|v\| \|u\| \cos \theta$, we have that

$$|v \cdot u| = |\|v\| \|u\| \cos \theta| = \|v\| \|u\| |\cos \theta| \le \|v\| \|u\|$$

since $|\cos \theta| \le 1$. Remarkably, this inequality, called the Cauchy-Schwartz inequality,[2] holds in any inner product space.

THEOREM 9.3   **Cauchy-Schwartz Inequality**   If v and u are vectors in an inner product space V, then

$$\boxed{|\langle v, u \rangle| \le \|v\| \|u\|.}$$

Proof   If either $v = 0$ or $u = 0$, then $\langle v, u \rangle = 0$ by part (1) of Theorem 9.1 and either $\|v\|$ or $\|u\|$ will be zero by part (4) of the definition of an inner product. Hence we in fact have equality of $|\langle v, u \rangle|$ and $\|v\| \|u\|$ when either $v = 0$ or $u = 0$. Suppose then that both $v \ne 0$ and $u \ne 0$. Dividing by $\|v\| \|u\|$, we have that the inequality we wish to obtain is equivalent to the inequality

$$\left| \left\langle \frac{v}{\|v\|}, \frac{u}{\|u\|} \right\rangle \right| \le 1.$$

Since

$$\left\| \frac{v}{\|v\|} \right\| = 1 \quad \text{and} \quad \left\| \frac{u}{\|u\|} \right\| = 1$$

(see Exercise 16), it suffices to assume that both v and u have length 1 and prove that

$$|\langle v, u \rangle| \le 1.$$

We have that

$$0 \le \langle v - u, v - u \rangle = \langle v, v - u \rangle - \langle u, v - u \rangle = \langle v, v \rangle - \langle v, u \rangle - \langle u, v \rangle + \langle u, u \rangle$$
$$= \|v\|^2 - 2\langle v, u \rangle + \|u\|^2 = 2 - 2\langle v, u \rangle.$$

Solving the inequality $0 \le 2 - 2\langle v, u \rangle$ for $\langle v, u \rangle$, we see

$$\langle v, u \rangle \le 1.$$

Carrying out similar steps with $\langle v + u, v + u \rangle$, the details of which we leave as an exercise (Exercise 17), we obtain

$$\langle v, u \rangle \ge -1.$$

Thus we have

$$|\langle v, u \rangle| \le 1$$

as needed.   ●

---

[2] Named for the French mathematician Augustin Louis Cauchy (1789–1857) and the German mathematician Herman Amandus Schwartz (1843–1921).

The next notion about inner product spaces we introduce in this section is that of the **angle between two vectors** v and u in an inner product space V. If either v or u is the zero vector, we take the angle θ between v and u to be zero just as we do in $\mathbb{R}^2$ or $\mathbb{R}^3$. Suppose that v and u are both nonzero. Recall that if v and u are in $\mathbb{R}^2$ or $\mathbb{R}^3$, the angle between them is $\theta = \cos^{-1}(v \cdot u / \|v\| \, \|u\|)$. Let us use the same type of equation in V; that is, we define the angle θ between v and u in the inner product space V to be

$$\theta = \cos^{-1} \frac{\langle v, u \rangle}{\|v\| \, \|u\|}.$$

There is, however, one potential flaw in this definition of angle between vectors. We need

$$\left| \frac{\langle v, u \rangle}{\|v\| \, \|u\|} \right| \le 1$$

for this inverse cosine to be defined. Do we have this for any nonzero vectors v and u? By the Cauchy-Schwartz inequality we do, and hence our definition does make sense.

As in $\mathbb{R}^2$ or $\mathbb{R}^3$, we will say that two nonzero vectors v and u in the inner product space V are **orthogonal** (or **perpendicular**) if the angle between v and u is $\pi/2$. We also consider the zero vector of V to be orthogonal to each vector in V. Since $\theta = \cos^{-1}(\langle v, u \rangle / \|v\| \, \|u\|)$ is $\pi/2$ if and only if $\langle v, u \rangle = 0$ in the case when both v and u are nonzero and $\langle v, u \rangle = 0$ when either v or u is the zero vector, we have Theorem 9.4.

THEOREM 9.4   Two vectors v and u in an inner product space V are orthogonal if and only if $\langle v, u \rangle = 0$.

EXERCISES 9.1

In Exercises 1–6, find $\langle v, u \rangle$, $\|v\|$, $\|u\|$, and the angle θ between v and u for the given vectors and the indicated inner product.

1. $v = \begin{bmatrix} 2 \\ -1 \end{bmatrix}, u = \begin{bmatrix} -3 \\ 1 \end{bmatrix}$; $\langle v, u \rangle = v \cdot u$

2. $v = \begin{bmatrix} 3 \\ 2 \\ 1 \end{bmatrix}, u = \begin{bmatrix} 2 \\ -1 \\ -4 \end{bmatrix}$; $\langle v, u \rangle = v \cdot u$

3. $v = \begin{bmatrix} 3 \\ -1 \\ 1 \\ 0 \end{bmatrix}, u = \begin{bmatrix} 6 \\ -2 \\ 1 \\ 3 \end{bmatrix}$; $\langle v, u \rangle = v \cdot u$

4. $v = \begin{bmatrix} -1 \\ 3 \\ -4 \\ 2 \\ 1 \end{bmatrix}, u = \begin{bmatrix} -6 \\ -1 \\ 2 \\ 1 \\ -3 \end{bmatrix}$; $\langle v, u \rangle = v \cdot u$

5. $v(x) = x^2 - x, u(x) = 3x + 5$; $\langle v, u \rangle = \int_0^1 v(x)u(x)\,dx$

6. $v(x) = \sin x, u(x) = \cos x$; $\langle v, u \rangle = \int_{-\pi}^{\pi} v(x)u(x)\,dx$

7. Verify that the function

$$\langle f, g \rangle = \int_a^b f(x)g(x)\,dx$$

in Example 2 satisfies the four properties of an inner product.

8. a) Let

$$v = \begin{bmatrix} a_1 \\ a_2 \\ \vdots \\ a_n \end{bmatrix} \quad \text{and} \quad u = \begin{bmatrix} b_1 \\ b_2 \\ \vdots \\ b_n \end{bmatrix}$$

be vectors in $\mathbb{R}^n$ and let $w_1, w_2, \ldots, w_n$ be positive real numbers. Define $\langle v, u \rangle$ to be

$$\langle v, u \rangle = a_1 w_1 b_1 + a_2 w_2 b_2 + \cdots + a_n w_n b_n.$$

Verify that this defines an inner product on $\mathbb{R}^n$. This is called a weighted inner product; $w_1, w_2, \ldots, w_n$ are called the weights.

b) For the weighted inner product on $\mathbb{R}^2$ with weights $w_1 = 2$, $w_2 = 3$, find $\langle v, u \rangle$, $\|v\|$, $\|u\|$, and the angle between v and u if

$$v = \begin{bmatrix} -1 \\ 3 \end{bmatrix} \quad \text{and} \quad u = \begin{bmatrix} 2 \\ -1 \end{bmatrix}.$$

9. a) Let A be an invertible $n \times n$ matrix. For vectors v and u in $\mathbb{R}^n$, define $\langle v, u \rangle$ to be

$$\langle v, u \rangle = (Av)^T Au = v^T A^T Au.$$

Verify that this defines an inner product on $\mathbb{R}^n$. This is called the inner product on $\mathbb{R}^n$ determined by A.

b) Let

$$A = \begin{bmatrix} 1 & -1 \\ 2 & 1 \end{bmatrix}.$$

Find $\langle v, u \rangle$, $\|v\|$, $\|u\|$, and the angle between v and u for the inner product on $\mathbb{R}^2$ determined by A if

$$v = \begin{bmatrix} 1 \\ 1 \end{bmatrix} \quad \text{and} \quad u = \begin{bmatrix} 2 \\ -3 \end{bmatrix}.$$

10. Find an $n \times n$ matrix A so that the weighted inner product in Exercise 8(a) is expressed as the inner product on $\mathbb{R}^n$ determined by A.

11. a) Let $x_1, x_2, \ldots, x_n$ be distinct real numbers. For $f(x)$ and $g(x)$ in P_{n-1}, define $\langle f(x), g(x) \rangle$ to be

$$\langle f(x), g(x) \rangle = f(x_1)g(x_1) + f(x_2)g(x_2) + \cdots + f(x_n)g(x_n).$$

Verify that this defines an inner product on the vector space P_{n-1}.

b) For the inner product space in part (a) with $x_1 = -1$, $x_2 = 0$, $x_3 = 1$, find $\langle f(x), g(x) \rangle$, $\|f(x)\|$, $\|g(x)\|$, and the angle between $f(x)$ and $g(x)$ if $f(x) = x$ and $g(x) = 1 - x^2$.

12. a) Suppose that $w(x)$ is a continuous nonnegative function on $[a, b]$ for which $\int_a^b w(x)\,dx > 0$. For f and g in $C[a, b]$, define $\langle f, g \rangle$ to be

$$\langle f, g \rangle = \int_a^b f(x)w(x)g(x)\,dx.$$

Verify that this defines an inner product on $C[a, b]$.

b) For the inner product space in part (a) with $w(x) = x$ and $[a, b] = [0, 1]$, find $\langle f, g \rangle$, $\|f\|$, $\|g\|$, and the angle between f and g if $f(x) = x^2$ and $g(x) = x^3$.

13. From Exercise 7 of Section 2.2 we have that the set of sequences that converge to zero forms a vector space. Let V denote this vector space.

a) Show that the set of sequences $\{a_n\}$ so that $\sum_{n=1}^{\infty} a_n^2$ converges is a subset of V. This subset of sequences of V is commonly denoted l_2.

b) Show that if $\{a_n\}$ and $\{b_n\}$ are sequences in l_2, then the series $\sum_{n=1}^{\infty} a_n b_n$ converges. (*Hint*: $|a_n b_n| \leq a_n^2 + b_n^2$.)

c) Show that l_2 is a subspace of V.

d) For $\{a_n\}$ and $\{b_n\}$ in l_2, define $\langle \{a_n\}, \{b_n\} \rangle$ to be

$$\langle \{a_n\}, \{b_n\} \rangle = \sum_{n=1}^{\infty} a_n b_n.$$

Show that this defines an inner product on l_2.

e) Verify that $\{a_n\} = \{1/2^n\}$ and $\{b_n\} = \{(-3/4)^n\}$ are vectors in l_2 and then find $\langle \{a_n\}, \{b_n\} \rangle$, $\|\{a_n\}\|$, $\|\{b_n\}\|$, and the angle between $\{a_n\}$ and $\{b_n\}$ for these vectors.

14. Prove the following parts of Theorem 9.1.

a) Part (3)

b) Part (4)

c) Part (5)

15. Prove the following parts of Theorem 9.2.

a) Part (1)

b) Part (2)

16. Show that if v is a nonzero vector in an inner product space V, then $v/\|v\|$ is a vector of length 1.

17. Complete the proof of Theorem 9.3 by showing that if v and u are vectors of length 1, then $\langle v, u \rangle \geq -1$.

18. Show that $\sin x$ and $\sin 2x$ define two orthogonal vectors in the inner product space $C[0, \pi]$ with inner product $\langle f, g \rangle = \int_0^\pi f(x)g(x)\,dx$.

19. Show that $2\sqrt{3}x$ and $3\sqrt{5}x^2 - \sqrt{5}$ define two orthogonal vectors in the inner product space $C[-1, 1]$ with inner product $\langle f, g \rangle = \int_{-1}^1 f(x)g(x)\,dx$.

20. Let V be an inner product space and let v and u be vectors in V. Prove that

$$\|v + u\| \leq \|v\| + \|u\|.$$

This is called the triangle inequality. (*Hint:* Consider $\|v + u\|^2$ and apply the Cauchy-Schwartz inequality.)

21. Prove the following generalization of the Pythagorean Theorem: If v and u are orthogonal vectors in an inner product space V, then

$$\|v + u\|^2 = \|v\|^2 + \|u\|^2.$$

22. Let w be a fixed vector in an inner product space V. Show that the set of all vectors in V orthogonal to w is a subspace of V.

23. Let w be a fixed vector in an inner product space V. Define the function $T : V \to \mathbb{R}$ by

$$T(v) = \langle v, w \rangle.$$

Verify that T is a linear transformation. What are the elements in $\ker(T)$?

In Exercises 24–26, V denotes an inner product space and u and v denote vectors in V with $v \neq 0$.

24. We say that u and v are parallel if the angle between u and v is 0 or π.

   a) Show that if u is a scalar multiple of v, then u and v are parallel.

   b) Show that if the angle between u and v is zero, then $u = (\|u\|/\|v\|)v$. (*Hint:* Suppose $w = u - (\|u\|/\|v\|)v \neq 0$. Consider $\langle w, w \rangle$.)

   c) Show that if the angle between u and v is π, then $u = -(\|u\|/\|v\|)v$.

   d) Show that if u is parallel to v, then u is a scalar multiple of v.

25. The vector projection of u onto v, $\operatorname{proj}_v(u)$, is

$$\operatorname{proj}_v(u) = \frac{\langle u, v \rangle}{\|v\|^2} v.$$

   a) Show that $\operatorname{proj}_v(u)$ is parallel to v.

   b) Show that for a fixed v the function $T : V \to V$ defined by $T(u) = \operatorname{proj}_v(u)$ is a linear transformation.

   c) Describe the vectors in $\ker(T)$.

   d) Show that the only eigenvectors of T with nonzero associated eigenvalues are nonzero vectors parallel to v and the eigenvalue associated with these eigenvectors is 1.

   e) The vector $w = u - \operatorname{proj}_v(u)$ is called the vector component of u orthogonal to v. Show that w is orthogonal to v and $u = \operatorname{proj}_v(u) + w$. (Consequently, for a given nonzero vector v, we can write any other vector as the sum of a vector parallel to v and a vector orthogonal to v.)

   f) The vector $w = 2\operatorname{proj}_v(u) - u$ is called the reflection of u about v. Show that $\|w\| = \|u\|$, that the angles between u and v and between w and v are the same, and that w is in $\operatorname{Span}\{u, v\}$.

26. Suppose that $u \neq 0$ and u and v are orthogonal. The function $T : V \to V$ given by $T(w) = w + \langle u, w \rangle v$ is called a shear parallel to the vector v.

   a) Show that T is a linear transformation.

   b) Show that $T(w) - w$ is parallel to v for all vectors w in V.

   c) Show that the only eigenvectors of T are nonzero vectors orthogonal to u and the eigenvalue associated with these eigenvectors is 1.

9.2 ORTHONORMAL BASES

Suppose that V is an inner product space. A set of vectors $v_1, v_2, \ldots, v_n$ in V is called an **orthogonal set of vectors** if each v_i is orthogonal to each v_j or, equivalently, $\langle v_i, v_j \rangle = 0$ for each $i \neq j$. Further, if each v_i has length 1, or $\|v_i\| = 1$ (this is described by saying v_i is a **unit vector**), then we say that $v_1, v_2, \ldots, v_n$ form an **orthonormal set of vectors**.

An orthonormal basis is a basis that forms an orthonormal set.

> **DEFINITION**   The vectors $v_1, v_2, \ldots, v_n$ in an inner product space V form an **orthonormal basis** for V if:
> 1. $v_1, v_2, \ldots, v_n$ form a basis for V;
> 2. $\langle v_i, v_j \rangle = 0$ for each $i \neq j$;
> 3. $\|v_i\| = 1$ for each i.

The standard basis

$$
e_1 = \begin{bmatrix} 1 \\ 0 \\ \vdots \\ 0 \end{bmatrix}, \quad
e_2 = \begin{bmatrix} 0 \\ 1 \\ \vdots \\ 0 \end{bmatrix}, \quad \ldots, \quad
e_n = \begin{bmatrix} 0 \\ 0 \\ \vdots \\ 1 \end{bmatrix}
$$

for $\mathbb{R}^n$ is easily seen to be an orthonormal basis for $\mathbb{R}^n$ when the dot product is used as the inner product. It is not the only one, however, as the following example illustrates.

EXAMPLE 1   Verify that

$$
v_1 = \begin{bmatrix} 1/\sqrt{2} \\ 1/\sqrt{2} \\ 0 \end{bmatrix}, \quad
v_2 = \begin{bmatrix} 1/\sqrt{6} \\ -1/\sqrt{6} \\ 2/\sqrt{6} \end{bmatrix}, \quad
v_3 = \begin{bmatrix} 1/\sqrt{3} \\ -1/\sqrt{3} \\ -1/\sqrt{3} \end{bmatrix}
$$

form an orthonormal basis for $\mathbb{R}^3$ with the dot product as the inner product.

Solution   Let us first show v_1, v_2, v_3 are linearly independent. If

$$c_1 v_1 + c_2 v_2 + c_3 v_3 = 0,$$

we have the system:

$$
\frac{c_1}{\sqrt{2}} + \frac{c_2}{\sqrt{6}} + \frac{c_3}{\sqrt{3}} = 0
$$

$$
\frac{c_1}{\sqrt{2}} - \frac{c_2}{\sqrt{6}} - \frac{c_3}{\sqrt{3}} = 0
$$

$$
\frac{2c_2}{\sqrt{6}} - \frac{c_3}{\sqrt{3}} = 0.
$$

Adding the first two equations gives us $2c_1/\sqrt{2} = 0$, and hence $c_1 = 0$. Once this is known, we see $3c_2/\sqrt{6} = 0$ by adding the first and third equations, and consequently $c_2 = 0$. It now follows that $c_3 = 0$, and hence v_1, v_2, v_3 are linearly independent. Thus these three linearly independent vectors form a basis for the three-dimensional space $\mathbb{R}^3$.

Next we observe that

$$v_1 \cdot v_2 = \frac{1}{\sqrt{12}} - \frac{1}{\sqrt{12}} = 0,$$

$$v_1 \cdot v_3 = \frac{1}{\sqrt{6}} - \frac{1}{\sqrt{6}} = 0,$$

and

$$v_2 \cdot v_3 = \frac{1}{\sqrt{18}} + \frac{1}{\sqrt{18}} - \frac{2}{\sqrt{18}} = 0,$$

and consequently v_1, v_2, v_3 form an orthogonal set of vectors.
Finally,

$$\|v_1\| = \sqrt{\frac{1}{2} + \frac{1}{2}} = 1,$$

$$\|v_2\| = \sqrt{\frac{1}{6} + \frac{1}{6} + \frac{4}{6}} = 1,$$

and

$$\|v_3\| = \sqrt{\frac{1}{3} + \frac{1}{3} + \frac{1}{3}} = 1. \qquad\bullet$$

The next theorem shows us that it is easy to express a vector in an inner product space as a linear combination of the vectors in an orthonormal basis.

THEOREM 9.5   Suppose that $v_1, v_2, \ldots, v_n$ form an orthonormal basis for an inner product space V. If v is a vector in V, then

$$\boxed{v = \langle v, v_1\rangle v_1 + \langle v, v_2\rangle v_2 + \cdots + \langle v, v_n\rangle v_n.}$$

Proof   Since $v_1, v_2, \ldots, v_n$ form a basis for V, there are scalars $c_1, c_2, \ldots, c_n$ so that

$$v = c_1 v_1 + c_2 v_2 + \cdots + c_n v_n.$$

For each $1 \le i \le n$,

$$\langle v, v_i\rangle = \langle c_1 v_1 + c_2 v_2 + \cdots + c_n v_n, v_i\rangle = c_1\langle v_1, v_i\rangle + c_2\langle v_2, v_i\rangle + \cdots + c_n\langle v_n, v_i\rangle.$$

Since $\langle v_j, v_i\rangle = 0$ for $j \ne i$ and $\langle v_i, v_i\rangle = \|v_1\|^2 = 1$, this equation reduces to

$$\langle v, v_i\rangle = c_i$$

and the proof is complete. $\bullet$

Let us look at an example using the result of Theorem 9.5.

EXAMPLE 2   Express the vector

$$v = \begin{bmatrix} 1 \\ 2 \\ 3 \end{bmatrix}$$

as a linear combination of the vectors

$$v_1 = \begin{bmatrix} 1/\sqrt{2} \\ 1/\sqrt{2} \\ 0 \end{bmatrix}, \qquad v_2 = \begin{bmatrix} 1/\sqrt{6} \\ -1/\sqrt{6} \\ 2/\sqrt{6} \end{bmatrix}, \qquad v_3 = \begin{bmatrix} 1/\sqrt{3} \\ -1/\sqrt{3} \\ -1/\sqrt{3} \end{bmatrix}$$

in Example 1.

Solution   We know from Example 1 that v_1, v_2, v_3 form an orthonormal basis for $\mathbb{R}^3$. Since

$$\langle v, v_1 \rangle = \frac{3}{\sqrt{2}},$$

$$\langle v, v_2 \rangle = \frac{5}{\sqrt{6}},$$

and

$$\langle v, v_3 \rangle = -\frac{4}{\sqrt{3}},$$

Theorem 9.5 then tells us that

$$v = \frac{3}{\sqrt{2}} v_1 + \frac{5}{\sqrt{6}} v_2 - \frac{4}{\sqrt{3}} v_3. \qquad \bullet$$

Orthonormal bases often are convenient bases to use for inner product spaces. The ease of expressing other vectors as linear combinations of the vectors in an orthonormal basis as illustrated in Example 2 is one reason for this. In fact, you may well have encountered cases where vectors are expressed as linear combinations of orthonormal basis vectors in previous courses. For instance, in calculus, physics, or engineering courses, you may have seen the acceleration vector of a particle moving along a curve expressed in terms of the unit tangent and unit normal vectors.

Given a basis of vectors $u_1, u_2, \ldots, u_n$ for an inner product space V, there is a procedure called the **Gram-Schmidt process**[3] for converting a basis $u_1, u_2, \ldots, u_n$ into an orthonormal basis $v_1, v_2, \ldots, v_n$. Here is how the steps of this procedure go. In the first step, we make u_1 into a unit vector v_1 by dividing u_1 by its length (this is called **normalizing** the vector).

Step 1. Let $v_1 = u_1/\|u_1\|$.

---

[3] Named for the Danish mathematician and actuary Jörgen Pederson Gram (1850–1916) and the German mathematician Erhardt Schmidt (1876–1959).

In the second step, we first use v_1 and u_2 to construct a vector w_2 orthogonal to v_1 by setting

$$k_1 = \langle v_1, u_2 \rangle \quad \text{and} \quad w_2 = u_2 - k_1 v_1.$$

The vector w_2 is orthogonal to v_1 since

$$\langle v_1, w_2 \rangle = \langle v_1, u_2 \rangle - k_1 \langle v_1, v_1 \rangle = k_1 - k_1 = 0.$$

We then normalize w_2 to obtain a unit vector v_2 orthogonal to v_1. In summary, our second step is:

Step 2. Let $k_1 = \langle v_1, u_2 \rangle$ and $w_2 = u_2 - k_1 v_1$. Then $v_2 = w_2 / \|w_2\|$.

In the third step, we set

$$k_1 = \langle v_1, u_3 \rangle, \qquad k_2 = \langle v_2, u_3 \rangle, \qquad \text{and} \qquad w_3 = u_3 - k_1 v_1 - k_2 v_2.$$

Since

$$\langle v_1, w_3 \rangle = \langle v_1, u_3 \rangle - k_1 \langle v_1, v_1 \rangle - k_2 \langle v_1, v_2 \rangle = k_1 - k_1 = 0,$$

w_3 is orthogonal to v_1. Likewise, w_3 is orthogonal to v_2. (Verify this.) Normalizing w_3 gives us v_3. In summary, our third step is:

Step 3. Let $k_1 = \langle v_1, u_3 \rangle$, $k_2 = \langle v_2, u_3 \rangle$, and $w_3 = u_3 - k_1 v_1 - k_2 v_2$. Then $v_3 = w_3 / \|w_3\|$.

Can you see the pattern developing here? Just for good measure, let us state the fourth step:

Step 4. Let $k_1 = \langle v_1, u_4 \rangle$, $k_2 = \langle v_2, u_4 \rangle$, $k_3 = \langle v_3, u_4 \rangle$, and $w_4 = u_4 - k_1 v_1 - k_2 v_2 - k_3 v_3$. Then $v_4 = w_4 / \|w_4\|$.

The mth step of the Gram-Schmidt process ($m \leq n$) produces a set of m orthonormal vectors. Since orthonormal sets of vectors are linearly independent (see Exercise 18), applying the Gram-Schmidt process with n steps will lead us to a linearly independent set of n orthonormal vectors $v_1, v_2, \ldots, v_n$, which will then be an orthonormal basis for the n-dimensional vector space V.

Here is an example illustrating the Gram-Schmidt process.

EXAMPLE 3   Use the Gram-Schmidt process to transform the basis of $\mathbb{R}^3$ consisting of

$$u_1 = \begin{bmatrix} 1 \\ 1 \\ 0 \end{bmatrix}, \qquad u_2 = \begin{bmatrix} 1 \\ 0 \\ 1 \end{bmatrix}, \qquad u_3 = \begin{bmatrix} 2 \\ -1 \\ 1 \end{bmatrix}$$

into an orthonormal basis where the inner product is the dot product.

Solution

Step 1:

$$v_1 = \frac{u_1}{\|u_1\|} = \frac{1}{\sqrt{2}} \begin{bmatrix} 1 \\ 1 \\ 0 \end{bmatrix} = \begin{bmatrix} 1/\sqrt{2} \\ 1/\sqrt{2} \\ 0 \end{bmatrix}$$

Step 2:

$$k_1 = v_1 \cdot u_2 = \frac{1}{\sqrt{2}}, \qquad w_2 = u_2 - \frac{v_1}{\sqrt{2}} = \begin{bmatrix} 1/2 \\ -1/2 \\ 1 \end{bmatrix}$$

$$v_2 = \frac{w_2}{\|w_2\|} = \frac{1}{\sqrt{6/4}} \begin{bmatrix} 1/2 \\ -1/2 \\ 1 \end{bmatrix} = \begin{bmatrix} 1/\sqrt{6} \\ -1/\sqrt{6} \\ 2/\sqrt{6} \end{bmatrix}$$

Step 3:

$$k_1 = v_1 \cdot u_3 = \frac{1}{\sqrt{2}}, \qquad k_2 = v_2 \cdot u_3 = \frac{5}{\sqrt{6}}$$

$$w_3 = u_3 - \frac{1}{\sqrt{2}}v_1 - \frac{5}{\sqrt{6}}v_2 = \begin{bmatrix} 2/3 \\ -2/3 \\ -2/3 \end{bmatrix}$$

$$v_3 = \frac{w_3}{\|w_3\|} = \frac{1}{\sqrt{12/9}} \begin{bmatrix} 2/3 \\ -2/3 \\ -2/3 \end{bmatrix} = \begin{bmatrix} 1/\sqrt{3} \\ -1/\sqrt{3} \\ -1/\sqrt{3} \end{bmatrix}$$

The orthonormal basis we have found is then

$$v_3 = \begin{bmatrix} 1/\sqrt{2} \\ 1/\sqrt{2} \\ 0 \end{bmatrix}, \qquad v_2 = \begin{bmatrix} 1/\sqrt{6} \\ -1/\sqrt{6} \\ 2/\sqrt{6} \end{bmatrix}, \qquad v_3 = \begin{bmatrix} 1/\sqrt{3} \\ -1/\sqrt{3} \\ -1/\sqrt{3} \end{bmatrix}. \qquad \bullet$$

Notice the orthonormal basis we found in Example 3 is the orthonormal basis given in Example 1. If you were wondering how we came up with the orthonormal basis in Example 1, you can now see how we got it.

If A is an $n \times n$ matrix,

$$A = \begin{bmatrix} a_{11} & a_{12} & \cdots & a_{1n} \\ a_{21} & a_{22} & \cdots & a_{2n} \\ \vdots & \vdots & & \vdots \\ a_{n1} & a_{n2} & \cdots & a_{nn} \end{bmatrix},$$

the ij-entry of the product $A^T A$ is

$$\sum_{k=1}^{n} a_{ki} a_{kj}.$$

This is the same as the dot product of column i of A and column j of A. To put it another way, if we let

$$v_i = \begin{bmatrix} a_{1i} \\ a_{2i} \\ \vdots \\ a_{ni} \end{bmatrix},$$

then

$$A^T A = [v_i \cdot v_j] = [v_i^T v_j].$$

Consequently, if the columns of A form an orthonormal basis for $\mathbb{R}^n$, then

$$A^T A = I.$$

Conversely, if $A^T A = I$, then the columns of A form an orthonormal set. Also, since A is an invertible matrix (with $A^{-1} = A^T$), the columns of A form a basis for $\mathbb{R}^n$. Thus if $A^T A = I$, the columns of A form an orthonormal basis for $\mathbb{R}^n$. An $n \times n$ matrix A with the property that $A^T A = I$ (or $A^{-1} = A^T$) is called an **orthogonal matrix.** We summarize this discussion with the following theorem.

THEOREM 9.6   A $n \times n$ matrix A is an orthogonal matrix if and only if the columns of A form an orthonormal basis for $\mathbb{R}^n$.

The Gram-Schmidt process gives us one way of finding orthonormal bases. Another procedure that is sometimes used is based on the following theorem.

THEOREM 9.7   Suppose that

$$v = \begin{bmatrix} a_1 \\ a_2 \\ \vdots \\ a_n \end{bmatrix}$$

is a unit vector with $a_1 \neq -1$. Let w denote the column

$$w = \begin{bmatrix} a_2 \\ a_3 \\ \vdots \\ a_n \end{bmatrix}.$$

Then the $n \times n$ matrix

$$A = \begin{bmatrix} a_1 & w^T \\ w & -I + \frac{1}{1+a_1} w w^T \end{bmatrix}$$

where I is the $(n-1) \times (n-1)$ identity matrix is an orthogonal matrix.

Proof   It suffices to show that $A^T A = I_n$. Keep in mind that since v is a unit vector, $v^T v = 1$. Because of this,

$$w^T w = 1 - a_1^2.$$

We have

$$A^T A = \begin{bmatrix} a_1 & w^T \\ w & -I + \frac{1}{1+a_1} w w^T \end{bmatrix} \begin{bmatrix} a_1 & w^T \\ w & -I + \frac{1}{1+a_1} w w^T \end{bmatrix}.$$

(Why?) It then follows that

$$A^T A = \begin{bmatrix} a_1^2 + w^T w & a_1 w^T - w^T + \frac{1}{1+a_1} w^T w w^T \\ a_1 w - w + \frac{1}{1+a_1} w w^T w & w w^T + I - \frac{2}{1+a_1} w w^T + \frac{1}{(1+a_1)^2} w w^T w w^T \end{bmatrix}$$

$$= \begin{bmatrix} 1 & a_1 w^T - w^T + \frac{1}{1+a_1}(1 - a_1^2) w^T \\ a_1 w - w + \frac{1}{1+a_1}(1 - a_1^2) w & w w^T + I - \frac{2}{1+a_1} w w^T + \frac{1}{(1+a_1)^2}(1 - a_1^2) w w^T \end{bmatrix}$$

$$= \begin{bmatrix} 1 & O_{1 \times (n-1)} \\ O_{(n-1) \times 1} & I \end{bmatrix} = I_n$$

●

The following example illustrates how, starting with a nonzero vector in $\mathbb{R}^n$, we can use this vector in conjunction with Theorem 9.7 to find an orthonormal basis for $\mathbb{R}^n$.

EXAMPLE 4   Starting with the vector

$$u = \begin{bmatrix} 1 \\ 0 \\ 1 \end{bmatrix},$$

find an orthonormal basis for $\mathbb{R}^3$.

Solution   We begin by normalizing u to get a unit vector v:

$$v = \frac{u}{\|u\|} = \begin{bmatrix} 1/\sqrt{2} \\ 0 \\ 1/\sqrt{2} \end{bmatrix}.$$

Calculating the entries of the matrix A of Theorem 9.7 with this vector v, in which case

$$w = \begin{bmatrix} 0 \\ 1/\sqrt{2} \end{bmatrix},$$

we find (try it)

$$A = \begin{bmatrix} 1/\sqrt{2} & 0 & 1/\sqrt{2} \\ 0 & -1 & 0 \\ 1/\sqrt{2} & 0 & -1/\sqrt{2} \end{bmatrix}.$$

The orthonormal basis we then obtain consists of

$$\begin{bmatrix} 1/\sqrt{2} \\ 0 \\ 1/\sqrt{2} \end{bmatrix}, \begin{bmatrix} 0 \\ -1 \\ 0 \end{bmatrix}, \begin{bmatrix} 1/\sqrt{2} \\ 0 \\ -1/\sqrt{2} \end{bmatrix}.$$

●

EXERCISES 9.2

1. a) Verify that the vectors

$$v_1 = \begin{bmatrix} 1/2 \\ 1/2 \\ \sqrt{2}/2 \end{bmatrix}, \quad v_2 = \begin{bmatrix} 1/2 \\ -5/6 \\ \sqrt{2}/6 \end{bmatrix},$$

$$v_3 = \begin{bmatrix} \sqrt{2}/2 \\ \sqrt{2}/6 \\ -2/3 \end{bmatrix}$$

form an orthonormal basis for $\mathbb{R}^3$ with the dot product as the inner product.

b) Express

$$v = \begin{bmatrix} -3 \\ 6 \\ -5 \end{bmatrix}$$

as a linear combination of v_1, v_2, v_3.

2. a) Verify that the vectors

$$v_1 = \begin{bmatrix} 0 \\ 1/\sqrt{3} \\ 1/\sqrt{3} \\ 1/\sqrt{3} \end{bmatrix}, \quad v_2 = \begin{bmatrix} 1/\sqrt{3} \\ -2/3 \\ 1/3 \\ 1/3 \end{bmatrix},$$

$$v_3 = \begin{bmatrix} 1/\sqrt{3} \\ 1/3 \\ -2/3 \\ 1/3 \end{bmatrix}, \quad v_4 = \begin{bmatrix} 1/\sqrt{3} \\ 1/3 \\ 1/3 \\ -2/3 \end{bmatrix}$$

form an orthonormal basis for $\mathbb{R}^4$ with the dot product as the inner product.

b) Express

$$v = \begin{bmatrix} 1 \\ -3 \\ 4 \\ -2 \end{bmatrix}$$

as a linear combination of v_1, v_2, v_3, v_4.

In Exercises 3–6, use the Gram-Schmidt process to convert the given basis vectors to an orthonormal basis where the inner product is the dot product.

3. $\begin{bmatrix} 1 \\ 2 \end{bmatrix}, \begin{bmatrix} 1 \\ -1 \end{bmatrix}$ **4.** $\begin{bmatrix} 3 \\ -2 \end{bmatrix}, \begin{bmatrix} 2 \\ -2 \end{bmatrix}$

5. $\begin{bmatrix} 3 \\ 0 \\ 0 \end{bmatrix}, \begin{bmatrix} 2 \\ -3 \\ 4 \end{bmatrix}, \begin{bmatrix} 1 \\ -15 \\ -5 \end{bmatrix}$

6. $\begin{bmatrix} 1 \\ 0 \\ 1 \end{bmatrix}, \begin{bmatrix} 1 \\ 1 \\ 0 \end{bmatrix}, \begin{bmatrix} 0 \\ 1 \\ 1 \end{bmatrix}$

7. Use the approach of Example 4 to find an orthonormal basis for $\mathbb{R}^3$ starting with the vector

$$u = \begin{bmatrix} 1 \\ -1 \\ 1 \end{bmatrix}.$$

8. Use the approach of Example 4 to find an orthonormal basis for $\mathbb{R}^4$ starting with the vector

$$u = \begin{bmatrix} 0 \\ 1 \\ 2 \\ -2 \end{bmatrix}.$$

Let P_n have the inner product

$$\langle p, q \rangle = \int_{-1}^{1} p(x)q(x)\, dx.$$

The polynomials obtained by applying the Gram-Schmidt process to the standard basis of P_n consisting of $1, x, x^2, \ldots, x^n$ are called normalized Legendre polynomials.[4]

9. Find the normalized Legendre polynomials when $n = 2$.

10. Find the normalized Legendre polynomials when $n = 3$.

11. Apply the Gram-Schmidt process to convert the standard basis of $\mathbb{R}^2$ to an orthonormal basis if the inner product on $\mathbb{R}^2$ is that of Exercise 8(b) in Section 9.1.

12. Do Exercise 11 if the inner product is that of Exercise 9(b) in Section 9.1.

13. Show that a matrix of the form

$$\begin{bmatrix} \cos\alpha & \pm\sin\alpha \\ \sin\alpha & \mp\cos\alpha \end{bmatrix}$$

is an orthogonal matrix.

14. Show that every 2×2 orthogonal matrix has the form of the matrix in Exercise 13.

15. Show that if A and B are $n \times n$ orthogonal matrices, then AB is an orthogonal matrix.

16. Suppose that A is an $n \times n$ orthogonal matrix. Prove that for any vector v in $\mathbb{R}^n$ with inner product the dot product, $\|Av\| = \|v\|$.

17. a) Apply the Gram-Schmidt process to find an orthonormal basis for the subspace of $\mathbb{R}^3$ spanned by the vectors

$$\begin{bmatrix} 1 \\ -1 \\ 1 \end{bmatrix}, \begin{bmatrix} 2 \\ 0 \\ 1 \end{bmatrix}$$

where the inner product is the dot product.

b) Use the result of part (a) in conjunction with the cross product of vectors you learned in calculus to find an orthonormal basis for $\mathbb{R}^3$.

18. Show that if $v_1, v_2, \ldots, v_m$ are orthonormal vectors in an inner product space V, then $v_1, v_2, \ldots, v_m$ are linearly independent. (*Hint:* Use an approach along the lines of the one in the proof of Theorem 9.5.)

19. The *GramSchmidt* command in Maple converts a set of vectors (the vectors do not have to be a basis or even linearly independent) in $\mathbb{R}^n$ into an orthogonal set of vectors but does not normalize them. Apply the *GramSchmidt* command or the corresponding command in another appropriate software package to the vectors in Exercise 5. Compare your software package's result with the answer to Exercise 5.

20. The *innerprod* command in Maple can be used to find dot products of vectors and hence lengths of vectors in $\mathbb{R}^n$. Use this command or the corresponding command in another appropriate software package to convert the orthogonal basis found in Exercise 19 to an orthonormal basis.

9.3   SCHUR'S THEOREM AND SYMMETRIC MATRICES

Suppose A is an $n \times n$ matrix. Recall from Chapter 5 that an $n \times n$ matrix B is similar to A if there is an invertible $n \times n$ matrix P so that

$$B = P^{-1}AP.$$

If the matrix P is an orthogonal matrix, in which case

$$P^{-1} = P^T \quad \text{and} \quad B = P^T AP,$$

---

[4] These normalized Legendre polynomials can also be obtained by normalizing the Legendre polynomials of Section 8.2.

we say that B is **orthogonally similar** to A. Also, recall that P is the change of basis matrix of the matrix transformation $T(X) = AX$ from the standard basis α for $\mathbb{R}^n$ to the basis β for $\mathbb{R}^n$ consisting of the columns of P when $B = P^{-1}AP$. If P is an orthogonal matrix, then the basis β is an orthonormal basis for $\mathbb{R}^n$ by Theorem 9.6.

Our first major result of this section is a result known as Schur's Theorem.[5] This theorem gives us a case in which a square matrix is orthogonally similar to a triangular matrix.

THEOREM 9.8   **Schur's Theorem**   Suppose that A is an $n \times n$ matrix. If all of the eigenvalues of A are real numbers, then A is orthogonally similar to an upper triangular matrix.

Proof   We use induction on n. The result is immediate if $k = 1$. Assume the result holds for all $k \times k$ matrices that have only real eigenvalues and suppose A is a $(k+1) \times (k+1)$ matrix with only real eigenvalues. Let $\lambda = r_1$ be one of the eigenvalues of A and let u_1 be an eigenvector of A associated with the eigenvalue r_1. By part (1) of Lemma 2.11, we can extend the linearly independent set consisting of the vector u_1 to a basis $u_1, u_2, \ldots, u_{k+1}$ of $\mathbb{R}^{k+1}$. Applying the Gram-Schmidt process to this basis, we obtain an orthonormal basis $v_1, v_2, \ldots, v_{k+1}$ with $v_1 = u_1/\|u_1\|$ also being an eigenvector associated with the eigenvalue r_1.[6] Letting P_1 denote the orthogonal matrix

$$P_1 = \begin{bmatrix} v_1 & v_2 & \cdots & v_{k+1} \end{bmatrix},$$

the matrix of the linear transformation $T(X) = AX$ with respect to the basis consisting of $v_1, v_2, \ldots, v_{k+1}$ has the form

$$P_1^{-1}AP_1 = P_1^T AP_1 = \begin{bmatrix} r_1 & C \\ O_{k \times 1} & A_1 \end{bmatrix}$$

where C is a $1 \times k$ matrix and A_1 is a $k \times k$ matrix. Let $p(\lambda)$ denote the characteristic polynomial of A and $p_1(\lambda)$ denote the characteristic polynomial of A_1. Using the fact that similar matrices have the same characteristic polynomial (see Exercise 31 of Section 5.5), it follows that

$$p(\lambda) = \det \begin{bmatrix} \lambda - r_1 & -C \\ O_{k \times 1} & \lambda I - A_1 \end{bmatrix} = (\lambda - r_1) \det(\lambda I - A_1) = (\lambda - r_1)p_1(\lambda).$$

This gives us that all of the eigenvalues of A_1 are real numbers. Hence by the induction hypothesis there is an orthogonal $k \times k$ matrix Q so that $Q^T A_1 Q$ is an upper triangular matrix. Let P_2 be the $(k+1) \times (k+1)$ matrix

$$P_2 = \begin{bmatrix} 1 & O_{1 \times k} \\ O_{k \times 1} & Q \end{bmatrix}.$$

---

[5] Issai Schur (1875–1941) was a student of Georg Frobenius and, as is the case with Frobenius, is best known for his many important contributions to group theory.

[6] If the first entry of $u_1/\|u_1\|$ is not -1, we could use Theorem 9.7 as an alternative method for finding this orthonormal basis.

We leave it as an exercise (Exercise 15) to verify that $P = P_1 P_2$ is an orthogonal matrix and that $P^T A P$ is an upper triangular matrix. ●

The theory of inner product spaces that we have developed for vector spaces over the real numbers can be extended to vector spaces over the complex numbers, but we will not do so fully here. We do mention, however, one special example of a complex inner product space: the vector space of $n \times 1$ column vectors over the complex numbers $\mathbb{C}$, denoted $\mathbb{C}^n$, with the dot product of two vectors

$$v = \begin{bmatrix} a_1 \\ a_2 \\ \vdots \\ a_n \end{bmatrix} \quad \text{and} \quad u = \begin{bmatrix} b_1 \\ b_2 \\ \vdots \\ b_n \end{bmatrix}$$

in $\mathbb{C}^n$ defined to be

$$v \cdot u = a_1 \overline{b_1} + a_2 \overline{b_2} + \cdots + a_n \overline{b_n} = v^T \overline{u}.$$

This dot product is then a function from pairs of vectors from $\mathbb{C}^n$ to $\mathbb{C}$. In the case when v and u are vectors in $\mathbb{R}^n$, their dot product in $\mathbb{C}^n$ is the same as it is in $\mathbb{R}^n$. Many, but not all, of the properties we have for dot products of vectors in $\mathbb{R}^n$ carry over to the dot product on $\mathbb{C}^n$. Perhaps the most notable exception is with scalars in the right-hand factor. Notice that because of the conjugate of u, we have

$$v \cdot (cu) = \overline{c}(v \cdot u)$$

rather than $c(v \cdot u)$ as in the real case.

With the introduction of the dot product on $\mathbb{C}^n$, we can extend Schur's Theorem to the case where A has imaginary eigenvalues. If C is a matrix whose entries are complex numbers, the **Hermitian conjugate**[7] of C, denoted C^*, is the conjugate tranpose of C; that is,

$$\boxed{C^* = \overline{C}^T.}$$

For instance, if

$$C = \begin{bmatrix} 2 & 4 + 3i \\ 1 - 2i & \sqrt{2}i \end{bmatrix},$$

then

$$C^* = \begin{bmatrix} 2 & 4 - 3i \\ 1 + 2i & -\sqrt{2}i \end{bmatrix}^T = \begin{bmatrix} 2 & 1 + 2i \\ 4 - 3i & -\sqrt{2}i \end{bmatrix}.$$

An $n \times n$ matrix P with complex entries is called a **unitary matrix** if

$$\boxed{P^* P = I.}$$

---

[7] Named for Charles Hermite (1822–1901).

Unitary matrices are the analogue of orthogonal matrices in $M_{n \times n}(\mathbb{C})$. An $n \times n$ matrix B is **unitarily similar** to an $n \times n$ matrix A if there is an $n \times n$ unitary matrix P so that $B = P^* A P$. In the case when A has complex eigenvalues, Schur's Theorem holds provided we replace orthogonally similar by unitarily similar.

THEOREM 9.9   **Schur's Theorem (continued)**   If A is an $n \times n$ matrix, then A is unitarily similar to an upper triangular matrix.

To prove Theorem 9.9, we adjust the proof of Theorem 9.8 to the complex setting. We will let you go through these details as an exercise (Exercise 16).

The inductive step in the proof of Schur's Theorem gives us an algorithm for triangularizing matrices. The following example illustrates how we do this in practice.

EXAMPLE 1   Find an orthogonal matrix P and a matrix $B = P^T A P$ so that B is a triangular matrix for

$$A = \begin{bmatrix} 6 & 4 & -3 \\ 1 & 4 & -1 \\ 4 & 5 & -1 \end{bmatrix}.$$

Solution   We will leave out many details and just outline how we arrive at our solution. In fact, we authors did many of the calculations that follow using Maple. When doing so, we sometimes applied Maple's *simplify* command to some of the matrices since Maple did not simplify the entries. The only eigenvalue of A is $\lambda = 3$. (This is real, so we now do know that A is orthogonally similar to a triangular matrix—otherwise we would have to do unitarily similar.) The vector

$$\begin{bmatrix} 1 \\ 0 \\ 1 \end{bmatrix}$$

forms a basis for E_3. Let us use this for u_1 in the proof of Theorem 9.8. We could use the approach in the proof of Theorem 9.8, which involves finding a basis for $\mathbb{R}^3$ containing u_1 and then applying the Gram-Schmidt process to find an othonormal basis forming the matrix P_1. However, the first entry of $u_1 / \|u_1\|$ is not -1 and it will be easier to apply the alternative approach involving Theorem 9.7 mentioned in the footnote to the proof of Theorem 9.8. In fact, we deliberately set up Example 4 of Section 9.2 so that the vector in this example is u_1. Consequently, we may take P_1 to be

$$P_1 = \begin{bmatrix} 1/\sqrt{2} & 0 & 1/\sqrt{2} \\ 0 & -1 & 0 \\ 1/\sqrt{2} & 0 & -1/\sqrt{2} \end{bmatrix}.$$

Calculating the product $P_1^T A P$, we find

$$A_1 = P_1^T A P = \begin{bmatrix} 3 & -9\sqrt{2}/2 & 7 \\ 0 & 4 & -\sqrt{2} \\ 0 & \sqrt{2}/2 & 2 \end{bmatrix}.$$

Now we have to triangularize (find Q) for the submatrix

$$B_1 = \begin{bmatrix} 4 & -\sqrt{2} \\ \sqrt{2}/2 & 2 \end{bmatrix}$$

of A_1. The only eigenvalue of B_1 is $\lambda = 3$ and E_3 has

$$\begin{bmatrix} \sqrt{2} \\ 1 \end{bmatrix}$$

as a basis vector. Applying the same set of steps we just applied to A to B_1, we find that

$$Q = \begin{bmatrix} \sqrt{6}/3 & \sqrt{3}/3 \\ \sqrt{3}/3 & -\sqrt{6}/3 \end{bmatrix}$$

and

$$Q^T B_1 Q = \begin{bmatrix} 3 & 3\sqrt{2}/2 \\ 0 & 3 \end{bmatrix}.$$

Setting

$$P_2 = \begin{bmatrix} 1 & 0 & 0 \\ 0 & \sqrt{6}/3 & \sqrt{3}/3 \\ 0 & \sqrt{3}/3 & -\sqrt{6}/3 \end{bmatrix}$$

and

$$P = P_1 P_2 = \begin{bmatrix} \sqrt{2}/2 & \sqrt{6}/6 & -\sqrt{3}/3 \\ 0 & -\sqrt{6}/3 & -\sqrt{3}/3 \\ \sqrt{2}/2 & -\sqrt{6}/6 & \sqrt{3}/3 \end{bmatrix},$$

we have

$$B = P^T A P = \begin{bmatrix} 3 & -2\sqrt{3}/3 & -23\sqrt{6}/6 \\ 0 & 3 & 3\sqrt{2}/2 \\ 0 & 0 & 3 \end{bmatrix}. \qquad \bullet$$

Our ability to triangularize square matrices gives us an alternative method to the methods of Chapter 6 for solving homogeneous systems of linear differential equations with constant coefficients.

EXAMPLE 2   Find the general solution to $Y' = AY$ where A is the matrix in Example 1.

Solution   Solving the triangular system $Y' = BY$ for the matrix B in the solution to Example 1, or equivalently, solving the system

$$y_1' = 3y_1 - \frac{2\sqrt{3}}{3}y_2 - \frac{23\sqrt{6}}{6}y_3$$

$$y_2' = 3y_2 + \frac{3\sqrt{2}}{2}y_3$$

$$y_3' = 3y_3,$$

we find its general solution to be

$$Z = \begin{bmatrix} c_1 e^{3x} - \frac{2\sqrt{3}}{3}c_2 x e^{3x} - c_3\left(\frac{\sqrt{6}}{2}x^2 + \frac{23\sqrt{6}}{6}x\right)e^{3x} \\ c_2 e^{3x} + \frac{3\sqrt{2}}{2}c_3 x e^{3x} \\ c_3 e^{3x} \end{bmatrix}.$$

It follows from Theorem 6.7 that the general solution to $Y' = AY$ is as follows.

$Y = PZ$

$$= \begin{bmatrix} \sqrt{2}/2 & \sqrt{6}/6 & -\sqrt{3}/3 \\ 0 & -\sqrt{6}/3 & -\sqrt{3}/3 \\ \sqrt{2}/2 & -\sqrt{6}/6 & \sqrt{3}/3 \end{bmatrix} \begin{bmatrix} c_1 e^{3x} - \frac{2\sqrt{3}}{3}c_2 x e^{3x} - c_3\left(\frac{\sqrt{6}}{2}x^2 + \frac{23\sqrt{6}}{6}x\right)e^{3x} \\ c_2 e^{3x} + \frac{3\sqrt{2}}{2}c_3 x e^{3x} \\ c_3 e^{3x} \end{bmatrix}$$

(The entries are long and complicated if we calculate this product, so let us leave our answer in this product form.)   ●

Of course, if the matrix A of a system $Y' = AY$ has imaginary eigenvalues, the type of work we did in Examples 1 and 2 would yield complex solutions which we would use to obtain real solutions in a manner similar to the one we used in Chapter 6.

In Chapter 5 we discussed the diagonalizability of square matrices and saw that not all square matrices are diagonalizable. Further, we saw in Chapter 6 that whether or not the matrix of a constant coefficient homogeneous system of linear differential equations is diagonalizable has a significant impact on how we find the general solution to such a system. Using Schur's Theorem, we will show that all symmetric matrices with real entries are diagonalizable. This then gives us a large class of matrices that we immediately know are diagonalizable. In order to obtain this, we first prove Theorem 9.10.

THEOREM 9.10   If A is an $n \times n$ symmetric matrix with real entries, then all the eigenvalues of A are real numbers.

Proof   Suppose that $\lambda = r$ is an eigenvalue of A. Let v be an eigenvector of A associated with the eigenvalue r. On the one hand, notice that $(Av) \cdot v$ (viewing this dot product in $\mathbb{C}^n$) is

$$(Av) \cdot v = (rv) \cdot v = r(v \cdot v).$$

On the other hand,

$$(Av) \cdot v = (Av)^T \overline{v} = v^T A^T \overline{v} = v^T A \overline{v} = v^T \overline{Av} = v \cdot (rv) = \overline{r}(v \cdot v).$$

Since $v \cdot v \neq 0$, these two equations give us that $r = \overline{r}$ and hence r is a real number. ●

Now we are ready to prove the aforementioned result about diagonalizability.

THEOREM 9.11   If A is a symmetric matrix with real entries, then A is diagonalizable.

Proof   Since all of the eigenvalues of A are real by Theorem 9.10, Schur's Theorem tells us that there is an orthogonal matrix P so that $P^T A P$ is an upper triangular matrix. Using the property $(AB)^T = B^T A^T$, we have

$$(P^T A P)^T = P^T A^T (P^T)^T = P^T A P$$

since A is symmetric. Hence $P^T A P$ is symmetric. However, the only way $P^T A P$ can be both upper triangular and symmetric is for it to be a diagonal matrix and our proof is complete. ●

We can even say a bit more for the matrix A in Theorem 9.11. From its proof we can see that not only is A diagonalizable, but it is orthogonally similar to a diagonal matrix. One way to obtain an orthogonal matrix P so that $P^T A P$ is a diagonal matrix is to use the approach of Example 1. However, there is a more commonly used approach. Before stating it, we prove the following theorem, which will be used to justify our technique.

THEOREM 9.12   If A is a symmetric matrix with real entries and v_1 and v_2 are eigenvectors of A with different associated eigenvalues, then v_1 and v_2 are orthogonal.

Proof   Suppose that $Av_1 = r_1 v_1$ and $Av_2 = r_2 v_2$. We proceed in a manner similar to the proof of Theorem 9.10. On the one hand,

$$(Av_1) \cdot v_2 = (r_1 v_1) \cdot v_2 = r_1 (v_1 \cdot v_2).$$

On the other hand,

$$(Av_1) \cdot v_2 = (Av_1)^T v_2 = v_1^T A^T v_2 = v_1^T A v_2 = v_1^T (r_2 v_2) = r_2 (v_1 \cdot v_2).$$

If $v_1 \cdot v_2 \neq 0$, these equations give us $r_1 = r_2$. Thus $v_1 \cdot v_2 = 0$ and hence v_1 and v_2 are orthogonal. ●

Now we state a commonly used set of steps for finding an orthogonal matrix P that diagonalizes an $n \times n$ symmetric matrix A with real entries.

Step 1. Find bases for the eigenspaces of A.

Of course, step 1 is nothing new. We have been doing this all along. But next:

Step 2. Apply the Gram-Schmidt process to the basis of each eigenspace to obtain an orthormal basis for the eigenspace.

Since we know A is diagonalizable, we have from Chapter 5 that altogether the vectors from the eigenspace bases in step 1 form a basis of eigenvectors of A for $\mathbb{R}^n$. The same then holds for all of the vectors we obtain in step 2. Because of Theorem 9.12, the vectors obtained in step 2 will be orthonormal. Thus they give us an orthonormal basis for $\mathbb{R}^n$, which forms the desired matrix P making our third and final step:

Step 3. Use for P the matrix whose columns are the vectors from step 2.

The following example illustrates how this procedure looks in practice.

EXAMPLE 3   Find an orthogonal matrix P so that $P^T A P$ is a diagonal matrix for

$$A = \begin{bmatrix} 3 & 0 & 1 \\ 0 & 4 & 0 \\ 1 & 0 & 3 \end{bmatrix}.$$

Solution

Step 1. The characteristic polynomial is

$$\det(\lambda I - A) = \begin{vmatrix} \lambda - 3 & 0 & -1 \\ 0 & \lambda - 4 & 0 \\ -1 & 0 & \lambda - 3 \end{vmatrix} = (\lambda - 4)((\lambda - 3)^2 - 1) = (\lambda - 4)^2(\lambda - 2).$$

Let us next find a basis for E_4.

$$\left[\begin{array}{ccc:c} 1 & 0 & -1 & 0 \\ 0 & 0 & 0 & 0 \\ -1 & 0 & 1 & 0 \end{array} \right] \rightarrow \left[\begin{array}{ccc:c} 1 & 0 & -1 & 0 \\ 0 & 0 & 0 & 0 \\ 0 & 0 & 0 & 0 \end{array} \right]$$

$$\begin{bmatrix} x \\ y \\ z \end{bmatrix} = \begin{bmatrix} z \\ y \\ z \end{bmatrix}.$$

The vectors

$$\begin{bmatrix} 0 \\ 1 \\ 0 \end{bmatrix}, \begin{bmatrix} 1 \\ 0 \\ 1 \end{bmatrix}$$

form a basis for E_4. To complete step 1, we find a basis for E_2.

$$\begin{bmatrix} -1 & 0 & -1 & \vdots & 0 \\ 0 & -2 & 0 & \vdots & 0 \\ -1 & 0 & -1 & \vdots & 0 \end{bmatrix} \rightarrow \begin{bmatrix} 1 & 0 & 1 & \vdots & 0 \\ 0 & 1 & 0 & \vdots & 0 \\ 0 & 0 & 0 & \vdots & 0 \end{bmatrix}$$

$$\begin{bmatrix} x \\ y \\ z \end{bmatrix} = \begin{bmatrix} -z \\ 0 \\ z \end{bmatrix}$$

The vector

$$\begin{bmatrix} -1 \\ 0 \\ 1 \end{bmatrix}$$

forms a basis for E_2.

Step 2. Applying the Gram-Schmidt process to the basis consisting of

$$u_1 = \begin{bmatrix} 0 \\ 1 \\ 0 \end{bmatrix}, \quad u_2 = \begin{bmatrix} 1 \\ 0 \\ 1 \end{bmatrix}$$

for E_4 we obtain:

$$v_1 = u_1, \quad k_1 = v_1 \cdot u_2 = 0, \quad w_2 = u_2, \quad v_2 = \begin{bmatrix} 1/\sqrt{2} \\ 0 \\ 1/\sqrt{2} \end{bmatrix}.$$

The vectors

$$\begin{bmatrix} 0 \\ 1 \\ 0 \end{bmatrix}, \begin{bmatrix} 1/\sqrt{2} \\ 0 \\ 1/\sqrt{2} \end{bmatrix}$$

form an orthonormal basis for E_4. Applying the Gram-Schmidt process to the basis consisting of

$$\begin{bmatrix} -1 \\ 0 \\ 1 \end{bmatrix}$$

for E_2, we quickly get the vector

$$\begin{bmatrix} -1/\sqrt{2} \\ 0 \\ 1/\sqrt{2} \end{bmatrix},$$

which forms an orthonormal basis for E_2.

$$\text{Step 3. } P = \begin{bmatrix} 0 & 1/\sqrt{2} & -1/\sqrt{2} \\ 1 & 0 & 0 \\ 0 & 1/\sqrt{2} & 1/\sqrt{2} \end{bmatrix}$$ ●

EXERCISES 9.3

In Exercises 1–4, find an orthogonal matrix P and a matrix $B = P^T A P$ so that B is a triangular matrix for the given matrix A.

1. $A = \begin{bmatrix} 1 & 1 \\ -1 & 3 \end{bmatrix}$ **2.** $A = \begin{bmatrix} -2 & 3 \\ -3 & 4 \end{bmatrix}$

3. $A = \begin{bmatrix} -2 & 1 & 0 \\ -2 & -1 & 1 \\ -3 & 1 & 0 \end{bmatrix}$

4. $A = \begin{bmatrix} 3 & 2 & -1 \\ 4 & 9 & -5 \\ 8 & 12 & -7 \end{bmatrix}$

5. Find the general solution to the system of linear differential equations $Y' = AY$ for the matrix A in Exercise 3.

6. Find the general solution to the system of linear differential equations $Y' = AY$ for the matrix A in Exercise 4.

In Exercises 7–12, find an orthogonal matrix P so that $P^T A P$ is a diagonal matrix for the given matrix A.

7. $A = \begin{bmatrix} 1 & 2 \\ 2 & 1 \end{bmatrix}$ **8.** $A = \begin{bmatrix} 7 & -24 \\ -24 & 7 \end{bmatrix}$

9. $A = \begin{bmatrix} 3 & -1 & 0 \\ -1 & 3 & 0 \\ 0 & 0 & 2 \end{bmatrix}$

10. $A = \begin{bmatrix} 2 & -1 & -1 \\ -1 & 2 & -1 \\ -1 & -1 & 2 \end{bmatrix}$

11. $A = \begin{bmatrix} \sqrt{2} & \sqrt{2} & 2 \\ \sqrt{2} & \sqrt{2} & 2 \\ 2 & 2 & 2\sqrt{2} \end{bmatrix}$

12. $A = \begin{bmatrix} 4 & 0 & 0 & 0 \\ 0 & 3 & 0 & 1 \\ 0 & 0 & 2 & 0 \\ 0 & 1 & 0 & 3 \end{bmatrix}$

13. Find the general solution to the system of linear differential equations $Y' = AY$ for the matrix A in Exercise 9.

14. Find the general solution to the system of linear differential equations $Y' = AY$ for the matrix A in Exercise 10.

15. Complete the proof of Theorem 9.8 by showing:

  a) $P = P_1 P_2$ is an orthogonal matrix;

  b) $P^T A P$ is an upper triangular matrix.

16. Prove Theorem 9.9.

17. Find a unitary matrix U and a matrix $B = U^* A U$ so that B is a triangular matrix for

$$A = \begin{bmatrix} -1+i & 1 & 0 \\ -1 & 1+i & 0 \\ -2+i & 1 & 1 \end{bmatrix}.$$

18. Prove that if an $n \times n$ matrix A is orthogonally similar to a diagonal matrix, then A is a symmetric matrix.

19. Prove that for any matrix A with real entries, $A^T A$ and AA^T are diagonalizable matrices.

Answers to Odd Numbered Exercises

CHAPTER 1

Exercises 1.1, pp. 15–17

1. $x = -14/3$, $y = 6$, $z = 4/3$ **3.** No solution   **5.** $x = -3z$, $y = 7z$ **7.** No solution

9. $x_1 = 7/2 - 3x_4/4$, $x_2 = -3/4 - 13x_4/8$, $x_3 = 0$ **11.** $x = 5 + 3z$, $y = -7/2 - 2z$

13. $x = 4/5$, $y = -3/5$ **15.** $x_1 = -7x_4 + 4x_5$, $x_2 = -9x_4 + 5x_5$, $x_3 = -4x_4 + 4x_5$

17. $c + b - a = 0$ **19.** Solutions for all a, b, c, d **21.** Does   **23.** Does

25. It will either have no solution or infinitely many solutions.

27. If the graphs of the two equations are distinct parallel planes, there will be no solution to the system. If the graphs of the two equations are the same plane, all the points on this plane will be solutions to the system. If the graphs of the equations are planes that are not parallel, the solutions to the system will be the points on the line of intersection of the planes.

29. No solution

31. $x_1 = t_1$, $x_2 = t_2$, $x_3 = -28830t_1/5657 - 38130t_2/5657$, $x_4 = -20181t_1/90512 - 26691t_2/90512$,
$x_5 = 145607t_1/67884 + 192577t_2/67884$, $x_6 = 13237t_1/271536 + 17507t_2/271536$,
$x_7 = 92659t_1/271536 + 122549t_2/271536$, $x_8 = 0$

Exercises 1.2, pp. 26–27

1. $\begin{bmatrix} 3 & 1 \\ 0 & -3 \\ 2 & 3 \end{bmatrix}$ **3.** $\begin{bmatrix} 4 & -2 \\ -6 & -4 \\ 0 & 8 \end{bmatrix}$ **5.** $\begin{bmatrix} -7 & 6 \\ 15 & 7 \\ 2 & -17 \end{bmatrix}$ **7.** $\begin{bmatrix} -3 & 3 \\ 15 & -4 \end{bmatrix}$ **9.** $\begin{bmatrix} 0 & 8 & -9 \\ 3 & -5 & 13 \\ 3 & -4 & 10 \end{bmatrix}$

11. Undefined   **13.** $\begin{bmatrix} 8 & -1 \\ 8 & 5 \\ 7 & 2 \end{bmatrix}$ **15.** $\begin{bmatrix} 12 & 27 \\ 15 & -24 \\ 9 & -21 \end{bmatrix}$ **17.** $\begin{bmatrix} 3 & -7 \\ 7 & 24 \end{bmatrix}$

19. $\begin{bmatrix} 2 & -1 & 4 \\ 1 & 1 & -1 \\ 0 & 1 & 3 \\ 1 & 1 & 0 \end{bmatrix} \begin{bmatrix} x \\ y \\ z \end{bmatrix} = \begin{bmatrix} 1 \\ 4 \\ 5 \\ 2 \end{bmatrix}$ **21.** $\begin{aligned} 2x_1 - 2x_2 + 5x_3 + 7x_4 &= 12 \\ 4x_1 + 5x_2 - 11x_3 + 3x_4 &= -3 \end{aligned}$

27. AB has a column of zeros; does not necessarily hold if A has a column of zeros.   **29.** (b) $\begin{bmatrix} 28 & -7 & 13 \end{bmatrix}$

31. −3   **33.** $\begin{bmatrix} 2 & -4 & 16 & 27 & -11 \\ 5 & 39 & 5 & -8 & -1 \\ 34 & 14 & -9 & -6 & 21 \\ -5 & 4 & -4 & -1 & 33 \\ 0 & 12 & -2 & -12 & -7 \end{bmatrix}$ **35.** Undefined

37. $\begin{bmatrix} 423 & 294 & 307 & 455 & 1409 \\ 964 & 2223 & 609 & -121 & -303 \\ 408 & 1265 & 779 & 830 & -94 \\ 236 & 953 & -38 & 113 & 1074 \\ 49 & 602 & 140 & -45 & 145 \end{bmatrix}$ **39.** $\begin{bmatrix} -52 & 234 & -114 & -230 & 516 \\ 278 & 258 & -28 & -23 & 323 \\ -155 & -200 & 183 & 344 & -83 \\ 223 & 31 & -106 & -27 & 545 \end{bmatrix}$

Exercises 1.3, pp. 36–37

1. $\begin{bmatrix} 1/7 & -2/7 \\ 3/7 & 1/7 \end{bmatrix}$ **3.** Not invertible   **5.** $\begin{bmatrix} 3/2 & 5/6 & -1/3 \\ -1 & -1/3 & 1/3 \\ -1 & -2/3 & 2/3 \end{bmatrix}$

7. $\begin{bmatrix} 11/12 & 3/4 & 1/12 & -1/4 \\ -1/4 & -1/4 & -1/4 & 1/4 \\ 1 & -1/2 & -1/2 & 0 \\ -7/12 & -1/4 & 1/12 & 1/4 \end{bmatrix}$ **9.** $x = 1/2, y = 0, z = 2$

11. (a) $\begin{bmatrix} 1 & 0 \\ 0 & 2 \end{bmatrix}$ (b) $\begin{bmatrix} 1 & 2 \\ 0 & 1 \end{bmatrix}$ (c) $\begin{bmatrix} 0 & 1 \\ 1 & 0 \end{bmatrix}$

23. $\begin{bmatrix} -4/19 & 973/247 & -648/247 & 27/247 & 8999/494 \\ -1/19 & -1215/247 & 864/247 & -36/247 & -11587/494 \\ 2/19 & -3519/494 & 2453/494 & -225/988 & -66429/1976 \\ 0 & -1/26 & 1/26 & 1/52 & -23/104 \\ -2/57 & 8419/741 & -1970/247 & 308/741 & 26659/494 \end{bmatrix}$

25. $\begin{bmatrix} 1/4 & -1/2\pi & -(-2+\pi)\sqrt{3}/12\pi & (2+2\sqrt{3}-\pi+\sqrt{3}\pi)\sqrt{3}/8\pi \\ -1/2 & 0 & \sqrt{3}/6 & -(-1+\sqrt{3})\sqrt{3}/4 \\ -11/16 & 3/8\pi & (-2+5\pi)\sqrt{3}/16\pi & -(6-15\pi+6\sqrt{3}+11\sqrt{3}\pi)\sqrt{3}/32\pi \\ 1/16 & -1/72\pi & -(-2+37\pi)\sqrt{3}/432\pi & (6+6\sqrt{3}-111\pi+11\sqrt{3}\pi)\sqrt{3}/864\pi \end{bmatrix}$

Exercises 1.4, pp. 41–43

1. $\begin{bmatrix} 1 & 0 & 0 \\ 0 & 4 & 0 \\ 0 & 0 & 1 \end{bmatrix}$ **3.** $\begin{bmatrix} -1 & 0 & 0 \\ 0 & -32 & 0 \\ 0 & 0 & 1 \end{bmatrix}$ **5.** $\begin{bmatrix} 6 & 7 & 1 \\ 0 & -4 & 7 \\ 0 & 0 & 2 \end{bmatrix}$ **7.** $\begin{bmatrix} 1 & 1 \\ 2 & -2 \\ -3 & 1 \end{bmatrix}$

9. $\begin{bmatrix} -7 & 5 \\ 14 & 18 \\ -19 & 5 \end{bmatrix}$ **11.** $\begin{bmatrix} 16 & -12 \\ 8 & -8 \end{bmatrix}$ **13.** $\begin{bmatrix} -1 & 8 & -3 \\ -6 & -4 & -10 \\ 7 & -4 & 13 \end{bmatrix}$ **15.** Is   **17.** Is not   **19.** Is

31. $\begin{bmatrix} 1 & 0 & 0 \\ 0 & 1 & 0 \\ -1 & -1 & 0 \end{bmatrix}$ **35. (a)** $\begin{bmatrix} 1 & 0 & -1 & 0 & 0 \\ 0 & 1 & -1 & 0 & 0 \\ 0 & 0 & 0 & 1 & 0 \\ 0 & 0 & 0 & 0 & 1 \\ 0 & 0 & 0 & 0 & 0 \end{bmatrix}$ **(b)** $\begin{bmatrix} 2 & 1 & -3 & 4 & 1 \\ 0 & 2 & -2 & 4 & 8 \\ 0 & 0 & 0 & -6 & 10 \\ 0 & 0 & 0 & 0 & -35/3 \\ 0 & 0 & 0 & 0 & 0 \end{bmatrix}$;

row-echelon form except for having leading entries 1.   **(c)** No

Exercises 1.5, pp. 50–51

1. 41   **3.** 41   **5.** 41   **7.** 31   **9.** 0   **11.** 28   **13.** 15   **15.** 177652

Exercises 1.6, pp. 57–58

1. Not invertible   **3.** Invertible   **5.** $\begin{bmatrix} 1/7 & -3/7 \\ 2/7 & 1/7 \end{bmatrix}$ **7.** $x = 11/17$, $y = 4/17$

9. $x = -2/5$, $y = -26/5$, $z = -3$ **11.** $x = (t^2 \cos 2t + 2t \sin 2t)/2e^t$, $y = (2t \cos 2t - t^2 \sin 2t)/2e^t$

15. (a) $\det(A) = 14$, $\det(B) = 7$ **(b)** $\det(AB) = 98$, $\det(A^{-1}) = 1/14$, $\det(B^T A^{-1}) = 1/2$
   (c) $\det(A + B) = 28$, $\det(A) + \det(B) = 21$

17. (a) $\begin{bmatrix} +34.24 & -143.355 & -27.260 & 51.102 \\ 4.00 & -28.395 & 7.828 & 17.406 \\ 1.424 & -90.780 & -6.640 & 32.664 \\ -13.92 & 26.301 & 1.764 & -2.562 \end{bmatrix}$ **(b)** $\begin{bmatrix} 72.5136 & 0 & 0 & 0 \\ 0 & 72.5136 & 0 & 0 \\ 0 & 0 & 72.5136 & 0 \\ 0 & 0 & 0 & 72.5136 \end{bmatrix}$
   (c) 72.5136

CHAPTER 2

Exercises 2.1, pp. 73–74

3. (a) Is not; properties 1, 3, 4, 6 do not hold   **(b)** Is not; properties 5, 6, 7, 8 do not hold
   (c) Is not; properties 1, 2, 6 do not hold   **7.** It is a vector space.

Exercises 2.2, pp. 81–83

1. (a) Is   **(b)** Is   **(c)** Is not   **(d)** Is   **3. (a)** Is   **(b)** Is not   **(c)** Is   **(d)** Is   **(e)** Is not   **5.** No

7. Yes; no   **9.** Yes   **11.** No   **13.** Yes   **15.** Do span   **17.** Do not span   **19.** Do span

21. Is in span   **23.** Do span   **25.** It cannot have a zero row.

Exercises 2.3, pp. 93–95

1. Linearly independent   **3.** Linearly dependent   **5.** Linearly independent   **7.** Linearly independent

9. Linearly dependent   **23. (a)** $\begin{bmatrix} 1/5 \\ -1 \\ 2/5 \end{bmatrix}$ **(b)** $\begin{bmatrix} -1 \\ 0 \\ 0 \end{bmatrix}$ **25. (a)** $\begin{bmatrix} 2 \\ -1 \\ 1 \end{bmatrix}$ **(b)** $6x^2 - x$

29. v_1 and v_2 are linearly independent if and only if v_2 is not parallel to L.

31. (b) $\begin{bmatrix} -3697930773/391460975 \\ 4130489956/391460975 \\ -1445251861/78292195 \\ -1926239514/391460975 \\ 3619315327/391460975 \\ 3387950282/391460975 \end{bmatrix}$

Exercises 2.4, pp. 104–106

1. (a) Do not   (b) Do   (c) Do not   (d) Do not   **3.** (a) Do   (b) Do not   (c) Do not   (d) Do not

5. (a) No basis   (b) $\begin{bmatrix} 1 & 0 \end{bmatrix}, \begin{bmatrix} 0 & 1 \end{bmatrix}$ (c) $\begin{bmatrix} 1 \\ 0 \end{bmatrix}, \begin{bmatrix} 0 \\ 1 \end{bmatrix}$ (d) 2

7. (a) $\begin{bmatrix} 1 \\ 1 \\ 0 \end{bmatrix}$ (b) $\begin{bmatrix} 1 & -1 & 0 \end{bmatrix}, \begin{bmatrix} 0 & 0 & 1 \end{bmatrix}$ (c) $\begin{bmatrix} 1 \\ 0 \\ 2 \end{bmatrix}, \begin{bmatrix} 0 \\ 1 \\ 1 \end{bmatrix}$ (d) 2

9. (a) $\begin{bmatrix} -1 \\ 1 \\ 0 \\ 0 \end{bmatrix}, \begin{bmatrix} -3 \\ 0 \\ 5 \\ 1 \end{bmatrix}$ (b) $\begin{bmatrix} 1 & 1 & 0 & 3 \end{bmatrix}, \begin{bmatrix} 0 & 0 & 1 & -5 \end{bmatrix}$ (c) $\begin{bmatrix} 1 \\ 0 \\ 1 \end{bmatrix}, \begin{bmatrix} 0 \\ 1 \\ 2 \end{bmatrix}$ (d) 2

11. (a) $\begin{bmatrix} -1/3 \\ 5/3 \\ 1 \\ 0 \end{bmatrix}, \begin{bmatrix} -1/3 \\ 2/3 \\ 0 \\ 1 \end{bmatrix}$ (b) $\begin{bmatrix} 1 & 0 & 1/3 & 1/3 \end{bmatrix}, \begin{bmatrix} 0 & 1 & -5/3 & -2/3 \end{bmatrix}$

(c) $\begin{bmatrix} 1 \\ 0 \\ -1 \\ -2 \\ 2 \end{bmatrix}, \begin{bmatrix} 0 \\ 1 \\ 1 \\ 1 \\ 1 \end{bmatrix}$ (d) 2   **13.** $\begin{bmatrix} 1 \\ 0 \\ 0 \end{bmatrix}, \begin{bmatrix} 0 \\ 1 \\ 0 \end{bmatrix}, \begin{bmatrix} 0 \\ 0 \\ 1 \end{bmatrix}$ **15.** $\begin{bmatrix} 1 \\ 0 \\ -7/3 \\ 3 \end{bmatrix}, \begin{bmatrix} 0 \\ 1 \\ -5/3 \\ 2 \end{bmatrix}$

21. (b) $x^2 - 1, x^2 + 1, x - 1$

25. (a) $\begin{bmatrix} -17/47 & -135/47 & -183/47 & 1 & 0 \end{bmatrix}, \begin{bmatrix} -102/47 & -58/47 & -64/47 & 0 & 1 \end{bmatrix}$,

(b) *gausselim:* $\begin{bmatrix} -2 & 3 & -1 & 4 & -2 \end{bmatrix}, \begin{bmatrix} 0 & 12 & -5 & 15 & 8 \end{bmatrix}, \begin{bmatrix} 0 & 0 & -47/24 & -61/8 & -8/3 \end{bmatrix}$

gaussjord: $\begin{bmatrix} 1 & 0 & 0 & 17/47 & 102/47 \end{bmatrix}, \begin{bmatrix} 0 & 1 & 0 & 135/47 & 58/47 \end{bmatrix}, \begin{bmatrix} 0 & 0 & 1 & 183/47 & 64/47 \end{bmatrix}$

rowspace: $\begin{bmatrix} 1 & 0 & 0 & 17/47 & 102/47 \end{bmatrix}, \begin{bmatrix} 0 & 0 & 1 & 183/47 & 64/47 \end{bmatrix}, \begin{bmatrix} 0 & 1 & 0 & 135/47 & 58/47 \end{bmatrix}$

rowspan: $\begin{bmatrix} 0 & -7 & -1 & -24 & -10 \end{bmatrix}, \begin{bmatrix} 0 & 0 & 47 & 183 & 64 \end{bmatrix}, \begin{bmatrix} -2 & 3 & -1 & 4 & -2 \end{bmatrix}$

(c) *gausselim:* $\begin{bmatrix} -2 \\ 3 \\ 1 \\ 0 \end{bmatrix}, \begin{bmatrix} 0 \\ 5 \\ 3 \\ 8 \end{bmatrix}, \begin{bmatrix} 0 \\ 0 \\ -29/5 \\ -29/5 \end{bmatrix}$ *gaussjord:* $\begin{bmatrix} 1 \\ 0 \\ 0 \\ 2 \end{bmatrix}, \begin{bmatrix} 0 \\ 1 \\ 0 \\ 1 \end{bmatrix}, \begin{bmatrix} 0 \\ 0 \\ 1 \\ 1 \end{bmatrix}$

$$colspace: \begin{bmatrix} 1 \\ 0 \\ 0 \\ 2 \end{bmatrix}, \begin{bmatrix} 0 \\ 1 \\ 0 \\ 1 \end{bmatrix}, \begin{bmatrix} 0 \\ 0 \\ 1 \\ 1 \end{bmatrix} \qquad colspan: \begin{bmatrix} 0 \\ -7 \\ -17 \\ -24 \end{bmatrix}, \begin{bmatrix} 0 \\ 0 \\ 47 \\ 47 \end{bmatrix}, \begin{bmatrix} -2 \\ 3 \\ 1 \\ 0 \end{bmatrix}$$

27. Apply the *gausselim* or *gaussjord* command to the matrix with rows $v_1^T, v_2^T, \ldots, v_k^T, v_{k+1}^T$. If this does not result in a matrix with a zero row, v_{k+1} will not be in Span$\{v_1, v_2, \ldots, v_k\}$.

CHAPTER 3

Exercises 3.1, pp. 119–120

1. Second order   **3.** Second order   **5.** Solves the differential equation. Does not satisfy the initial condition.
7. Solves the differential equation. Does not satisfy the initial condition.   **9.** 0, 16, 8   **11.** 3, 24, −24
13. (1) $y = -2$, (2) $y' > 0$ for $y > -2$, $y' < 0$ for $y < -2$; $y'' = 2y' = 4y + 8$, $y'' > 0$ for $y > -2$ and $y'' < 0$ for $y < -2$
15. (1) $y = 0, 4$, (2) $y' > 0$ for $y > 4$ and $y < 0$, $y' < 0$ for $0 < y < 4$; $y'' = 2yy' - 4y' = (2y - 4)(y^2 - 4y)$, $y'' > 0$ for $y > 4$ and $0 < y < 2$, $y'' < 0$ for $2 < y < 4$ and $y < 0$.
17. (1) $y = 0, 3, -3$, (2) $y' > 0$ for $y > 3$ and $-3 < y < 0$, $y' < 0$ for $0 < y < 3$ and $y < -3$; $y'' = (3y^2 - 9)y' = (3y^2 - 9)(y^3 - 9y)$

Exercises 3.2, p. 124

1. Separable   **3.** Not separable   **5.** $\ln|y| = 2x^3/3 - 4x + C$ **7.** $2 \arctan y = \ln(x^2 + 1) + C$
9. $\ln|y - 4| - \ln|y| = -e^x + C$ **11.** $y = -\ln(\ln|t + 1| - t + c)$ **13.** $3y^2 + 2x^3 = 3$
15. $y^2 + 8\ln|y| = 4x^2 - 3$ **17.** (a) $y = \sqrt{9 - 6x^3}/3$ (b) $0 \le x \le (3/2)^{1/3}$ **21.** (c) $y = \tan(x + c) - x - 1$

Exercises 3.3, pp. 129–130

1. Exact; $x^3 - 4y^2x + 3y^4 = C$ **3.** Exact; $x^2y + e^xy = C$ **5.** Not exact   **7.** $x^2y - x = C$
9. $e^s + t^2 \cos s - e^t = C$ **11.** $x^2y + y^3 = 2$ **13.** $2\cos(xy) - y^3 = 1$
19. $c = -b$; $-ax^2/2 - bxy + dy^2/2 = k$

Exercises 3.4, pp. 135–136

1. $e^{-1/x}$; $y = ce^{1/x}$ **3.** e^{-x^2}; $y = -1/2 + Ce^{x^2}$ **5.** $(1 + 2x)^{-1/2}$; $y = 1 + 2x + C(1 + 2x)^{1/2}$
7. x; $y = x^2/3 + C/x$ **9.** $1/x$; $y = -1/(2x) + Cx$ **11.** e^{e^t}; $y = 1 + Ce^{-e^t}$ **13.** $y = 1/2 + (3/2)e^{4-4x}$
15. $y = (x^2 + 2x + 2)/(x + 1)$ **17.** $y = (\cos(2x) + 1)/(4x) + \sin(2x)/2$

Exercises 3.5, p. 143

1. $2x^2 \ln|x| - y^2 = Cx^2$ **3.** $2x^{-1/2}y^{1/2} - 2x^{1/2}y^{3/2} = C$ **5.** $-2x^{-1} - x^{-3}y^2 + 2x^{-3}/3 = C$
9. $x^2(y + 1)^2/2 + y^4/4 + y^3/3 - y^2/2 - y = C$ **11.** $y^2 = -2x^2 \ln|x| + Cx^2$
13. $e^{-y/x} + \ln|x| = C$ **17.** $y = 1/(1 + cx)$

Exercises 3.6, pp. 151–153

1. ≈ 48 years   **3.** ≈ 7567 amoeba   **5.** ≈ 21.64 pounds   **7.** (a) 13.48 gal   (b) 47.12 gal
9. (a) ≈ 46.2 hours   (b) No solution   **11.** ≈ 55.09 minutes   **13.** ≈ 1.12 hours
15. ≈ 4.91 meters per second   **17.** $20R/(R - 20)$ where R is the earth's radius   **19.** $y = mx$

21. $(\alpha Ce^{kt})/(1 + \alpha e^{kt})$ where $\alpha = B/(C - B)$ and $B = 10$ billion, $C = 1$ trillion
23. (c) $(1 + R/c_V) \ln V + \ln P = K$, $K -$ a constant

Exercises 3.7, p. 157

1. $y = -x^2/18 - x/81 + C_1 e^{-9x} + C_2$ **3.** $y = \pm(C_1 - 2x)^{1/2} + C_2$ **5.** $y^3 = 1/(C_1 x + C_2)$
7. $y = C_1 \sin 2x + C_2 \cos 2x$ **9.** $y = ((2x + 1)^{3/2} - 1)/3$ **11.** $y = \ln x$ **13.** $y = 2/(1 - x)^2$
15. $y = (x^3 + x)/2$ **17.** $dx/dt = \pm\sqrt{2gR^2/(R + x) + v_0^2 - 2gR}$

Exercises 3.8, pp. 167–168

1. $y(x) = \int_0^x (s + 2y(s))\, ds$ **3.** Maclaurin and Picard are the same: $1 + 2x + 2x^2 + 4x^3/3 + 2x^4/3$
5. Maclaurin: $1 + 2x + 4x^2 + 8x^3 + 16x^4$, Picard: $1 + 2x + 4x^2 + 8x^3 + 16x^4 + (416/15)x^5 + (128/3)x^6$
  $+ (3712/63)x^7 + (4544/63)x^8 + (44032/567)x^9 + (22528/315)x^{10} + (10240/189)x^{11} + (2048/63)x^{12}$
  $+ (8192/567)x^{13} + (16384/3969)x^{14} + (32768/59535)x^{15}$
7. $-11/42 - x/9 + x^3/3 + x^4/18 - x^7/63$ **9.** $\int_0^x \cos(\sin s)\, ds$
13. $1 + (x - 1)^2/2 - (x - 1)^3/3 + (x - 1)^4/6 - (x - 1)^5/20 + 13(x - 1)^6/180 - (x - 1)^7/63 + (x - 1)^8/160$
  $- (x - 1)^9/270 - (x - 1)^{11}/4400$

Exercises 3.9, pp. 176–177

1. Euler values: 1.2, 1.43, 1.962, 2.3144   **3.** Euler values: 1.2, 1.61472, 2.63679, 6.1966
5. Taylor values: 1.215, 1.4663, 2.102124, 2.5156   **7.** Taylor values: 1.3067, 2.20126, 7.9066, 730.5869
11. Runge-Kutta values: 1.21605, 1.46886, 1.76658, 2.11914, 2.53869
13. Runge-Kutta values: 0, 0.000333, 0.0026669, 0.00900, 0.021359, 0.041791

CHAPTER 4

Exercises 4.1, pp. 188–189

1. Linear, third order   **3.** Linear, fourth order   **5.** $(-\infty, \infty)$ **7.** No interval
9. $y = c_1 e^{-x} + c_2 e^{2x}$, $y = (2/3)e^{-x} + (1/3)e^{2x}$
11. $y = c_1 e^x + c_2 e^{2x} + c_3 e^{-3x}$, $y = (3/2)e^x - (3/5)e^{2x} + (1/10)e^{-3x}$
13. $y = c_1 e^{-x} + c_2 e^{2x} - 2$, $y = 2e^{-x} + e^{2x} - 2$
15. $y = c_1 e^x + c_2 e^{2x} + c_3 e^{-3x} + e^{-x}$, $y = (1/5)e^{2x} - (1/5)e^{-3x} + e^{-x}$
17. $g(x) = e^x(x^2/2 - x + 1)$, $y = c_1 x + c_2 x^2 + e^x$, $y = (2 - e)x - x^2 + e^x$

Exercises 4.2, pp. 201–203

1. $y = c_1 e^{-8x} + c_2$ **3.** $y = c_1 e^{(-2+\sqrt{3})x} + c_2 e^{-(2+\sqrt{3})x}$ **5.** $y = c_1 e^{-4x} + c_2 e^{-x} + c_3 e^x$
7. $y = c_1 e^{-3x} + c_2 x e^{-3x}$ **9.** $y = c_1 e^{-2x} + c_2 x e^{-2x} + c_3 x^2 e^{-2x}$ **11.** $y = c_1 \cos 2x + c_2 \sin 2x$
13. $y = c_1 e^{-2x} + e^x(c_2 \cos x + c_3 \sin x)$ **15.** (a) $y = c_1 e^x + c_2 e^{-x}$ (b) $y = -e^x/(2e) + e^{1-x}/2$
17. (a) $y = e^{-2x}(c_1 \sin 2x + c_2 \cos 2x)$ (b) $y = -(e^{-2x} \sin 2x)/2$ **19.** (a) $(c_1 + c_2 x)e^{-1/2x}$ (b) $-xe^{-1/2x}$
21. (a) $c_1 + c_2 x + c_3 e^{-5x}$ (b) $33/25 + (2/5)x + 2e^{-5x}/(25e^5)$
23. (a) $y = c_1 + c_2 e^{-2x} + c_3 e^{3x}$ (b) $y = (2/3) + (1/5)e^{-2x} + (2/15)e^{3x}$
25. (a) $y = c_1 e^{-x} + (c_2 + c_3 x)e^{3x}$ (b) $y = -(7/16)e^{3x} + (7/16)e^{-x} + (3/4)xe^{3x}$
27. $y = (A + Bx)\sin 2x + (C + Dx)\cos 2x + (E + Fx + Gx^2)e^{4x}$

29. $y = (A + Bx + Cx^2)\cos 3x + (D + Ex + Fx^2)\sin 3x + Ge^{-2x} + (H + Ix)e^{-3x} + (J + Kx)e^{3x}$
31. $y_2 = 1$ **33.** $y_2 = x$ **45.** $y = c_1/x + c_2 x^5$ **47.** $y = c_1 x + c_2 x \ln x$

Exercises 4.3, p. 211

1. $y = c_1 e^{-2x} + c_2 e^{3x} - (3/4)e^{2x}$ **3.** $y = c_1 \sin 5x + c_2 \cos 5x + 623/625 + x^2/25$
5. $y = c_1 e^{2x} \sin 3x + c_2 e^{2x} \cos 3x + (4/37)e^x \sin 3x + (24/37)e^x \cos 3x$
7. $y = c_1 e^{-x} + c_2 + x - x^2/2 + e^x/2 + (1/10)\sin 2x - (1/5)\cos 2x$
9. $y = c_1 \sin 2x + c_2 \cos 2x + (3/16)x \sin 2x$ **11.** $y = c_1 e^{-8x} + c_2 + x^3/12 - (15/32)x^2 + (63/128)x$
13. $y = c_1 e^{2x} \cos 3x + c_2 e^{2x} \sin 3x - (2/3)xe^{2x} \cos 3x$
15. $y = c_1 e^{-x} + c_2 e^{4x} + c_3 - x^2/8 + (3/16)x - (1/20)xe^{4x}$
17. $y = 1/32 - x/8 - (31/96)e^{4x} - (41/24)e^{-2x}$
19. $y = (5/2)xe^x - (15e + 2)e^x/(4e) + ((5/4)e^2 + e/2)e^{-x}$
21. $y = -(1/4)\cos x - (1/4)\sin x - (127/100)e^{2x} \cos x + (139/100)e^{2x} \sin x + (7/5)x + (63/25)$
23. $y_P = Ax^2 + Bx + C + (Dx^2 + Ex)\sin x + (Fx^2 + Gx)\cos x$
25. $y_P = (Ax^2 + Bx)\sin 3x + (Cx^2 + Dx)\cos 3x + Exe^{-x}$

Exercises 4.4, p. 217

1. $y_P = -5 - 3x^2$ **3.** $y = c_1 e^{-3x} + c_2 xe^{-3x} + e^{3x}/12$
5. $y = c_1 \cos 2x + c_2 \sin 2x - (5/2)x \cos 2x + (5/4)\sin 2x \ln(|\sin 2x|)$
7. $y = c_1 e^{-x} + c_2 e^{2x} + e^x((1/10/\sin x - (3/10)\cos x)$ **9.** $y = c_1 e^x + c_2 e^{-4x} + c_3 e^{-x} + (1/6)e^{-2x}$
11. $y_P = x^3/3$ **13.** $y_P = (2x + 1)e^x/4$ **15.** $y_P = -(5/8)x^3$

Exercises 4.5, pp. 228–229

1. $u = 0.5 \cos 10t$, amp $= 0.5$, $\omega = 10$ **3.** $u = 0.25 \sin 8t$, amp $= 0.25$, $\omega = 8$
5. $u = (e^{-2t}/2)(\cos 2t + \sin 2t)$ **7.** $u = 5e^{-2t} - 2e^{-5t}$
9. $u = 3/400 + (197/400)\cos 10t$
11. $u = (4/5)\sin 2t + (2/5)\cos 2t - (7/10)e^{-2t} \sin 2t + (1/10)e^{-2t} \cos 2t$
13. $u = (14/3)e^{-2t} - (5/3)e^{-5t} + te^{-2t}$
15. (a) $f = \sqrt{4mk}$ (b) (i) $f = 0$ (ii) $f^2 < 4mk$ (iii) $f^2 = 4mk$ (iv) $f^2 > 4mk$
17. $e^{-t/5}((6/7)\sin(7t/5) + \cos(7t/5))$ **19.** $R = 0$, $\omega = \frac{1}{\sqrt{LC}}$

CHAPTER 5

Exercises 5.1, pp. 243–245

1. Is   **3.** Is not   **5.** Is   **7.** Is   **9.** Is not   **11.** Is   **15.** $\begin{bmatrix} 3 & 2 \\ 2 & 3 \end{bmatrix}$ **17.** $\begin{bmatrix} 1 & -1 & 1 \\ 2 & 1 & 2 \\ 3 & 3 & 3 \\ 1 & 2 & 1 \end{bmatrix}$

19. (a) $\begin{bmatrix} 0 \\ 3 \\ 3 \end{bmatrix}$ (b) $\begin{bmatrix} -y + z \\ x - y + z \\ x \end{bmatrix}$ **21.** (a) $2x$ (b) $2ax + b$

Note that the answers in Exercises 23 and 25 are not unique.

23. No basis for kernel

Basis for range: $\begin{bmatrix} 1 \\ 0 \end{bmatrix}, \begin{bmatrix} 0 \\ 1 \end{bmatrix}$

25. Basis for kernel: $\begin{bmatrix} -1 \\ 0 \\ 1 \end{bmatrix}$

Basis for range: $\begin{bmatrix} 1 \\ 0 \\ -1 \\ -1 \end{bmatrix}, \begin{bmatrix} 0 \\ 1 \\ 2 \\ 1 \end{bmatrix}$

27. Constant functions

Exercises 5.2, pp. 252–253

1. $\begin{bmatrix} 3x + 2y \\ 2x + y \end{bmatrix}$ **3.** $\begin{bmatrix} 2x + 6y \\ 2x - 2y \end{bmatrix}$ **5.** $\begin{bmatrix} x + 7y \\ 3x + y \end{bmatrix}$ **7.** $4ax + 4a + 4b$

9. $ax + 4a + b$ **11.** e^{3x}, e^{-x} **13.** $1, x, \cos x, \sin x$

23. (a) $A = \begin{bmatrix} 1 & 3 \\ 1 & -1 \end{bmatrix}, B = \begin{bmatrix} 2 & -1 \\ 1 & 2 \end{bmatrix}$ (b) $C = \begin{bmatrix} 1 & 7 \\ 3 & 1 \end{bmatrix}$ (c) $D = \begin{bmatrix} 5 & 5 \\ 1 & -3 \end{bmatrix}$

Exercises 5.3, pp. 267–269

1. (a) $\begin{bmatrix} 1 & 1 \\ 1 & -1 \end{bmatrix}$ (b) $\begin{bmatrix} 1 & -2 \\ -1 & 1 \end{bmatrix}$ (c) $\begin{bmatrix} -1 & -2 \\ -1 & -1 \end{bmatrix}$ (d) $\begin{bmatrix} -4 & 7 \\ -2 & 4 \end{bmatrix}$

(e) $\begin{bmatrix} -4 \\ -1 \end{bmatrix}$ (f) $\begin{bmatrix} 9 \\ 4 \end{bmatrix}$ (g) $\begin{bmatrix} 1 \\ -5 \end{bmatrix}$

3. (a) $\begin{bmatrix} 17 & -8 & -12 \\ 16 & -7 & -12 \\ 16 & -8 & -11 \end{bmatrix}$ (b) $\begin{bmatrix} 1 & 1 & 1 \\ 1 & 2 & -1 \\ 1 & 0 & 2 \end{bmatrix}$ (c) $\begin{bmatrix} -4 & 2 & 3 \\ 3 & -1 & -2 \\ 2 & -1 & -1 \end{bmatrix}$ (d) $\begin{bmatrix} -3 & 0 & 0 \\ 0 & 1 & 0 \\ 0 & 0 & 1 \end{bmatrix}$

(e) $\begin{bmatrix} 2 \\ -1 \\ 1 \end{bmatrix}$ (f) $\begin{bmatrix} -6 \\ -1 \\ 1 \end{bmatrix}$ (g) $\begin{bmatrix} -6 \\ -9 \\ -4 \end{bmatrix}$

5. (a) $\begin{bmatrix} 1 & 0 & 0 \\ 2 & 1 & 0 \\ 1 & 1 & 1 \end{bmatrix}$ (b) $\begin{bmatrix} 1 & 1 & 0 \\ 0 & 0 & 1 \\ -1 & 1 & 1 \end{bmatrix}$ (c) $\begin{bmatrix} 1/2 & 1/2 & -1/2 \\ 1/2 & -1/2 & 1/2 \\ 0 & 1 & 0 \end{bmatrix}$

(d) $\begin{bmatrix} 3/2 & 1/2 & -1/2 \\ -1/2 & 1/2 & 1/2 \\ 2 & 2 & 1 \end{bmatrix}$ (e) $\begin{bmatrix} 1/2 \\ 1/2 \\ 1 \end{bmatrix}$ (f) $\begin{bmatrix} 1/2 \\ 1/2 \\ 3 \end{bmatrix}$ (g) $x^2 + 3x + 3$

7. (a) $\begin{bmatrix} 0 & -1 \\ 1 & 0 \end{bmatrix}$ (b) $\begin{bmatrix} 1 & 1 \\ 1 & -1 \end{bmatrix}$ (c) $\begin{bmatrix} 1/2 & 1/2 \\ 1/2 & -1/2 \end{bmatrix}$ (d) $\begin{bmatrix} 0 & 1 \\ -1 & 0 \end{bmatrix}$

(e) $\begin{bmatrix} 1/2 \\ -5/2 \end{bmatrix}$ (f) $\begin{bmatrix} -5/2 \\ -1/2 \end{bmatrix}$ (g) $-3\sin x - 2\cos x$

9. (a) $\begin{bmatrix} 1 & 0 & -1 \\ -1 & 1 & 0 \\ 0 & -1 & 1 \end{bmatrix}$ (b) $\begin{bmatrix} -2 \\ -3 \\ 5 \end{bmatrix}$ (c) $-2v_1 - 3v_2 + 5v_3$

13.
$$\begin{bmatrix} a_{33} & a_{31} & a_{32} \\ a_{13} & a_{11} & a_{12} \\ a_{23} & a_{21} & a_{22} \end{bmatrix}$$
; rearrange the row and column entries of $[T]_\alpha^\alpha$ in the same way to get $[T]_\beta^\beta$.

17. (a)
$$\begin{bmatrix} 1 & -1 & 1 & -1 \\ 2 & -1 & 2 & 1 \\ 3 & 1 & -1 & 1 \\ -1 & 3 & -5 & 1 \end{bmatrix}$$
(b)
$$\begin{bmatrix} 1 & 1 & 1 & 1 \\ -1 & 0 & 0 & -1 \\ 1 & 1 & -1 & 1 \\ 0 & 0 & 1 & 1 \end{bmatrix}$$
(c)
$$\begin{bmatrix} 1/2 & -1 & -1/2 & -1 \\ 1/2 & 1 & 1/2 & 0 \\ 1/2 & 0 & -1/2 & 0 \\ -1/2 & 0 & 1/2 & 1 \end{bmatrix}$$

(d)
$$\begin{bmatrix} 5 & 2 & -9 & 2 \\ 7 & 6 & 3 & 8 \\ 1 & 0 & -3 & 0 \\ -10 & -6 & 8 & -8 \end{bmatrix}$$
(e)
$$\begin{bmatrix} 2 \\ 3 \\ -1 \\ -2 \end{bmatrix}$$
(f)
$$\begin{bmatrix} 21 \\ 13 \\ 5 \\ -30 \end{bmatrix}$$
(g)
$$\begin{bmatrix} 9 \\ 9 \\ -1 \\ -25 \end{bmatrix}$$

19. (a)
$$\begin{bmatrix} 1 & 1 & 0 & 0 \\ 0 & 1 & -1 & 0 \\ 1 & 1 & 1 & 0 \\ 0 & 1 & -1 & 0 \end{bmatrix}$$
(b)
$$\begin{bmatrix} 1 & 0 & 1 & 0 \\ -1 & 1 & 0 & 0 \\ 0 & 1 & 0 & 1 \\ 1 & 1 & -1 & -1 \end{bmatrix}$$
(c)
$$\begin{bmatrix} 1/4 & -1/2 & 1/4 & 1/4 \\ 1/4 & 1/2 & 1/4 & 1/4 \\ 3/4 & 1/2 & -1/4 & -1/4 \\ -1/4 & -1/2 & 3/4 & -1/4 \end{bmatrix}$$

(d)
$$\begin{bmatrix} 1/4 & 3/4 & 1/2 & 1/2 \\ -3/4 & 3/4 & 1/2 & -1/2 \\ -1/4 & 1/4 & 1/2 & -1/2 \\ 3/4 & 5/4 & 1/2 & 3/2 \end{bmatrix}$$
(e)
$$\begin{bmatrix} 7/2 \\ -3/2 \\ -1/2 \\ 7/2 \end{bmatrix}$$
(f)
$$\begin{bmatrix} 5/4 \\ -23/4 \\ -13/4 \\ 23/4 \end{bmatrix}$$
(g) $-2x^3 - 7x^2 - 7$

Exercises 5.4, pp. 277–278

1. $-1, 3; E_{-1} : \begin{bmatrix} 0 \\ 1 \end{bmatrix}, E_3 : \begin{bmatrix} 1/2 \\ 1 \end{bmatrix}$ **3.** $4; E_4 : \begin{bmatrix} 3/2 \\ 1 \end{bmatrix}$

5. $2\sqrt{3}, -2\sqrt{3}; E_{2\sqrt{3}}: \begin{bmatrix} \sqrt{3}/2 \\ 1 \end{bmatrix}, E_{-2\sqrt{3}}: \begin{bmatrix} -\sqrt{3}/2 \\ 1 \end{bmatrix}$

7. $2, 3, 1; E_2: \begin{bmatrix} -1/2 \\ 1 \\ 1 \end{bmatrix}, E_3 : \begin{bmatrix} -1 \\ 1 \\ 1 \end{bmatrix}, E_1 : \begin{bmatrix} 0 \\ 1 \\ 0 \end{bmatrix}$ **9.** $4; E_4 : \begin{bmatrix} 0 \\ 0 \\ 1 \end{bmatrix}$

11. $3, -4; E_3 : \begin{bmatrix} 5 \\ -2 \\ 1 \end{bmatrix}, E_{-4} : \begin{bmatrix} -2 \\ 8/3 \\ 1 \end{bmatrix}$ **13.** $4, -2; E_4 : \begin{bmatrix} 1 \\ 1 \\ 1 \end{bmatrix}, E_{-2} : \begin{bmatrix} -1 \\ 1 \\ 0 \end{bmatrix}, \begin{bmatrix} -1 \\ 0 \\ 1 \end{bmatrix}$

15. $1 + 2i, 1 - 2i; E_{1+2i} : \begin{bmatrix} 1/2 + i/2 \\ 1 \end{bmatrix}, E_{1-2i} : \begin{bmatrix} 1/2 - i/2 \\ 1 \end{bmatrix}$

17. $1, i, -i; E_1 : \begin{bmatrix} 1 \\ 0 \\ 0 \end{bmatrix}, E_i : \begin{bmatrix} 0 \\ -i \\ 1 \end{bmatrix}, E_{-i} : \begin{bmatrix} 0 \\ i \\ 1 \end{bmatrix}$

27. $3 + \sqrt{3}, 3 - \sqrt{3}, 0; E_{3+\sqrt{3}} : \begin{bmatrix} 1 \\ 1/2 - \sqrt{3}/2 \\ 1 \end{bmatrix}, E_{3-\sqrt{3}} : \begin{bmatrix} 1 \\ 1/2 + \sqrt{3}/2 \\ 1 \end{bmatrix}, E_0 : \begin{bmatrix} 1/4 \\ 13/8 \\ 1 \end{bmatrix}$

29. $r_1 = 4.732050808$, $r_2 = 1.267949199$, $r_3 = -0.2 \times 10^{-8}$;

$$E_{r_1} : \begin{bmatrix} -1.038008500 \\ 0.3799374765 \\ -1.038008498 \end{bmatrix}, E_{r_2} : \begin{bmatrix} 0.562305740 \\ 0.7681239226 \\ 0.5623057380 \end{bmatrix}, E_{r_3} : \begin{bmatrix} -0.3297195484 \\ -2.143177059 \\ -1.318878189 \end{bmatrix}$$

31. $3, -2, i, -i$; $E_3 : \begin{bmatrix} 1 \\ 1 \\ 1 \\ 1 \end{bmatrix}$, $E_{-2} : \begin{bmatrix} 1 \\ 1/2 \\ 1 \\ 1 \end{bmatrix}$, $E_i : \begin{bmatrix} 1/2 + i/2 \\ -1/2 + 5i/2 \\ 1 \\ 1/2 + i \end{bmatrix}$, $E_{-i} : \begin{bmatrix} 1/2 - i/2 \\ -1/2 - 5i/2 \\ 1 \\ 1/2 - i \end{bmatrix}$

Exercises 5.5, pp. 286–287

1. Diagonalizable, $\begin{bmatrix} -1 & 0 \\ 0 & 3 \end{bmatrix}$, $P = \begin{bmatrix} 0 & 1/2 \\ 1 & 1 \end{bmatrix}$ **3.** Not diagonalizable

5. Diagonalizable, $\begin{bmatrix} 2\sqrt{3} & 0 \\ 0 & -2\sqrt{3} \end{bmatrix}$, $P = \begin{bmatrix} \sqrt{3}/2 & -\sqrt{3}/2 \\ 1 & 1 \end{bmatrix}$

7. Diagonalizable, $\begin{bmatrix} 2 & 0 & 0 \\ 0 & 3 & 0 \\ 0 & 0 & 1 \end{bmatrix}$, $P = \begin{bmatrix} -1/2 & -1 & 0 \\ 1 & 1 & 1 \\ 1 & 1 & 0 \end{bmatrix}$ **9.** Not diagonalizable

11. Not diagonalizable   **13.** Diagonalizable, $\begin{bmatrix} 4 & 0 & 0 \\ 0 & -2 & 0 \\ 0 & 0 & -2 \end{bmatrix}$, $P = \begin{bmatrix} 1 & -1 & -1 \\ 1 & 1 & 0 \\ 1 & 0 & 1 \end{bmatrix}$

15. Diagonalizable, $\begin{bmatrix} 1 + 2i & 0 \\ 0 & 1 - 2i \end{bmatrix}$, $P = \begin{bmatrix} 1/2 + i/2 & 1/2 - i/2 \\ 1 & 1 \end{bmatrix}$

17. Diagonalizable, $\begin{bmatrix} 1 & 0 & 0 \\ 0 & i & 0 \\ 0 & 0 & -i \end{bmatrix}$, $P = \begin{bmatrix} 1 & 0 & 0 \\ 0 & -i & i \\ 0 & 1 & 1 \end{bmatrix}$ **19.** $\begin{bmatrix} 1 & 0 \\ 0 & 2 \end{bmatrix}$

21. $\begin{bmatrix} 0 & 0 & 0 \\ 0 & 0 & 0 \\ 0 & 0 & 1 \end{bmatrix}$, $\begin{bmatrix} 0 & 1 & 0 \\ 0 & 0 & 0 \\ 0 & 0 & 1 \end{bmatrix}$

23.
$$\begin{bmatrix} 5 & 0 & 0 & 0 & 0 & 0 \\ 0 & 1 & 0 & 0 & 0 & 0 \\ 0 & 0 & 1 & 0 & 0 & 0 \\ 0 & 0 & 0 & -2 & 0 & 0 \\ 0 & 0 & 0 & 0 & -2 & 0 \\ 0 & 0 & 0 & 0 & 0 & -2 \end{bmatrix}, \begin{bmatrix} 5 & 0 & 0 & 0 & 0 & 0 \\ 0 & 1 & 1 & 0 & 0 & 0 \\ 0 & 0 & 1 & 0 & 0 & 0 \\ 0 & 0 & 0 & -2 & 0 & 0 \\ 0 & 0 & 0 & 0 & -2 & 0 \\ 0 & 0 & 0 & 0 & 0 & -2 \end{bmatrix}, \begin{bmatrix} 5 & 0 & 0 & 0 & 0 & 0 \\ 0 & 1 & 0 & 0 & 0 & 0 \\ 0 & 0 & 1 & 0 & 0 & 0 \\ 0 & 0 & 0 & -2 & 1 & 0 \\ 0 & 0 & 0 & 0 & -2 & 0 \\ 0 & 0 & 0 & 0 & 0 & -2 \end{bmatrix},$$

$$\begin{bmatrix} 5 & 0 & 0 & 0 & 0 & 0 \\ 0 & 1 & 0 & 0 & 0 & 0 \\ 0 & 0 & 1 & 0 & 0 & 0 \\ 0 & 0 & 0 & -2 & 1 & 0 \\ 0 & 0 & 0 & 0 & -2 & 1 \\ 0 & 0 & 0 & 0 & 0 & -2 \end{bmatrix}, \begin{bmatrix} 5 & 0 & 0 & 0 & 0 & 0 \\ 0 & 1 & 1 & 0 & 0 & 0 \\ 0 & 0 & 1 & 0 & 0 & 0 \\ 0 & 0 & 0 & -2 & 1 & 0 \\ 0 & 0 & 0 & 0 & -2 & 0 \\ 0 & 0 & 0 & 0 & 0 & -2 \end{bmatrix}, \begin{bmatrix} 5 & 0 & 0 & 0 & 0 & 0 \\ 0 & 1 & 1 & 0 & 0 & 0 \\ 0 & 0 & 1 & 0 & 0 & 0 \\ 0 & 0 & 0 & -2 & 1 & 0 \\ 0 & 0 & 0 & 0 & -2 & 1 \\ 0 & 0 & 0 & 0 & 0 & -2 \end{bmatrix}$$

25. $\begin{bmatrix} 2 & 0 & 0 \\ 0 & 3 & 0 \\ 0 & 0 & 1 \end{bmatrix}$ **27.** $\begin{bmatrix} 4 & 1 \\ 0 & 4 \end{bmatrix}$ **29.** $\begin{bmatrix} 3 & 1 & 0 \\ 0 & 3 & 0 \\ 0 & 0 & -4 \end{bmatrix}$

39. $\begin{bmatrix} -1 & 0 & 0 & 0 & 0 \\ 0 & 3 & 1 & 0 & 0 \\ 0 & 0 & 3 & 1 & 0 \\ 0 & 0 & 0 & 3 & 0 \\ 0 & 0 & 0 & 0 & -1 \end{bmatrix}$, $P = \begin{bmatrix} -1 & -3 & 18 & 1 & -1 \\ 3359/9 & -3 & 15 & 7 & 3422/9 \\ -970/9 & -3 & 18 & -2 & -988/9 \\ 467/9 & -3 & 18 & 1 & 476/9 \\ 476/9 & -3 & 18 & 1 & 485/9 \end{bmatrix}$

Exercises 5.6, pp. 291–292

1. (a) $2, -3$; $V_2 : \begin{bmatrix} 1 \\ 0 \\ 1 \end{bmatrix}, \begin{bmatrix} 0 \\ 1 \\ 0 \end{bmatrix}$, $V_{-3} : \begin{bmatrix} 1/2 \\ 3/2 \\ 1 \end{bmatrix}$ (b) Is diagonalizable (c) $\begin{bmatrix} 2 & 0 & 0 \\ 0 & 2 & 0 \\ 0 & 0 & -3 \end{bmatrix}$

3. (a) $-1, 2$; $V_{-1} : -2x + 1$, $V_2 : x + 1$ (b) Is diagonalizable (c) $\begin{bmatrix} -1 & 0 \\ 0 & 2 \end{bmatrix}$

5. (a) $1, -1$; $V_1 : e^x$, $V_{-1} : e^{-x}$ (b) Is diagonalizable (c) $\begin{bmatrix} 1 & 0 \\ 0 & -1 \end{bmatrix}$ **9.** -2

CHAPTER 6

Exercises 6.1, pp. 301–302

5. $y_1 = c_1 e^x$
$y_2 = c_2 e^{-2x}$

7. $y_1 = c_1 e^{-x}$
$y_2 = c_2$
$y_3 = c_3 e^{4x}$

9. $y_1 = 2e^x$
$y_2 = e^{-2x}$

11. $y_1 = 2e^{-x}$
$y_2 = 1$
$y_3 = 0$

13. $y_1 = c_1 e^x - 2$
$y_2 = c_2 e^{-2x} + (1/2)x - 1/4$

15. $y_1 = c_1 e^{-x} - 2x + 3$
$y_2 = c_2 - e^{-x}(x + 1)$
$y_3 = c_3 e^{4x} - (1/5)\cos 2x + (1/10)\sin 2x$

Exercises 6.2, p. 311

1. $y_1 = c_1 e^{3x}$
$y_2 = 2c_1 e^{3x} + c_2 e^{-x}$

3. $y_1 = c_1 e^{2\sqrt{3}x} + c_2 e^{-2\sqrt{3}x}$
$y_2 = (2/\sqrt{3})c_1 e^{2\sqrt{3}x} - (2/\sqrt{3})c_2 e^{-2\sqrt{3}x}$

5. $y_1 = c_2 e^{2x} - c_3 e^{3x}$
$y_2 = c_1 e^x - 2c_2 e^{2x} + c_3 e^{3x}$
$y_3 = -2c_2 e^{2x} + c_3 e^{3x}$

7. $y_1 = -c_1 e^{-2x} - c_2 e^{-2x} + c_3 e^{4x}$
$y_2 = c_1 e^{-2x} + c_3 e^{4x}$
$y_3 = c_2 e^{-2x} + c_3 e^{4x}$

9. $y_1 = e^x(c_1(\cos 2x + \sin 2x) + c_2(\cos 2x - \sin 2x))$
$y_2 = 2e^x(c_1 \cos 2x + c_2 \sin 2x)$

11. $y_1 = c_1 e^x$
$y_2 = c_2 \cos x + c_3 \sin x$
$y_3 = -c_2 \sin x + c_3 \cos x$

13. $y_1 = 2e^{3x}$
$y_2 = 4e^{3x} - 3e^{-x}$

15. $y_1 = e^{2x} - e^{3x}$
$y_2 = 2e^x - 2e^{2x} + e^{3x}$
$y_3 = -2e^{2x} + e^{3x}$

17. $y_1 = e^x(2\cos 2x + \sin 2x)$
$y_2 = e^x(\cos 2x + 3\sin 2x)$

19. $y_1 = c_1 e^{-2x} + 2c_2 e^{3x}$
$y_2 = c_1 e^{-2x} + c_2 e^{3x}$

21. $y_1 = c_1 e^{(2+\sqrt{10})x} + c_2 e^{(2-\sqrt{10})x}$

$y_2 = (-1 + \sqrt{10})c_1 e^{(2+\sqrt{10})x} - (1 + \sqrt{10})c_2 e^{(2-\sqrt{10})x}$

23. $y_1 = -c_1 e^{5x} + c_2 e^{2x} + c_3 e^{-x}$

$y_2 = 2c_1 e^{5x} + c_2 e^{2x} + 2c_3 e^{-x}$

$y_3 = c_1 e^{5x}$

25. $y_1 = e^x (c_1 \cos 3x + c_2 \sin 3x)$

$y_2 = e^x (c_1 \sin 3x - c_2 \cos 3x)$

27. $x = \cos t - 3 \sin t$

$y = \cos t - \sin t$

Exercises 6.3, pp. 314–315

1. $P = \begin{bmatrix} 6 & 1 \\ 4 & 0 \end{bmatrix}$; $\begin{aligned} y_1 &= (6c_1 + c_2 + 6c_2 x)e^{4x} \\ y_2 &= (4c_1 + 4c_2 x)e^{4x} \end{aligned}$

3. $P = \begin{bmatrix} 0 & 0 & 1 \\ 0 & 1 & 0 \\ 1 & 0 & 0 \end{bmatrix}$; $\begin{aligned} y_1 &= c_3 e^{4x} \\ y_2 &= (c_2 + c_3 x)e^{4x} \\ y_3 &= (1/2)(2c_1 + 2c_2 x + c_3 x^2)e^{4x} \end{aligned}$

5. $P = \begin{bmatrix} 6/49 & 20/7 & 43/49 \\ -8/49 & -8/7 & 8/49 \\ -3/49 & 4/7 & 3/49 \end{bmatrix}$; $\begin{aligned} y_1 &= 6c_1 e^{-4x} + 140(c_2 + c_3 x)e^{3x} + 43c_3 e^{3x} \\ y_2 &= -8c_1 e^{-4x} - 56(c_2 + c_3 x)e^{3x} + 8c_3 e^{3x} \\ y_3 &= -3c_1 e^{-4x} + 28(c_2 + c_3 x)e^{3x} + 3c_3 e^{3x} \end{aligned}$

7. $P = \begin{bmatrix} 2 & 2 & -3/2 + (1/2)i & -3/2 - (1/2)i \\ 1 & 2 & -3/2 + (11/2)i & -3/2 - (11/2)i \\ 2 & 2 & -2 - i & -2 + i \\ 2 & 2 & -2 + (3/2)i & -2 - (3/2)i \end{bmatrix}$

$y_1 = 2c_1 e^{-2x} + 2c_2 e^{3x} + (-3/2 + (1/2)i)c_3 e^{-ix} + (-3/2 - (1/2)i)c_4 e^{ix}$

$y_2 = c_1 e^{-2x} + 2c_2 e^{3x} + (-3/2 + (11/2)i)c_3 e^{-ix} + (-3/2 - (11/2)i)c_4 e^{ix}$

$y_3 = 2c_1 e^{-2x} + 2c_2 e^{3x} + (-2 - i)c_3 e^{-ix} + (-2 + i)c_4 e^{ix}$

$y_4 = 2c_1 e^{-2x} + 2c_2 e^{3x} + (-2 + (3/2)i)c_3 e^{-ix} + (-2 - (3/2)i)c_4 e^{ix}$

9. $y_1 = (2 + 3x)e^{4x}$

$y_2 = (1 + 2x)e^{4x}$

11. $y_1 = e^{3x}(-30/49 + (100/7)x) + (30/49)e^{-4x}$

$y_2 = e^{3x}(138/49 - (40/7)x) - (40/49)e^{-4x}$

$y_3 = e^{3x}(-34/49 + (20/7)x) - (15/49)e^{-4x}$

17. $x = (1 + t)e^{-6t}$

$y = te^{-6t}$

Exercises 6.4, p. 318

$c_1 - c_4 = 2.$

1. $y_1 = c_1 e^{3x} - 2/3$

$y_2 = 2c_1 e^{3x} + c_2 e^{-x} + x - 19/3$

3. $y_1 = e^x((c_1(\cos 2x + \sin 2x) + c_2(\cos 2x - \sin 2x)) + 19/25 + (2/5)x$

$y_2 = 2e^x (c_1 \cos 2x + c_2 \sin 2x) + 86/25 + (3/5)x$

5. $y_1 = c_2 e^{2x} - c_3 e^{3x} - 2/9 + x/3$

$y_2 = c_1 e^x - 2c_2 e^{2x} + c_3 e^{3x} + 2/9 + (2/3)x - e^{-x}/4 - (x/2)e^{-x}$

$y_3 = -2c_2 e^{2x} + c_3 e^{3x} + 2/9 + (2/3)x$

7. $y_1 = 6(c_1 + c_2 x)e^{4x} + c_2 e^{4x} - (9/16)x - 1/32$

$y_2 = (4c_1 + 4c_2 x)e^{4x} - (5/8)x + 1/4$

9. $y_1 = 6c_1 e^{-4x} + 140(c_2 + c_3 x)e^{3x} + 43c_3 e^{3x}$

$y_2 = -8c_1 e^{-4x} - 56(c_2 + c_3 x)e^{3x} + 8c_3 e^{3x} - e^{-2x}/25$

$y_3 = -3c_1 e^{-4x} + 28(c_2 + c_3 x)e^{3x} + 3c_3 e^{3x} + (3/25)e^{-2x}$

11. $y_1 = (8/3)e^{3x} - 2/3$

$y_2 = (16/3)e^{3x} + 2e^{-x} + x - 19/3$

13. $y_1 = e^{2x} - (7/9)e^{3x} - 2/9 + x/3$

$y_2 = -2e^{2x} + (7/9)e^{3x} + 2/9 + (2/3)x + (9/4)e^x - (1/4)e^{-x} - (x/2)e^{-x}$

$y_3 = -2e^{2x} + (7/9)e^{3x} + 2/9 + (2/3)x$

15. $x = 1 + t - (2/3)e^t + (2/3)e^{4t}$

$y = -1 - t + (2/3)e^t + (4/3)e^{4t}$

Exercises 6.5, p. 322

1. $y_1 = c_1 e^{-x} + c_2 e^{-2x}$

$y_2 = -c_1 e^{-x} - 2c_2 e^{-2x}$

$y = y_1, y' = y_2$

3. $y_1 = c_1 e^{-x} + c_2 e^{2x} + (1/10)\cos x - (3/10)\sin x$

$y_2 = -c_1 e^{-x} + 2c_2 e^{2x} - (1/10)\sin x - (3/10)\cos x$

$y = y_1, y' = y_2$

5. $y_1 = c_1 e^x + c_2 e^{-x} + c_3 e^{-4x}$

$y_2 = c_1 e^x - c_2 e^{-x} - 4c_3 e^{-4x}$

$y_3 = c_1 e^x + c_2 e^{-x} + 16c_3 e^{-4x}$

$y = y_1, y' = y_2, y_3 = y''$

7. $y_1 = e^x (c_1 \cos x + c_2 \sin x) + c_3 e^{-2x}$

$y_2 = e^x (c_1 (\cos x - \sin x) + c_2 (\cos x + \sin x)) - 2c_3 e^{-2x}$

$y_3 = 2e^x (-c_1 \sin x + c_2 \cos x) + 4c_3 e^{-2x}$

$y = y_1, y' = y_2, y_3 = y''$

9. $y_1 = c_1 + c_2 e^{-8x} + (63/128)x - (15/32)x^2 + x^3/12$

$y_2 = -8c_2 e^{-8x} + (63/128) - (15/16)x + x^2/4$

$y = y_1, y' = y_2$

13. $y_1 = c_1 - (2/3)c_2 + c_2 x + (1/2)c_3 e^{-x} + (1/6)c_4 e^{3x}$

$y_2 = -2c_2 + (3/2)c_3 e^{-x} + (1/2)c_4 e^{3x}$

Exercises 6.6, pp. 331–334

1. $q_1 = 9/2 e^{-t/10} + 11/2$

$q_2 = -9/2 e^{-t/10} + 11/2$

3. $q_1 = (-95/2 - (50/3)\sqrt{6})e^{((-3+\sqrt{6})/50)x} + (-95/2 + (50/3)\sqrt{6})e^{((-3-\sqrt{6})/50)x} + 100$

$q_2 = (-95/2 - (50/3)\sqrt{6})(\sqrt{6}/2)e^{((-3+\sqrt{6})/50)x} - (-95/2 + (50/3)\sqrt{6})(\sqrt{6}/2)e^{((-3-\sqrt{6})/50)x} + 100$

5. $i_1 = (1 - 2t)e^{-4t}$

$i_2 = 2te^{-4t}$

7. $i_1 = e^{-t}(3\cos t - \sin t) + 4\sin t - 3\cos t$

$i_2 = e^{-t}(-\cos t - 3\sin t) + \cos t + 2\sin t$

9. $x = 2e^{-5t} + 2e^{5t} - e^{10t} - e^{-10t}$

$y = 3e^{-5t} + 3e^{5t}$

11. $x_1 = (3/5)\sin t + (\sqrt{6}/15)\sin\sqrt{6}t$

$x_2 = (6/5)\sin t - (\sqrt{6}/30)\sin\sqrt{6}t$

13. $x_1 = (3/50)\cos 2\sqrt{3}t + (1/25)\cos\sqrt{2}t$

$x_2 = (-1/50)\cos 2\sqrt{3}t + (3/25)\cos\sqrt{2}t$

15. $x_1 = e^{-t/2}((2\sqrt{3}/5)\sin\sqrt{3}t/2 + (4\sqrt{23}/115)\sin\sqrt{23}t/2)$

$x_2 = e^{-t/2}((4\sqrt{3}/5)\sin\sqrt{3}t/2 - (2\sqrt{23}/115)\sin\sqrt{23}t/2)$

17. $x_1 = (3/5)\sin t + (1/15)\cos t - (2/5)\cos\sqrt{6}t + (\sqrt{6}/15)\sin\sqrt{6}t + (1/3)\cos 2t$

$x_2 = (6/5)\sin t + (2/15)\cos t + (1/5)\cos\sqrt{6}t - (\sqrt{6}/30)\sin\sqrt{6}t - (1/3)\cos 2t$

21. $250(e^{-20} + (13/4)e^{-10} + 2e^{-5} - 2e^{-15} + e^{20} - 2e^{15} + 2e^5 + (13/4)e^{10} - 17/2)$

23. $x = -(5/2)e^{-2t} + (25/2)e^{6t}$
$y = (5/2)e^{-2t} + (25/2)e^{6t}$

Exercises 6.7, pp. 342–343

1. (a) $(0, 0)$, $(2, -1)$ (b) $A(0, 0) = \begin{bmatrix} 1 & 0 \\ 0 & 2 \end{bmatrix}$, $A(2, -1) = \begin{bmatrix} 0 & 2 \\ 1 & 0 \end{bmatrix}$

(c) $(0, 0) \begin{bmatrix} c_1 e^t \\ c_2 e^{2t} \end{bmatrix}$, $(2, -1) \begin{bmatrix} (c_1/2)e^{\sqrt{2}t} + (c_2/2)e^{-\sqrt{2}t} \\ (\sqrt{2}c_1/4)e^{\sqrt{2}t} - (\sqrt{2}c_2/4)e^{-\sqrt{2}t} \end{bmatrix}$ (d) $(0, 0)$ unstable, $(2, -1)$ unstable

3. (a) $(0, 0)$, $(-1, 1)$ (b) $A(0, 0) = \begin{bmatrix} 1 & 0 \\ 0 & 1 \end{bmatrix}$, $A(-1, 1) = \begin{bmatrix} 0 & 0 \\ 1 & 0 \end{bmatrix}$

(c) $(0, 0) \begin{bmatrix} c_1 e^t \\ c_2 e^t \end{bmatrix}$, $(-1, 1) \begin{bmatrix} c_2 \\ c_1 + c_2 t \end{bmatrix}$ (d) $(0, 0)$ unstable, $(-1, 1)$ unstable

5. (a) $(0, 0)$ (b) $A(0, 0) = \begin{bmatrix} 1 & 4 \\ 1 & -1 \end{bmatrix}$ (c) $(0, 0) \begin{bmatrix} (1/2 + \sqrt{5}/10)c_1 e^{\sqrt{5}t} + (1/2 - \sqrt{5}/10)c_2 e^{-\sqrt{5}t} \\ (\sqrt{5}/10)c_1 e^{\sqrt{5}t} - (\sqrt{5}/10)c_2 e^{-\sqrt{5}t} \end{bmatrix}$

(d) $(0, 0)$ unstable

9. (a) $(0, 0)$, $(2/3, 2)$ (b) $A(0, 0) = \begin{bmatrix} 1/2 & 0 \\ 0 & 0 \end{bmatrix}$, $A(2/3, 2) = \begin{bmatrix} 0 & -1/6 \\ 3/2 & -1/2 \end{bmatrix}$

(c) $(0, 0) \begin{bmatrix} c_1 e^{t/2} \\ c_2 \end{bmatrix}$, $(2/3, 2) \begin{bmatrix} e^{-t/4}(c_1(\cos \sqrt{3}t/4 + c_2 \sin \sqrt{3}t/4)) \\ (3e^{-t/4}/2)(c_1(\cos \sqrt{3}t/4 + \sqrt{3} \sin \sqrt{3}t/4)) \\ \quad + c_2(-\sqrt{3} \cos \sqrt{3}t/4 + \sin \sqrt{3}t/4) \end{bmatrix}$

(d) $(0, 0)$ unstable, $(2/3, 2)$ stable

11. (a) $(0, 0)$, $(3, 3)$ (b) $A(0, 0) = \begin{bmatrix} 3/4 & 0 \\ 0 & 0 \end{bmatrix}$, $A(3, 3) = \begin{bmatrix} 0 & -3/4 \\ 3/4 & -3/4 \end{bmatrix}$

(c) $(0, 0) \begin{bmatrix} c_1 e^{3t/4} \\ c_2 \end{bmatrix}$, $(3, 3) \begin{bmatrix} e^{-3t/8}(c_1((1/2) \cos 3\sqrt{3}t/8 + (\sqrt{3}/6) \sin 3\sqrt{3}t/8) + \\ c_2(-(\sqrt{3}/6) \cos 3\sqrt{3}t/8 + (1/2) \sin 3\sqrt{3}t/8)) \\ (\sqrt{3}/3)e^{-3t/8}(c_1 \sin 3\sqrt{3}t/8 - c_2 \cos 3\sqrt{3}t/8) \end{bmatrix}$

(d) $(0, 0)$ unstable, $(3, 3)$ stable

13. $y^2 - (2g/l) \cos x = c, c = -64\sqrt{2}$

15. (a) $(0, 0)$, $((2k + 1)\pi, 0)$, $(2k\pi, 0)$, $k = \pm 1, \pm 2, \ldots$ (b) $A(0, 0) = \begin{bmatrix} 0 & 1 \\ -(g/l) & -c/(ml) \end{bmatrix}$,

$A((2k + 1)\pi, 0) = \begin{bmatrix} 0 & 1 \\ g/l & -c/(ml) \end{bmatrix}$, $A(2k\pi, 0) = \begin{bmatrix} 0 & 1 \\ -(g/l) & -c/(ml) \end{bmatrix}$

17. (a) $(0, 0)$, $(4, 3)$ (b) $A(0, 0) = \begin{bmatrix} -3/4 & 0 \\ 0 & 1 \end{bmatrix}$, $A(4, 3) = \begin{bmatrix} 0 & 1 \\ -3/4 & 0 \end{bmatrix}$

(c) $(0, 0) \begin{bmatrix} c_1 e^{-3t/4} \\ c_2 e^t \end{bmatrix}$, $(4, 3) \begin{bmatrix} (1/2)(c_1 \cos \sqrt{3}t/2 + c_2 \sin \sqrt{3}t/2) \\ (\sqrt{3}/4)(c_2 \cos \sqrt{3}t/2 - c_1 \sin \sqrt{3}t/2) \end{bmatrix}$

19. (a) $(0, 0)$, $(0, 4)$, $(4, 0)$, $(4/3, 4/3)$ (b) $A(0, 0) = \begin{bmatrix} 1 & 0 \\ 0 & 1 \end{bmatrix}$, $A(0, 4) = \begin{bmatrix} -1 & 0 \\ -2 & -1 \end{bmatrix}$,

$A(4, 0) = \begin{bmatrix} -1 & -2 \\ 0 & -1 \end{bmatrix}$, $A(4/3, 4/3) = \begin{bmatrix} -1/3 & -2/3 \\ -2/3 & -1/3 \end{bmatrix}$ (c) $(0, 0) \begin{bmatrix} c_1 e^t \\ c_2 e^t \end{bmatrix}$,

$$(0,4) \begin{bmatrix} c_2 e^{-t} \\ -2e^{-t}(c_1 + c_2 t) \end{bmatrix}, (0,4) \begin{bmatrix} -2e^{-t}(c_1 + c_2 t) \\ c_2 e^{-t} \end{bmatrix}, (4/3, 4/3) \begin{bmatrix} c_1 e^{t/3} + c_2 e^{-t} \\ -c_1 e^{t/3} + c_2 e^{-t} \end{bmatrix}$$

CHAPTER 7

Exercises 7.1, pp. 351–352

1. $4/s$ **3.** $4/s^3$ **5.** $1/(s+4)$ **7.** $s/(s^2+9)$ **9.** $2s/(s^2+1)^2$ **11.** $2/(s-3)^3$
13. $3/((s-2)^2+9)$ **15.** $(s-a)/((s-a)^2+b^2)$ **17.** $s/(s^2-9)$ **19.** $4s/(s^2-1)^2$
21. $(s-a)/((s-a)^2-b^2)$ **35.** 4 **37.** $4e^{6t}$ **39.** $(2/3)e^{2t} - (2/3)e^{-4t}$ **41.** $2\cos 3t$

Exercises 7.2, pp. 355–356

1. $y = e^{2t}$ **3.** $y = -1$ **5.** $y = e^{-3t} + 5te^{-3t}$ **7.** $y = (1/2)\sin 2t$
9. $y = -(3/10)e^t + (2/15)e^{-4t} - (5/6)e^{-t}$ **11.** $y = (2/3) - (2/3)e^{-3t}$
13. $y = (2/3) - e^{3t}/24 + e^{3t}/4 - 7e^{-t}/8$
15. $y = e^{2t}((139/100)\sin t - (127/100)\cos t) - (1/4)\sin t - (1/4)\cos t + 7t/5 + 63/25$
17. $\approx 328{,}056$ **19.** $70 + 110e^{5\ln 9/11}$
21. $u = 60e^{-t/10}/3961 + e^{-2t}((3847/23766)\sin 6t + (3721/15844)\cos 6t)$
23. $(96/65)\sin 2t - (12/65)\cos 2t + 18e^{-3t}/13 - 6e^{-t}/5$

Exercises 7.3, pp. 364–366

1. (a) $t - tu_2(t)$ (c) $1/s^2 - e^{-2s}/s^2 - 2e^{-2s}/s$
3. (a) $e^4 e^{2(t-2)}u_2(t) - e^8 e^{2(t-4)}u_4(t)$ (c) $e^{-2(s-2)}/(s-2) - e^{-4(s-2)}/(s-2)$
5. (a) $u_\pi(t)\cos 2(t-\pi) - u_{2\pi}(t)\cos 2(t-2\pi)$ (c) $e^{-\pi s}s/(s^2+4) - e^{-2\pi s}s/(s^2+4)$
7. (b) $e^{-2s}(2/s + 1/s^2) - 2e^{-4s}(4/s + 1/s^2)$ **9.** (b) $-e^{-\pi s}s/(s^2+1)$ **11.** $u_2(t)$
13. $u_1(t)(e^{4(t-1)} - 1)/4$ **15.** $e^{-3t}/9 + t/3 - 1/9 - u_2(t)(-5e^{-3(t-2)}/9 + t/3 - 1/9)$
17. $u_1(t)(e^{3(t-1)}/24 + e^{-3(t-1)}/12 - e^{1-t}/8)$
19. $-(39/20)e^t + 7e^{3t}/12 + (1/30)\cos 3t + (1/15)\sin 3t + 7/3 + u_{2\pi}(t)(69e^{t-\pi}/20 - 13e^{3(t-2\pi)}/12 - (1/30)\cos 3t - (1/15)\sin 3t - 7/3)$
21. $e^{2t}/6 - e^{-4t}/6$ **25.** $\approx 503{,}660$
27. $60/3961e^{-t/10} + e^{-2t}((3721/15844)\cos 6t + (3847/23766)\sin 6t) - e^{-3/10}u_3(t)/3961(60e^{-(t-3)/10} - 60e^{-2(t-3)}\cos(6(t-3)) - 19e^{-2(t-3)}\sin(6(t-3)))$
29. $4 - 6e^{-t} + 2e^{-3t} - u_2(t)(4 + 2e^{-3(t-2)} - 6e^{-(t-2)})$ **31.** $6u_\pi(t)(e^{-(t-\pi)} - e^{-3(t-\pi)})$

Exercises 7.4, p. 369

1. $e^t - 1$ **3.** $e^{-t}/2 - (1/2)\cos t + (1/2)\sin t$ **5.** $1/((s-1)s^2)$ **7.** t
9. $\cos 2t + (1/8)\sin 2t - (t/4)\cos 2t$ **11.** $t^4/24$ **13.** $(3/2)(1 + e^{2t})$ **15.** $(1/3)(e^t + 2e^{4t})$
17. $5e^t/3 + ((4\sqrt{3}/3)\sin\sqrt{3}t/2 - (2/3)\cos\sqrt{3}t/2)e^{(-t/2)}$

Exercises 7.5, pp. 372–373

1. $y_1 = e^{2t} - e^{3t} - 2u_2(t)/3(1 - 3e^{2(t-2)} + 2e^{3(t-2)})$
   $y_2 = 2e^{3t} - e^{2t} - 2u_2(t)/3(1 + 3e^{2(t-2)} - 4e^{3(t-2)})$
3. $y_1 = (1/5)(4e^{2t} + e^{-3t}) + u_2(t)(-e^t/4 + e^{2t-2}/5 + e^{8-3t}/20)$
   $y_2 = (1/5)(4e^{2t} - 4e^{-3t}) + u_2(t)(e^{2t-2} - e^{8-3t})/5$

5. $x_1 = u_1(t)(5/48 - 35t/256 + 5t^3/96 - (25/2048)e^{4(t-1)} - (15/2048)e^{-4(t-1)})$
$x_2 = u_1(t)(5/24 - 43t/128 + 5t^3/48 + (15/1024)e^{4(t-1)} + (9/1024)e^{-4(t-1)})$

7. $x_1 = (2\sqrt{2}/3)\sin\sqrt{2}t - (\sqrt{5}/15)\sin\sqrt{5}t$ **9.** $q_1 = (1/3)(4 + 8e^{-9t/100})$, $q_2 = (1/3)(8 - 8e^{-9t/100})$
$x_2 = (\sqrt{2}/3)\sin\sqrt{2}t + (\sqrt{5}/15)\sin\sqrt{5}t$

11. $Q_1 = -25 + 50t + 25e^{-t}(\cos t - \sin t) + u_1(t)(75 - 50t - 25e^{1-t}(\cos(t - 1) - \sin(t - 1))$
$Q_2 = 25 - 25e^{-t}(\cos t + \sin t) + u_1(t)(-25 + 25e^{1-t}(\cos(t - 1) + \sin(t - 1))$

13. $x_1 = u_2(t)((2/5)\sin(t - 2) - (\sqrt{6}/15)\sin\sqrt{6}(t - 2))$
$x_2 = u_2(t)((4/5)\sin(t - 2) + (\sqrt{6}/30)\sin\sqrt{6}(t - 2))$

CHAPTER 8

Exercises 8.1, pp. 383–384

1. $2\sum_{n=0}^{\infty}(-x)^n/n!$ **3.** $\sum_{n=0}^{\infty}(x - 1)^{2n}/n!$ **5.** $\sum_{n=0}^{\infty}x^{2n}/(2n)!$ **7.** $\sum_{n=0}^{\infty}(-x)^n/n!$
9. $\sum_{n=0}^{\infty}x^{2n}/(2^n(2n)!)$ **11.** $\frac{1}{2}\sum_{n=0}^{\infty}(-1)^n(2x)^{2n+1}/(2n + 1)!$ **13.** $1 + x^2/2 + x^4/2 + x^6/6$
15. $1 - 4x^2 + 2x^3/3 + 8x^4/3$

Exercises 8.2, pp. 392–393

1. $1 + x^3/3! + 4/6!x^6 + \sum_{n=2}^{\infty}(1)(4)\cdots(3n + 1)x^{3(n+1)}/(3(n + 1))!$
3. $2(x - x^3/3! - x^5/5! - 3/7!x^7 - \sum_{n=2}^{\infty}(1)(3)\cdots(2n + 1)x^{2n+5}/(2n + 5)!$
5. $2x^2 + x^3 + x^4/2 - x^6/15 - 3/70x^7 + \cdots$ **7.** $1 + x^2/2 + x^3/3 + x^5/40 + x^6/90 + \cdots$
9. $a_0(1 + x^3/3! + 4/6!x^6 + \sum_{n=2}^{\infty}(1)(4)\cdots(3n + 1)x^{3(n+1)}/(3(n + 1))!) + a_1(x + 2x^4/4! + (2)(5)/7!x^7 +$
$\sum_{n=2}^{\infty}(2)(5)\cdots(3n + 2)x^{3n+4}/(3n + 4)!)$
11. $a_0(1 - x^2) + a_1(x - x^3/3! - x^5/5! - \sum_{n=1}^{\infty}(1)(3)\cdots(2n + 1)x^{2n+5}/(2n + 5)!)$
13. $a_0 + a_1x + (a_0/2)x^2 + (a_1/6)x^3$ **15.** $a_0 + a_1x - (a_0/6)x^3 + ((a_0 - a_1)/12)x^4$
17. $a_0 + a_1x - (3a_0/8)x^2 + (5a_1/24)x^3 - (21/128)a_0x^4 + \cdots$ **19.** $a_1x + a_0(1 - (1/2)x\ln((1 + x)/(1 - x)))$
21. $(a_0/3)(35x^4 - 30x^2 + 3)$
23. (b) $\lambda = 0, y = 1; \lambda = 2, y = x; \lambda = 4, y = 1 - 2x^2; \lambda = 6, y = x - 2x^3/3; \lambda = 8, y = 1 - 4x^2 + (4/3)x^4$
25. $p(x_0) = b_0, p'(x_0) = b_1, p''(x_0) = 2b_2, \cdots, n!b_n = p^{(n)}(x_0)$

Exercises 8.3, p. 397

1. $c_1x + c_2x^3$ **3.** $c_1\cos(\ln x) + c_2\sin(\ln x)$ **5.** $c_1/x^2 + c_2\ln x/x^2$ **7.** $c_1\cos(\ln x)/x + c_2\sin(\ln x))/x$
9. $x\cos(\ln x) - x\sin(\ln x)$ **11.** $y = c_1x + c_2x^2 - (1/2)x\ln x$ **13.** $(c_1(x^2 - 2x) + c_2)/(x - 1)^3$

Exercises 8.4, pp. 410–411

1. $x = 0$ **3.** $r^2 = 0$ **5.** $x^{-2}(c_1\cos x^2 + c_2\sin x^2)$
7. $c_1x^{1/3}(1 - x^2/14 + x^4/728 - x^6/82992 + \cdots) + c_2(1 - x^2/10 + x^4/440 - x^6/44880 + \cdots)$
9. (b) $J_0(x)\ln x + x^2/4 - (3/128)x^4 + (11/13824)x^6 + \cdots$ (d) $x^{-1/2}(c_1\cos x + c_2\sin x)$
11. $x^{-3/2}(c_1(\cos x + x\sin x) + c_2(\sin x - x\cos x))$
13. (b) $x^{-1/2}\cos x, x^{-1/2}\sin x$ (c) $x^{-1/2}(c_1\cos x + c_2\sin x) + x^{-1/2}$
15. $x^2 + x^3 + x^4/3 + x^5/36 - (7/720)x^6 + \cdots$ **17.** $x^2 - x^4/12 + x^6/384 - x^8/23040 + \cdots$
19. (b) $r(r - 1) + \alpha r = 0$ (c) $r = 0; a_{n+1} = \dfrac{(n + \beta)(n + \gamma)}{(n + 1)(n + \alpha)}a_n$

$r = 1 - \alpha; a_{n+1} = \dfrac{(n + 1 - \alpha + \beta)(n + 1 - \alpha + \gamma)}{(n + 2 - \alpha)(n + 1)}a_n$

CHAPTER 9

Exercises 9.1, pp. 419–421

1. $-7, \sqrt{5}, \sqrt{10}, \cos^{-1}\left(-\frac{7}{5\sqrt{2}}\right)$ **3.** $21, \sqrt{11}, 5\sqrt{2}, \cos^{-1}\left(\frac{21}{5\sqrt{22}}\right)$

5. $-13/12, 1/\sqrt{30}, \sqrt{43}, \cos^{-1}\left(-\frac{13\sqrt{30}}{12\sqrt{43}}\right)$ **9. (b)** $3, 3, \sqrt{26}, \cos^{-1}\left(\frac{1}{\sqrt{26}}\right)$

11. (b) $0, \sqrt{2}, 1, \pi/2$ **13. (e)** $-3/11, 1/\sqrt{3}, 3/\sqrt{7}, \cos^{-1}\left(-\frac{\sqrt{21}}{11}\right)$

Exercises 9.2, pp. 429–430

1. (b) $\frac{3-5\sqrt{2}}{2}v_1 - \frac{39+5\sqrt{2}}{6}v_2 + \frac{20-3\sqrt{2}}{6}v_3$ **3.** $\begin{bmatrix} 1/\sqrt{5} \\ 2/\sqrt{5} \end{bmatrix}, \begin{bmatrix} 2/\sqrt{5} \\ -1/\sqrt{5} \end{bmatrix}$

5. $\begin{bmatrix} 1 \\ 0 \\ 0 \end{bmatrix}, \begin{bmatrix} 0 \\ -3/5 \\ 4/5 \end{bmatrix}, \begin{bmatrix} 0 \\ -4/5 \\ -3/5 \end{bmatrix}$ **7.** $\begin{bmatrix} 1/\sqrt{3} \\ -1/\sqrt{3} \\ 1/\sqrt{3} \end{bmatrix}, \begin{bmatrix} -1/\sqrt{3} \\ (-3-\sqrt{3})/6 \\ (-3+\sqrt{3})/6 \end{bmatrix}, \begin{bmatrix} 1/\sqrt{3} \\ (-3+\sqrt{3})/6 \\ (-3-\sqrt{3})/6 \end{bmatrix}$

9. $1/\sqrt{2}, \sqrt{3}x/\sqrt{2}, \frac{3\sqrt{5}x^2-\sqrt{5}}{2\sqrt{2}}$ **11.** $\begin{bmatrix} 1/\sqrt{2} \\ 0 \end{bmatrix}, \begin{bmatrix} 0 \\ 1/\sqrt{3} \end{bmatrix}$

17. (a) $\begin{bmatrix} 1/\sqrt{3} \\ -1/\sqrt{3} \\ 1/\sqrt{3} \end{bmatrix}, \begin{bmatrix} 1/\sqrt{2} \\ 1/\sqrt{2} \\ 0 \end{bmatrix}$ **(b)** $\begin{bmatrix} 1/\sqrt{3} \\ -1/\sqrt{3} \\ 1/\sqrt{3} \end{bmatrix}, \begin{bmatrix} 1/\sqrt{2} \\ 1/\sqrt{2} \\ 0 \end{bmatrix}, \begin{bmatrix} -1/\sqrt{6} \\ 1/\sqrt{6} \\ 2/\sqrt{6} \end{bmatrix}$

19. $\begin{bmatrix} 3 \\ 0 \\ 0 \end{bmatrix}, \begin{bmatrix} 0 \\ -3 \\ 4 \end{bmatrix}, \begin{bmatrix} 0 \\ -12 \\ -9 \end{bmatrix}$

Exercises 9.3, p. 439

1. $P = \begin{bmatrix} \sqrt{2}/2 & \sqrt{2}/2 \\ \sqrt{2}/2 & -\sqrt{2}/2 \end{bmatrix}, B = \begin{bmatrix} 2 & -2 \\ 0 & 2 \end{bmatrix}$

3. $P = \begin{bmatrix} \sqrt{6}/6 & \sqrt{2}/2 & -\sqrt{3}/3 \\ \sqrt{6}/6 & -\sqrt{2}/2 & -\sqrt{3}/3 \\ \sqrt{6}/3 & 0 & \sqrt{3}/3 \end{bmatrix}, B = \begin{bmatrix} -1 & -2\sqrt{3} & 3\sqrt{2}/2 \\ 0 & -1 & -\sqrt{6}/2 \\ 0 & 0 & -1 \end{bmatrix}$

5. $\begin{bmatrix} \sqrt{6}/6 & \sqrt{2}/2 & -\sqrt{3}/3 \\ \sqrt{6}/6 & -\sqrt{2}/2 & -\sqrt{3}/3 \\ \sqrt{6}/3 & 0 & \sqrt{3}/3 \end{bmatrix}\begin{bmatrix} c_1e^{-x} - 2\sqrt{3}c_2xe^{-x} + \frac{3\sqrt{2}}{2}c_3xe^{-x} + \frac{3\sqrt{2}}{2}c_3x^2e^{-x} \\ c_2e^{-x} - \frac{\sqrt{6}}{2}c_3xe^{-x} \\ c_3e^{-x} \end{bmatrix}$

7. $\begin{bmatrix} -1/\sqrt{2} & 1/\sqrt{2} \\ 1/\sqrt{2} & 1/\sqrt{2} \end{bmatrix}$ **9.** $\begin{bmatrix} -1/\sqrt{2} & 1/\sqrt{2} & 0 \\ 1/\sqrt{2} & 1/\sqrt{2} & 0 \\ 0 & 0 & 1 \end{bmatrix}$ **11.** $\begin{bmatrix} 1/2 & -1/\sqrt{2} & -1/2 \\ 1/2 & 1/\sqrt{2} & -1/2 \\ 1/\sqrt{2} & 0 & 1/\sqrt{2} \end{bmatrix}$

13. $\begin{bmatrix} -\frac{1}{\sqrt{2}}c_1e^{4x} + \frac{1}{\sqrt{2}}c_2e^{2x} \\ \frac{1}{\sqrt{2}}c_1e^{4x} + \frac{1}{\sqrt{2}}c_2e^{2x} \\ c_3e^{2x} \end{bmatrix}$ **17.** $\begin{bmatrix} 0 & \sqrt{2}/2 & \sqrt{2}/2 \\ 0 & \sqrt{2}/2 & -\sqrt{2}/2 \\ 1 & 0 & 0 \end{bmatrix}$

Index of Maple Commands

Index

461